Ethology and Behavioral Ecology of Marine Mammals

Series Editor

Bernd Würsig, Department of Marine Biology, Texas A&M University at Galveston, Galveston, TX, USA

The aim of this series is to provide the latest ethological information on the major groupings of marine mammals, in six separate books roughly organized in similar manner. These groupings are the 1) toothed whales and dolphins, 2) baleen whales, 3) eared seals and walrus, 4) true seals, 5) sea otter, marine otter and polar bear, and 6) manatees and dugong, the sirens. The scope shall present 1) general patterns of ethological ways of animals in their natural environments, with a strong bent towards modern behavioral ecology; and 2) examples of particularly well-studied species and species groups for which we have enough data. The scope shall be in the form of general and specific reviews for concepts and species, with an emphasis especially on data gathered in the past 15 years or so. A final 7th book was added since the beginning of this series, on "The Evolving Human Factor" to explore the effects that humans had, are having and will have (unless we change our ways) on these magnificent mammals of the seas. The editors and authors are all established scientists in their fields, even though some of them are quite young.

More information about this series at https://link.springer.com/bookseries/15983

Christopher W. Clark · Ellen C. Garland
Editors

Ethology and Behavioral Ecology of Mysticetes

Springer

Editors
Christopher W. Clark
K. Kisa Yang Center for Conservation Bioacoustics, Cornell Lab of Ornithology
Cornell University
Ithaca, NY, USA

Ellen C. Garland
Sea Mammal Research Unit, School of Biology
University of St Andrews
St Andrews, UK

ISSN 2523-7500 ISSN 2523-7519 (electronic)
Ethology and Behavioral Ecology of Marine Mammals
ISBN 978-3-030-98448-9 ISBN 978-3-030-98449-6 (eBook)
https://doi.org/10.1007/978-3-030-98449-6

Cover Front-Photos: Baleen-Whales: right side: Mother and calf humpback (Megaptera novaeangliae) whales from the Ha'apai group in Tonga. The calf is about 4 weeks old; at this bold, inquisitive and playful age it approached to within 2m of diver and photographer, Rob Harcourt. The mother is in the background, about 15m away. (Photo credit Rob Harcourt)
left side: An Omura's whale (*Balaenoptera omurai*) mother and calf near Banc du Goliath, north of Nosy Be, Madagascar, in November 2018. This female was first seen on Nosy Be without a calf in November 2015. Drone photo collected under research permit 266/18/MEEF/SG/DGF/DSAP/SCB issued by la Direction Générale de l'Environnement et des Forêts, and la Direction de la Gestion des Ressources Naturelles Renouvelables et des Écosystèmes, Madagascar. (Photo credit and © Salvatore Cerchio)

This Springer imprint is published by the registered company Springer Nature Switzerland AG
The registered company address is: Gewerbestrasse 11, 6330 Cham, Switzerland

Shooting Stars

The shooting stars are out tonight
I've seen at least a half a dozen
more than I did at this same time last
The night is calm
The twinkling sight
Of sights is lovely lovelier than I've ever seen
I've never seen a night like this
before you go and say goodnight
Please tell me tell me tell me
Have you ever seen have you ever seen
A night like this before

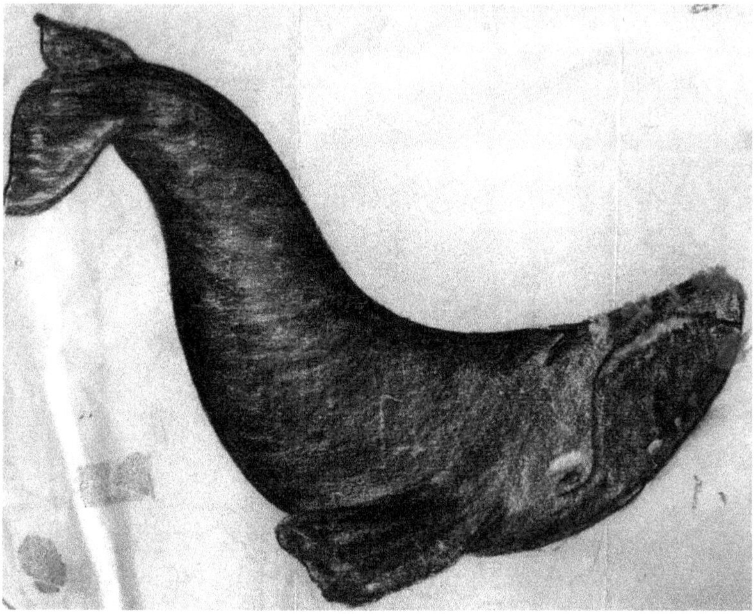

Christopher W. Clark
First song and first drawing
after arriving at Whale Camp
Golfo San José,
Chubut, Argentina
September 1972
An adventure that became my life

Introduction to the Series

We—multiple topic editors and authors—are pleased to provide a series on the ethology and behavioral ecology of marine mammals. We define ethology as "the science of animal behavior" and behavioral ecology as "the science of the evolutionary basis for animal behavior because of ecological pressures." Those ecological pressures include us, the humans.

We determine, somewhat arbitrarily but with some background, that "marine mammals" habitually feed in the sea, but also include several mammals that went from saltwater oceans back into rivers, as seen in the chapter by Sutaria et al., in the first book on odontocetes. Polar bears represent a somewhat outlier "marine mammal," as they are quite at home in the sea but can also feed on terrestrial mammals, birds, berries, lichens, and mosses.

In six books, we include toothed whales (the odontocetes); baleen whales (the mysticetes); sea lions and fur seals (the otariids) as well as the walrus; true seals (the phocids); the special cases of the sea otter and polar bear; and manatees and the dugong (the sirens). Each of our chosen editors and their chapter authors has their own schedules, so the series is not arriving in the order given above, but within the 4 years of 2019 through 2022, all six marine mammal books on Ethology and Behavioral Ecology of Marine Mammals are seeing light of day, and you the readers will ascertain their worth and promise as to current knowledge and to accumulating data while our fields of science advance.

Since the first book on odontocetes came out in 2019, we added a seventh final book, on The Evolving Human Factor, with chapters on present knowledge of behavioral capabilities and societal ways of marine mammals, past assaults on marine mammals, continuing assaults on the marine and other environments, dawning of awareness of assaults, and perhaps ways that we humans can and must do better. Several of us simply felt that to detail modern science of marine mammal ethology and behavioral ecology was not enough—we need to be aware of the amazingly destructive Anthropocene epoch in which we live and try to improve for all of nature, and therefore also for us. While topics of human influence run throughout each of the first six books, a concentration on human actions and potential solutions is supplied in Book 7.

Not all mammals that occur in marine waters are represented, nor all that have gone back to freshwater. Thus, there is nary a mention of marine-feeding bats, marine-feeding river otters, those aspects of beluga whales that foray way up into major rivers, seals living in land-locked lakes at times thousands of kilometers from the ocean, and other species that occasionally make the marine environment or—as generally accepted marine mammals—adjacent freshwater systems their home. Such are the ways of a summary, and we apologize that we do not fully encompass all. As a series editor, I have been a science partner to all major taxonomic entities of this series, but to this only because I have been in the marine mammal field for 50+ years now, with over 65 graduate students who—in aggregate—have conducted research on all seven continents. In no manner do I pretend to have kept up with all aspects of diverse fields of modern enquiry. It is a special privilege (and delight) to have multiple up-to-date editors and their fine authors involved in this modern compilation and am extremely grateful (and humbled) for this, still learning, and ever so.

Each chapter is reviewed by the book editors, peer-reviewed by other scientists as chosen by the editors, and perused and commented on by me. If you learned something new and imparted that to your colleagues, students, or your own mentors, then the series and sections of it shall have been worthwhile.

Tortolita Desert, AZ, USA Bernd Würsig
January 2022

Preface

This book presents an extensive and in-depth compendium of chapters on baleen whale ethology and behavioral ecology. Nine of the 15 chapters are led by leading female scientists in the field, and eight chapters are all-female teams, a reflection of the reality that women are significant contributors to baleen whale science.

The task of co-editing this book started out like a blind date. Initially, I (CWC) had accepted Bernd Würsig's offer to edit this book on baleen whales, but soon realized that I would need help. When I asked Phil Clapham if he would consider working with me as a co-editor, he emphatically said no, but suggested that I seriously consider co-editing with Ellen C. Garland (ECG). Speaking of an apparent cultural separation, how was this going to work? Ellen is advancing through the beginning of her stellar career, while I am slowly and at times reluctantly sliding into the end of mine. She is a female scientist from New Zealand growing her scientific garden at the University of St Andrews, while I am a male scientist from the USA and recently retired from Cornell University. It was one of those situations in which we sort of "knew" each other by scientific reputation, but not as individuals, each with our own personal and professional experiences, histories, and idiosyncrasies. Furthermore, I was representative of the old-white-male days when men, especially those from the USA, dominated the waterfront of marine mammal science. These conditions posed a challenge, especially given that our means of communicating and learning to work together were constrained by meeting virtually because of the global COVID-19 pandemic. Could we find a balance in our shared responsibilities without the biases and attitudes that are the silent manifestations of different experiences and ways of thinking? Could we trust each other to leave ego aside? Would we listen to each other with open minds? We are delighted to assert that working together as co-editors has been an amazing, at times challenging, but very rewarding experience. We learned. We adapted. We grew.

This book aspires to present the current state of knowledge on baleen whale ethology and behavioral ecology, while highlighting important challenges and future directions for study. Through its chapters, several proximate themes emerge: the diversity of the authors' technical and scientific backgrounds, the benefits from applied combinations of emergent technologies, and the critical importance of long-term datasets. These lead to insights as to how whale behaviors change through time over ecologically appropriate spatial scales and the evolutionary implications of such observations. An underlying theme is our collective concern over the influences of man-made activities on ocean acoustic environments and their ultimate impacts on individuals and populations.

The layout of this book follows others in the series. It is divided into two parts. The first consists of eight chapters, the first of which (Chap. 1, Clark and Garland) showcases a variety of photographs demonstrating some of the biological themes and behaviors discussed in the next seven chapters. The second part of the book consists of seven chapters that are species specific with an emphasis on acoustic ecology, reflecting the strong bioacoustics focus of both co-editors. Chapter 2 (Clark and Gagnon) serves as an overview of baleen whale acoustic ecology and how the field has benefited from contributions by physical acousticians and acoustic oceanographers. By utilizing U.S. Navy large-scale arrays as telescopes, Clark and Gagnon peer deep into ocean acoustic environments and specifically into the acoustic habitats and behaviors of baleen whales at ocean basin scales. The lessons gleaned from these experiences have been profound. They have provided a continuous stream of novel insights and discrepant events at unique scales of space, time, and frequency. The chapter highlights the implications of acoustic observations of baleen whale behaviors on global scales based on small-scale recorders and large-scale acoustic arrays.

Chapter 3 (Reidenberg) provides an overview of baleen whale anatomical and physiological adaptations, with a focus on traits adapted for listening to and producing sounds in a marine acoustic environment. The chapter highlights how underlying anatomical structures can influence biological function, which is a challenging task given the difficulties of studying baleen whales. Chapter 4 (Horton et al.) provides an overview on the current knowledge of spatial and temporal distributions of baleen whales and their migratory behaviors. It highlights how technological innovations (e.g., satellite tags, photo-ID, and passive acoustics) have helped to revolutionize our understandings of baleen whale behavioral ecology and ethology. Chapter 5 (Friedlaender) explores the feeding strategies that baleen whales exhibit through an evolutionary foundation of baleen and filter feeding, focusing on how and why these strategies developed. Such a foundation is imperative to contextualize feeding behaviors that are observed across a range of species in varying habitats. These insights, driven by new technologies, are fundamentally changing our understanding of the roles of baleen whales in ocean ecosystems. Chapter 6 (Carroll and Garland) highlights genetic and genomic tools that provide unique insights into the hidden social lives of baleen whales. By integrating genetic information with other data, new aspects of whale ecology are explored including information on mating systems, population recovery, identification of previously unknown prey species, and identification of units to conserve. Chapter 7 (Tyack) investigates the social organization of baleen whales. It explores the consequences of an annual reproduction cycle, which creates spatiotemporally separated breeding and feeding grounds and seasons, and long-distance migrations between them. It discusses social groupings and the few examples of long-term associations in some foraging groups. It concludes by questioning the traditional assumption of what constitutes a "group" given baleen whale abilities to maintain acoustic contact over much greater distances than are usually assessed by human observers. As originally stated by Roger Payne, baleen whales may form long-range "heards" in addition to shorter range "herds." Chapter 8 (Garland and Carroll) explores the concept of social learning and culture in baleen whales. Some of the strongest evidence for culture in animals has come from cetacean studies, as culture dominates cetaceans' lives. The chapter highlights a number of studies including examples of maternally directed (and thus culturally transmitted) site fidelity to breeding and feeding areas and migratory routes; dynamic cultural transmission of song; and social transmission of novel feeding techniques.

The second part of this book (Chaps. 8–15) focuses on species-specific examples of acoustic behavioral ecology in baleen whales and delves deeply into understandings of acoustic communication, with an emphasis on songs and singing behaviors. Chapter 9 (Širović and Oleson) investigates diversity and behavioral variability in blue whale vocalizations, including song. The chapter explores some of the variability in blue whale acoustic behaviors and the possible influences of habitat and prey on blue whale population trends. Chapter 10 (Dunlop) investigates the mysteries of humpback whale social communication, which has until recently been a neglected part of humpback whale vocal behavior. It explores their large and complex acoustic repertoire and concludes by suggesting that humpback whales may have a complex social structure, contrary to previous assumptions. Chapter 11 (Cholewiak and Cerchio) provides a historical synthesis and global overview based on five decades of research into humpback whale song. It explores common themes and highlights gaps that remain in our understanding of this complex, structured, and culturally transmitted male reproductive behavior.

Chapter 12 (Stafford) explores the acoustic behavior of the bowhead whale, the only baleen whale endemic to the Arctic. Acoustic communication is critical in bowhead whale life history, and many unknowns remain; the chapter explores the year-round calls and the complex, two-voiced song displays produced during the winter–spring season. Chapter 13 (Parks) explores the acoustic communication of the three currently recognized species of right whales. It highlights the graded nature of the acoustic repertoire and integrates behavioral context into understandings of sound type usage, which may also reflect the sex and age class of the calling individual. Chapter 14 (Risch) explores the acoustic diversity and variability in the elusive minke whale. It highlights how recent advances in technology have aided in discovering that minke whales are the source of several sounds including the "boing" and "bioduck," which baffled seafarers for half a century. It also explores minke whale song and suggests that, like other baleen whales, most minke whale populations produce long duration song sequences presumably in a reproductive context. Chapter 15 (Cerchio) synthesizes current understandings of Omura's whale global distribution, song, and vocal behavior. Omura's whale was first described in 2003, and there remains a paucity of information on this species. This chapter presents valuable new information that leads to inferences about the species' behavioral patterns at a population level and explores what these observations may suggest about its mating system.

We would like to thank all authors for completing their chapters during a historic human event—the COVID-19 pandemic. The challenges were many, devastating, and very personal. We dedicate this book to all those we lost.

Finally, as the co-editors, we want to thank each other for the many energetic and enlightening conversations about the finer points of bioacoustics, English grammar, and life. To Chris, thank you for sharing your knowledge and wisdom; it has been a privilege and a pleasure. To Ellen, your youthful energy has been contagious, and your novel ways of thinking about vertebrate social systems have unambiguously motivated me to expand my ways of thinking about baleen whale ethology and behavioral ecology. There is so much more to learn and so much more to share. Thank you.

Christopher W. Clark at the edge of the North Atlantic Ocean; Longnook Beach, Cape Cod, MA USA, where he was born and raised

Ellen C. Garland on fieldwork in the Southern Lagoon, New Caledonia, in 2018, listening for singing humpback whales

Truro, MA, USA
St Andrews, Scotland
December 2021

Christopher W. Clark
Ellen C. Garland

Contents

Part I
Patterns of Mysticete Ethology and Behavioral Ecology

Chapter 1
Observing Baleen Whales

Christopher W. Clark and Ellen C. Garland

Abstract There is a saying that goes: "A picture is worth a thousand words." This book is a simple storybook of sorts, full of thousands of words about baleen whales, how they live their lives today, and how their many remarkable adaptations might have come to be. The idea for this opening chapter arose out of an undeniable desire to have more pictures, to *show you*, the reader, the diversity of species and some of their observable adaptive behaviors. By combining photographs with captions, we further elucidate adaptations and behaviors that are explored in depth throughout this book. In the coming decade, we should all be prepared for a dramatic increase in scientifically based, multimedia storybooks that explore and reveal the full spectrum of biology on this singing planet.

Keywords Antarctic minke whale (*Balaenoptera bonaerensis*) · Baleen whale · Behavior · Behavioral context · Blue whale (*Balaenoptera musculus*) · Bowhead whale (*Balaena mysticetus*) · Fin whale (*Balaenoptera physalus*) · Gray whale (*Eschrichtius robustus*) · Humpback whale (*Megaptera novaeangliae*) · Minke whale (*Balaenoptera acutorostrata*) · North Atlantic right whale (*Eubalaena glacialis*)

See Figs. 1.1, 1.2, 1.3, 1.4, 1.5, 1.6, 1.7, 1.8, 1.9, 1.10, 1.11, and 1.12.

C. W. Clark (✉)
K. Lisa Yang Center for Conservation Bioacoustics, Cornell Lab of Ornithology, Cornell University, 159 Sapsucker Woods Road, Ithaca, NY 14850, USA
e-mail: cwc2@cornell.edu

E. C. Garland
Sea Mammal Research Unit, School of Biology, University of St. Andrews, St. Andrews, Fife KY16 8LB, UK

C.W. Clark and E. C. Garland (eds.), *Ethology and Behavioral Ecology of Mysticetes*, Ethology and Behavioral Ecology of Marine Mammals,
https://doi.org/10.1007/978-3-030-98449-6_1

Fig. 1.1 Two gray whales (*Eschrichtius robustus*) off the west coast of Vancouver Island, likely diving to feed on dense swarms of planktonic mysids that are near the shallow bottom (15 m). Although gray whales are best known as bottom feeders (e.g., sifting the benthos for ampeliscid amphipods), they will also feed in the water column and at the surface (Moore et al. 2007). (*Photo Credit* Jim Darling)

Fig. 1.2 Large group of Antarctic minke whales (*Balaenoptera bonaerensis*) in the sea ice of Wilhelmina Bay on the western side of the Antarctic Peninsula. Among the group is a single whale instrumented with a suction-cup, motion-sensing, recording tag. Data from the tag are used to study subsurface behavior including acoustic communication, feeding and diving patterns, and habitat use (Chaps. 5 and 14). As a species that is dependent on sea ice, Antarctic minkes are at risk because the waters of the Antarctic Peninsula are rapidly warming. Data from tagging technology provide insights to better understand how minke whales behave and interact with their environment, and how baleen whales are impacted by anthropogenic disturbances. (*Photo Credit* Ari Friedlaender, photo taken under the authority of NMFS and ACA permits as well as OSU IACUC)

Fig. 1.3 Bowhead whale (*Balaena mysticetus*) surfacing in a spring lead, surrounded by *sikuliaq* (the Inupiat word for young ice) about 5 miles Northwest of Utqiaġvik, Alaska, May 2021 (Chap. 12). This whale is navigating a half-mile offshore through a partially open-water lead system. The following day, heavy sea ice completely closed the lead, and no open water was visible. Note the pronounced height of the whale's head above the waterline. This is where its nares (nostrils/blow holes) are located. The height of the nares is an adaptation for breathing in small openings in the ice. The elasticity of the tissue around the nares is a physiological adaptation for breaking up through the ice as the whale comes to the surface to breathe (George et al. 1989; Hillmann et al. 2021). Also note the small white scar on the right side of the whale's back, a result of ice cutting through the whale's skin and exposing its blubber. After migrating through the Bering Strait and swimming past Utqiaġvik in the spring, bowheads turn east and migrate for 100 s of miles through mostly 100% ice cover into the Canadian Beaufort Sea where they spend the summer feeding. (*Photo Credit* John Craighead George)

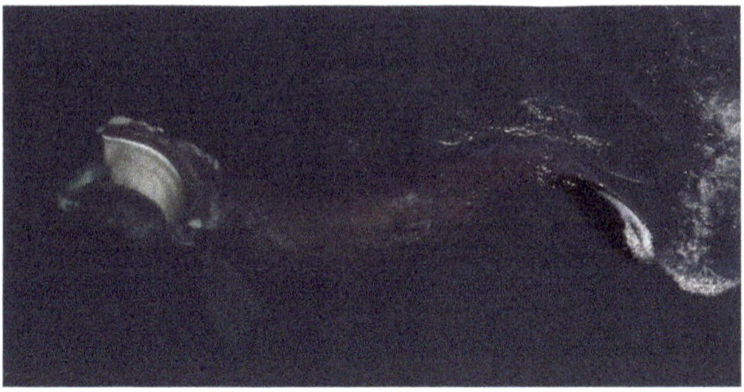

Fig. 1.4 Aerial view of a North Atlantic right whale (*Eubalaena glacialis*, Eg #4520) subsurface feeding on its right side in Cape Cod Bay, MA, USA, May 2021. The whale's mouth is open revealing a row of whitish baleen plates suspended from its upper left jaw. The curved rim of its lower left jaw starts at the front of the mouth, arches upward, across, and downward, and ends with a slight curl just below the right eye. Above the eye is a white patch made white by whale lice attached to a raised area of roughened skin called a callosity. Directly above this white patch, at the top of the head, are the whale's nostrils, around which there are also several callosities. Each whale has a unique assemblage of callosities that are used to identify, recognize, and distinguish between individuals (Seger and Rowntree 2018; Chap. 13). (*Photo Credit* Center for Coastal Studies, taken under NOAA permit #19,315–01)

Fig. 1.5 Minke whales (*Balaenoptera acutorostrata,* dwarf type) on Great Barrier Reef (Ribbon Reef #10), Australia. These four whales were part of a larger group of 14 that were present for eleven hours around Curt & Micheline Jenner's research vessel, *Whale Song*, and interacted with divers in the water. This population and location offer opportunities to apply combinations of methodologies (e.g., acoustic recording, biopsy sampling, photo-identification, satellite tagging, and visual observation) to study individual and group behaviors within a known ecological context (Chap. 14). (*Photo Credit* Wayne Osborn)

Fig. 1.6 Aerial view of a surface active group of at least four North Atlantic right whales (*Eubalaena glacialis*) in Cape Cod Bay, MA, USA, January 2020. Three individual whales are at the surface, and the white belly patch of a fourth individual is evident beneath them. (*Photo Credit* Center for Coastal Studies, taken under NOAA permit #19,315–01)

Fig. 1.7 Two humpback whales (*Megaptera novaeangliae*) off Isla Socorro, México, displaying high-energy surface activity in the midst of a competitive group. The whale in the background is breaching, while the one in the foreground is tail-slapping on the water's surface (Chaps. 10 and 11). (*Photo Credit* Danielle Cholewiak)

Fig. 1.8 Bowhead whale (*Balaena mysticetus*) mother and calf offshore Cumberland Sound, Canada, photographed from an unmanned aerial vehicle (drone), August 2019. Note the white chins of both whales, the lack of any callosities (compared to right whales, another balaenid, which have callosities), and the numerous gray and white marks, most likely from scraping against ice. The mother's blowholes are barely evident as twin slits near the front of her head beneath a cloud of exhalation mist (Chap. 12). (*Photo Credit* Ricky Kilabuk of Pangnirtung, NU, Canada)

Fig. 1.9 Aerial view of North Atlantic right whale (*Eubalaena glacialis*, Eg #3942) feeding below the sea surface next to her 2021 calf in Cape Cod Bay, MA, USA, May 2021. The mother's mouth is open revealing the rows of whitish baleen plates suspended from the left and right sides of her narrow upper jaw. The two smallish white dots at the back sides of her open mouth are callosity patches above the left and right eyes: The whitish characteristic is the result of whale lice (cyamids) attached to each patch. At the top of the head, midway between the eye patches are two very small white callosities. These are just behind the whale's two blowholes, which appear as a mirrored pair of curved grayish slits. This female's unique assemblage of callosities along her narrow, upper jaw, as well as the callosity patches on her calf's head, is used to identify and recognize this mother and calf (Chaps. 7 and 13). (*Photo Credit* Center for Coastal Studies, taken under NOAA permit #19,315–01)

Fig. 1.10 Pair of fin whales (*Balaenoptera physalus*) swimming from left to right while surface feeding in tandem through a dense patch of euphausiids (*Nyctiphanes simplex*) in the Gulf of California, Baja Sur, Mexico. This tandem formation allows for greater water column coverage when feeding (Chap. 5). In addition, a fin whale's head region is asymmetrically colored such that the left lower jaw is dark and the right lower jaw is whitish, an adaptive asymmetry that aids in corralling prey. Ventral pleats extend two-thirds of the body length back to the umbilical scar, enabling the whales to engulf large amounts of food-laden water. Rows of baleen plates suspended from both sides of the upper jaw and an enormous muscular ventral pouch and tongue are used to expel the water and filter the food. In this photo, each whale is swimming on its right side while plowing through the food patch such that only parts of the lower left jaw, upper left jaw, ventral pleats, and baleen are visible (Ramp et al. 2015). (*Photo Credit* Richard Sears, Mingan Island Cetacean Study)

Fig. 1.11 North Atlantic right whale (ID, Eg #4633) turns while skim feeding at the surface in Cape Cod Bay, MA, USA, March 2020. Its open mouth reveals the comb-like structure of baleen plates attached to its upper jaw, an adaptation that enables the whale to filter and consume copepods and other tiny zooplankton from the ocean (Chap. 5 and 13). (*Photo Credit* Center for Coastal Studies, taken under NOAA permit #19,315–01)

Fig. 1.12 Blue whale (*Balaenoptera musculus*) lunge feeding at the surface off Redondo Beach, California, USA, as viewed from the front right side of the whale (Chap. 9). The blow holes (i.e., nares or nostrils) are in a barely visible indentation on the top of the head. The right eye is visible on the side of the head below the nostrils where there is a slightly slanted, rectangular patch of white water. The black baleen is suspended from the whale's upper jaw and appears fuzzy due to the hair-like endings of the baleen plates. Baleen is a keratinaceous material, similar to the composition of fingernails (Chap. 3). (*Photo Credit* John Calambokidis)

References

George JC, Clark C, Carroll GM, Ellison WT (1989) Observations on the ice-breaking and ice navigation behavior of migrating bowhead whales (*Balaena mysticetus*) near Point Barrow, Alaska, Spring 1985. Arctic 42(1):24–30

Hillmann D, Tarpley RJ, George J, Nader P, Thewissen J (2021) Anatomy of skull and mandible. In: The bowhead whale. Elsevier, pp 127–136

Moore SE, Wynne KM, Kinney JC, Grebmeier JM (2007) Gray whale occurrence and forage southeast of Kodiak, Island, Alaska. Mar Mamm Sci 23(2):419–428
Ramp C, Delarue J, Palsbøll PJ, Sears R, Hammond PS (2015) Adapting to a warmer ocean—seasonal shift of baleen whale movements over three decades. PLoS One 10 (3):e0121374
Seger J, Rowntree VJ (2018) Whale lice. In: Encyclopedia of marine mammals. Elsevier, pp 1051–1054

Chapter 2
Baleen Whale Acoustic Ethology

Christopher W. Clark and George J. Gagnon

Abstract There has been enormous growth in technical mechanisms for collecting, analyzing, and visualizing baleen whale acoustic behaviors. Organizing and synthesizing the import of these behaviors remain a challenge, as is the placement of such efforts within the broader framework of adaptation, selective advantage, and behavioral and evolutionary ecology. Synthesis based on bioacoustic behavior includes consideration of low-frequency, physical acoustic propagation in the marine environment and the resultant potential beneficial opportunities for baleen whales to communicate, forage, navigate, orient, and maintain social organization. Observations of baleen whale bioacoustic behaviors range from singing as a male reproductive advertisement display occurring over periods of many months within a potentially enormous communication space to non-singing events associated with short duration social contexts involving both sexes and multiple age groups. Such observations are helpful as a starting framework but should be recognized as simplifications given the high levels of behavioral variability and complexity inherent in these long-lived, large-brained species. A variety of observations, referred to as discrepant events, suggest that present understandings of baleen whale behavioral ecology are insufficient to explain the spatial and temporal scales over which baleen whales engage in bioacoustic behaviors. Such behaviors and behavioral variability at ocean basin scales have promoted the concepts of acoustic environment and acoustic habitat, and motivated concerns over the biological influences of anthropogenic sounds on species-specific habitats, behaviors, and survival.

Keywords Baleen whales · Behavioral ecology · Acoustic ecology · Acoustic habitat · Acoustic behavior

---

C. W. Clark (✉)
K. Lisa Yang Center for Conservation Bioacoustics, Cornell Lab of Ornithology, Cornell University, 159 Sapsucker Woods Road, Ithaca, NY 14850, USA
e-mail: cwc2@cornell.edu

G. J. Gagnon
Marine Acoustics, Inc., 2417 Camino Real So., Virginia Beach, VA 23456, USA

C.W. Clark and E. C. Garland (eds.), *Ethology and Behavioral Ecology of Mysticetes*, Ethology and Behavioral Ecology of Marine Mammals,
https://doi.org/10.1007/978-3-030-98449-6_2

Background (CWC)

The first time I heard the idea that some whales could have communicated across an ocean basin was in 1972. The idea was hypothetical. Roger Payne and Doug Webb, a biologist and an acoustic oceanographer, postulated that prior to modern shipping and the collective noise from all those engines and propellers, fin whales (*Balaenoptera physalus*) could have communicated across an ocean basin (Payne and Webb 1971). Roger and Katy Payne, along with their four young children, Bernd and Mel Würsig, and I, were sitting around a campfire, beneath the Southern Cross on the shores of Golfo San José, Argentina. We could hear the blows of right whales (*Eubalaena australis*) just a few hundred meters away. Roger was expounding on the theoretical possibility that under naturally quiet conditions, the extremely low-frequency sounds of fin whales could propagate unbelievably great distances. This, he explained, resulted from a special combination of physical realities and biological adaptations, such that some whale sounds that were extremely intense and below the threshold of human hearing (infrasonic) could theoretically propagate across an ocean basin. Although I was curiously incredulous, that notion remained in my mind like a dormant seed for two decades. In 1992, I observed the truth of that seedling first-hand during an initial visit to a high-security US Navy facility in which exceptionally skilled acoustic analysts detect, identify, and track submarines and surface vessels. It was there that I realized that the Payne and Webb hypothesis was true: the song of a blue whale (*Balaenoptera musculus*) singing on the edge of the Grand Banks of Canada, could be detected off Bermuda 1600 km away! That was the beginning of the "Whales' 93" project (Costa 1993; Nishimura 1994) and a lifetime friendship with Lt. Commander (LCDR) George "Chuck" Gagnon, who became my teacher and very close friend in our expeditions using the US Navy's Integrated Undersea Surveillance System (IUSS) to observe acoustically active whales at ocean basin scales (Gagnon and Clark 1993; Clark 1995b; Clark and Gagnon 2002; Clark et al. 2019). I often refer to this system as an acoustic telescope because it allows one to peer deep into the ocean's acoustic environment and specifically into the acoustic habitats and behaviors of baleen whales. The lessons from these experiences have been profound; a continuous stream of novel insights and discrepant events at scales of space, time, and frequency that I believe are better matched to the sensory systems, voices, behaviors, movements, and lifespans of these whales, not the unintentionally restricted perspectives resulting from the limited observational mechanisms presently available in the non-military, unclassified domain. Biological insights from the acoustic telescopes in combination with the rapidly growing observations of acoustic behaviors by scientists on a global scale have significant implications for how we synthesize and merge these observations within the pantheon of behavioral ecology. The voices of these whales are neither arbitrary nor whimsical, just as the suite of behaviors that accompany these acoustic performances are the living reflections of evolution's economy; the costs and benefits imposed by the Earth's physical and biological environments.

2.1 Introduction

This chapter is not only about baleen whales. They are the focus, the cast of characters in the play, but the play is but one of many in the anthology of plays about acoustic ecology, whether species-specific, phylogenetic, or taxonomically agnostic.

Peter Marler in "On innateness: Are sparrow songs 'learned' or 'innate'?" (Marler 1999, p. 293) wrote:

> A layperson might be excused for wondering if the nature-nurture debate will ever be resolved to everyone's satisfaction. There is no unanimous agreement on this theme even among serious students of animal behavior. Many still seem to be either unaware of the contemporary revolution in molecular genetics and developmental biology, or reluctant to admit its relevance. No sooner does a biologically satisfactory consensus seem to have emerged than the assertion resurfaces that any invocation of genetic influences on the development of behavior implies complete predestination, eliminating all prospects of mutability and adaptive ontogenetic change. Outdated attitudes such as these are not the only obstacles to scientific progress. There is also a problem with the terminology that many of us use, and that is the issue this paper seeks to address.
>
> Even those of us who are all too aware of the maze of interactive, often probabilistic steps to be taken in the journey from a gene to a complex behavioral trait sometimes still apply, albeit somewhat casually, terms like innate and learned to a behavior. We use them as useful adjectives that provide an apparently objective and instructive way of classifying behavioral traits. This classificatory usage of pairs of dichotomous terms, like instinctive and acquired, learned and innate, or learned and unlearned, is deeply ingrained, apparently continuing to be useful as a way of distinguishing between categories of behavior with very different ontogenetic histories. It is true that some refrain completely from using the term innate, but they still label behaviours as unlearned without any hesitation.

Although perhaps not obvious, the translation of this nature-nurture perspective into a similar dynamic regarding the combination of physical acoustics, biological inheritance, and behavioral acquisition through selective adaptation is especially appropriate when discussing the acoustic ecology of baleen whales. Science involves and requires a level of self-awareness, which can be acquired but must certainly be self-taught, that every attempt at acquiring knowledge has the potential to compromise further advancement unless one is willing to sacrifice one's self-imposed conditions, especially when the phenomenon of our curious attention is biological and occurring at scales beyond most presently available observational mechanisms. To paraphrase teachers of Darwinian selection and evolution: variability is grist for the evolutionary mill. We can expand the term evolutionary to embrace the full spectrum of possibilities from genetic to cultural evolution, while recognizing that evolution can appear categorical as well as a Mobius strip of possibilities (Starostin and Van Der Heijden 2007). Constraining our embrace of variability for the sake of simplification often leads to paths that satisfy our proximate motivations, yet takes us into places often devoid of ultimate understandings. The science of baleen whale acoustic ecology is very much a case in point: although still at the base of the mountain, we are looking upward. Our excitement, rather than concern of being only at the base, arises from the remarkably rapid increase of the accumulated understandings, accelerated by an equally remarkable rise in observational and analytical technologies. It is now timely and appropriate to shift the collective scientific focus onto questions of behavioral, ecological, and evolutionary importance (Slabbekoorn and Smith 2002).

2.2 How We Got Here

Scientific research on baleen whales was initially driven by direct exploitation of whales as a commercial resource and the need to know more about remaining populations and how many whales were available for continued harvesting (Ross 1835a; b; Scammon 1874; Brown 1954; Holt 1974; Lockyer 1984; Tormosov et al. 1998; Bannister 2009). The 1960–80s initiated a shift from exploitation to scientific investigations (Nishiwaki 1966; Payne 1972; Katona et al. 1979; Herman et al. 1980; Winn et al. 1983). This primarily coincided with the advent of novel and mostly visual observation tools (e.g., aerial survey, photo-identification, tagging, theodolite tracking), but also included dabs of bioacoustic research driven by the Cold War and the need to understand biological contributions to the ocean's low-frequency (<1000 Hz), underwater acoustic environment (Schevill and Watkins 1962; Patterson and Hamilton 1964; Schevill et al. 1964; Cummings and Philippi 1970; Cummings and Thompson 1971; Northrop 1974; Watkins et al. 1981). After the Cold War, in the mid-1990s, the US Navy produced an Environmental Impact Statement for their Low Frequency Active Sonar (LFAS) system. This coincided with and fueled the growing scientific and public concern over potential adverse impacts from non-lethal, manmade sound-making activities in the ocean, both intentional (e.g., active sonars, seismic exploration explosions) and non-intentional (e.g., vessel engine noise), on individuals and populations (National Research Council 2003, 2005).

The more recent focus on anthropogenic noise impacts represents a form of scientific conservation where the underlying motivation for increased knowledge is species and ecosystem protection. In some western societies, these recent concerns have been increasingly driven by socially driven, legally mandated requirements about acoustic exposure levels and possible resultant biological impacts from anthropogenic acoustic activities (National Research Council 2003, 2005; Wartzok et al. 2004). Scientific progress has been driven by a broadening of understandings of physical acoustic environments, increases in acoustic data collection and analysis technologies, and a subsequent rise in descriptive observations on bioacoustic behaviors. Such observations have led to initial efforts at species-specific levels of syntheses and acknowledgment that biological understandings at ecosystem scales are essential for assessing possible negative consequences of human activities (e.g., Southall et al. 2019).

A focus on deriving anthropogenic risk assessments can be a mixed blessing. On the one hand, it has resulted in an influx of scientist-engineers trained in physical acoustics, acoustic oceanography, and data analytics, along with the co-evolving investments in analytical and computational technologies. On the other hand, it has led to an underinvestment in scientists trained in such fields as behavioral and evolutionary ecology, anatomy, endocrinology, and neuroethology. That said, research on marine mammal acoustic ecology although embryonic by comparison, is undergoing a rapid change that is placing baleen whale acoustic behaviors within a foundation of ecology as well as within the broader frameworks of biological evolution and cultural transmission.

In this chapter, we emphasize several features of sound propagation in the low-frequency range (approximately 10–1000 Hz),[1] within which baleen whales typically produce their sounds, and are known to have specialized auditory (Ketten 1994; Parks et al. 2007; Yamato et al. 2012) and sound production capabilities (Reidenberg and Laitman 2007; Damien et al. 2019) adapted for the oceans' low-frequency acoustic environment. On the broad ocean basin scale, there is much variability in when and where different species produce different types of low-frequency sounds (Chaps. 9–15). Associating the nuances of these relatively scant and often disparate observations within a behavioral ecological context is difficult and typically defaults to existing assumptions about what animals should or might be doing (e.g., communicating, foraging, mating, migrating), which can lead to circular reasoned conclusions about behavioral function. At the same time and on a much finer scale, there are intentional efforts by the research community to obtain bioacoustic observations during specific behavioral contexts (e.g., Southall and Novacek 2009; Goldbogen et al. 2013). Such observations, although extraordinary, don't yet include the extremely large spatial–temporal ecological scales within which baleen whale acoustic behaviors most likely occur.

How does one assign a function (e.g., communication, navigation, foraging) or an ecological context to a baleen whale acoustic behavior when the spatial and temporal scales over which that behavior can occur are still beyond the scales of our observational capabilities? Over what distances and over what time spans do acoustic behaviors such as communication occur? How does one recognize an acoustic communication event when the potential communication range and audience is many orders of magnitude greater than our present conceptual understandings and includes, for example, auditory time delays of up to tens of minutes? Is it possible to conduct experiments to test hypotheses about the behavioral significance of different baleen whale acoustic behaviors at these large scales? Presently, the intra- and interspecific variability in possible scales of bioacoustic behaviors is presently under-appreciated. This condition should be taken as a phenomenal opportunity for exploration and discovery, not as an impossibility. Our community is gathering huge amounts of raw data, developing a wide variety of technological and analytical mechanisms for extracting information, and increasingly able to synthesize some biologically meaningful knowledge from these efforts. The challenge is to combine these advancements to match the behavioral and ecological scales over which different species engage in bioacoustic behaviors that span the breadth of biological and cultural evolution. If we are devoted to lifelong advocations on behalf of ocean life, we must embrace this

---

[1] The adverb "approximately" is used because singing blue whales produce sounds in the 6–10 Hz frequency range that have only been detected when the animal is very close to the sensor (i.e., <100–500 m). I have assumed these were made when the whale is shunting air from its ventral sac back into the lungs (Chapter 3). It is also known that some baleen whales produce sounds in a frequency range > 1000 Hz (Chapter 11).

challenge, knowing that the power of discovery and the achievements of scientific knowledge through coherent syntheses can lead to inspirational conceptualizations as to life and baleen whales, as well as possibly our places in those histories.

This chapter is about placing observations of baleen whale acoustic behaviors into an ecological framework that is conceptually matched to the spatial, temporal, and acoustic scales of the biology (i.e., auditory neurophysiology, laryngeal anatomy, ethology). Getting the scales right is like finding and combining the edge pieces in a giant puzzle so that one starts to comprehend the dimensions and contexts of behavioral variability as well as the existing limitations of the mechanisms by which we make our observations. At this point, the puzzle is about acoustic ecology in the broadest sense, with an emphasis on acoustic observation as the mechanistic modality, but with a clear-eyed understanding that the acoustic modality is but one of many in an organism's sensory behavioral response arsenal wrapped within the larger complexities of biological evolution and cultural adaptation. The use of acoustics as our primary observational modality at least brings us into a multi-dimensional space matched to the one in which baleen whales are by all accounts exceptionally well adapted. We might not know much about the biological significance (i.e., function) of a particular acoustic behavior let alone the actual contexts in which it occurs, but at least we are starting to observe behaviors at ecologically appropriate spatiotemporal scales, and we are learning how to enhance our understandings by merging such observations with those from broader suites of scientific tools and concepts, many of which have already demonstrated their essential values as applied to other taxa (e.g., multi-taxonomic, Bradbury and Vehrencamp 1998; birds, Dantzker et al. 1999; anurans, Feng et al. 2006; bats, Simmons 2012).

2.3 Acoustic Ecology

Ecology is the branch of biology that deals with the relations of organisms to their physical surroundings and to one another. In a similar line of reasoning, habitat is that portion of the biotic and abiotic environment that influences an organism's survival and reproduction. Habitat is partially constrained and thereby defined by an organism's abilities to perceive and interact with those portions of the environment that influence its ability to survive and reproduce.

Following these definitions guided by insights from chemical ecology (Eisner and Meinwald 1995), we define:

1. *Acoustic ecology* is the branch of biology that deals with the relationships, mediated through sound, of organisms to their physical surroundings and to one another.
2. *Acoustic environment* is the physical acoustic milieu, independent of an organism's auditory or sound production capabilities. It includes biotic and abiotic contributors to the auditory scene (Bass and Clark 2003), increasingly

referred to as the "soundscape" (Schafer 1977). In the ocean, acoustic environment is dependent on hard oceanographic features, such as bathymetry and seafloor composition, that influence sound velocity and acoustic reflectivity, as well as ocean features, such as water temperature and salinity, that influence sound velocity and refraction.

3. *Acoustic habitat* is that portion or subset of the acoustic environment that influences an organism's abilities to survive and reproduce. An organism's acoustic habitat is partially constrained and thereby defined by an organism's abilities to both perceive (passive bioacoustic habitat) and produce sounds (active bioacoustic habitat), and thereby take advantage of those portions of the acoustic environment that influence its ability to survive and reproduce. The distinction between the acoustic environment and an acoustic habitat is important, because it provides terms by which to refer to the physical acoustic scene without qualifications for species or context. Thus, for example, one can define features of the active acoustic environment simply by collecting, analyzing, and describing the content of acoustic recordings. In contrast, the acoustic habitat of a species is defined by species-specific auditory (i.e., passive listening) and active sound production capabilities, is exploited by the species' acoustic signals and signaling behaviors, and can be further segregated by the features of the behavioral ecological context, which are typically categorized (e.g., communicating, foraging, migrating), but should be done so with caution. By this distinction between environment and habitat, all the marine mammals living in the same ocean area live in the same acoustic environment, but most live in different acoustic habitats. Thus, the acoustic habitat of a blue whale foraging in the Southern California Bight (Calambokidis et al. 2008) is very different from the acoustic habitat of Baird's beaked whale (*Berardius bairdii*) foraging in that same region (Stimpert et al. 2014).[2]

2.4 Physical Acoustics

The transmission properties of sound in water and the many factors that influence sound propagation are well-studied (Medwin and Clay 1998), as are the resultant dependencies of sound fields on an ocean's physical properties and the properties of its surrounding topography (Kuperman and Lynch 2004; Jensen et al. 2011). By the mid-twentieth century, it was well known within the national defense communities responsible for deploying active and passive sonars that whales contributed to the ocean's low-frequency acoustic environment (Medwin and Clay 1998). In fact, it was that defense community that provided most of the initial observations on low-frequency and very-low-frequency (10–100 Hz) whale sounds (Walker 1963; Patterson and Hamilton 1964; Cummings and Thompson 1971). Over the

[2] An acoustic habitat, which can include multiple species, is similar to but not the same as an acoustic niche, which is species-specific. Thus, an acoustic environment contains multiple acoustic habitats, and an acoustic habitat can include multiple acoustic niches.

last 30 years, the physical acoustic and acoustic oceanography communities have continued to make significant contributions regarding baleen whale acoustic behaviors, in some cases by directly providing new approaches and analytical observations over longer time scales than previously observed (Kuperman et al. 1993; Curtis et al. 1999; Watkins et al. 2000; Thode et al. 2000; Wilcock et al. 2014; Au and Lammers 2016; Ellison et al. 2016; Weirathmueller et al. 2017b) and behavioral observations with unique insights (McDonald and Moore 2002; McDonald et al. 2006; Ellison et al. 2012; Weirathmueller et al. 2017a; Helble et al. 2020). Today, there is an incredibly talented suite of biologists–engineers–oceanographers working along the continuum of the technical to biological communities making invaluable contributions on baleen whale bioacoustics (Mellinger and Clark 1994; Stafford et al. 1999; Nieukirk et al. 2004; Delarue et al. 2009; Baumgartner et al. 2011; Klinck et al. 2012; Rice et al. 2014; Wilcock et al. 2014; Erbe et al. 2015; Miksis-Olds et al. 2018).[3] Here, we merge insights from these advancements and our own experiences to explore some of their biological implications and to highlight a number of hypotheses about baleen whale acoustic behaviors.

2.5 Acoustic Opportunities for Bioacoustic Exploitation

The publication by Payne and Webb (1971) was the first in which a biologist and an ocean acoustician explicitly combined their expertise to hypothesize that baleen whales could exploit a unique feature of an ocean's very-low-frequency sound transmission environment, a feature referred to as the deep sound channel, to communicate across an ocean basin. This initial hypothesis, based on physical attributes of an ocean's acoustic environment in combination with the literature on terrestrial acoustic environments (e.g., Slabbekoorn and Smith 2002), motivates a simple question. Are there regions and periods of time in an ocean where and when physical acoustic conditions offer beneficial opportunities that whales could exploit either through passive listening alone or in combination with active sound production and other forms of behavior, such as diving and swimming? Diving vertically down and back through the water column is consistent with the hypothesis that the animal is sampling water temperature and the occurrence and features of a thermocline. Swimming at a constant speed and heading while also sampling water temperature is consistent with the hypothesis that the animal is listening for changes in the received level of a distant singer to estimate the singer's range and bearing by taking advantage of physical acoustic phenomena known as a convergence zone (Fig. 2.1) in a biological form of synthetic aperture array tracking (Hayes and Gough 2009). Performing these behaviors near an island or along a shelf break where the sound channel often

[3] See all other chapters in this book. The references we include here are a representative subset of publications by biologists, engineers, and acoustic oceanographers who have made important contributions. We sincerely apologize to all those we have not included, but who have also made equally important contributions.

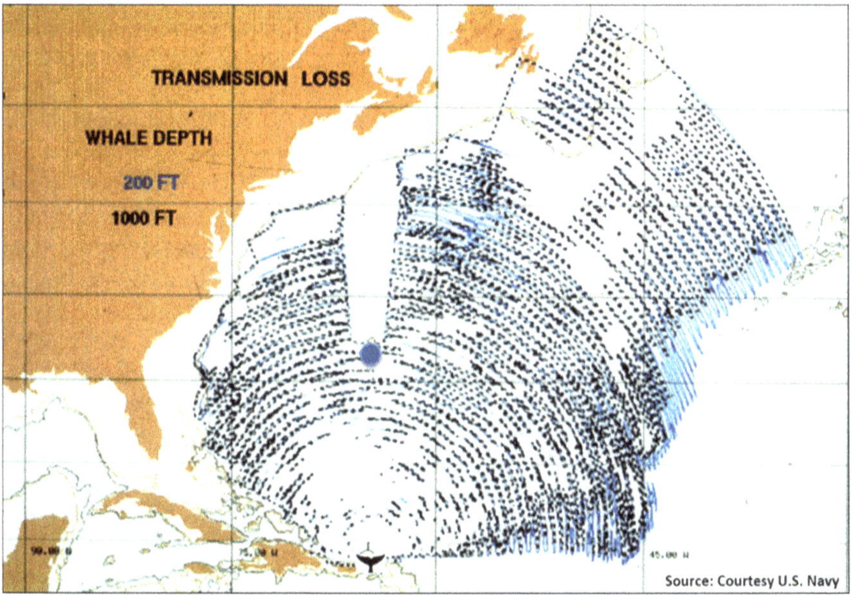

Fig. 2.1 Image based on a sound transmission model using features of North Atlantic blue whale singer (20 Hz center frequency, 2–6 Hz bandwidth, 185 decibel (dB re 1 μPA) source level, 20 s duration, 200 ft, and 1000 ft depth) located on the shelf break north of the Virgin Islands (whale tail). Notice areas of higher sound level at continental shelf edge of eastern North America, mid-Atlantic Ridge (far right comb-like edging), and convergence zone (CZ) rings, as well as the sound shadow from the island of Bermuda (blue dot)

comes closer to the surface would further increase a listening and/or an acoustically active animal's access to a larger audience. A condition of essentially all such opportunities is that they are frequency-dependent and time-bandwidth-dependent. The lower the frequency band and center frequency of the signal, the greater the physical propagation space, and in some ways, the greater the temporal space available for exploitation, while acknowledging the condition that the natural ambient noise level in the very-low-frequency range increases inversely with frequency (Urick 1983). The greater the product of signal duration and bandwidth (a unitless value referred to as time-bandwidth product), the greater the likelihood of both detection and recognition (Oppenhein et al. 1999). These considerations are driven by the assumption that exploitation of any such opportunity increases an individual's access to certain types and amounts of information and provides a selective advantage to the individual.

For simplicity, we separate these opportunities into two general forms: passive listening and active sound production. Passive listening opportunities increase the space and time span over which an individual can effectively listen to the ocean, while active sound production opportunities increase the space (i.e., volume of water) within and time over which an acoustically active individual can effectively hear the reflections of its own sounds (intra-animal communication aiding in navigation-orientation during multiple behavioral contexts) or be heard by other conspecifics (inter-animal communications). Although the order in which these two opportunities are written is intended to reflect the assumption that passive listening evolved prior to the evolution of active sound production, we do not know of any actual evidence that supports this assumption (Thewissen et al. 2009). Furthermore, our categorical separation into the passive listening and active production opportunity types is certainly a simplification, knowing that in all likelihood these opportunities are derived from co-evolved combinations of both passive listening and active sound production in concert with other adaptive behaviors (e.g., diving and swimming) and reside along some form of an evolutionary continuum (i.e., most likely punctuated (Eldredge 2014)). The important point is that the potential opportunities for biological exploitation are a result of ocean acoustic propagation conditions and an ocean's resultant acoustic environment. Multiple marine taxa, including invertebrates (Kaifu et al. 2007; Gervaise et al. 2019), seabirds (Crowell 2016), fishes (Ladich 2008; Lugli 2010; Rice et al. 2021), and sea turtles (Bartol and Ketten 2006) have passive and active adaptations that exploit opportunities in the ocean's low-frequency acoustic environment.

2.5.1 Passive Listening

By listening to an ocean's acoustic environment, including ambient sound fields resulting from physical acoustic phenomena such as reflection, refraction, and reverberation, an animal can perceive clues that could serve as indirect and direct indicators of possible foraging or reproductive opportunities. For example, late winter and early spring storms along the Aleutian Islands in the North Pacific lead to increases in very-low-frequency noise levels (Farrell et al. 2016) that are indirect indicators, i.e., potential predictors, of the onset of wind conditions along the west coast of North America. These offshore winds lead to coastal upwelling events that drive vertical migrations of benthic organisms, some of which are a major food resource (e.g., euphausiids) for blue whales (Croll et al. 2005). Could whales be using these seasonal changes in ambient noise levels as cues that motivate their movements to the coast in search of potential nutrient-rich resources and could this be an additional factor driving the timing and destinations of whale migrations to the western coast of North America (Abrahms et al. 2019)? Was this chain of causal relationships between a set of physical atmospheric-oceanic phenomena and an adaptive response from an invertebrate community reliable enough to contribute to auditory adaptations and behaviors beneficial to a vertebrate megafauna community? As an example of an auditory adaptation related to migration, could the nearly continual, very-low-frequency rumblings from the motions of tectonic plates in such places as the Caribbean (North Atlantic) and Bonin Islands (North Pacific) serve as reliable acoustic beacons for whales during their seasonal migrations between feeding areas and calving- breeding areas (e.g., humpback whales (*Megaptera novaeangliae*); see Chap. 4)? Various other types of natural marine hydroacoustic events occur in association with seafloor spreading, volcanic eruptions, and hydrothermal vents (Dziak

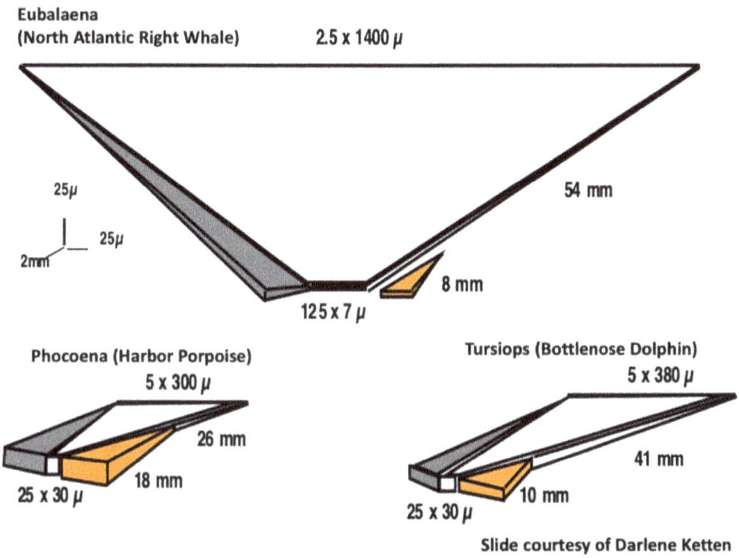

Fig. 2.2 Schematic illustrations diagraming some of the anatomically measured features and thus differences between the inner ears of a baleen whale, North Atlantic right whale (*Eubalaena glacialis*); a bottlenose dolphin (*Tursiops truncatus*), and a harbor porpoise (*Phocoena phocoena*). These differences illustrate the mechanical differences and imply the frequency response differences of the three basilar membranes, and in this case, the baleen whale is a low-frequency receiver, the dolphin is a high-frequency receiver, and the porpoise is a very high-frequency receiver. Slide courtesy of Darlene Ketten

and Fox 2002) and could also serve as natural acoustic beacons. For such hypothetical responses and benefits to become possible, one necessary condition is an auditory system adapted for perception of very-low-frequency sounds. There are unambiguous anatomical and neuroanatomical characteristics of the baleen whale inner ear that are consistent with this hypothesis, as illustrated in Fig. 2.2 (see Ketten 1994, 1997).

In a more subtle, dynamic sensing condition, ocean temperature gradients and currents can result in partially reflective boundaries that can change ambient noise conditions. Thus, for example, a cold water ocean eddy in a warmer ocean environment can have an acoustic signature in the very-low or infrasonic frequency ranges, as does a warmer water current in a colder ocean (e.g., the Gulf Stream moving through the colder waters of the North Atlantic) (Stommel 2020). Are whales adapted to sense differences in ambient noise conditions as a result of these natural physical oceanographic phenomena? If so, are there observations of their behaviors that are consistent with this consideration and its suggested hypothesis?

2.5.2 Active Sound Production

Many of the ocean's physical acoustic conditions offer opportunities for an individual or groups of individuals to produce sounds to probe and communicate within expanded ocean regions throughout longer portions of the year. Are baleen whales adapted (i.e., predisposed) to producc low-frequency sounds? Do they produce sounds that enable them to assess ocean conditions associated with opportunities to communicate (i.e., a social–cultural benefit), navigate (move between breeding and feeding areas), and find food (acquire energy)? Are baleen whales evolved specialists that use active acoustics for communicating, foraging, and navigating in the low-frequency band, just as odontocetes are highly evolved specialists that use active acoustics for communicating, foraging, and navigating in high-frequency and ultrasonic-frequency bands (Jones et al. 2005; Tyack 2019)?

In the early 1990s, there was no adequate explanation as to how baleen whales produced low-frequency sounds. Since then, Reidenberg and Laitman (2007) have published a comparative anatomical description of baleen whale laryngeal anatomy that demonstrates that baleen whales are endowed with a specialized laryngeal sac and a laryngeal structure homologous to the vocal folds in terrestrial mammals (Chap. 3). These mysticete anatomical features are dramatically different from those in odontocetes (Fig. 2.2). More recently, detailed morphologies of laryngeal structures in several baleen whale species offer evidence that baleen whales are anatomically adapted for producing low-frequency sounds leading to hypotheses as to how bowhead whales (*Balaena mysticetus*) and several rorquals might produce these sounds (Schoenfuss et al. 2014; Damien et al. 2019).

2.6 Baleen Whale Acoustic Environments, Acoustic Behaviors, and Bioacoustic Features

The highly variable suite of baleen whale acoustic behaviors is typically separated into singing and calling, with the outcomes of those behaviors referred to as songs and calls, respectively. This is a convenient, but overly simplistic bifurcation that obscures deeper levels of exploration about acoustic form and function. Because songs tend to be sung at high source levels, singing is the most obvious baleen whale acoustic behavior, and there are pronounced species-specific differences in the sound units (e.g., units or notes) from which songs are composed and in the syntax governing the composition of notes into higher levels of organization (e.g., phrases, themes, songs). Baleen whale species that are known to sing produce species-specific combinations of units within a species-specific syntax, and most have geographically distinct song types, which can be used to distinguish between different populations of singers and could be considered dialects (Chaps. 9, 11, 12, 14, 15). These basic characteristics of baleen whale singing behaviors and songs have some remarkable similarities to avian species as has been observed and so richly described for songbirds (Catchpole and

Slater 1995; Kroodsma 1996; Marler 1997). It also offers fertile ground for higher levels of examination of singing as a cultural phenomenon (see Chap. 8).

A synthesis of baleen whale song concludes that singers are males, and songs are typically composed of various combinations of frequency-modulated (FM) and amplitude-modulated (AM) units. There are two exceptions to this conclusion. The first involves bowhead whales, which are known to sing with two voices (Chap. 12) and in which there are observations that suggest females produce simple forms of song (Tervo et al. 2011). The second involves North Pacific right whales (*Eubalaena japonica*) in which it has been suggested that songs are composed of broadband, impulsive sounds known as gunshots (Crance et al 2019, Chap. 13). These exceptions should be considered evidence that there is variability in our commonly held opinions about baleen whale songs and singing behaviors, and there is much more to discover about this bioacoustic behavior.

Some studies of calling behavior have yielded interpretations on the biological significance (i.e., behavioral function or "meaning") of certain call types (Chap. 10; Clark 1982, southern right whales (*Eubalaena australis*); Chap. 13). Overall, baleen whale calling behaviors are not well described and poorly understood, and a full comparative, multi-species analysis of baleen whale calls has yet to be completed. There are various collections of sounds attributed to baleen whales, but species attribution of many of these sounds is uncertain. In some cases, this is because the only evidence is an acoustic record in a known location on a known date and time, but it is unaccompanied by contextual information and unassociated with visual observations (e.g., minke whale (*Balaenoptera acutorostrata*), Chap. 14; Omura's whale (*Balaenoptera omurai*), Chap. 15). Research into the function of baleen whale calls has advanced with the coordinated application of multimodal tools (e.g., acoustic arrays, tags, biopsy sampling), which has enabled detailed behavioral descriptions of individuals and groups within reasonably well-described social and ecological contexts. These research opportunities often include different combinations of males, females, calves, sub-adults, and adults, and occur in accessible areas before, during, or after periods of migration (Chap. 4), foraging (Chap. 5) and breeding (Tyack 1981; Darling and Berube 2001; Chaps. 7 and 11). These opportunities in concert with the growing sophistication of multimodal observation technologies and behavioral understandings offer a bounty of opportunities to study and understand baleen whale social behaviors (Chaps. 7, 8, 10, and 13).

Along with low-frequency auditory perception (Ketten 1994, 1997; Parks et al. 2007; Yamato et al. 2012; Fig. 2.2) and sound production (Reidenberg and Laitman 2007; Chap. 3; Fig. 2.3), there are physical acoustic factors and bioacoustic observations that are consistent with the hypothesis that baleen whale songs are adapted for long-range communication (Payne and Webb 1971; Chap. 9). Under natural ambient noise conditions, there are places where and times when the ambient noise level in a particular frequency band is lower than an adjacent band. Such bands of low ambient noise have been documented in terrestrial acoustic environments and are referred to as "quiet windows" (Morton 1975; Wiley and Richards 1978; Wiley 1992). Different quiet windows have been documented for different baleen species depending upon their preferred acoustic habitats in either shallow water (e.g., Clark 1982, southern

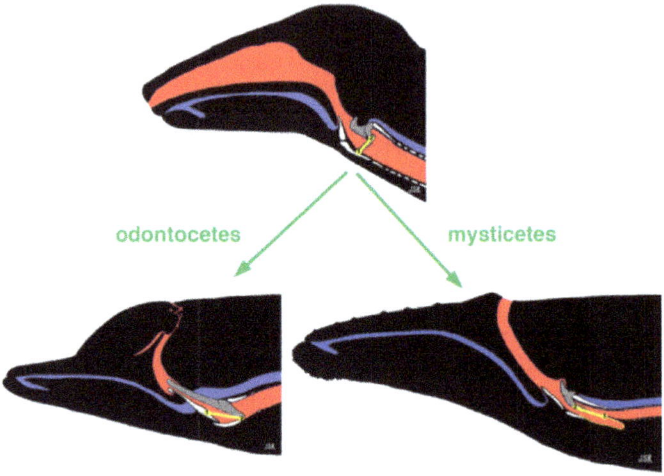

Fig. 2.3 Schematic diagram showing divergence of laryngeal regions in toothed whales (odontocetes, lower left) and baleen whales (mysticetes, lower right) relative to a modern deer (top) that represents a hypothetical baseline terrestrial ancestral condition. The respiratory tracts are indicated in red, digestive tracts in blue, midline cartilages in white, and parasagittal cartilages in gray. Courtesy of J. Reidenberg (Reidenberg and Laitman 2007)

right whales, *Eubalaena australis*; Chap. 13) or pelagic acoustic environments (Urick 1983; Clark and Ellison 2004 Fig. 73.4; Tyack 2008).

Low-frequency baleen whale sounds have been reliably detected with a single pressure sensor out to distances of many tens of kilometers (Clark and Clapham 2004), and their very-low and infrasonic frequency sounds have been detected out to distances of many hundreds of kilometers (Stafford et al. 1998; Clark and Gagnon 2002; Watkins et al. 2004). Under certain unique conditions, sound transmission models demonstrate that a 20 Hz source (e.g., blue whale song) would be detectable out to distances of several thousands of kilometers (Miksis-Olds 2015), and such enormous distances have been empirically verified (pers. obs.). A serious challenge in efforts to determine distances over which very long-range communication might occur is knowing where an acoustically active whale is located. For us that challenge has been overcome by access to large passive arrays, which are explicitly designed to operate over ocean basin areas and can be used to simultaneously locate and track individual singing and calling whales (Costa 1993; Gagnon and Clark 1993). These observations enable exploration as to whether or not there are baleen whale acoustic behaviors consistent with the hypothesis that such behaviors are adapted for different behavioral contexts such as long-range communication (e.g., Payne and Webb 1971; Clark et al. 2019), migration, or navigation. Testing the validity of this hypothesis seems nearly impossible given the ranges over which some of these physical acoustic opportunities occur. But they gain greater viability through the accumulation and synthesis of observations collected at ocean basin scales over decades and after careful consideration of discrepant events.[4]

Biological observations include both characteristics of the auditory system (Ketten 1992, 1997; Fig. 2.2) and characteristics of the sound production system (e.g., laryngeal anatomy; Chap. 3; Fig. 3.2) that are consistent with hypotheses regarding adaptive advantages for low-frequency, long-range communication (Damien et al. 2019). These considerations become the basis for the following questions. Are there places and times in the ocean where and when physical acoustic conditions offer beneficial opportunities that whales could exploit through passive listening, active sound production, and other forms of behavior (e.g., diving through thermoclines to sample temperature changes in the acoustic environment) so as to gain some advantage, and what would that advantage be? In this case, the terms "exploit" and "advantage" explicitly imply that whales that engage in any such behavior or combination of behaviors would gain a "selective advantage" through the processes of selection and Darwinian evolution. If so, what, when, and where are these conditions and circumstances in the ocean? What are the potential advantages for a whale, what are some of the expected adaptations that one would expect baleen whales to have (e.g., anatomical features, neurophysiological characteristics, and behaviors) and are there observations that are consistent with or contradictory to this general proposition?

Different combinations of physical ocean conditions (water depth; depth-dependent water temperature, pressure and salinity; and subsea composition) can result in situations in which certain types of sound remain above background noise at great distances at various depths. The relationships between the characteristics of the original sound source (e.g., frequency content, source level, time-bandwidth product) and how ocean conditions modify those acoustic features are complex (Jensen et al. 2000). That complexity aside, one relationship is unambiguous: The ocean acts as a low-pass filter such that the further away one is from a very-low-frequency, narrow-band sound source, the lower the loss of acoustic energy, while the further away one is from a broadband sound source, the greater the loss of higher frequencies (Clark and Ellison 2004, Fig. 73.5).

The ocean refracts sounds such that a sound's original characteristics can become significantly transformed as different frequencies in a signal propagate over different pathways and increasingly greater distances from the source. The physical acoustics of sound propagation are highly dependent upon frequency content and sound speed as a function of depth. As a result, sound travels through this refracted frequency-dependent path of low sound speed, not necessarily over the shortest distance between the source and a receiver. Several obvious and pronounced refractive transformations, especially in the low-frequency and very-low-frequency bands include combinations of time dispersion, in which the duration of the original sound can become compressed or lengthened, and frequency dispersion, in which the frequency content of the sound can become modified. Thus, for example, the combination of frequency and time dispersion can result in the transformation of a simple FM event into a family

[4] A discrepant event is a surprising and seemingly paradoxical observation that is not what an observer would normally expect.

of FM events (e.g., a right whale contact call, Chap. 13; McDonald and Moore 2002) or a nearly instantaneous impulsive event can be transformed into a FM event (Guerra et al. 2011; Erbe et al. 2016). Although frequency-dependent refraction leads to signal distortion, if the original signal is known, has low variability and is redundant, that distortion can provide information about where the sender is located and some of the characteristics of the ocean between the sender and receiver. Might baleen whales take advantage of such information, and what types of behaviors can be observed that would or would not be consistent with that hypothesis?

2.7 Convergence

In the mid-1990s, acoustic oceanographers conducted the Acoustic Tomography of Ocean Climate (ATOC) experiment to determine heat content throughout the Pacific Ocean (Munk et al. 1995). To probe the ocean for variations in sound speed, they designed and synthesized a very-low-frequency signal that had no biologically salient features. It consisted of a phase-modulated m-sequence with a 75 Hz center frequency, 30 Hz bandwidth, 195 dB (re 1 μPa at 1 m) source level, and a 27 s duration (a time-bandwidth product of 810) that was repeated (i.e., redundant) in transmissions lasting 10-, 20-, and 40-min (Worcester et al. 1999; Howe et al. 2019). Using their own acoustic receiving systems and the US Navy Sound Surveillance System (SOSUS)[5] receivers, these scientists collected and processed recordings of the received signal to produce detailed renderings of sound speed and heat content as functions of space and time; essentially vertical slices of ocean temperature over many thousands of kilometers from the sound source to the receivers. While listening to several of the ATOC engineer–scientists discussing the acoustic features designed into the signal, another thought emerged. Although this signal and transmission system were impressive, humans were not able to build a system that could produce sounds as low and as powerful as those produced by singing blue whales: 15–30 Hz center frequency, 5–10 Hz bandwidth, 180–185 dB (re 1 μPa at 1 m) source level, a 15–30 s duration (a time-bandwidth product of 75–300), repeated *ca.* every 70–130 s; sounds that were routinely observed on those same Navy receivers throughout ocean basin areas. During a half serious conversation with some of the acoustic oceanographers, CWC suggested that the ATOC engineer–scientists figure out a way to use the high intensity, very-low-frequency, and infrasonic sounds produced by free-ranging blue whales as their sound source (We then mused that we could call this BTOC for Blue Whale Tomography of Ocean Climate!). It was during such conversations with these brilliant oceanographers, in combination with our growing experiences using those same Navy receivers to observe the different behaviors of acoustically active baleen whales, that in some ways humans were applying signal processing theory to design very-low-frequency sounds to probe the ocean just as thousands of

[5] SOSUS is part of the IUSS, both of which are operational, and security restrictions as to the system's specifications apply.

baleen whales were doing every day, and had most likely been doing for hundreds of thousands if not millions of years. We could not help but see possible parallels between these recent physical acoustic scientific endeavors and how baleen whale bioacoustic behaviors were adapted to take advantage of ocean acoustic opportunities in the low-frequency band within the ocean's naturally quiet acoustic environment (Chap. 9).

During the first months as we were starting to work together in the SOSUS facility, CWC asked GJG (then LCDR Gagnon) what he would do if he had an underwater sonar system that could produce the sounds of a singing blue whale: Without hesitating, he responded: "I could illuminate the ocean." Fig. 2.1 is the illustration that LCDR Gagnon soon produced in support of his words (see also Miksis-Olds 2015). Which is to say, that based on blue whale song characteristics and an empirically informed low-frequency sound propagation model, a blue whale song could propagate throughout an ocean basin and create patterns of occurrence as a result of the refractive and reflective influences of sound speed gradients and bathymetric features (i.e., the depths and shapes of the underwater terrain).

Can these patterns derived from the low-frequency physical acoustics of an ocean basin be exploited by whales to communicate with other whales in the ocean's very-low-frequency acoustic environment, which in this case is part of the whales' low-frequency acoustic habitat? Do whales take advantage of these opportunities for "illuminating" the ocean in order to communicate throughout enormous ocean areas as hypothesized for fin whales by Payne and Webb (1971), and might whales also sense ocean features that could be beneficial for migrating and finding food? Can their songs actually be detected over such distances? This last question was at least partially answered a few months later when GJG showed CWC his 43 day, 3100 km track of a blue whale singer (aka Ol' Blue) that was first detected 750 km away from receivers off the island of Bermuda (Fig. 2.4).

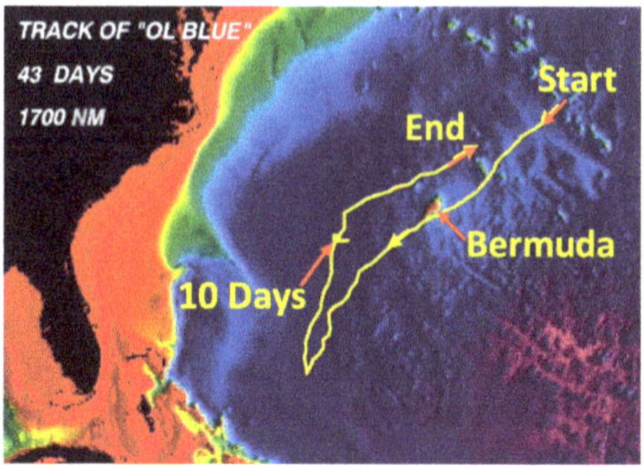

Fig. 2.4 Track of singing blue whale (aka Ol' Blue) made by Clyde Nishimura based on data collected by LCDR Gagnon in 1993 using US Navy receivers during the Whales '93 Project. The "10 Days" red arrow points to the area in which the whale stopped swimming at a steady speed and meandered for 10 days (for fin whale singers; see Helble et al. 2020)

Under certain ambient noise and sound propagation conditions in the very-low-frequency band, an ocean's acoustic environment offers opportunities for an animal to expand the space over which its sounds can propagate through the refractive features of the ocean (e.g., currents, thermoclines) so as to return reflections off the hard features of the ocean (e.g., seafloor, shelf breaks, seamounts). This type of opportunity is referred to as echo-ranging, a form of echolocation in which the animal attends to the reflective–refracted–reverberant acoustic environment generated by its own sounds, as well as by the sounds of conspecifics, a form of bistatics (Tyack 1997). An important characteristic of this phenomenon in the ocean is that echoes of low-frequency baleen whale sounds can have time delays of many minutes rather than seconds,[6] which poses some interesting questions about possible neurophysiological attributes of baleen whale auditory processes.

The primary potential benefit of echo-ranging comes from the acoustically active animal attending to the sound field generated by reflections of its sounds off hard features in the ocean when inhomogeneities in the medium are minimal, much like one would experience in a canyon or valley after making a loud whoop on a cold windless night. In the ocean, resultant arrival times of reflections provide a sense of static ocean features (i.e., topography) that would be beneficial for orientation and navigation, especially during migrations. A simple, short-range example of an echo-ranging behavior was observed during the spring migration of bowhead whales off Pt. Barrow, AK and is consistent with the hypothesis that bowheads use reflection of their calls off ice keels to echo-navigate under heavy ice conditions, often in complete darkness (George et al. 1989; Clark and Ellison 2004, Figs. 73.7 and 73.8).

Given the positive and relatively high target strengths of islands, shelf breaks, underwater mountains and ocean ridges, the reflections of very-low-frequency baleen whale sounds off such structures can be detected hundreds of miles away (pers. obs.). Could whales attend to these echoes to aid in navigation? Are there acoustic features of species-specific sounds and their associated behaviors (e.g., swimming speed and directivity, diving pattern) that are at least consistent with this hypothesis of echo-ranging, and what would those be? Such benefits would be especially valuable for energy-efficient navigation during migrations between areas of high nutrient productivity (i.e., "feeding grounds") and havens for calving and mating (i.e., "breeding grounds"; see Chap. 4).

A second potential benefit of low-frequency echo-ranging comes from the acoustically active animal attending to the refracted and volume reverberation distortions in the reflective sound field (see Tyack 1997 for excellent discussion re odontocetes), similar to what one might experience after making a loud whoop that is then refracted as a result of an atmospheric thermocline and air movements (Larom et al. 1997). In the ocean, attending to these low-frequency echo distortions could provide information about dynamic ocean features (i.e., cold water eddies, ocean currents,

[6] A simple rule of thumb when converting travel time in seawater to distance is that a minute of travel time is *ca.* equivalent to a distance of 50 nmi (93 km). Thus, the echo off a reflective feature that is 25 nmi away from a calling whale arrives back at that whale about 1 min later.

upwellings) associated with foraging opportunities (Croll et al. 2005; Johnston et al. 2005; Keiper et al. 2005).

What beneficial information could an animal derive by listening to a familiar biological or naturally occurring abiotic sound that has been transformed during its passage through the ocean and seafloor? Could an animal that attended to the attributes of acoustic reflections, reverberations, and refractions gain an advantage that increased its ability to orient and navigate (see George et al. 2004, Clark and Ellison 2004)? Using passive acoustics, one can readily observe the acoustic signatures of tectonic seismic events along the edges of the Caribbean Island (Bryan and Nishimura 1995) and Aleutian Island chains (pers. obs.). Could these events serve as very broadscale acoustic beacons? Might the same be posited for the high latitude, late winter–spring storms that pump energy from atmospheric pressure oscillations into the North Atlantic and North Pacific Oceans; could the resultant changes in very-low-frequency ambient noise provide a signal of pending changes in upwellings along coastal regions of North America? Could temperature interfaces associated with density differences between ocean currents (e.g., warmer Gulf Stream in the North Atlantic; warmer North Pacific Current, colder California Current in the North Pacific) and internal wave signatures provide cues that would be beneficial for migrating, foraging, and communicating (see Worcester et al. 1999; Howe et al. 2018)?

How is it that a singing blue whale leaves a northern North Atlantic feeding habitat and heads on a 2400 km track directly toward the Flemish Cap off Newfoundland, Canada, where he turns and swims 1700 km at a constant speed (4.8 km/h) toward an underwater seamount in the New England Seamount chain, where he turns again and swims 800 km directly toward the eastern side of Bermuda (pers. obs.)? Similarly, at a population level, how is it that the thousands of acoustic tracks of singing humpbacks leaving their high latitude summer feeding areas to appear as a grand procession, a highway of whales that is repeated year after year (pers. obs.)? The courses and speeds of the individuals as well as the migrating population seem to follow great circle routes that appear well-organized, as if the animals were following some sort of global map. Could these long-lived, large-brained, low-frequency specialists be using combinations of genetic predisposition and passive–active acoustic behaviors to acquire and maintain acoustic maps and teach the nuances of migrating through their ocean acoustic habitats to other members of the population (Chap. 4) or is all of this behavior learned (Chap. 8)?

From a simple biological perspective, if there is a benefit from an acoustic behavior that reaches a greater number of conspecific listeners (i.e., the largest audience within a certain range and depth), this can be achieved by producing a sound that is low in frequency, powerful (intense) and simple (redundant and easily recognizable). The song of a North Atlantic blue whale singer is typically composed of simple 10–20 s, infrasonic (15–20 Hz), high intensity (185 dB) note types repeated *ca.* every 70–75 s (Mellinger and Clark 2003; Chap. 9) and is recognizable at both close (*ca.* 1 *km*) and distant receivers (*ca.* 100 km) (Clark and Ellison 2004, Fig. 73.5 upper panels). In contrast, the complex, broadband song of a North Atlantic humpback singer (Chap. 11), although still obviously recognizable as a humpback at 100 km, is low-pass filtered during its propagation through the ocean such that at 100 km only sounds below approximately 400 Hz remain (Clark and Ellison 2004, Fig. 73.5, lower panels).

Distortions of baleen whale sounds occur in the form of time dispersion and frequency dispersion in which frequency-dependent propagation modes act to modify a sound's original temporal and frequency characteristics. Fin whale songs in the North Atlantic have been analyzed to demonstrate that their very-low, FM 20-Hz song notes excite multiple acoustic modes (D'spain et al. 1995), and both features are visually apparent in spectrograms and aurally discerned to the human ear when played back at higher speeds. Blue whale songs from the Eastern North Pacific have been combined with matched-field processing and global inversion techniques to examine seafloor sediment properties and to demonstrate that the application of these tools can reveal intriguing observations of baleen whale bioacoustic behaviors (Thode et al. 2000). For example, the authors noted the co-occurrence of a blue whale producing an "unusual diversity of signals" in association with a "sudden change in oceanographic conditions." There have been observations indicating that very-low-frequency synthetic signals can reveal features of the seafloor depending on the transmission path. As noted in Heaney (2005), a "*primary result of this test was the indication that there may be significant energy in the first bottom bounce, comparable to the direct-path (purely refracted) ray. This implies that the seafloor near the source influences the arrival times and patterns of the field at long distances, and should be taken into account.*" FM calls from North Pacific right whales have been observed with evidence of frequency and time dispersion (McDonald and Moore 2002). What is particularly intriguing in these various observations of time–frequency dispersion is that if one knows some basic features of the acoustic environment (e.g., water depth and depth of thermocline), one can estimate the range to the source as well as features of the intervening ocean (G. D'Spain, pers. comm.). Given the likelihood that baleen whales can sense both water depth by listening to direct-path echoes of their sounds from the seafloor and changes in water temperature as a function of depth (i.e., the thermocline) while diving, it seems reasonable to assume that baleen whales can listen for and perceive time–frequency dispersions of songs and calls. By doing so, a listening whale could recognize dispersion features (e.g., arrive time differences for different frequencies in a FM signal) indicating both distances to singing and calling whales and features of the ocean medium between itself and those acoustically active whales.

2.8 Discrepant Events

Early on in the Whales'93 project, CWC well remembers being admonished by a physical acoustician after showing GJG's "Ol' blue" whale track (Fig. 2.4). In that presentation, CWC had implied that the singing blue whale could use the reflections of its song off underwater features to navigate within the deep ocean and learn bathymetric features in a form of echo-ranging, and that other listening whales could follow the movements of the singer using a form of low-frequency directional tracking (Webb and Worthington 1968; Gould 2005). The acoustician claimed that this was impossible because the song is composed of CW and FM units with wavelengths on the order of 80–100 m,[7] but the distance between the whale's inner ears was only about 1 m, too short to allow any such directional hearing capability. CWC then explained the reasoning for his hypothesis; stating that the target strength of the underwater slope of Bermuda was >25 dB (pers. comm. W. T. Ellison), a level of reflection that would provide considerable sound level in the echo of the whale's sounds. Next, CWC told the acoustician about a discrepant event and its recent scientific explanation. This event concerned the discovery that a 2 mm dipterid female fly homes in on a male cricket's 5 kHz song (68 mm wavelength) to find and lay her egg inside the unsuspecting singer (Robert et al. 1992). The simple physics indicated that this fly was doing something acoustically impossible; detecting a signal with a wavelength 34 times greater than its body length, yet observations, experiments and the eventual discovery of the mechanism confirmed that it was a biological reality (Miles et al. 1995). The paths and highways of science are littered with the remnants from such claims of biological impossibilities. One might thereby conclude that it is often wise to question such statements about biological impossibilities when confronted with a contradictory observation. This is now especially true given the research evidence that baleen whales are well adapted for both low-frequency sound reception and production (Ketten 1992; Reidenberg and Laitman 2007; Chap. 3) and the recognition that science has yet to fully observe baleen whales performing many of their acoustic behaviors within the actual scales and contexts of their acoustic habitats and sensory capabilities.

Throughout our many decades together using Navy receivers to observe baleen whales at ocean basin scales, there have been multiple observations that can be considered discrepant events[8]. For security reasons, we can only give very general details as to where in an ocean these observations occurred and their spatial and temporal resolutions, but we are most certain of their authenticity. These events involve the most obvious baleen whale singers: blue (Chap. 9), fin (Watkins et al. 1987), humpback (Chap. 11), and minke (Chap. 14) whales[9] and are based on observations of thousands of singing whales that were tracked and recorded as they moved throughout an ocean basin. Some observations involve combinations of singing behaviors that result in a dramatic increase in the singer's potential audience. For example, when a singer positions himself over a shelf break where the water depth drops rapidly and the deep sound channel is closer to the surface (Clark and Ellison 2004; Miksis-Olds 2015). In this case, a song's acoustic energy refracts downward, reflects off the shelf and enters the deep sound channel such that it can propagate to many hundreds, and

---

[7] Some bat species produce and use their CW-FM signals in the > 20 kHz frequency band to echolocate (Jones 2005). In sonar, CW signals can be used to calculate the speed of an acoustic source, while FM signals are used to estimate both speed and range of the source.

[8] A discrepant event is a surprising and seemingly paradoxical observation that is not what an observer would normally expect (e.g., a 2 mm dipterid fly homing in on a 68 mm cricket sound).

[9] This likely holds true for other species as well, including sei whale (*Balaenoptera borealis*), Bryde's whale (*Balaenoptera edeni*), and Omura's whales (*Balaenoptera omurai*)(Chapter 15).

sometimes thousands of kilometers and still be above the ambient noise level at a distant non-Navy recording site. In these cases, the songs of a singer that were only detected a few tens of kilometers away when it was singing on the continental shelf in shallow water, were suddenly detected great distance away once the whale was singing above the slope of that continental shelf. This observation is consistent with observations showing that some of the highest densities of blue and fin whale singers can be found seasonally along shelf breaks.

Because of their low-density and high-intensity songs, individual blue whale singers are the easiest baleen whale to unambiguously detect, locate, and track, often for weeks at a time. Individual singers often appear to slalom from one underwater feature to another, as if they have memorized bathymetric and oceanographic acoustic features of the ocean, acquired over decades, and enabled by an auditory sensory system evolved and devoted to listening in the ocean's three-dimensional low-frequency acoustic environment. It's as if they have learned the spatial and seasonal nuances of their acoustic habitat. This habitat includes the acoustic cacophony originating from earthquakes, atmospheric weather systems, ocean currents, and internal waves, as well as all the sounds of ocean life, especially sounds produced by other baleen whales in all their various forms. These observations are consistent with the hypothesis that, given their present suite of inherited and learned capabilities, blue whales are adapted to exploit the features of their different acoustic habitats for communicating over extremely large ocean areas. This chapter's initial passage from Marler (1999) reminds us to carefully attend to differences between the concepts and the realities of innate and acquired behaviors, while remaining cognizant of any limitations of our observational mechanisms.

Before the Whales'93 project, fin whale song variability was assumed to be relatively small, especially when compared to humpback whales that had become the standard by which baleen whale acoustic behaviors were judged. Watkins et al. (1987) was an impressive *tour de force* that clearly showed variation in song features based on the analysis of 20-Hz song units over a span of several decades, although the authors could not adequately explore spatial variation because the data were mostly limited to Bermuda and the Gulf of Maine, USA. Spatial variation in song features was immediately evident during the Whales'93 project because data were available from throughout the North Atlantic (Nishimura 1994; Clark 1995a; Clark 1996; Clark and Gagnon 2002). During the 1993–1996 period, we noticed that fin whale songs in the more northerly parts of the North Atlantic were often composed of repeated three-note syllables that included an initial low intensity, 140–142 Hz, FM unit lasting < 1 s. We referred to this note type as the "precursor" because it always occurred first, but slightly overlapped the signature 20-Hz note. Although this precursor sound had sometimes been observed by previous researchers, its center frequency of around 140 Hz had been erroneously assumed to be the 7th harmonic of the 20-Hz unit (i.e., the 20-Hz "pulse" á la Watkins et al. 1987). On closer examination, this higher frequency sound was clearly a separate note type, not a harmonic of the 20-Hz note. Its identification, along with the lower intensity, 18 Hz "backbeat" note lasting <1 s (Clark et al. 2002; Castellote et al. 2012) expanded the note repertoire in North Atlantic fin whale songs to three. Although fin whale song syllables can be composed of combinations of only three notes, there is a great deal of intra- and inter-song variability, as well as spatial (i.e., where) and temporal (i.e., when) variability in song syntax and singing behavior. What is particularly interesting relative to the center frequencies of the three note types from which North Atlantic fin whale songs can be composed is that the center frequency of the precursor note has been steadily declining, such that it is now (in 2021) at 126 Hz, while the center frequencies of the 20-Hz and backbeat notes have not been in a steady decline. This suggests that the laryngeal mechanism by which these notes and syllables are produced is more complex than previously considered (Chap. 3).

The levels of variability in North Atlantic fin whale singing behavior and song syntax are known to exist in other ocean regions (Hatch 2004, North Atlantic and North Pacific; Hatch and Clark 2004; Mediterranean Sea, Castellote et al. 2012; Sea of Cortez, Mexico, Croll et al. 2002; Chukchi Sea, Delarue et al. 2013; Scotia Sea, Širovič et al. 2006; Archer, Rankin et al. 2020). As bioacoustic observations continue to expand, more details of spatial and temporal variation in fin whale singing behaviors are beginning to emerge (Clark et al. 2019; Helble et al. 2020), but a comprehensive descriptive analysis and synthesis of variability in fin whale song and certainly a behavioral explanation of different singing behaviors across multiple ocean regions remain major challenges.

It is now well-established that the center frequency in blue whale songs has been decreasing in different populations for decades (Chap. 9). Based on this global observation within different blue whale song populations and the observations that blue whale populations hybridize with fin whales (Bérubé and Aguilar 1998), one might expect to find that fin whale song notes have also been decreasing in frequency. This is not necessarily the case for fin whale very-low-frequency 20-Hz song notes throughout the North Atlantic for which the center frequency has remained the same (approx. 20 Hz center frequency with a 17–24 Hz 3 dB bandwidth) over the 26-year span from 1995 to 2021 (pers. obs.). Any hypothesis explaining multi-decadal decreases in blue whale song frequencies must also account for decreasing center frequency in the fin whale precursor note yet constancy in the center frequency of the 20-Hz note.

Humpbacks are arguably the most well-studied baleen whale species to date (Chaps. 4– 8, 10, and 11). Two fundamental concepts in their natural history are the existence of higher latitude feeding grounds and lower latitude breeding-calving grounds, and these concepts are interwoven into understandings of foraging, migratory, and reproductive behaviors (Chaps. 4, 5, 7, and 11). The generalized seasonal patterns of humpback singing behavior, based on acoustic observations (i.e., thousands of acoustic tracks over several decades) using Navy receivers, are consistent with existing knowledge, while some observations expand on that incomplete knowledge, or collectively highlight the incompleteness of or a possible contradiction with that knowledge. An example of consistent knowledge is that a major portion of humpback singers migrate from higher latitude summer feeding areas into the Caribbean region during the late fall through winter and singers remain in the Caribbean region until early spring, but relatively few singers are observed migrating back toward higher latitude summer feeding areas. A second example is that there is a second migratory wave of fewer singers moving from higher latitude summer feeding areas on the eastern side of the North Atlantic (e.g., Wenzel et al. 2009).

Two examples of expanded knowledge in the North Atlantic include acoustic observations indicating that singers occur in bands many hundreds of kilometers wide and thousands of kilometers long, and these bands of singers are so spatially large and consistent within and between years that we refer to them as "humpback highways." An interesting detail of singer tracks in these highways is that one often observes multiple singers in a line with individuals separated by many tens of kilometers as if following a common path, and this procession continues for days to weeks. During periods of the year when singing humpbacks are migrating to and spending time in their lower latitude Caribbean breeding–calving areas, many singers do not move out of the region and remain in their higher latitude feeding areas.

An example of incomplete and possibly contradictory knowledge has been observed during the winter period when females are known to calve and males are known to sing in nearshore waters throughout much of the Caribbean region (e.g., Winn et al. 1975; Levenson and Leapley 1978; Chap. 11), and yet little to no attention has been given to song and singing behavior in offshore areas. In the first months of the Whales'93 project, it was evident that nearshore singers were not the only singers in the Caribbean region as singers were observed throughout offshore, deepwater areas (Gagnon and Clark 1993; Clark et al. 1993; Clark and Gagnon 2002). Offshore singing in deeper water is not really surprising, given that recordings from US Navy hydrophones off the Island of Bermuda were used to describe humpback song (Payne and McVay 1971). We believe that this imbalance between observations of and attention to humpback song and singing behavior in nearshore and offshore areas is leading to a scientific blind spot and a form of a discrepant event. Our observations of offshore singing are consistent from year to year and occur seasonally throughout a period of four to six months (Clark and Gagnon 2002, Fig. 10). Furthermore, song durations are long (15–30 min), high intensity, structurally stereotyped and maintain a consistent syntax throughout multiple bouts, which last for many hours to days. Research on offshore singing and a biological explanation of these offshore observations need to be included in any comprehensive synthesis of humpback song function and singing behavior, and this synthesis needs to include our similar observations in other ocean basins (i.e., this observation is not unique to just the Western North Atlantic population of singing humpback whales). Are these offshore, deepwater singers high-quality males who establish and advertise their dominance by where they sing and the persistence of their singing behavior and the characteristics of their songs? Is this an example of a lek mating system as initially suggested by Herman and Tavolga (1980) (Chap. 11)? Are there obvious differences between the songs and singing behaviors of inshore and offshore singers? Does singing in deeper water along shelf breaks give them a selective advantage by expanding a singer's potential audience of receptive females while intimidating other males? Are there differences between songs and singing behaviors from the entire Caribbean regions compared to singers that remain in their high latitude feeding areas?

Minke whales have always been a bit of an enigma. Early on in the Whales'93 Program, it became clear that minkes were known within the Navy community as an acoustically distinctive species in the North Atlantic (Clark and Gagnon 2002). Their two most common song types were referred to as the "A-train" and "B-train." These terms for two different minke whale song types in the North Atlantic most likely coincide with an interesting singing behavior that we have consistently observed each year throughout the minke whale's annual multi-month singing season[10]. This behavior has not yet been described in the published literature. During the peak of the singing season, when the highest numbers and densities of singers are detected, song bouts often consist of two alternating song types. Both types are pulse trains typical of minke songs but differ in center frequency and bandwidth such that one pulse train type, the A-train, has a higher center frequency (e.g., 220 Hz) and a broader bandwidth (e.g., 700 Hz), while the other type, the B-train, has a lower center frequency (80 Hz) and a narrower bandwidth (e.g., 200 Hz). What has been particularly interesting from these multi-decadal, ocean basin observations of minke singing behavior is that the broadly distributed population of singers occurs in a seasonally consistent spatial pattern that moves clockwise through the western North Atlantic over a period of six to seven months. Our explanation for the discrepancies between the seasonal and spatial variability in minke whale singing behavior and song composition as observed through the published and classified observations is primarily based on the pronounced differences in the spatial and temporal scales and resolutions of the two observation technologies, and this statement certainly applies to all baleen whale species.

2.9 Conclusions

Throughout the past 20–30 years, there have been dramatic increases in physical acoustic technologies and resultant increases in spatial, temporal, and spectral scales of marine physical acoustic and bioacoustic observations. Separately, these observations rarely occur at the same rates, but at their intersections, the cross-fertilized levels of expertise are beginning to expose some intriguing patterns of baleen whale acoustic behaviors. Such patterns, whether revealed through relatively simple visualizations, multi-dimensional analyses, or intellectual synthesis, indicate that we are entering a new phase of scientific discovery. Progress depends on increased attention to evolutionary and behavioral ecology, fully integrated with physical acoustics and signal design theory. This switch in emphasis from technology and physics to behavioral

[10] We never heard a complete explanation for the origin of these names, but we did know that the Navy determined the source of these sounds as coming from minkes using a combination of simultaneous visual sightings from aircraft and localizations via sonobuoys deployed from aircraft. An interesting observation based on simultaneous visual observation and acoustic monitoring was that the pulse trains occurred after the whale dove from the surface and did not occur once the animal returned to the surface.

ecology and biology are likely to reveal biological adaptations (i.e., genetic, anatomical, neurophysiological, signal form-function, cultural), their selective advantages and relationships (i.e., behavioral ecological context), and the spatiotemporal scales of their occurrence. Progress will also identify the risks that human activities impose on marine life forms that are dependent on a naturally quiet, low-frequency acoustic environment.

Acknowledgements Generous mentors with curious minds, many of whom became lifelong friends, all of whom shaped our thinking, taught us how to be and remain curious, shared their ideas whether seemingly impossible or obviously simple once explained, opened doors, and were always there - thank you with more than thanks while acknowledging that you also taught us to pay it forward. In particular, I (CWC) am eternally grateful to Roger and Katy Payne for taking me to the whales in Argentina and showing me that one could both sing beneath the stars and dedicate one's mindful life to scientific discovery. Equally as well, I sing my loudest praises and unwavering gratitude to Charlie Walcott, who without a hint of fanfare or self-directed gratification, somehow recognized my conscious bent for combining biology and engineering in novel ways. Charlie, you pointed me onto paths that I did not know were there and gave me support without which I would have perished. Also, I wish to acknowledge my co-author and best friend, Chuck Gagnon. Your unbelievable curiosity about everything, especially about anything that makes a sound in the ocean, in combination with your exceptional practical understanding of underwater acoustics has been my inspiration and guiding light. Lastly, I devote this chapter to my sweetheart and wife, Janie, without whom I might well be living in multiple forms of poverty.

References

Abrahms B, Hazen EL, Aikens EO, Savoca MS, Goldbogen JA, Bograd SJ, Jacox MG, Irvine LM, Palacios DM, Mate BR (2019) Memory and resource tracking drive blue whale migrations. Proc Natl Acad Sci 116(12):5582–5587

Archer FI, Rankin S, Stafford KM, Castellote M, Delarue J (2020) Quantifying spatial and temporal variation of North Pacific fin whale (Balaenoptera physalus) acoustic behavior. Mar Mamm Sci 36(1):224–245. https://doi.org/10.1111/mms.12640

Au WW, Lammers MO (2016) Listening in the Ocean. Springer, Berlin

Bannister JL (2009) Baleen whales (mysticetes). In: Encyclopedia of marine mammals. Elsevier, pp 80–89

Bartol SM, Ketten DR (2006) Turtle and tuna hearing. In: Swimmer Y, Brill R (eds) Sea turtle and pelagic fish sensory biology: Developing techniques to reduce sea turtle bycatch in longline fisheries, vol NOAA Tech. Memo., NOAA-TM-NMFS-PIFSC-07. U.S. Dep. Commer., pp 98–103

Bass AH, Clark CW (2003) The physical acoustics of underwater sound communication. In: Simmons A, Fay R, Popper A (eds) Springer handbook of auditory research. pp 15–64

Baumgartner MF, Lysiak NSJ, Schuman C, Urban-Rich J, Wenzel FW (2011) Diel vertical migration behavior of Calanus finmarchicus and its influence on right and sei whale occurrence. Mar Ecol-Prog Ser 423:167–184. https://doi.org/10.3354/meps08931

Bérubé M, Aguilar A (1998) A new hybrid between a blue whale, *Balaenoptera musculus*, and a fin whale, *B. physalus*: frequency and implications of hybridization. Mar Mamm Sci 14(1):82–98

Bradbury JW, Vehrencamp SL (1998) Principles of animal communication. Sinauer Associates, Sunderland

Brown SG (1954) Dispersal in blue and fin whales. Cambridge University Press, Cambridge

Bryan C, Nishimura C (1995) Monitoring oceanic earthquakes with SOSUS: An example from the Caribbean. Oceanography 8(1):4–10

Calambokidis J, Schorr GS, Steiger GH, Francis J, Bakhtiari M, Marshall G, Oleson EM, Gendron D, Robertson K (2008) Insights into the underwater diving, feeding, and calling behavior of blue whales from a suction-cup-attached video-imaging tag (CRITTERCAM). Mar Technol Soc J 41 (1):19–29

Castellote M, Clark CW, Lammers MO (2012) Acoustic and behavioural changes by fin whales (*Balaenoptera physalus*) in response to shipping and airgun noise. Biol Conserv 147(1):115–122. https://doi.org/10.1016/j.biocon.2011.12.021

Catchpole CK, Slater PJB (1995) Bird Song: Biological themes and variations, 1st edn. Cambridge University Press, Cambridge

Clark CW (1982) The acoustic repertoire of the Southern right whale, a quantitative analysis. Anim Behav 30(4):1060–1071

Clark CW (1995a) Acoustic behaviors of blue and finback whales: insights from the Navy's dual uses program (abstract). In: XXIV international ethological conference. Honolulu, HI

Clark CW (1995b) Application of US Navy underwater hydrophone arrays for scientific research on whales. In: European Cetacean Society

Clark CW (1996) Blue whales' 93; Fin whales' 93; Minke Whales' 93, Humpback Whales' 93, North Atlantic. Whales' 93: Hot tracks and greatest hits, vol A seven-CD set Cornell Lab of Ornithology, Cornell University

Clark CW, Clapham PJ (2004) Acoustic monitoring on a humpback whale (*Megaptera novaeangliae*) feeding ground shows continual singing into late spring. Proc R Soc Lond B Biol Sci 271(1543):1051–1057

Clark CW, Ellison WT (2004) Potential use of low-frequency sounds by baleen whales for probing the environment: evidence from models and empirical measurements. In: Thomas JA, Moss CF, Vater M (eds) Advances in the study of echolocation in bats and dolphins. University of Chicago Press, pp 564–589

Clark CW, Gagnon GC (2002) Low-frequency vocal behaviors of baleen whales in the North Atlantic: Insights from IUSS detections, locations and tracking from 1992 to 1996. J Underwater Acoust 52(3):609–640

Clark CW, Gagnon CJ, Mellinger DK Whales '93 (1993): The application of the Navy IUSS for low-frequency marine mammal research (abstract). In: Tenth biennial conference on the biology of marine mammals, Galveston, TX, p 3

Clark CW, Borsani JF, Notarbartolo di Sciara G (2002) Vocal activity of fin whales, *Balaenoptera physalus*, in the Ligurian sea. Mar Mamm Sci 18(1):286–295

Clark CW, Gagnon GJ, Frankel AS (2019) Fin whale singing decreases with increased swimming speed. Roy Soc Open Sci 6(6). https://doi.org/10.1098/rsos.180525

Costa DP (1993) The secret life of marine mammals: Novel tools for studying their behavior and biology at sea. Oceanography 6(3):120–128

Crance JL, Berchok CL, Wright DL, Brewer AM, Woodrich DF (2019) Song production by the North Pacific right whale, Eubalaena Japonica. J Acoust Soc Am 145(6):3467–3479. https://doi.org/10.1121/1.5111338

Croll DA, Clark CW, Acevedo A, Tershy B, Floress S, Gedamke J, Urban J (2002) Only male fin whales sing loud songs. Nature 417:809

Croll DA, Marinovic B, Benson S, Chavez FP, Black N, Ternullo R, Tershy BR (2005) From wind to whales: trophic links in a coastal upwelling system. Mar Ecol Prog Ser 289:117–130

Crowell SC (2016) Measuring in-air and underwater hearing in seabirds. In: The effects of noise on aquatic life II. Springer, pp 1155–1160

Cummings WC, Philippi LA (1970) Whale phonations in repetitive stanzas. Naval Undersea Research and Development Center

Cummings WC, Thompson PO (1971) Underwater sounds from the blue whale, *Balaenoptera musculus*. J Acoust Soc Am 50 (4 (Part 2)):1193–1198

Curtis KR, Howe BM, Mercer JA (1999) Low-frequency ambient sounds in the North Pacific: long time series observations. J Acoust Soc Am 106(6):3189–3200

D'Spain G, Kuperman W, Clark C, Mellinger D (1995) Simultaneous source ranging and bottom geoacoustic inversion using shallow water, broadband dispersion of fin whale calls. J Acoust Soc Am 97(5):3353–3353

Damien J, Adam O, Cazau D, White P, Laitman JT, Reidenberg AJS (2019) Anatomy and functional morphology of the mysticete rorqual whale larynx: Phonation positions of the U-Fold. Anat Rec-Adv Integr Anat Evol Biol 302(5):703–717. https://doi.org/10.1002/ar.24034

Dantzker MS, Deane GB, Bradbury JW (1999) Directional acoustic radiation in the strut display of male sage grouse Centrocercus urophasianus. J Exp Biol 202(21):2893–2909

Darling JD, Bérubé M (2001) Interactions of singing humpback whales with other males. Mar Mamm Sci 17(3):570–584

Delarue J, Todd SK, Van Parijs SM, Di Iorio L (2009) Geographic variation in Northwest Atlantic fin whale (*Balaenoptera physalus*) song: Implications for stock structure assessment. J Acoust Soc Am 125(3):1774–1782

Delarue J, Martin B, Hannay D, Berchok CL (2013) Acoustic occurrence and affiliation of fin whales detected in the Northeastern Chukchi sea 10(66):159–172, July to Oct 2007

Dziak RP, Fox CG (2002) Evidence of harmonic tremor from a submarine volcano detected across the Pacific Ocean basin. J Geophys Res Solid Earth 107 (B5). https://doi.org/10.1029/2001jb000177

Eisner T, Meinwald J (1995) Chemical ecology. Proc Natl Acad Sci USA 92(1):1–1. https://doi.org/10.1073/pnas.92.1.1

Eldredge N (2014) Punctuated equilibria: an alternative to phyletic gradualism. In: Time frames. Princeton University Press, pp 193–224

Ellison WT, Southall BL, Clark Christopher W, Frankel AS (2012) A new context-based approach to assess marine mammal behavioral responses to anthropogenic sounds. Conserv Biol 26(1):21–28

Ellison WT, Racca R, Clark CW, Streever B, Frankel AS, Fleishman E, Angliss R, Berger J, Ketten D, Guerra M, Leu M, McKenna M, Sformo T, Southall B, Suydam R, Thomas L (2016) Modeling the aggregated exposure and responses of bowhead whales Balaena mysticetus to multiple sources of anthropogenic underwater sound. Endangered Species Res 30:95–108

Erbe C, Verma A, McCauley R, Gavrilov A, Parnum I (2015) The marine soundscape of the Perth Canyon. Prog Oceanogr 137:38–51. https://doi.org/10.1016/j.pocean.2015.05.015

Erbe C, Reichmuth C, Cunningham K, Lucke K, Dooling R (2016) Communication masking in marine mammals: a review and research strategy. Mar Pollut Bull 103(1–2):15–38. https://doi.org/10.1016/j.marpolbul.2015.12.007

Farrell WE, Berger J, Bidlot JR, Dzieciuch M, Munk W, Stephen RA, Worcester PF (2016) Wind sea behind a cold front and deep ocean acoustics. J Phys Oceanogr 46(6):1705–1716. https://doi.org/10.1175/jpo-d-15-0221.1

Feng AS, Narins PM, Xu C-H, Lin W-Y, Yu Z-L, Qiu Q, Xu Z-M, Shen J-X (2006) Ultrasonic communication in frogs. Nature 440(7082):333–336

Gagnon GJ, Clark CW (1993) The use of U.S. Navy IUSS passive sonar to monitor the movement of blue whales. In: Tenth biennial conference on the biology of marine mammals. Galveston, TX, p 50

George JC, Clark C, Carroll GM, Ellison WT (1989) Observations on the ice-breaking and ice navigation behavior of migrating bowhead whales (balaena-mysticetus) near point barrow, Alaska, Spring 1985. Arctic 42(1):24–30

Gervaise C, Lossent J, Valentini-Poirier C, Boissery P, Noel C, Di Iorio L (2019) Three-dimensional mapping of the benthic invertebrates biophony with a compact four-hydrophones array. Appl Acoust 148:175–193

Goldbogen JA, Friedlaender AS, Calambokidis J, Mckenna MF, Simon M, Nowacek DP (2013) Integrative approaches to the study of baleen whale diving behavior, feeding performance, and foraging ecology. Bioscience 63(2):90–100. https://doi.org/10.1525/bio.2013.63.2.5

Gould WJ (2005) From swallow floats to Argo—The development of neutrally buoyant floats. Deep Sea Res Part II 52(3–4):529–543

Guerra M, Thode AM, Blackwell SB, Michael Macrander A (2011) Quantifying seismic survey reverberation off the Alaskan North Slope. J Acoust Soc Am 130(5):3046–3058. https://doi.org/10.1121/1.3628326

Hatch LT (2004) Male genes and male song: integrating genetic and acoustic data in defining fin whale, Balaenoptera physalus, management units. Cornell University

Heaney K (2005) The Kauai Near-Source Test (KNST): modeling and measurements of downslope propagation near the North Pacific Acoustic Laboratory (NPAL) Kauai source. JASA

Helble TA, Guazzo RA, Alongi GC, Martin CR, Martin SW, Henderson EE (2020) Fin whale song patterns shift over time in the central north pacific. Frontiers in Marine Science 7. https://doi.org/10.3389/fmars.2020.587110

Herman LM, Tavolga WN (1980) The communication systems of cetaceans. In: Herman LM (ed) Cetacean behavior; mechanisms and function. Wiley, New York, pp 149–209

Herman LM, Forestell PH, Antinoja RC (1980) The 1976/77 migration of humpback whales into Hawaiian waters: composite description. Marine Mammal Commission Report No. 77/19,

Holt S (1974) Whales: conserving a resource. Nature 251(5474):366–367

Howe BM, Miksis-Olds J, Rehm E, Sagen H, Worcester PF, Haralabus G (2019) Observing the oceans acoustically. Front Mar Sci 6. https://doi.org/10.3389/fmars.2019.00426

Jensen BA, Kuperman W, Porter MB, Schmidt H (2000) Computational ocean acoustics. Springer, New York

Jensen FB, Kuperman WA, Porter MB, Schmidt H (2011) Wave propagation theory. In: FBJe Al (ed) Computational ocean acoustics. Springer, Berlin. https://doi.org/10.1007/978-1-4419-8678-8

Johnston DW, Thorne LH, Read AJ (2005) Fin whales *Balaenoptera physalus* and minke whales *Balaenoptera acutorostrata* exploit a tidally driven island wake ecosystem in the Bay of Fundy. Mar Ecol Prog Ser 305:287–295

Jones G (2005) Echolocation. Curr Biol 15(13):R484–R488

Jones BA, Stanton TK, Lavery AC, Johnson MP, Madsen PT, Tyack PL (2005) A scattering analysis of echoes due to biosonar signals emitted by foraging beaked whales. JASA 118 (3):1909

Kaifu K, Segawa S, Tsuchiya K (2007) Behavioral responses to underwater sound in the small benthic octopus Octopus ocellatus. J Marine Acoust Soc Jpn 34(4):266–273

Katona S, Baxter B, Brazier O, Kraus S, Perkins J, Whitehead H (1979) Identification of humpback whales by fluke photographs. In: Winn HE, Olla BL (eds) Volume 3: cetaceans. Behavior of marine animals: current perspectives in research. Plenum Press, New York, pp 33–44

Keiper CA, Ainley DG, Allen SG, Harvey JT (2005) Marine mammal occurrence and ocean climate off central California, 1986 to 1994 and 1997 to 1999. Mar Ecol Prog Ser 289:285–306

Ketten DR (1992) The Cetacean Ear: Form, Frequency, and Evolution. In: Thomas J, Kastelein RA, Supin AY (eds) Marine mammal sensory systems. Plenum Press, New York, pp 53–75

Ketten DR (1994) Functional analyses of whale ears: adaptations for underwater hearing. IEEE 264–270

Ketten DR (1997) Structure and function in whale ears. Bioacoustics 8(1–2):103–135

Klinck H, Nieukirk SL, Mellinger DK, Klinck K, Matsumoto H, Dziak RP (2012) Seasonal presence of cetaceans and ambient noise levels in polar waters of the North Atlantic. J Acoust Soc Am 132 (3):EL176–181. https://doi.org/10.1121/1.4740226

Kroodsma DE (1996) Ecology of passerine song development. Kroodsma, D E and E H Miller 6 (paper):0–8014

Kuperman WA, Hildebrand JA, D'Spain G (1993) Acoustic environment and source characterization using marine mammal vocalizations. Mar Physical Lab/SIO

Kuperman WA, Lynch JF (2004) Shallow-water acoustics. Phys Today 57(10):55–61. https://doi.org/10.1063/1.1825269

Ladich F (2008) Sound communication in fishes and the influence of ambient and anthropogenic noise. Bioacoustics 17(1–3):34–38. https://doi.org/10.1080/09524622.2008.9753755

Larom D, Garstang M, Lindeque M, Raspet R, Zunckel M, Hong Y, Brassel K, O BS, Sokolic F (1997) Meteorology and elephant infrasound at Etosha National Park, Namibia. J Acoust Soc Am 101(3):1710–1717
Levenson C, Leapley W (1978) Distribution of Humpback whales (Megaptera novaeangliae) in the caribbean determined by a rapid acoustic method. J Fish Res Board Can 35(8):1150–1152
Lockyer C (1984) Review of baleen whale (*Mysteceti*) reproduction and implications for management. In: Perrin WF, Brownell RLJ, DeMaster DP (eds) Reproduction in whales, dolphins, and porpoises. Reports of the International Whaling Commission, Special Issue 6 edn. International Whaling Commission Special Issue 6, Cambridge, England, pp 27–50
Lugli M (2010) Sounds of shallow water fishes pitch within the quiet window of the habitat ambient noise. J Comp Physiol A Neuroethol Sens Neural Behav Physiol 196(6):439–451. https://doi.org/10.1007/s00359-010-0528-2
Marler P (1997) Three models of song learning: Evidence from behavior. J Neurobiol 33(5):501–516
Marler P (1999) On innateness: Are sparrow songs "learned" or "innate"? Design of Animal Communication
McDonald MA, Moore SE (2002) Calls recorded from North Pacific right whales (*Eubalaena japonica*) in the eastern Bering Sea. J Cetac Res Manage 4(3):261–266
McDonald MA, Mesnick SL, Hildebrand JA (2006) Biogeographic characterisation of blue whale song worldwide: using song to identify populations. J Cetacean Res Manage 8(1):55–65
Medwin H, Clay CS (1998) Chapter 2—Sound Propagation. In: Medwin H, Clay CS (eds) Fundamentals of acoustical oceanography. Academic Press, San Diego, pp 17–69. https://doi.org/10.1016/B978-012487570-8/50004-0
Mellinger DK, Clark CW (1994) Methods for automatic detection of mysticete calls
Mellinger DK, Clark CW (2003) Blue whale (*Balaenoptera musculus*) sounds from the North Atlantic. J Acoust Soc Am 114(2):1108–1119
Miksis-Olds JL (2015) Unraveling Soundscapes: Learning to be good ocean listeners. Paper presented at the 3rd Underwater Acoustics Conference, Crete, Greece
Miksis-Olds JL, Nieukirk SL, Harris DV (2018) Two unit analysis of Sri Lankan pygmy blue whale song over a decade. J Acoust Soc Am 144(6):3618–3626. https://doi.org/10.1121/1.5084269
Miles R, Robert D, Hoy R (1995) Mechanically coupled ears for directional hearing in the parasitoid fly Ormia ochracea. J Acoust Soc Am 98(6):3059–3070
Morton ES (1975) Ecological sources of selection on avian sounds. Am Nat 109:17–34
Munk WP, Worcester PF, Wunsch C (1995) Ocean acoustic tomography. Journal of Fluid Mechanics, 2006/04/26 edn. Cambridge University Press
National Research Council (2003) Ocean noise and marine mammals. National Academies Press, Washington, D.C.
National Research Council (2005) Marine mammal populations and ocean noise: Determining when noise causes biologically significant effects. National Academies Press, Washington, D.C.
Nieukirk SL, Stafford KM, Mellinger DK, Dziak RP, Fox CG (2004) Low-frequency whale and seismic airgun sounds recorded in the mid-Atlantic Ocean. J Acoust Soc Am 115(4):1832–1843
Nishimura CE (1994) Monitoring whales and earthquakes by using SOSUS. Naval Research Laboratory
Nishiwaki M (1966) Distribution and migration of the larger cetaceans in the North Pacific as shown by Japanese whaling results. In: Norris KS (ed) Whales, dolphins, and porpoises. University of California Press, Berkeley, CA, pp 171–191
Northrop J (1974) Detection of low-frequency underwater sounds from a submarine volcano in the Western Pacific. J Acoust Soc Am 56(3):837–841
Oppenhein AV, Schafer RW, Buck JR (1999) Discrete-time signal processing. Prince Hall, Sec 11
Parks SE, Ketten DR, O'Malley JT, Arruda J (2007) Anatomical predictions of hearing in the North Atlantic right whale. Anat Rec Adv Integr Anat Evol Biol 290(6):734–744
Patterson B, Hamilton GR (1964) Repetitive 20 cycle per second biological hydroacoustic signals at Bermuda. In: Tavolga WN (ed) Marine bio-acoustics. Pergamon Press, Oxford, pp 125–145
Payne RS, McVay S (1971) Songs of humpback whales. Science 173:585–597

Payne R, Webb D (1971) Orientation by means of long range acoustic signaling in baleen whales. Ann N Y Acad Sci 188:110–142
Payne R (1972) Report from Patagonia: The right whales. New York Zoological Society (Sprong 1972):1–4
Reidenberg JS, Laitman JT (2007) Discovery of a low frequency sound source in mysticeti (Baleen whales): Anatomical establishment of a vocal fold homolog. Anat Rec-Adv Integr Anat Evol Biol 290(6):745–759. https://doi.org/10.1002/ar.20544
Rice AN, Tielens JT, Estabrook BJ, Muirhead CA, Rahaman A, Guerra M, Clark CW (2014) Variation of ocean acoustic environments along the western North Atlantic coast: a case study in context of the right whale migration route. Eco Inform 21:89–99. https://doi.org/10.1016/j.ecoinf.2014.01.005
Robert D, Amoroso J, Hoy RR (1992) The evolutionary convergence of hearing in a parasitoid fly and its cricket host. Science 258(5085):1135–1137
Ross J (1835a) Narrative of a second voyage in search of a North-West passage and of a residence in the arctic regions during the years 1829, 1830, 1831, 1832, 1833: Including the reports of Capt. JC Ross... and the Discovery of the Northern Magnetic Pole. Baudry
Ross J (1835b) Appendix to the narrative of a second voyage in search of north-west passage, and of a residence in the arctic regions during the years 1829, 1830, 1831, 1832, 1833: Including the reports of commander, now captain, james clark ross and the discovery of the Northern magnetic pole, vol 2. AW Webster
Scammon CM (1874) The marine mammals of the north-western coast of North America: described and illustrated; together with an account of the American whale-fishery. JH Carmany
Schafer RM (1977) The Tuning of the world, vol 1st edn. Knopf, New York
Schevill WE, Watkins WA (1962) Whale and porpoise voices: a phonograph record. Woods Hole Oceanographic Institute
Schevill WE, Watkins WA, Backus RH (1964) The 20-cycle signals and *Balaenoptera* (fin whales). In: Tavolga WN (ed) Marine bio-acoustics. Pergamon Press, Oxford, pp 147–152
Schoenfuss HL, Bragulla HH, Schumacher J, Henk WG, Craig George J, Hillmann DJ (2014) The anatomy of the larynx of the bowhead whale, Balaena mysticetus, and its sound-producing functions. Anatomical record (Hoboken, NJ : 2007) 297:1316–1330. https://doi.org/10.1002/ar.22907
Simmons JA (2012) Bats use a neuronally implemented computational acoustic model to form sonar images. Curr Opin Neurobiol 22(2):311–319
Širovič A, Hildebrand JA, Thiele D (2006) Baleen whales in the Scotia Sea during January and February 2003. J Cetac Res Manage 8(2):161–171
Slabbekoorn H, Smith TB (2002) Habitat-dependent song divergence in the little greenbul: an analysis of environmental selection pressures on acoustic signals. Evolution 56(9):1849–1858
Southall BL, Novacek DP (2009) Acoustics in marine ecology: innovation in technology expands the use of sound in ocean science. Mar Ecol Prog Ser 395:1–3. https://doi.org/10.3354/meps08393
Southall BL, Finneran JJ, Reichmuth C, Nachtigall PE, Ketten DR, Bowles AE, Ellison WT, Nowacek DP, Tyack PL (2019) Marine mammal noise exposure criteria: updated scientific recommendations for residual hearing effects. Aq Mamm 45(2):125–232. https://doi.org/10.1578/am.45.2.2019.125
Stafford KM, Fox CG, Clark DS (1998) Long-range acoustic detection and localization of blue whale calls in the northeast Pacific Ocean. J Acoust Soc Am 104(6):3616–3625
Stafford KM, Nieukirk SL, Fox CG (1999) An acoustic link between blue whales in the eastern tropical Pacific and the Northeast Pacific. Mar Mamm Sci 15(4):1258–1268
Starostin E, Van Der Heijden G (2007) The shape of a Möbius strip. Nat Mater 6(8):563–567
Stimpert AK, DeRuiter SL, Southall BL, Moretti DJ, Falcone EA, Goldbogen JA, Friedlaender A, Schorr GS, Calambokidis J (2014) Acoustic and foraging behavior of a Baird's beaked whale, Berardius bairdii, exposed to simulated sonar. Sci Rep 4:7031. https://doi.org/10.1038/srep07031
Stommel H (2020) The gulf stream. University of California Press, Berkeley

Tervo OM, Parks SE, Christoffersen MF, Miller LA, Kristensen RM (2011) Annual changes in the winter song of bowhead whales (Balaena mysticetus) in Disko Bay. W Greenl Mar Mamm Sci 27(3):E241–E252. https://doi.org/10.1111/j.1748-7692.2010.00451.x
Thewissen J, Cooper L, George J, Bajpai S (2009) From land to water: the origin of whales, dolphins, and porpoises. evolution: education and outreach 2 (2):272–288
Thode AM, D'Spain GL, Kuperman WA (2000) Matched-field processing, geoacoustic inversion, and source signature recovery of blue whale vocalizations. JASA 107(3):1286–1300
Tormosov D, Mikhaliev Y, Best P, Zemsky V, Sekiguchi K, Brownell Jr R (1998) Soviet catches of southern right whales Eubalaena australis, 1951–1971. Biological data and conservation implications. Biol Conserv 86 (2):185–197
Tyack P (1981) Interactions between singing Hawaiian humpback whales and conspecifics nearby. Behav Ecol Sociobiol 8:105–116
Tyack PL (1997) Studying how cetaceans used sound ot explore their environment. In: Communication. Springer, pp 251-297
Tyack PL (1998) Acoustic communication under the sea. In: Owren MJ, Evans CS (eds) Hopp SL. Sound analysis and research methods. Springer-Verlag, Animal acoustic communication, pp 163–220
Tyack PL (2008) Implications for marine mammals of large-scale changes in the marine acoustic environment. J Mammal 89(32):549–558
Tyack P (2019) Communication by Sound and by visual, tactile, and chemical sensing. In: Würsig B (ed) Ethology and behavioral ecology of odontocetes. Springer International Publishing, Cham, pp 25–50. https://doi.org/10.1007/978-3-030-16663-2_2
Urick RJ (1983) Principles of underwater sound, 3rd edn. McGraw-Hill Book Company, New York
Walker RA (1963) Some intense, low-frequency, underwater sounds of wide geographic distribution, apparently of biological origin. J Acoust Soc Am 35(11):1816 1824
Wartzok D, Poppper A, Gordon J, Merrill J (2004) Factors affecting the responses of marine mammals to acoustic disturbance. Mar Technol Soc J 37 (4)
Watkins WA, Moore KE, Wartzok D, Johnson JH (1981) Radio tracking of finback (*Balaenoptera physalus*) and humpback (*Megaptera novaengliae*) whales in Prince William Sound, Alaska. Deep-Sea Res 28A(6):577–588
Watkins WA, Tyack P, Moore KE (1987) The 20-Hz signals of finback whales (*Balaenoptera physalus*). J Acoust Soc Am 82(6):1901–1912
Watkins WA, Daher MA, Reppucci GM, George JE, Martin DL, DiMarzio NA, Gannon DP (2000) Seasonality and distribution of whale calls in the North Pacific. Oceanography 13(1):62–67
Watkins WA, Daher MA, George JE, Rodriguez D (2004) Twelve years of tracking 52-Hz whale calls from a unique source in the North Pacific. Deep Sea Res I 51(12):1889–1901
Webb D, Worthington L (1968) Measurements of vertical water movement in the Cayman Basin. In: Deep sea research and oceanographic abstracts, vol 5. Elsevier, pp 609–612
Weirathmueller MJ, Stafford KM, Wilcock WSD, Hilmo RS, Dziak RP, Trehu AM (2017a) Spatial and temporal trends in fin whale vocalizations recorded in the NE Pacific Ocean between 2003–2013. PLoS One 12(10). https://doi.org/10.1371/journal.pone.0186127
Weirathmueller MJ, Wilcock WSD, Hilmo RS (2017b) Estimating range to a vocalizing fin whale using the timing and amplitude of multipath arrivals. J Acoust Soc Am 142(4):2101–2120. https://doi.org/10.1121/1.5005494
Wenzel FW, Allen J, Berrow S, Hazevoet CJ, Jann B, Seton RE, Steiner L, Stevick P, Su, rez PL, pez, Whooley P (2009) Current knowledge on the distribution and relative abundance of humpback whales (*Megaptera novaeangliae*) off the Cape Verde Islands, Eastern North Atlantic. Aq Mamm 35(4):502–510
Wilcock WSD, Stafford KM, Andrew RK, Odom RI (2014) Sounds in the Ocean at 1–100 Hz. In: Carlson CA, Giovannoni SJ (eds) Annual review of marine science, vol 6, Annu Rev Mar Sci pp 117–140. https://doi.org/10.1146/annurev-marine-121211-172423
Wiley RH (1992) Evolutionary adaptations for communication—a citation-classic commentary on physical constraints on acoustic communication in the atmosphere—implications for the evolution

of animal vocalizations by Wiley, R.H. and Richards, D.G. Curr Contents/agric Biol Environ Sci 48:8–8

Wiley RH, Richards DG (1978) Physical constraints on acoustic communication in the atmosphere: Implications for the evolution of animal vocalizations. Behav Ecol Sociobiol 3:69–94

Winn HE, Edel RK, Taruski AG (1975) Population estimate of the humpback whale (*Megaptera novaengliae*) in the West Indes by visual and acoustic techniques. J Fish Res Board Can 32(4):499–506

Winn HE, Price CA, Sorensen PW (1983) The distributional biology of the right whale *Eubalaena glacialis* in the western north atlantic. Brownwell, R L Jr, P B Best And J H Prescott 0 (0):129–138

Worcester PF, Cornuelle BD, Dzieciuch MA, Munk WH, Howe BM, Mercer JA, Spindel RC, Colosi JA, Metzger K, Birdsall TG, Baggeroer AB (1999) A test of basin-scale acoustic thermometry using a large-aperture vertical array at 3250-km range in the eastern North Pacific Ocean. J Acoust Soc Am 105(6):3185–3201

Yamato M, Ketten DR, Arruda J, Cramer S, Moore K (2012) The auditory anatomy of the minke whale (*Balaenoptera acutorostrata*): a potential fatty sound reception pathway in a baleen whale. Anat Rec (hoboken) 295(6):991–998. https://doi.org/10.1002/ar.22459

Chapter 3
Anatomy of Sound Production and Reception

Joy S. Reidenberg

Abstract Mysticetes (baleen whales) have an amazing array of adaptations of mammalian traits and functions enhanced for aquatic submersion. In this chapter, special emphasis is placed on anatomy and physiology of adaptions that enable communication (sound production and reception) in water, including during diving. Whenever possible, data revealing anatomical differences will be pointed out at the family level for balaenids (bowhead whale *Balaena mysticetus*, and right whales *Eubalaena glacialis, E. japonica, E. australis*), neobalaenid (pygmy right whale, *Caperea marginata*), eschrichtiid (gray whale *Eschrichtius robustus*), and balaenopterids (rorqual whales, i.e., whales with throat pleats, including humpback *Megaptera novaeangliae*, and the many species of the genus *Balaenoptera*, including minke whale *B. acutorostrata*, Antarctic minke whale *B. bonaerensis*, blue whale *B. musculus*, fin whale *B. physalus*, sei whale *B. borealis*, Bryde's whale *B. edeni*, and Omura's whale *B. omurai*). Some species groupings (e.g., the various right whales) or subspecies groupings (e.g., subspecies of blue whales) will not be addressed if there is no literature to indicate a difference, and the assumption is made that their anatomy is similar enough to be generalized for the whole group. In many cases, an anatomical feature may be discussed in generalities because it is present in all mysticetes.

Keyword Anatomy · Mysticete · Sound · Production · Generation · Hearing · Diving · Adaptations

3.1 Overview of Mysticete Aquatic Adaptations

Mysticetes are grouped into families based largely on the morphology of the mouth. The parvorder name Mysticeti (literally "mustached" + "whales") refers to the brushy appearance of baleen plates in the whale's mouth that resemble a mustache.

J. S. Reidenberg (✉)
Icahn School of Medicine at Mount Sinai, Center for Anatomy and Functional Morphology, mail box 1007, One Gustave L. Levy Place, New York, NY 10029-6574, USA
e-mail: joy.reidenberg@mssm.edu

C.W. Clark and E. C. Garland (eds.), *Ethology and Behavioral Ecology of Mysticetes*, Ethology and Behavioral Ecology of Marine Mammals,
https://doi.org/10.1007/978-3-030-98449-6_3

Mysticetes do not erupt teeth, although remnant tooth buds can be seen in the upper jaw of fetal whales (Deméré et al. 2008; Thewissen et al. 2017). Instead, they sport baleen plates made of keratin that are attached to the upper jaw. The baleen frames the gape on either side of the oral cavity and is used to separate prey from water in a process called filter-feeding using a mechanism of cross-flow filtration (Pinto 2011). Balaenids (bowhead and right whales), and presumably also the neobalaenid (pygmy right whale), feed using a strategy called "skim feeding" or "ram feeding" in which water flows through the mouth and prey is trapped on long baleen plates as the whale continuously swims (Chap. 5). The eschrichtiid (gray whale) can employ a benthic feeding strategy called "suction feeding" that involves sucking in and sifting muddy substrate to catch prey on the baleen. Balaenopterids (rorqual whales) use a different feeding strategy called "lunge feeding" or "gulp feeding." This batch-loading technique involves seeking (or actively corralling) prey in a tight cluster (bait ball), swimming directly at the cluster, opening the mouth, stretching the pleated throat region (engulfing prey-laden water), closing the mouth, and squeezing water out while retaining prey caught on the baleen.

All mysticetes have a very large head that can be up to one-third of the length of the whole body. Most of the head is comprised of the large mouth. Balaenids have a highly arched upper jaw. The blunted front of a balaenid is not a very streamlined shape, and this is associated with their relatively slow swimming speeds compared with the sleeker arrowhead shape of the balaenopterids. The only neobalaenid, the pygmy right whale, has many head features that resemble the balaenids (narrow rostrum, baleen covered by raised lips). However, it is unknown whether their swimming or feeding style is similar to balaenids. Balaenopterids, on the other hand, have a very wide, V-shaped head. The tip of the rostrum is pointed, and the top of the head is broad and flat. The only eschrichtiid, the gray whale, has head features that appear somewhat intermediate between balaenids and balaenopterids. The gray whale rostrum is only mildly arched, but the face is more streamlined because it is relatively narrow and conical. Unlike balaenids, the gray whale lower jaw lips are not raised and the baleen is short, and unlike balaenopterids, there are only a few deep throat grooves (usually only three).

Mysticete species can also be distinguished by unique shapes and color patterns of their caudal (tail) flukes. Each fluke contains no bones and is approximately a triangular shape with a gentle swept-back curve along the leading edge. The trailing edge is highly variable in texture, but there is always a notch that separates where the left and right flukes meet in the midline. The sleeker, swept-back flukes of balaenopterids appear related to hydrodynamic efficiency at higher speeds, compared with the broader flukes of the slower balaenids. Humpback whales have the most distinct flukes. The trailing edge is highly corrugated and may impart an advantage in maneuverability (van Nierop et al. 2008). Flukes often bear unique features, such as scars or notches from injuries (e.g., killer whale rake marks or bites, encounters with fishing gear entanglement, propeller cuts), distributions of parasites such as barnacles and pigmentation variations (e.g., complex black and white patterns of humpback whales). These specific features help identify individual whales.

Pectoral flippers also exhibit species-specific variations in shape. The most distinctive are the extremely elongated and lumpy flippers of humpback whale. Flipper shapes vary, including highly swept-back (balaenopterids), paddle-shaped (gray whale, pygmy right whale), or squarish (balaenids). The swept-back shape of balaenopterids may impart wing-like functions for improved maneuverability for prey capture. The squarish flippers of balaenids may relate to stable, straight-line swimming during ram feeding. Flippers contain nearly all the bones homologous to the forelimb bones of other vertebrates, but are often missing only one digital ray (Cooper et al. 2007a, 2007b).

The mysticete body is a fusiform shape that allows the whale to slip through the high friction aquatic environment with reduced resistance. The body is covered with very smooth and flexible integument (skin) that helps to reduce drag. Cutaneous ridges may sense flowing water (Shoemaker and Ridgway 1991). The skin readily deforms and recoils, channeling water across the body in the most efficient flow directions to reduce drag (Pavlov 2006). A subdermal connective tissue layer of fibers provides support for dermal projections (dorsal fin, caudal peduncle) and adds spring-action that stores and releases energy to aid locomotion (Pabst 1996).

Mysticetes have the mammalian trait of hairs, but these are sparse and found mainly on the head. These whisker-like hairs may be homologous to vibrissae (Bauer et al. 2018; Berta et al. 2015). Pits may remain in adults where the hairs fall out (especially visible on the rostrum of gray whales), but their function in receiving sensation is still not well understood. Many mysticetes retain hairs into adulthood on top of the head or the chin (Reeb et al. 2007). Chin hairs may transmit information about prey density and water flow during feeding (Drake et al. 2015; Nakai and Shida 1948). Sensory receptors are also suspected in what look like vibrissal pits on either side of the tip of the lower lip (Pyenson et al. 2012). Hairs are also found in the callosities of right whales and may also mediate feeding-associated sensations (Reeb et al. 2007). Adult humpback whales have a single hair in each raised bump (tubercle) on the dorsum of the head and the front and lateral surfaces of the lower jaw. Bending these hairs or stimulating the nerve endings in the tubercles could deliver a host of mechanical or electrical sensory information, such as detecting swimming speed, water currents, water's surface, contact with another whale, vibrations or electrical fields of prey, or sound pressures (Mercado 2014). The tubercles may also cushion impacts during aggressive encounters or disrupt the whale's outline (camouflage) deterring detection by echolocation.

Mysticetes migrate from productive high-/low-latitude feeding grounds to mid-latitude locations where they give birth away from predators (Corkeron and Connor 1999, Chap. 4). A recent study posits that whale migrations to warmer water increase metabolic rate, promoting shedding and rejuvenation of skin (Pitman et al. 2020). The large mysticete body has a higher surface-to-volume ratio that conserves heat. The blubber provides insulation and stores extra energy used during migration. Whales also use some of this fat as an energy source for behaviors relevant to reproduction (e.g., fighting, chasing, fleeing, singing). For females, gestation followed by lactation are additional long-term strains that require extra fat reserves. In humpback whales, larger females can produce larger calves and are suggested to be preferred by males for mating (Pack et al. 2009).

Mysticete brains are among the largest of any animal's, but are less impressive when measured relative to head size or body size. Mysticetes have unusually large heads adapted to hold their enormous feeding apparatus, skewing the brain-to-body size ratio. This ratio and other similar measures such as encephalization quotient appear to be poor methods for assessing intelligence in large whales (Marino 2009). Size is irrelevant to social complexity, and large bodies do not require much more brain tissue than small bodies to coordinate muscle and organ functions. Another method to assess complexity is the degree of gyrification (convolutions). Mysticete brains are highly convoluted, providing more surface area for gray matter (for processing signals), as opposed to white matter (for transmission of signals). Complexity of convolutions does not always equal intelligence, as some regions are specialized for other functions (e.g., sound production and reception). However, sound production and hearing are necessary not only for navigation-orienting and foraging, but also for complex social communication, and thus indicate a high level of cortical brain function. Mysticetes (specifically humpback and fin whales) have been documented to have spindle (*von economo*) neurons (Hof and VanDer Gucht 2007). These neurons are associated with complex social skills and emotions (Butti et al. 2013), but are not restricted to highly encephalized species (Raghanti et al. 2015).

3.2 Anatomy of Sound Production: Respiratory Tract

The respiratory tract begins at the crown of the head, where paired blowholes regulate the passage of air.[1] They are guarded by a chevron-shaped ridge that acts as a splashguard, diverting water away from the blowholes. This location on top of the head reduces the energetic cost of swimming, as the whole head does not need to be lifted out of the water to enable breathing. Opening and closing the blowholes are regulated by moving cartilage plates located at the lateral aspect of the openings (alar cartilages) (Buono et al. 2015). Immediately below the opening(s) are the paired nasal plugs, which obstruct the nasal lumen to protect the respiratory tract from incursions of water while submerged (Fig. 3.1). The paired nasal plugs are covered by pigmented epithelium, indicating they are derived from invaginations of the external skin—most likely the tissue that was previously the upper lip. These bulbous structures are largely comprised of fat, with strands of skeletal muscle inserted into them that contract to retract the plugs rostrally (Gil et al. 2020). Muscle action is only required for the momentary opening of the blowholes, while excluding water from the respiratory tract (the position for the majority of time) does not require energy

[1] Note the use of the colloquial term "air" to refer to the gas inside the respiratory tract, even though it may have a different mix of chemical concentrations compared with the ambient air at 1 atmosphere of pressure (sea level).

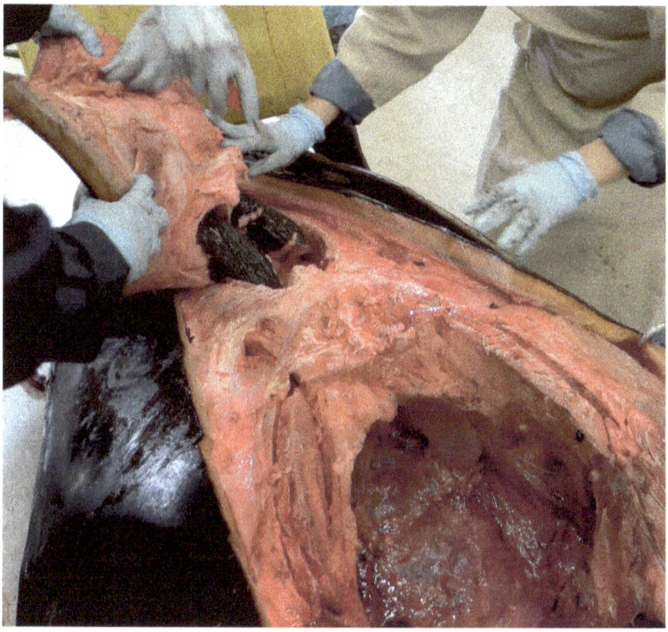

Fig. 3.1 Oblique caudal-cranial view of the nasal plugs being dissected away from the dorsum of a deceased neonate fin whale (*Balaenoptera physalus*). Note the paired nasal plugs have a black pigmented epithelium that matches the outer black skin (visible on the whale's left side). The lower right region of the image shows the cranial cavity (brain has been removed)

(Maust-Mohl et al. 2019). Nasal plug geometry appears to be species-specific and likely contributes to determining the unique shapes of the blow in each species.

Below the paired nasal plugs, the paired nasal passageways continue along a diagonal slope leading caudo-inferiorly through the skull (Fig. 3.2; Chap. 2, Fig. 3.3). These passageways exit the skull at the posterior choanae, where the two tubes are united into a common space called the nasopharynx. The nasopharynx is a soft-walled chamber and can be contracted by the surrounding skeletal muscle (superior pharyngeal constrictor). The nasopharynx has a pitted surface lined with glands that secrete mucus and may also expel excess salt. The bottom of this chamber is regulated by a muscular ring called the palatopharyngeal sphincter comprised of the palatal skeletal muscles. The top of the larynx (voice box) overlaps the palatopharyngeal sphincter. When the sphincter contracts, it closes the narrow gap through which the larynx is inserted, ensuring a tighter seal around the larynx and thus separating the respiratory tract from the digestive tract (Reidenberg and Laitman 2007a).

Lateral diverticulae emerge off the respiratory tract in the region of the nasopharynx. They are called pterygoid sacs (or sinuses). These are paired, air-filled chambers located under, but outside, the skull. Unlike true sinuses, the pterygoid sacs are not completely encased in bone, thus allowing flexibility to respond to ambient

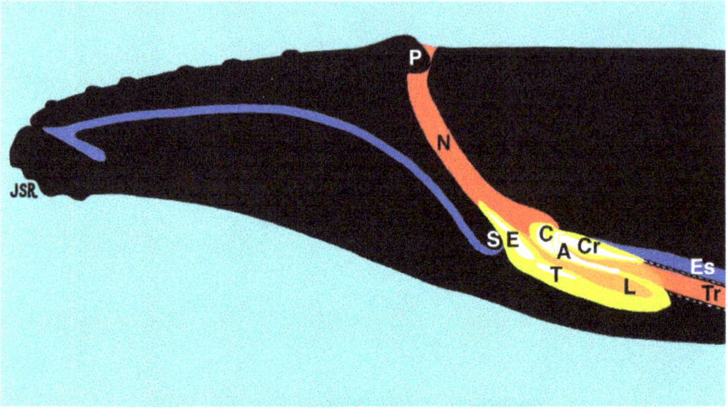

Fig. 3.2 Diagram of midsagittal anatomy of the upper respiratory system in a typical mysticete. Yellow = larynx, Red = respiratory passageways, Blue = digestive passageways. A = arytenoid cartilage, C = corniculate cartilage, Cr = cricoid cartilage, E = epiglottic cartilage, Es = esophagus, L = laryngeal sac, N = nasal passageways (paired), P = nasal plugs (paired), S = soft palate (palatopharyngeal sphincter), T = thyroid cartilage, Tr = trachea

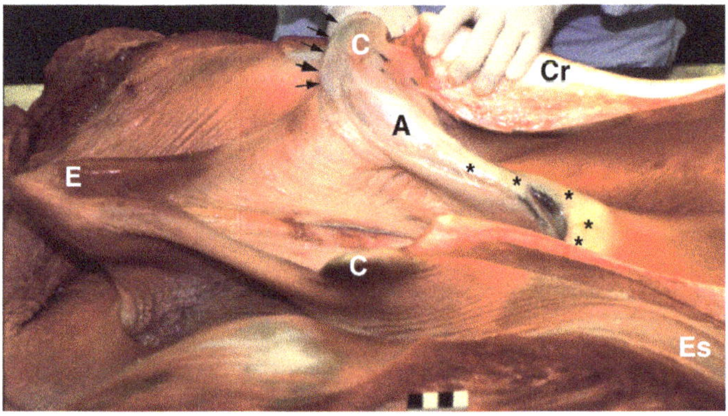

Fig. 3.3 Left latero-dorsal view of the larynx of a juvenile humpback whale (*Megaptera novaeangliae*) that has been cut along the dorsal midline and pulled open (note hand at top of figure). A = arytenoid (paired, only right is visible), C = corniculate (paired, left is projecting toward viewer), Cr = cricoid (unpaired, cut in midline), E = epiglottis (unpaired), Es = esophagus, arrows = corniculate flap, asterisks (*) = U-shaped vocal fold (only right side is visible). Scale bar: 1 square = 1 cm

pressure changes. They appear to be expansions of the Eustachian (pharyngotympanic) tubes connecting the respiratory tract to the middle ear. Similarly to terrestrial mammals, they are used for equilibrating pressure changes in the airspace of the middle ears. In mysticetes, this is a critical function for hearing, as ossicles in the middle ear must have space to vibrate in order to transmit sounds to the inner ears at depth. The sacs hold excess air that can be sacrificed to collapse at depth, thus protecting the remaining small but critical volume of residual air in the middle ear.

The hyoid apparatus is comprised of an unpaired basihyal in the midline and laterally projecting paired thyrohyals (the "greater horns"), and paired stylohyals that connect them to the skull. They provide a large surface area for the attachment of several robust skeletal muscles that can cause it to be elevated/depressed, or protracted/retracted. Depression/retraction lowers (opens) the mandibles and depresses the floor of the mouth, while elevation/protraction lifts the floor of the mouth and assists closing the gape. The hyoid apparatus also supports movements of attached lingual (tongue) muscles (Kienle et al. 2015). The hyoid apparatus also protects the airway during swallowing by elevating the larynx and opening the lateral food channels of the pharynx (Reidenberg and Laitman 1994, and Werth 2007 for detailed descriptions of the hyoid apparatus and associated skeletal muscles). Asymmetric placement of the hyoid apparatus in odontocetes assists with engulfing large prey (Reidenberg and Laitman 1994; MacLeod et al. 2007), but mysticetes do not appear to have any asymmetry of the hyoid apparatus or associated lateral food channels for swallowing prey.

The larynx is comprised of the epiglottic, thyroid, cricoid, and fused arytenoid–corniculate cartilages (the latter being paired cartilages). The larynx is positioned under the skull base and maintains an intranarial position through its interlock of the epiglottis and corniculates being positioned above the palatopharyngeal sphincter (Fig. 3.2). This ensures that the opening of the larynx is contiguous with the space of the nasopharynx, maintaining an interrupted connection between the nasal tract above and the trachea and lungs below. There are differences in the shapes of the laryngeal cartilages among the various species (Reidenberg and Laitman 2007a for details).

Balaenopterid whales have a slender and elongated epiglottis, with a pointy tip (Fig. 3.3). The epiglottis is deeply concave, having a trough-shaped channel for its entire length. It resembles a canoe, with the concavity facing the lumen of the larynx. In the rest position, the epiglottis lies over the soft palate, creating a bridge that channels air in this concavity in either direction. If the epiglottis is inserted into the oral cavity, the concavity faces the palate and together they form a half-cylinder tube. This position enables shunting air into the mouth (for bubble cloud production). Water in the mouth may be excluded if the epiglottis is forced dorsally against the ventral aspect of the soft palate, sealing the edges and allowing air to exit only at the epiglottic tip (Reidenberg and Laitman 2007b).

The corniculate portion of the arytenoid–corniculate complex in balaenopterids curls dorso-caudally away from the epiglottis to interlock with the posterior edge of the palatopharyngeal sphincter (Fig. 3.2). These paired cartilages are not able to oppose the epiglottis tightly to close the larynx. Rather, the epiglottis can be lifted dorsally to oppose the leading edges of the corniculates. Each corniculate has a flap of tissue (corniculate flap) on the rostral edge that extends toward the epiglottis (Fig. 3.3). It is unknown whether the flaps rest inside the concavity of the epiglottis, or only approach it to form a lateral boundary that helps enclose the air passageway. The size of the flap is species-specific, with humpback whales having the smallest flaps and minke whales having the largest flaps (relative to the size of the whole larynx). The function of these flaps is unknown, but as they are opposed against each other it may be possible for them to clap together and passively create pulse-like vibrations with airflow, with species-specific frequency characteristics related to their geometry (Damien et al. 2019).

The neobalaenid has an unusual appearance of the epiglottic and arytenoid–corniculate cartilages (Fig. 3.4). While it has a similar shape to balaenopterid laryngeal cartilages, the mucosal surface is unusually pitted (Reeb and Best 1999). The function of these pits is still unknown.

In balaenids, the epiglottis is shorter, thicker, has a rounded tip, and a ridge in the midline of the luminal surface (Reidenberg and Laitman 2007a) (Fig. 3.5). The epiglottis appears less mobile compared with balaenopterids. It is directed superiorly toward the skull, rather than forward on a diagonal as in balaenopterids. The arytenoid–corniculate cartilages are straight cartilages that extend in the same direction as the epiglottis, and slightly curl caudally only at the very tip. The corniculates are very thick and do not have any flaps of tissue extending from the leading edge, as in balaenopterid whales (Reidenberg and Laitman 2007a).

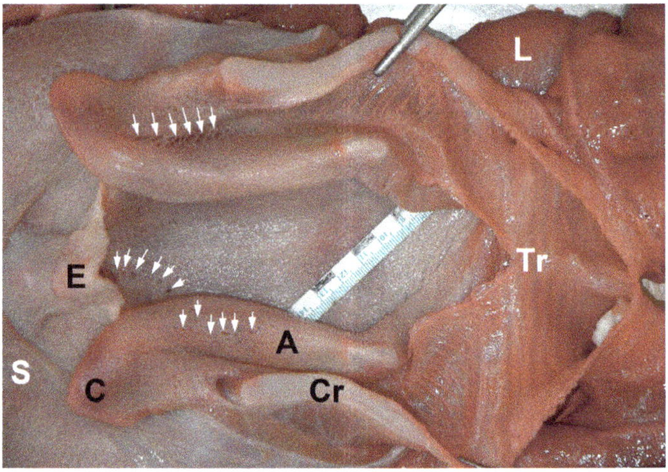

Fig. 3.4 Dorsal view of the larynx of a pygmy right whale (*Caperea marginata*) that has been cut along the dorsal midline and opened laterally to reveal the lumen. Note the many pits in the mucosa (some indicated by arrows). A = arytenoid cartilage (paired) supporting the U-shaped vocal fold (cut caudally), C = corniculate cartilage (paired) supporting the corniculate flap, Cr = cricoid cartilage (unpaired, but cut in the midline), E = epiglottis (unpaired, curled caudally and projecting dorsally), L = laryngeal sac (deviated to the right), S = soft palate (palatopharyngeal sphincter), Tr = trachea (just above the bifurcation or carina). Note the vocal fold is cut caudally in the midline to reveal passageway to laryngeal sac. The laryngeal sac begins where the scalebar ruler lies and deviates to the right side. The ruler is placed diagonally so that it is inserted fully into the sac. Note white paper towels were placed in the left main bronchus and under the carina to absorb fluids and hold those spaces open

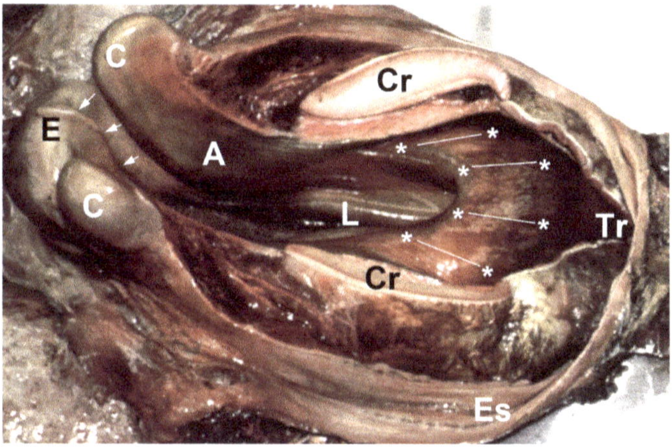

Fig. 3.5 Left latero-dorsal view of the larynx of a North Atlantic right whale (*Eubalaena glacialis*) that has been cut along the dorsal midline and opened laterally to reveal the lumen. A = arytenoid cartilage (paired but only right is visible) supporting the U-shaped vocal fold (asterisks), C = corniculate cartilage (paired) supporting thick "flap" tissue, Cr = cricoid cartilage (unpaired, but split in half in midline), E = epiglottis (unpaired) with a midline ridge (arrows), Es = esophagus (unpaired, but split in the midline), L = laryngeal sac (the lateral walls are collapsing medially), Tr = trachea. Note the depth of the very thick U-shaped vocal fold, indicated by the paired asterisks with connecting lines

Two shallow, concave channels lie on either side of the midline ridge of the balaenid epiglottis (Fig. 3.5). These channels are opposed by the corniculate cartilages (located at the rostral end of the paired arytenoid–corniculate complex cartilages). The three cartilages (epiglottis and paired corniculates) are more rigid and can nest into each other. The midline ridge helps form a seal in its position inserted between the two corniculate cartilages, obstructing any remaining lumen. Unlike the balaenopterid condition, the balaenid larynx appears capable of being tightly close at the tip.

The arytenoid–corniculate cartilages support the vocal folds of the mysticete larynx (note: the lay term "vocal cords" for these folds is erroneous, because they are not string-like cords, but rather, shelves of tissue that project into the laryngeal cavity). The paired vocal folds form a U-shape when viewed dorsally (Figs. 3.3 and 3.5). The vocal fold begins rostrally at the free edge of the corniculate flap and extends caudally to where it meets its pair in the midline. Here, the cartilages are anchored together by a strong ligament, together forming the U-shaped fold. The vocal fold has different characteristics along its length, each capable of vibrating at a different frequency (Damien et al. 2019). At rest, two vocal folds lie in opposition to each other. In balaenopterids, the opposing surfaces are rounded caudally, crinkled where they meet in the middle of the vocal fold's length, and become flatter toward the rostral end, terminating at a lip-like thickening at the base of each opposed corniculate flap (Reidenberg and Laitman 2007a; Damien et al. 2019). The vocal fold is very elongated in balaenopterids (Damien et al. 2019), but is shorter in balaenids (Schoenfuss et al. 2014). The arms of the U-shaped fold are also much thicker in balaenid whales (bowhead whales, Schoenfuss et al. 2014; North Atlantic right whales, Reidenberg and Laitman 2007a) compared with neobalaenid, eschrichtiid, and balaenopterid whales (Fig. 3.5). As the U-shaped fold appears to be the source of sounds, these differences likely determine the types of sounds each group can make.

In all mysticetes, the arytenoid cartilage is attached to the cricoid cartilage at a large synovial joint, indicating this is a highly mobile cartilage (Reidenberg and Laitman 2007a). Several intrinsic laryngeal skeletal muscles, all attached to the arytenoid cartilage, are responsible for these movements. Regulating how much and where air flows over or between the vocal folds can cause vibrations at different points along the tissue and is likely the reason whales can generate such a large repertoire of sounds (Damien et al. 2019).

Immediately ventral to the vocal folds is a large muscular diverticulum, called the laryngeal sac (Figs. 3.5 and 3.6). It lies in the midline, with its rostral end beginning at the base of the epiglottis, immediately dorsal to the thyroid cartilage of the larynx. It extends out from underneath the caudal edge of the thyroid cartilage and terminates caudally as far as the beginning of the trachea. It is unprotected by the thyroid cartilage for much of its length, allowing it to expand ventrally. Interestingly, the pygmy right whale's laryngeal sac is asymmetrically placed on the right side (Reeb and Best 1999) (Fig. 3.4). The reason for this neobalaenid asymmetry is unknown. It is particularly unusual because mysticete anatomy is otherwise symmetrical.

The opening into the laryngeal sac is regulated by movements of the vocal folds, controlled by muscular action on the arytenoid cartilages (Fig. 3.6). It is not known whether the sac itself can generate or propagate sounds, possibly through muscular

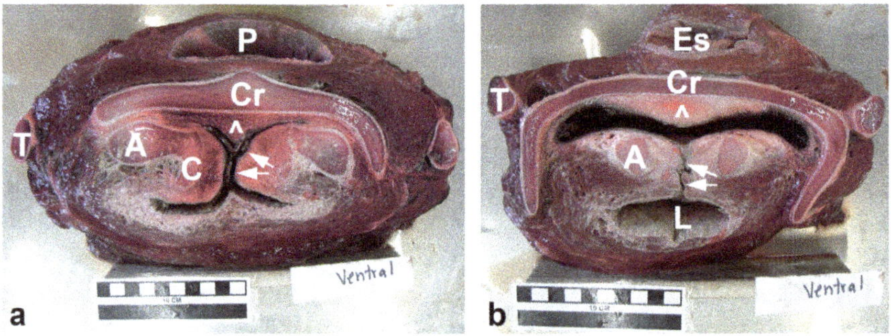

Fig. 3.6 Cross sections of the larynx of a minke whale (*Balaenoptera acutorostrata*), viewed from caudal to cranial. a, more rostral section taken at the junction between the arytenoid and corniculate cartilages and before the enlargement of the laryngeal sac. b, more caudal section taken from the region where the two "arms" of the U-shaped vocal folds meet and regulate the opening to the laryngeal sac. A = arytenoid cartilage (paired), C = corniculate cartilage (paired), Cr = cricoid cartilage (unpaired), Es = esophagus, L = laryngeal sac, P = pharynx, T = thyroid cartilage (unpaired, but two paired projecting caudal horns are visible laterally), arrows = surfaces of a vocal fold (note the rostral region is smooth, while the caudal region is corrugated), ^ = "cushion" of tissue under the cricoid cartilage that may seal the gap between he vocal folds

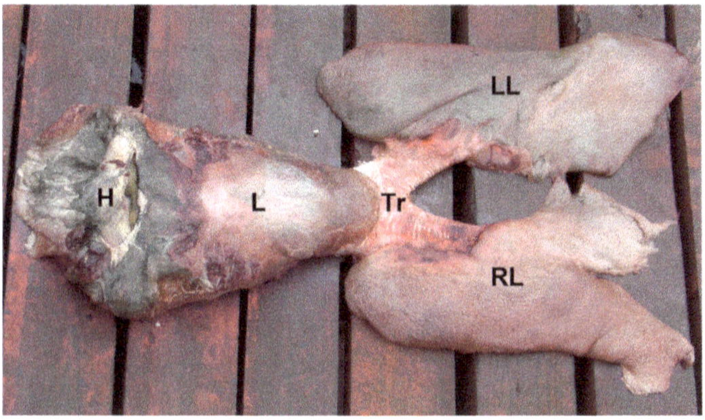

Fig. 3.7 Ventral view of the lower respiratory tract of a minke whale (*Balaenoptera acutorostrata*). H = hyoid bone (basihyal and thyrohyals), L = laryngeal sac immediately ventral to the larynx, LL = left lung, RL = right lung, Tr = trachea. Note that the laryngeal sac (L) is as long as the lungs (LL and RL)

movements along its walls or vibrations transferred to it from movements of the attached vocal folds at the sac's entrance. The sac volume is proportionately larger in balaenopterids compared to balaenids, but the sac walls are proportionately thicker in balaenids compared with balaenopterids. Differences in relative size and thickness of the sac may contribute to the species-specific characteristics of sounds produced.

The sac can distend to a volume that rivals the total volume of one fully inflated lung, and the circumferential skeletal muscle in the walls can contract to expel air from the sac (Reidenberg and Laitman 2008) (Fig. 3.7). The sac may also be used for buoyancy compensation (see below, Diving Adaptations), or to provide excess air volume to accommodate for the volume collapse that occurs when diving (Reidenberg 2018).

It is possible that this sac serves as a sound transducer, sending vibrations from its walls to the overlying blubber and skin, and then into the water (Reidenberg 2017). It is also likely that the sac is an air recycling chamber, catching air used in generating sounds at the vocal folds and then sending it back to be used again for another cycle of sound production (Reidenberg 2018; Damien et al. 2019). This recycling of air lets the whale make multiple sounds in series without having to return to the surface for more air, thus allowing it to stay submerged longer to continue singing. It has been hypothesized that some faint very low-frequency sounds (<10–12 Hz), that have been recorded in between song units from blue and fin whales, may be caused by air movement during air recycling from the sac back to the lungs (C. Clark, pers. comm. 2021).

The laryngeal sac may have multiple roles in all of the above functions. If so, the impact of one function on another may introduce limitations depending upon sac volume as related to depth. For example, as the sac compresses with higher dive pressures it contributes to negative buoyancy, but provides less air volume to be used for sound production. This may explain the observation of balaenopterids producing most sounds in the shallow phases of their dives (depths of only <30–50 m) (Lewis et al. 2018, Chap. 9). Compression or decompression of the laryngeal sac during descent or ascent may oscillate air through the vocal folds to generate sound, thus explaining the long-duration "B"-type calls blue whales make as they descend or ascend (Aroyan et al. 2000). Note that calls are not made below 90 m, the depth at which total lung collapse is predicted and thus airflow between the sac and lungs ceases.

The respiratory passageway continues caudally inside the lumen of the larynx. A thick cushion of fat bulges into the lumen dorsally (Fig. 3.6). It is attached to the ventral aspect of the cricoid cartilage of the larynx. This cricoid cushion is positioned immediately above the vocal folds and is shaped to fit directly into the slight gap between the folds that leads into the laryngeal sac (Damien et al. 2019). This cricoid cushion may function as a valve to help seal the gap between the U-shaped vocal folds, preventing airflow between the sac and the lungs.

Progressing caudally, the larynx transitions into the trachea. This transition is not clearly demarcated, as the cricoid cartilage often breaks into segmented, C-shaped extensions caudally that resemble tracheal rings (Reidenberg and Laitman 2007a). In this region, the C-shaped cartilages are incomplete ventrally, possibly allowing the laryngeal sac to inflate into this soft area and help collapse the trachea (Damien et al. 2019). There is a bifurcation of the trachea (carina), but also an accessory bronchus (epibronchus or eparterial bronchus) above the carina on the right side. Cartilage "rings" may be incomplete C-shapes (proximally near the larynx), complete O-shapes (mid-trachea), or spiraling interlocking shapes (distally in lung). Cartilage is found supporting the airways to the level of the tertiary bronchi. The various cartilage shapes contribute to airway compliance (Moore et al. 2014). The tracheobronchial tree will collapse under extreme pressure from diving, beginning caudally and progressing cranially (Denk et al. 2020).

The lungs are paired, smooth-surfaced organs, bilaterally positioned in the dorsal region of the thorax. Their anatomy is unusual because the lungs are not divided into lobes (Fig. 3.7). The function of this feature is unknown, but perhaps it enables rapid ventilation with less obstruction from angled bronchial branches. This may be related to the adaptive advantage of exchanging all the air in the lungs in one short sneeze-like breath as the whale approaches and briefly breaks the water's surface. The lungs have anatomy that supports pressure-induced collapse and re-expansion (Piscitelli et al 2013). Microscopic examination shows a large number of smooth muscles, particularly around the alveolar openings. Their function is also a puzzle, but perhaps relates to re-expansion of collapsed lungs after a deep dive.

3.3 Sound Generation

All mysticetes studied to date generate underwater sounds. Some sounds are infrasonic (<20 Hz) or in the very low-frequency range (ca. <100 Hz), such as the songs and calls of blue and fin whale singers (Chap. 9). Many sounds are in the low-frequency band (<1000 Hz), such as the songs and calls of humpback, bowhead, right, minke, and Omura's whales (Chaps. 10–15). Mysticetes are specialists in generating low-frequency sounds. These sounds have long wavelengths, suffer little to no attenuation due to frequency-dependent absorption, and can be refracted into pathways with cylindrical, not spherical spreading loss, such that they can travel great distances before attenuating below detection levels (Payne and Webb 1971). In some ways, this long-distance sound transmission phenomenon is similar to that experienced when hearing only the bass beat from the sound system of a distant car or the aggregate noise from a distant highway. Only the low frequencies are perceptible as they travel more efficiently between the distant sound system or highway, while the higher treble frequencies of the song or the collective whines of traffic attenuate before reaching your ears.

The physics of sound propagation in the ocean, from shallow waters along continental shelves or abyssal regions in open ocean environs, has demonstrated the mechanisms by which sounds of different frequency content can travel different distances (Jensen and Kuperman 1983; Jensen et al. 2000). Thus, in the deep ocean, the infrasonic songs of blue and fin whales can be detected on seafloor hydrophones at ranges of many thousands of kilometers, and the very low-frequency portions of humpback songs can be detected at many hundreds of kilometers. In contrast, in relatively shallow waters of a continental shelf, the higher-frequency (> 1 kHz) portions of humpback songs are only detectible out to many tens of kilometers (Clark and Clapham 2004).

All blue, fin, and humpback singers that have been sexed have been males (Glockner 1983; Croll et al. 2002; Oleson et al. 2007; Lewis et al. 2018). Male singing is thought to be important for mate selection and/or male competition (Smith et al. 2008; Herman 2017). Humpback whales are particularly famous for their songs because of their complexity, as well as the social exchange and learning of song components between groups, allowing for song evolution (Chaps. 8, 10 and 11). There are some remarkable species-specific differences in the types of sounds produced by mysticetes. This includes the ability of blue whales to produce intense long sequences of very low-frequency units lasting tens of seconds (>180 dB, Chap. 9), the complex songs of humpback whales (Chap. 10), the "megapclicks" of humpback whales (Stimpert et al. 2007), the "boing" and "star-wars" sounds of minke whales (Rankin and Barlow 2005, Chap. 14), the suggested simultaneous production of two sounds by humpback whales (Jeff Jacobsen, pers. comm.) or bowhead whales (Chap. 12), and the unusual "gunshot" sound of right whales (Crance et al. 2017, Chap. 13).

There are several unique features of mysticete larynges that may support the generation of their unique sounds (Chaps. 6, 9–15). As the air flows out from the lungs and trachea, it passes between the arytenoid–corniculate complexes, crossing over the freely mobile tissue of the vocal folds. This causes them to vibrate, generating variations in pressure at the vibration rate. There are differences in the size and shape of the vocal folds between balaenopterid and balaenid whales. Balaenopterids have long vocal folds with regionally specialized tissues that may generate many different types of sounds as air is directed in different paths across the vocal folds (Damien et al.

2019). This could account for the complex sounds of some species (e.g., humpback whales, Chap. 11). The longer folds also hold the possibility of generating much lower frequencies compared with those produced by shorter vocal folds.

Balaenids have shorter, thicker vocal folds, with a smooth surface that does not appear to have regional specializations, and lack corniculate flaps (Reidenberg and Laitman 2007a) (Fig. 3.5). Thicker folds may vibrate at a slower rate (producing lower-frequency sounds), but this is tempered by a short vocal fold length (limiting fold excursion during vibration). This anatomy is predicted to correlate with a more restricted vocal range in right whales (Chap. 13), but does not explain the great variety of sounds produced by singing bowhead whales (Chap. 12).

There is great variation in corniculate flap sizes among mysticetes. Flaps are very small in humpbacks whales and very large in minke whales. These flaps may generate pulses as they clap together, or fluttering sounds ("thrums," "purrs") as they passively vibrate (as a flag does in the wind). Larger flaps are presumed to have a greater role in contributing to the repertoire of sounds. One could speculate that they are perhaps responsible for the unique "boing" sounds of North Pacific minke whales and "star-wars" sounds of dwarf minke whales (Chap. 14).

Balaenids have a unique structure at the tip of the larynx (Fig. 3.5). The epiglottic and corniculate cartilages are much thicker than in other mysticetes, and they can be opposed against each other to form a complete seal of the laryngeal aditus (the opening connecting the larynx and pharynx). In the closed position, the epiglottis and paired corniculates may act together as one unit, completely plugging the opening of the nasopharynx. The trio of epiglottis and corniculates can be elevated into, or lowered out of, the nasopharyngeal space—acting as a piston. Lowering the larynx expands the nasopharyngeal lumen, in turn lowering the pressure in that space. If the larynx is withdrawn to the point of disconnecting from the nasopharynx, the seal of the soft palate tissues (palatopharyngeal sphincter) would be broken. This could create a sudden "pop" sound from air rushing into the space to equalize the pressure (similar to the sounds made when humans click the tongue or finger pop the mouth). This unique laryngeal feature may underlie the ability of balaenids to make unusual sounds, such as the "gunshot" sounds of right whales (Chap. 13) and "two-voiced singing" by bowhead whales (Würsig and Clark 1990, Chap. 12).

Vocal fold and/or corniculate flap vibrations generate sound energy that radiates out from the whale into the surrounding water. The pathway by which sound propagates through and radiates out of the whale's body is not known. One pathway could involve vibrations transferred along the rostral laryngeal cartilages to the skull. Vibrations from both/either the arytenoid-supported vocal folds and/or the corniculate-supported flaps could pass to the opposed epiglottis. When the larynx is elevated, the epiglottis contacts the vomer, a bone in the skull that cradles the mesorostral cartilage. The vomer is very thin, and often has a hole in this region, exposing the cartilage. This cartilage (homologous to the cartilaginous nasal septum) runs a straight path in balaenopterids (Fig. 3.8), a strongly curved path in balaenids, and a moderately curved path in neobalaenid and eschrichtiid whales. If sounds are transmitted along this cartilage, it could serve as a waveguide to send vibrations in a narrow path with no energy loss. The vibrations may travel directly to water

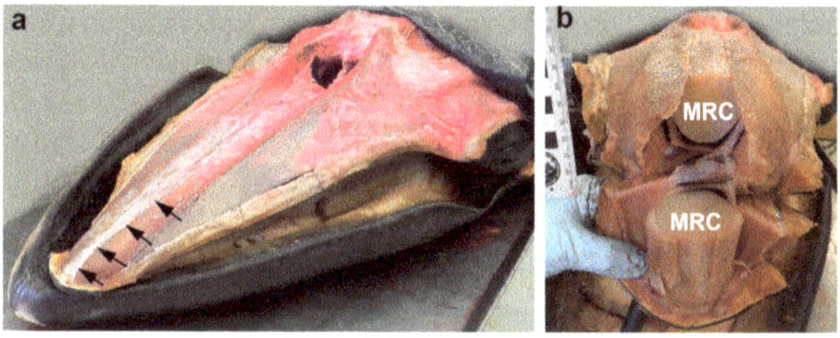

Fig. 3.8 a Left latero-rostro-dorsal view of a deceased juvenile fin whale (*Balaenoptera physalus*) head, dissected to show the mesorostral cartilage (arrows). **b** Same fin whale, rostral view, with the mesorostral cartilage (MRC) cut in cross section and the distal tip reflected inferiorly (ventrally)

at the rostral tip or through the dorsal gap where it is incompletely covered by the premaxillae, both being places where the cartilage meets a thin layer of overlying blubber. As cartilage is mostly comprised of water, there should be very little transmission loss as the energy is transferred to water. There could be great differences in the type of sound emitted, though, as the bowed cartilage should have different resonance properties compared to the straight cartilage rod. In addition, the balaenid whales should have the best coupling of the laryngeal cartilages to the mesorostral cartilage due to higher stiffness of the cartilages and tighter opposition between the epiglottis and corniculate cartilages. This topic needs further investigation, but it is predicted that the two types of mesorostral cartilage geometries will correlate with vocal differences between the balaenopterids and balaenids, while neobalaenid and eschrichtiid whales may possess characteristics shared with both groups.

Another sound path may utilize tissue vibrations from the arytenoid cartilages supporting the vocal folds, to the attached laryngeal sac, and from there to the overlying tissues of the throat including the blubber, and eventually into water. If this throat transmission pathway is used, there should be great diversity in the types of sounds that can be transmitted, as the overlying tissues are quite different between the groups of whales. Balaenopterids have very pliable throat grooves, so the sound vibration would have to pass through highly elastic, expandable tissues of the throat pouch. Perhaps this anatomy is correlated with singing talent, as balaenopterids are hypothesized to seek singing locations where they take advantage of ocean features to broadcast their sounds over greater areas (e.g., use vertical walls to reflect sounds in the manner of a band shell, or sing along the edge of the continental shelf where sound is refracted downward, reflects on the shelf and enters the deep sound channel; Chap. 2). Balaenids, on the other hand, may send their sounds a short distance to nearby whales, and thus may not utilize an elaborate sound transfer mechanism. Balaenid, neobalaenid, and eschrichtiid throats are less pliable, having at most only three short throat grooves. Balaenids have a much thicker blubber layer, and the tissues between the throat grooves do not expand to the same degree as in balaenopterids. If the throat transmission pathway is used, the sounds are likely to be dampened in the balaenid, neobalaenid, and eschrichtiid whales compared with balaenopterid whales, therein consistent with the expectation that sound source levels in balaenids would be lower than those in balaenopterids. The purpose of the neobalaenid's asymmetrically placed laryngeal sac is unclear. Perhaps the position is lateralized due to the absence of highly flexible throat pleats, instead channeling vibrations in a different pathway along the side of the head or neck to minimize dampening.

Another factor influencing sound generation and transmission is the degree of laryngeal sac expansion. As the sac enlarges, the walls become tighter, changing its resonance properties (the sac walls may vibrate as a drum does, with variations in resonance correlated to drum barrel shape and size and drumhead tension). The laryngeal sac of balaenopterids is larger than the sac of balaenids. Differences in laryngeal sac size, shape, and tension are expected to contribute to the qualities of the generated and transmitted sound.

During sound production, the mouth and nose are closed, and air does not escape from those openings. Instead, air flows through the gap between the free edges of each vocal fold into the ventrally placed laryngeal sac. As the sac expands, it captures the air used to generate sound. Then the sac contracts and sends the air back to be reused. This recycling of air is the mechanism by which whales can produce multiple sequential or a continual series of sounds over long periods without having to return to the surface for more air. Balaenopterids have the largest laryngeal sac volume, relative to the size of the larynx. Studies are ongoing to determine whether this difference is sexually dimorphic. A relatively large laryngeal sac may correlate with longer song cycles performed on a single breath.

3.4 Sound Reception

Mysticetes do not have external ears (pinnae), giving a streamlined contour. The external ear canals are filled with a wax plug, avoiding a cavity that would otherwise retain air that dampens underwater sound transmission, or could become a reservoir for developing an infection. In addition the Eustachian tubes are wide, allowing an easy mechanism for regulating diving-related pressure changes in the middle ear. Despite these aquatic adaptations, the anatomy of the mysticete inner ear (e.g., low basilar membrane thickness: width ratios, large organ of Corti mass) indicates they hear particularly well in the low-frequency range (Wartzok and Ketten 1999, Chap. 2). Although hearing ranges are often correlated with the sound production frequency ranges, one does not always predict the other. Rather, hearing ranges can be estimated through measurements of the cochlea's basilar membrane or reconstructed through physiologic measurements of brainstem responses to auditory evoked potentials (Auditory Brainstem Response, ABR). The ABR method has been applied successfully on odontocetes and is only applicable to temporarily restrained, very small mysticetes (e.g., it was attempted on a gray whale calf in a captive rehabilitation situation with equivocal results, Ridgway and Carder 2001). It cannot be performed on free-swimming wild mysticetes.

The ear has many of the same components as that of land mammals including a highly modified tympanic membrane, tympano-periotic complex (sometimes called the petrotympanic complex, comprised of the tympanic bulla, petrosal bone, and the ossicles of the middle ear), and a fluid-filled labyrinth system consisting of semicircular canals (for balance) and a cochlea (for hearing). The middle ear chamber is air-filled and maintained as such even at depth due to the adjacent pterygoid air sac that can collapse at depth and shunt all remaining air to the middle ear. The semicircular canals are very small relative to the whole labyrinth system. It has been proposed that this reduction may be related to decreased head rotation, particularly as mysticetes filter feed rather than pursue highly mobile individual prey (Ekdale and Racicot 2015). Head movements are restricted due to the flattened cervical vertebrae that compress the neck region and greatly curb its flexibility. This, in turn, stabilizes the head and neck from lateral movements (and thus undue friction), particularly at high speeds (VanBuren and Evans 2017).

Unlike the semicircular canals, the cochlea is hypertrophied, perhaps as an adaptation for high background noise (Ketten 2012). Cochlear apical width is greater than odontocetes, and similar to elephants that are also known to hear in the infrasonic range (Payne et al 1986). Unlike odontocetes, mysticetes do not have extensive outer bony laminae (a feature associated with high frequency hearing). Inside the cochlea, mysticetes have a broad, flaccid basilar membrane—an adaptation that responds well to infrasonics (Ketten 1997). The innervation densities of auditory ganglion cells are three times higher than that of humans, likely enabling the central nervous system to process multiple response sets in parallel. Interestingly, both odontocetes and mysticetes have similar high ratios of ganglion cells to hair cells, indicating both groups have equally complex signal processing even though the perceived sounds are on opposite ends of the frequency spectrum (Wartzok and Ketten 1999).

The petrotympanic bone is unusual because it is very large and extremely dense—more so than the petrous bone, or any other bone, in any other mammal (Cozzi, et al. 2009). It likely functions as a lever, bringing force (pressure) amplification of low amplitude sounds to the ossicular chain (Tsur et al. 2019). Both bone and tissue conduction of sound to the ear are likely (Ketten 2012). Modeling has shown that low-frequency sound transmission primarily occurs through bony conduction in the fin whale, and likely in all mysticetes (Cranford and Krysl 2015). Bone conduction enhances hearing at lower frequencies, and this mechanism is more sensitive than pressure transmission through soft tissues (Cranford and Krysl 2015). However, bone conduction transfers vibrations to both ears simultaneously and thus limits localization and discrimination of the sound source (Tubelli et al. 2018). The tympano-periotic complexes are attached to the skull at the pedicles, enabling bone conduction, but most of the complex is actually detached from the skull. Perhaps this detachment limits bone conduction or directs the sound in specific pathways to increase auditory localization processes. The bony conduction pathway has merit because the receiver is so large (the whole skull) that it may be ideal for perceiving the longer wavelengths of the low-frequency sounds. These sounds are generally used for communication, but have been suggested by some to serve also as coarse sonar for navigating around large underwater obstacles or using them as navigational aids (e.g., ice keels, George et al. 1989; seamounts or canyons, Clark and Ellison 2004). Precise localization is probably unnecessary when the target is so large, and thus bony conduction reaching both ears at approximately the same time may not be a serious detriment. Alternatively, the brain may be tuned to detect microsecond differences in arrival times between the bony conduction to each ear, enabling directional hearing.

The evolutionary retention of an ossicular chain implies it is also used for sound transfer. If bone conduction is the preferred mechanism for low-frequency sound transfer, perhaps tissue transfer is the mechanism for higher frequencies. Two mechanisms have been proposed for transferring sounds to the ossicular chain. Vibrations may be transferred by the tympanic membrane, although not from the external ear canal as in terrestrial mammals. This is because the tympanic membrane is not in contact with an air-filled external acoustic meatus. Rather, it is a highly modified structure (also called the "glove finger" due to its invagination into an elongated sac-like structure) that lies in a fluid-filled chamber adjacent to a large wax earplug. Modeling and experimental studies in humpback and minke whale ears predict the apex of the tympanic membrane receives vibrations passing through tissues in this region and regulates the vibration transfer to the ossicles of the middle ear (Tubelli and Zosuls 2012; Tubelli et al. 2018). In minke whales, sound may be received by the fat bodies located adjacent to the lower jaw and directly transferred through their connection to the middle ear ossicles (Yamato et al. 2012). Conversely, these fat pads were not found in the bowhead whale (unpublished data, pers. comm. D. Hillmann, Louisiana State University), although this may be because the specimens were only fetal.

3.5 Diving Adaptations

Fluids such as blood are relatively incompressible and may be used to compensate for pressure-related volume changes of air-containing spaces during diving. Blood may be shunted into these spaces to occupy volumes (that were expanded with air before diving compression) to prevent damage to structural tissues surrounding these spaces. There are two major regions where blood-filled tissue expands adjacent to collapsing air-filled spaces: the lungs and the pterygoid sacs.

As the lungs shrink during diving, the vacated negative space is filled by an expanding vascular tissue called the retia mirabilia thoracica that swells with blood (Moore et al. 2011). This is not a simple erectile tissue comprised of blood sinuses, but rather, a complex network of blood vessels in the thoracic cavity located on the dorsal, paraspinal region. These vessels may also function to trap bubbles that might form in the blood on ascent, preventing the bubbles from circulating to vital organs and thus mitigating for decompression sickness (Reidenberg and Laitman 2015).

The pterygoid sacs have a complex network of vessels running through and along pterygoid sac walls. The arterial system of this region has a considerable surface area in contact with both the fat of the mandibular region and the air in the pterygoid sacs. This may facilitate nitrogen gas exchange to prevent decompression sickness

(Costidis and Rommel 2016a). There are also thin-walled, venous lakes (large open blood sinuses through which arteries pass) in the pterygoid sac that may participate in nitrogen exchange (Costidis and Rommel 2016b). The pterygoid sacs are soft-walled and respond to ambient pressure changes by collapsing during a dive and re-expanding upon ascent. Filling the sacs with air before diving ensures that there is extra volume associated with the middle ear, thus mitigating the effect of volume collapse at depth. As the soft-walled pterygoid sacs deflate, the residual air volume is preferentially preserved in the rigid space of the middle ear (surrounded by bone). This arrangement preserves hearing at depth because it ensures that the ossicles have a space in which to vibrate so they can transmit sounds to the inner ear.

While diving, whales, particularly balaenopterids, approach neutral buoyancy as the positive buoyancy of blubber is countered by the negative buoyancy of the relatively incompressible tissues (bones, muscles, and organs) (Gray et al. 2007; George et al. 2016). These factors cannot be changed, and thus comprise the static component of buoyancy. Dynamic buoyancy, however, is regulated by the air spaces of the respiratory tract (lungs, laryngeal sac, trachea, larynx, pharynx, sinuses, nasal passageways), as they can contain variable amounts of air. During inhalation, a fully inflated respiratory tract makes the whale positively buoyant. However, as the whale begins its dive, and ambient pressure increases, the air in the respiratory tract begins to compress. As these air spaces collapse, the whale's body becomes denser and its buoyancy becomes negative, thus aiding descent. The whale's head would be pointing down during descent, so air will naturally rise to the highest point in the respiratory tract: the lungs. Evacuating air out of the nasal passageways, nasopharynx, pterygoid air sacs, larynx, and laryngeal sac should make the head heavier (particularly since it is mostly comprised of dense skull bone), and this would help the whale maintain a head-down orientation. After a few initial beats of the tail to begin the descent, the negatively buoyant whale could now save energy and glide the rest of the way down (Williams et al. 2000)—following its own heavy head as if following a sinking stone. Balaenopterids spend more energy swimming during ascent to overcome negative buoyancy. Achieving negative buoyancy is more difficult in balaenids, as they tend to float at the surface due to the thicker blubber layer. They must exert more energy in powerful tail strokes when initiating a dive until lung collapse contributes to achieving negative buoyancy, enabling continued descent with less effort.

Whales appear able to vocalize at any depth, but there can be advantages to emitting sound during deep dives. However, the collapse of air spaces should make vocalizing nearly impossible at extreme depths. The laryngeal sac may be just the diving adaptation necessary for sound production to proceed. Extra air carried in this sac can be added to the air volume available for generating sounds. As the whale descends, the lungs would only have a fraction of the original volume they contained at one atmosphere of pressure, according to Boyle's law (i.e., for a fixed amount of gas kept at steady temperature, pressure and volume are inversely proportional—that is, when one doubles, the other is reduced by half). The nasal region will also collapse its volume in the same proportion, sucking in the nasal plugs down the tubular nasal passageway's length and shrinking the walls of the nasopharynx and pterygoid sacs. The larynx and trachea would similarly collapse, squeezing the epiglottis toward the corniculates and contracting the tracheal rings similarly to a compressed accordion. These spaces may not be able to contract any further, or may be insulated from doing so in order to maintain some air in the pterygoid sac to preserve a space for ossicular vibration for hearing.

The laryngeal sac can inflate to approximately the volume of one lung at the surface. This is an expendable space that would have a lot of extra volume available if inflated before descent, although it would be reduced by the same percentage at depth. Contraction of the laryngeal sac muscles could evacuate this air to the lungs. Perhaps it would be enough to increase the lung volume to initiate sound production. The flattened rib cage that accommodated the lung collapse at depth would now partially re-expand. Contraction of the intercostal muscles would re-flatten the ribs, compressing the thorax, and drive air from the lungs into the trachea, larynx, and through the vocal folds. This should vibrate the folds and generate the fundamental frequency needed to create sounds. This small amount of air would be captured by the expanding laryngeal sac. The muscular walls of the sac could then compress this air volume and expel it back to the lungs to be recycled for repeated rounds of sound production.

As the whale ascends, the whole process reverses. Contraction of the intercostal muscles would compress the thorax and shunt lung air cranially to the trachea, larynx, laryngeal sac, pterygoid sacs, nasopharynx, and nasal passageways. This makes the head lighter than the rest of the body, and the whale now follows the head up, as if following a rising balloon. After a few initial tail strokes to begin the upward ascent, the whale could save energy and simply glide effortlessly to the surface. This is particularly true for balaenids, as the positively buoyant effect of the blubber becomes greater during ascent, helping to power their rapid glides back to the surface (Nowacek et al. 2001).

Whales are champion breath-holders, but this means less oxygen in the blood is delivered to the tissues during diving. There are many adaptations that accommodate for this, including the previously mentioned energy conservation during descent and ascent, and the slowed metabolism evidenced by the very low heart rate while diving. This means their bodies use less oxygen, so the muscle contractions can last longer before building up oxygen debt. Another adaptation lies in the skeletal muscles themselves. They contain high levels of myoglobin, the substance that makes the muscle appear very dark. Myoglobin helps bind extra oxygen where it is needed most for muscle contractions, thus compensating for the lack of circulating oxygenated blood during the dive. Myoglobin concentrations appear to correlate with average duration and depth of dives, thus enhancing dive capacity (Helbo and Fago 2012).

The bones have also evolved to be adapted for buoyancy control with anatomy specific to diving behavior. The skull is devoid of any bony paranasal air sinuses, likely as an adaptation for diving. Rigid-walled sinuses would crack under pressure as the contained volume of air shrinks during a dive. Bowhead whales (and presumably all balaenids) have dense bones as calves, but lose this density with age (George et al. 2016). The heavy bones impart ballast (negative buoyancy) that helps counter-balance flotation (positive buoyancy) caused by the fat tissues (e.g., blubber). Balaenids have comparatively more blubber (compared with other mysticetes), and there is proportionally more in calves. Thus, despite dense bones, balaenids tend to be positively buoyant near the surface. This causes them to spend more energy as they begin a dive (compared with non-balaenids) to overcome the positive buoyancy provided by the thicker blubber layer (Nowacek et al. 2001). In contrast, most other cetaceans have osteoporotic (less dense) bones in the ribs and flippers that balance the buoyancy of the thinner blubber layer, thus contributing to an overall neutral buoyancy (Gray et al. 2007). This enables dynamic buoyancy regulation, achieved by controlling the amount of air in the respiratory tract.

3.6 Conclusion

Mysticetes are a very diverse group of whales. Major differences are found between the anatomy and physiology of balaenopterids versus balaenids, with the neobalaenid and eschrichtiid whales often exhibiting an intermediate morphology and function between these two extremes. Unique features of balaenopterids describe a faster, more agile, and deeper diving whale, with a gulping feeding style, and (in some species) lower frequency calls/songs. Unique features of balaenids describe a slower swimming, shallower diving, less agile whale, with a continuous flow feeding style, and a broader vocal range with some unique sound production abilities (e.g., bowhead two-voiced singing, Chap. 12). Less well understood are the intermediate morphologies exhibited by the neobalaenid and eschrichtiid whales. Further research is needed to relate the variations in laryngeal anatomy to the species-specific sounds produced by the four different mysticete families.

References

Aroyan JL, McDonald MA, Webb SC, Hildebrand JA, Clark D, Laitman JT, Reidenberg JS (2000) Acoustic models of sound production and propagation. In: Au W, Fay R, Popper A (eds) Hearing by Whales and Dolphins. Springer-Verlag, New York, pp 409–469

Bauer GB, Reep RL, Marshall CD (2018) The tactile senses of marine mammals. Int J Comp Psychol 31

Berta A, Ekdale EG, Zellmer NT, Deméré TA, Kienle SS, Smallcomb M (2015) Eye, nose, hair, and throat: External anatomy of the head of a neonate gray whale (Cetacea, mysticeti, eschrichtiidae). Anat Rec 298(4):648–659

Buono MR, Reidenberg JS, Fernández MS (2015) Anatomy of the nasal complex in the southern right whale, *Eubalaena australis* (Cetacea, Mysticeti). J Anat 226:81–92

Butti C, Santos M, Uppal N, Hof P (2013) Von Economo neurons: Clinical and evolutionary perspectives. Cortex 49(1):312–326

Clark CW, Clapham PJ (2004) Acoustic monitoring on a humpback whale (Megaptera novaeangliae) feeding ground shows continual singing into late spring. Proc R Soc B 1051–1057

Clark CW, Ellison WT (2004) Potential use of low-frequency sounds by baleen whales for probing the environment: evidence from models and empirical measurements. In: Thomas J, Moss C, Vater M (eds) Echolocation in bats and dolphins. The University of Chicago Press, pp 564–582

Corkeron PJ, Connor RC (1999) Why do baleen whales migrate? Mar Mamm Sci 15:1228–1245

Cooper LN, Berta A, Dawson SD, Reidenberg JS (2007a) Evolution of hyperphalangy and digit reduction in the cetacean Manus. Anat Rec 290:654–672
Cooper LN, Dawson SD, Reidenberg JS, Berta A (2007b) Neuromuscular anatomy and evolution of the cetacean forelimb. Anat Rec 290:1121–1137
Costidis AM, Rommel SA (2016a) The extracranial arterial system in the heads of beaked whales, with implications on diving physiology and pathogenesis. J Morphol 277:5–33
Costidis AM, Rommel SA (2016b) The extracranial venous system in the heads of beaked whales, with implications on diving physiology and pathogenesis. J Morphol 277:34–64
Cozzi B, Mazzariol S, Podesta M, Zotti A (2009) Diving adaptations of the cetacean skeleton. Open Zool J 2:24–32
Crance JL, Berchok CL, Keatings JL (2017) Gunshot call production by the North Pacific Right Whale *Eubalaena japonica* in the Southeastern Bering Sea. Endangered Species Res 34:251–267
Cranford TW, Krysl P (2015) Fin whale sound reception mechanisms: skull vibration enables low-frequency hearing. PLoS ONE 10(1):e0116222
Croll DA, Clark CW, Acevedo A, Tershy B, Floress S, Gedamke J, Urban J (2002) Only male fin whales sing loud songs. Nature 417:809
Damien J, Adam O, Cazau D, White P, Reidenberg JS, Laitman J (2019) Anatomy and functional morphology of the mysticete rorqual whale larynx: phonation positions of the Ufold. Anat Rec 302:703–717
Deméré TA, McGowen MR, Berta A, Gatesy J (2008) Morphological and molecular evidence for a stepwise evolutionary transition from teeth to baleen in mysticete whales. Syst Biol 57(1):15–37
Denk M, Fahlman A, Dennison-Gibby S, Song Z, Moore M (2020) Hyperbaric tracheobronchial compression in cetaceans and pinnipeds. J Exp Biol 223:jeb217885
Drake SE, Crish SD, George JC, Stimmelmayr R, Thewissen JGM (2015) Sensory hairs in the bowhead whale, *Balaena mysticetus* (Cetacea, Mammalia). Anat Rec 298(7):1327 1335
Ekdale EG, Racicot RA (2015) Anatomical evidence for low frequency sensitivity in an archaeocete whale: comparison of the inner ear of *Zygorhiza kochii* with that of crown Mysticeti. J Anat 226:22–39
George JC, Clark CW, Carroll GM, Ellison WT (1989) Observations on the ice- breaking and ice navigation behavior of migrating bowhead whales (Balaena mysticetus) near Point Barrow, Alaska, spring 1985. Arctic 42:24–30
George JC, Stimmelmayr R, Suydam R, Usip S, Givens G, Sformo T et al (2016) Severe bone loss as part of the life history strategy of bowhead whales. PLoS ONE 11(6):e0156753
Gil KN, Lillie MA, Vogl AW, Shadwick RE (2020) Rorqual whale nasal plugs: protecting the respiratory tract against water entry and barotrauma. J Exp Biol 223:jeb219691
Glockner DA (1983) Determining the sex of humpback whales (*Megaptera novaeangliae*) in their natural environment. In: Payne R (ed) Communication and behavior of whales. Westview Press, Colorado, pp 447–466
Gray N-M, Kainec K, Madar S, Tomko L, Wolfe S (2007) Sink or swim? Bone density as a mechanism for buoyancy control in early cetaceans. Anat Rec 290:638–653
Helbo S, Fago A (2012) Functional properties of myoglobins from five whale species with different diving capacities. J Exp Bio 215:3403–3410
Herman LM (2017) The multiple functions of male song within the humpback whale (*Megaptera novaeangliae*) mating system: review, evaluation, and synthesis. Biol Rev 92:1795–1818. https://doi.org/10.1111/brv.12309
Hof PR, Van Der Gucht E (2007) Structure of the cerebral cortex of the humpback whale, *Megaptera novaeangliae* (Cetacea, Mysticeti, Balaenopteridae). Anat Rec 290:1–31
Jensen FB, Kuperman WA (1983) Optimum frequency of propagation in shallow water environments. J Acoust Soc Am 73:813–820
Jensen FB, Kuperman WA, Porter MB, Schmidt H (2000) Computational Ocean acoustics. AIP series in modern acoustics and signal processing, Amer Inst Phys 612
Ketten DR (1997) Structure and function in whale ears. Bioacoustics 8:103–135
Ketten D (2012) Structure and function in whale ears. Bioacoustics 8(102):103–135

Kienle SS, Ekdale EG, Reidenberg JS, Deméré TA (2015) Tongue and hyoid musculature and functional morphology of a neonate gray whale (Cetacea, Mysticeti, *Eschrichtius robustus*). Anat Rec 298:660–674

Lewis LA, Calambokidis J, Stimpert AK, Fahlbusch J, Friedlaender AS, McKenna MF, Mesnick SL, Oleson EM, Southall BL, Szesciorka AR, Sirovic A (2018) Context-dependent variability in blue whale acoustic behaviour. Roy Soc Open Sci 5(8) https://doi.org/10.1098/rsos.180241

MacLeod CD, Reidenberg JS, Weller M, Santos MB, Herman J, Goold J, Pierce GJ (2007) Breaking symmetry: The marine environment, prey size, and the evolution of asymmetry in cetacean skulls. Anat Rec 290:539–545

Marino L (2009) Brain size evolution. In: Würsig B, Perrin WF, Thewissen JGM (eds) Encyclopedia of marine mammals. Academic, Press, pp 149–152

Maust-Mohl M, Reidenberg JS, Reiss D (2019) A comparison of hippopotamus and cetacean nostrils: An open and shut case. Anat Rec 302:693–702

Mercado E III (2014) Short note: Tubercles: What sense is there? Aquat Mamm 40(1):95–103

Moore MJ, Hammar T, Arruda J, Cramer S, Dennison S, Montie E, Fahlman A (2011) Hyperbaric computed tomographic measurement of lung compression in seals and dolphins. J Exp Biol 214:2390–2397

Moore C, Moore M, Trumble S, Niemeyer M, Lentell B, McLellan W, Costidis A, Fahlman A (2014) A comparative analysis of marine mammal tracheas. J Exp Biol 217:1154–1166

Nakai J, Shida T (1948) Sinus-hairs of the sei whale (Balaenoptera borealis). Sci Rept Whales Res Inst 1:41–47

Nowacek DP, Johnson MP, Tyack PL, Shorter KAWA, Pabst DA (2001) Buoyant balaenids: the ups and downs of buoyancy in right whales. Proc Royal Soc B 268(1478):1811–1816

Oleson EM, Calambokidis J, Burgess WC, McDonald MA, LeDuc CA, Hildebrand A (2007) Behavioral context of call production by eastern North Pacific blue whales. Mar Ecol Prog Ser 330:269–284

Pabst DA (1996) Morphology of the subdermal connective tissue sheath of dolphins: a new fibre-wound, thin-walled, pressurized cylinder model for swimming vertebrates. J Zool 238(1):35–52

Pack AA, Herman LM, Spitz SS, Deakos MH, Herman EYK (2009) Male humpback whales in the Hawaiian breeding grounds preferentially associate with larger females. Anim Behav 77(3):653–662

Pavlov VV (2006) Dolphin skin as a natural anisotropic compliant wall. Bioinspir Biomim 1:31–40

Payne R, Webb D (1971) Orientation by means of long range acoustic signaling in baleen whales. Ann N Y Acad Sci 188(1):110–41

Payne KB, Langbauer WR, Thomas EM (1986) Infrasonic calls of the Asian elephant (*Elephas maximus*). Behav Ecol Sociobiol 18:297–301

Pinto SJD (2011) On the filtration mechanism and oral anatomy of lunge-feeding baleen whales. Dissertation, Univ. British Columbia. http://hdl.handle.net/2429/37849

Piscitelli MA, Raverty SA, Lillie MA, Shadwick RE (2013) A review of cetacean lung morphology and mechanics. J Morphol 274:1425–1440

Pitman RL, Durban JW, Joyce T, Fearnbach H, Panigada S, Lauriano G (2020) Skin in the game: Epidermal molt as a driver of long-distance migration in whales. Mar Mam Sci. 36:565–594

Pyenson N, Goldbogen J, Vogl A, Szathmary G, Drake R, Shadwick R (2012) Discovery of a sensory organ that coordinates lunge feeding in rorqual whales. Nature 485:498–501

Raghanti MA, Spurlock LB, Treichler RF, Weigel SE, Stimmelmayr R, Butti C, Thewissen JGMH, Hof P (2015) An analysis of von economo neurons in the cerebral cortex of cetaceans, artiodactyls, and perissodactyls. Brain Struct Funct 220:2303–2314

Rankin S, Barlow J (2005) Source of the North Pacific "boing" sound attributed to minke whales. J Acoust Soc Am 118:3346–3351

Reeb D, Best PB (1999) Anatomy of the laryngeal apparatus of the pygmy right whale, Caperea marginata (Gray 1846). J Morphol Oct 242(1):67–81

Reeb D, Best PB, Kidson SH (2007) Structure of the integument of southern right whales. *Eubalaena australis*. Anat Record 290(6):596–613

Reidenberg JS (2017) Terrestrial, semiaquatic, and fully aquatic mammal sound production mechanisms. Acoust Today 13(2):35–43

Reidenberg JS (2018) Where does the air go? Anatomy and functions of the respiratory tract in the humpback whale (*Megaptera novaeangliae*). Special Issue on Humpback Whales. Madagascar Cons. & Dev. 13(01) Dec 91–100

Reidenberg JS, Laitman JT (1994) Anatomy of the hyoid apparatus in Odontoceti (toothed whales): Specializations of their skeleton and musculature compared with those of terrestrial mammals. Anat Rec 240:598–624

Reidenberg JS, Laitman JT (2007a) Discovery of a low frequency sound source in mysticeti (baleen whales): Anatomical establishment of a vocal fold homolog. Anat Rec 290:745–760

Reidenberg JS, Laitman JT (2007b) Blowing bubbles: an aquatic adaptation that risks protection of the respiratory tract in humpback whales (*Megaptera novaeangliae*). Anat Rec 290:569–580

Reidenberg JS, Laitman JT (2008) Sisters of the sinuses: cetacean air sacs. Anat Rec 291:1389–1396

Reidenberg JS, Laitman JT (2015) Beating the bends: Dolphin thoracic rete as a natural vascular filter for trapping nitrogen bubbles and preventing diving decompression sickness. In: The FASEB journal american association of anatomists: experimental biology, Boston

Ridgway SH, Carder DA (2001) Assessing hearing and sound production in cetaceans not available for behavioral audiograms: experiences with sperm, pygmy sperm, and gray whales. Aquat Mamm 27(3):267–276

Schoenfuss HL, Bragulla HH, Schumacher J, Henk WG, Craig George J, Hillmann DJ (2014) The anatomy of the larynx of the bowhead whale, *Balaena mysticetus*, and its sound-producing functions. Anat Rec 297:1316–1330

Shoemaker PA, Ridgway SH (1991) Cutaneous ridges in odontocetes. Mar Mammal Sci 7:66–74

Smith JN, Goldizen AW, Dunlop RA, Noad MJ (2008) Songs of male humpback whales, Megaptera novaeangliae, are involved in intersexual interactions. Anim Behav 76(2):467–477

Stimpert AK, Wiley DN, Au WWL, Johnson MP, Arsenault R (2007) "Megapclicks": Acoustic click trains and buzzes produced during night-time foraging of humpback whales (Megaptera novaeangliae). Biol Let 3(5):467–470

Thewissen JGM, Hieronymus TL, George JC, Suydam R, Stimmelmayr R, McBurney D (2017) Evolutionary aspects of the development of teeth and baleen in the bowhead whale. J Anat 230(4):549–566

Tsur I, Shaviv N, Bronstein I, Elmakis D, Knafo O, Werner YL (2019) Topography of vibration frequency responses on the bony tympano-periotic complex of the pilot whale *Globicephala macrorhynchus*. Hear Res 384:107810

Tubelli A, Zosuls A (2012) A prediction of the minke whale (*Balaenoptera acutorostrata*) middle-ear transfer function. J Acoust Soc Am 132(5):3263–3272

Tubelli A, Zosuls A, Ketten DR, Mountain DC (2018) A model and experimental approach to the middle ear transfer function related to hearing in the humpback whale (*Megaptera novaeangliae*). J Acoust Soc Am 144:525–535

VanBuren CS, Evans DC (2017) Evolution and function of anterior cervical vertebral fusion in tetrapods. Biol Rev 92(1):608–626

van Nierop EA, Alben S, Brenner MP (2008) How bumps on whale flippers delay stall: an aerodynamic model. Phys Rev Lett 101(5):054502

Wartzok D, Ketten DR (1999) Marine mammal sensory systems. In: Reynolds JE, Rommel SA (eds) Biology of marine mammals. Smithsonian institution Press, pp 117–175

Werth AJ (2007) Adaptations of the cetacean hyolingual apparatus for aquatic feeding and thermoregulation. Anat Rec 290:546–568

Williams TM, Davis RW, Fuiman LA, Francis J, Le Boeuf BJ, Horning M, Calambokidis J, Croll DA (2000) Sink or swim: strategies for cost-efficient diving by marine mammals. Science 288:133–136

Würsig B, Clark CW (1990) Behavior of bowhead whales in the Western Arctic. In: Burns J, Montague J (eds) The bowhead whale book. Allen Press, Lawrence

Yamato M, Ketten DR, Arruda J, Cramer S, Moore K (2012) The auditory anatomy of the minke whale (*Balaenoptera acutorostrata*): a potential fatty sound reception pathway in a baleen whale. Anat Rec 295:991–998

Chapter 4
Baleen Whale Migration

Travis W. Horton, Daniel M. Palacios, Kathleen M. Stafford, and Alexandre N. Zerbini

T. W. Horton (✉)
Te Kura Aronukurangi/School of Earth and Environment, University of Canterbury, Private Bag, 4800, Christchurch 8140, New Zealand
e-mail: travis.horton@canterbury.ac.nz

D. M. Palacios
Marine Mammal Institute, Hatfield Marine Science Center, Oregon State University, 2030 SE Marine Science Dr, Newport, OR 97365, USA
e-mail: daniel.palacios@oregonstate.edu

K. M. Stafford
Applied Physics Laboratory, University of Washington, 1013 NE 40th Street, Box 355640, Seattle, WA 98105-6698, USA
e-mail: kate.stafford@oregonstate.edu

A. N. Zerbini
The Cooperative Institute for Climate, Ocean, and Ecosystem Studies (CICOES), University of Washington (UW), John M. Wallace Hall, 3737 Brooklyn Ave NE, Seattle, WA 98105, USA
e-mail: alex.zerbini@noaa.gov

Marine Mammal Laboratory, Alaska Fisheries Science Center, National Marine Fisheries Service, National Oceanic and Atmospheric Administration (NOAA), 7600 Sand Point Way, NE, Seattle, WA 98115, USA

Marine Ecology and Telemetry Research, 2468 Camp McKenzie Tr NW, Seabeck, WA 98380, USA

Instituto Aqualie, Av. Dr. Paulo Japiassú Coelho, 714/202, Juiz de Fora, MG 36033-310, Brazil

C.W. Clark and E. C. Garland (eds.), *Ethology and Behavioral Ecology of Mysticetes*, Ethology and Behavioral Ecology of Marine Mammals,
https://doi.org/10.1007/978-3-030-98449-6_4

> *Paikea is an ancestor of many* [Māori] *iwi (tribal groups) of the eastern seaboard of Aotearoa New Zealand. The Paikea story is known in other parts of the Pacific and provides an explanation for how this particular ancestor reached Aotearoa from the ancestral and spiritual home-land of Hawaiki. There are several versions of the story, but it is commonly accepted that he was the sole survivor of a marine disaster and through his endeavors reached shore at a place called Ahuahu. This was achieved through the mobilization of his marine ancestors, his family of whales, who helped him reach Aotearoa. Paikea is described as riding on the back of a whale, or transforming into a whale, and is referred to accordingly as he tahito, he tipua, he taniwha, he tohorā, he tangata, he tekoteko—an ancient being, an extraordinary being, a denizen of the deep, a whale, a man, a sentinel for his people. Paikea is also the Māori name for southern humpback whales [i.e. Oceania population of Megaptera novaeangliae].*
>
> *The Paikea narrative underpins a certain type of relationship with whales, one of kaitiakitanga—care or stewardship. This is conceptualized in whakapapa [i.e. genealogy] terms, whereby whales are identified as ancestors and kin. The kaitiaki [i.e. guardian] relationship underpins voyaging knowledge contained in oral histories. Whales guide waka (canoes/vessels) to land, through dangerous seas and channels, and are called upon to smooth rough waters for safe passage. Ocean-going waka hourua are double-hulled to replicate the physical qualities of a pair of whales cresting waves in tandem.*
>
> *Tere tohorā, tere tangata—where whales journey, people follow—is a whakataukī (proverb) that encapsulates the essence of this synergy.*
>
> Billie Lythberg and Wayne Ngata (Te Aitanga a Hauiti), New Histories of Pacific Whaling, Rachel Carson Center Perspectives, No. 5, pp. 105–106, 2019

Abstract Baleen whale migration emerges as a foundational theme of cetacean behavioral ecology and the relationships that bind humans and whales together. From facilitating the culmination of the great human migration many centuries ago, to their roles as ecosystem service providers, baleen whales have influenced the path of human history. With a focus on modern technologically enabled insights, we provide an overview of what scientists currently know about the spatial and temporal distribution of baleen whales and their migratory behaviors. Although a coarse model of seasonally paced north–south migration generally applies, a deeper analysis reveals the remarkable diversity of baleen whale migrations. Some species, including gray whales (*Eschrichtius robustus*), migrate relatively close to shore; others, including humpback whales (*Megaptera novaeangliae*), tend to migrate across ocean basins. Some species, including Bryde's whales (*Balaenoptera edeni*), appear to largely reside in middle to low-latitude ecosystems, relatively removed from cold, high-latitude water. In contrast, bowhead whales (*Balaena mysticetus*) remain within Arctic ecosystems all year, and others, including Omura's whales (*Balaenoptera* *omurai*), may not migrate at all. The scientific focus to date has largely been on population-specific studies of where whales go, what their behaviors are, and when they undertake their migrations. Thus, there remains much to be learned, particularly regarding why baleen whales migrate and how they navigate during their long-distance migrations. Technological innovations such as satellite tags and passive acoustics have revolutionized our understanding of baleen whale behavioral ecology and ethology, and technology will continue to play a critical role in advancing the science of baleen whale migration.

Keywords Baleen whale · Migration · Telemetry · Satellite tracking · Passive acoustics · Migratory behavior · Movement behavior · Acoustic behavior · Orientation · Navigation

4.1 Introduction

The powerful and ancient synergy between Paikea, the migratory humpback whales (*Megaptera novaeangliae*) of the Pacific, and Polynesian tūpuna/kūpuna/ancestors* continues to tie humans and whales together. In many places, indigenous rights to stranded whales are protected by law, and local communities like Kaikōura, in Aotearoa New Zealand, and Vava'u, in Tonga, benefit from whale watching activities that attract tens of thousands of visitors every year. Perhaps the most significant connection between whales and humans is the role whales play in removing carbon from the atmosphere. By spreading nutrients around, whales increase oceanic primary production (e.g., Lavery et al. 2014), and by virtue of their large body sizes and long life spans, growing whale populations act as a net sink in the global carbon cycle, locking away tons of carbon for extended periods of time (e.g., Pershing et al. 2010). As another whakataukī goes: *Ko ahau te tohorā, te tohorā ko ahau/I am the whale, and the whale is me.*

Emergent themes in the whale–human nexus include guardianship, sense of place, observation, connectivity, and unity. Yet, perhaps the most salient theme is migration. For example, 'People of the Whales', Iñupiat of Arctic Alaska, have accumulated vast knowledge of bowhead whale migrations over centuries of observation, and this knowledge is fundamental to Iñupiaq life and culture (Sakakibara 2017). Elsewhere, Paikea, an ancestor of many Māori, was guided by Paikea, the humpback whale, during the great human migration from Hawaiki to Aotearoa. Centuries old oral traditions underscore the more recent scientific observation that Aotearoa New Zealand is a global hotspot for whales and whale strandings at the intersection of the migratory corridors utilized by at least five different baleen whale populations. As the ancestors knew, a stranded migrating whale provides sacred resources to its descendants, while a live whale grows the ocean's bounty. Modern economies and local job markets where whale watching has a significant presence are paced by the phenology (seasonal timing) of great whale migrations. The epic migrations of baleen whales have connected us, supported us, protected us, and unified us for centuries.

From all perspectives, baleen whale migrations are remarkable. Generally defined, migration is a natural phenomenon that involves movement between different locations for specific purposes such as feeding or reproduction (Dingle and Drake 2007). Migration facilitates ecosystem services and the sustainability of marine resources, including small fish and krill (Roman and McCarthy 2010). Although many different animals migrate, migratory behaviors are diverse, and baleen whales are no exception.

This chapter provides a high-level overview, presented in two parts, of what is currently known about baleen whale migration. Part one considers the migratory patterns of baleen whales. Several central questions serve as a framework for this section, including: Why do baleen whales migrate? Where do baleen whales go? How fast do baleen whales swim? How long do the migrations last? When do the migrations occur? What marine ecosystems and environments do whales migrate between? Part two focuses on the impact of technology on our understandings of baleen whale migrations. We highlight some of the technological advances and innovations that have helped steer the scientific community toward potential answers to the central questions of marine megafauna movement ecology. The central questions we address in this section include: How has baleen whale recovery and conservation been facilitated by scientific insights? How might baleen whales orient and navigate during long-distance migration? How have baleen whale hotspots been identified? We conclude the chapter with a brief look at the horizon.

4.2 Part I: Migratory Patterns of Baleen Whales

Whale migrations stand out as some of the most incredible behaviors in nature. Most baleen, and some toothed, whales migrate. These journeys include the longest known migrations of any mammal (Fig. 4.1). For example, individual whales, including the gray whale (*Eschrichtius robustus*) 'Varvara' (Mate et al. 2015) and the humpback whale 'AHWC #2950' (Robbins et al. 2011), completed round-trip annual migrations of approximately 18,000–21,000 km. Although many school-age children know that blue whales (*Balaenoptera musculus*) are the largest animals ever to have lived, fewer likely know that despite their unprecedented size blue whales are capable of swimming at ~6 km h^{-1} over distances greater than 6000 km during a ~6-week-long migration across an open ocean (Hucke-Gaete et al. 2018). Although they have a more restricted geographic range in comparison, individual bowhead whales (*Balaena mysticetus*) swim more than 10,000 km in a single year, the equivalent of one-quarter the way around the world (Fortune et al. 2020a).

Despite many nuances and unknowns, a generalized pattern of migratory behavior has been recognized over the past century. Much of our early knowledge of migrations of pelagic species arose from reports of commercial whaleships and the extensive discovery marking program (Kellogg 1929; Mackintosh and Wheeler 1929; Mackintosh 1972). Arguably, the patterns described may have been more reflective of the seasonal migration of whalers from hemisphere to hemisphere, rather than that of whales. The discovery marking program, in which labeled steel cylinders were

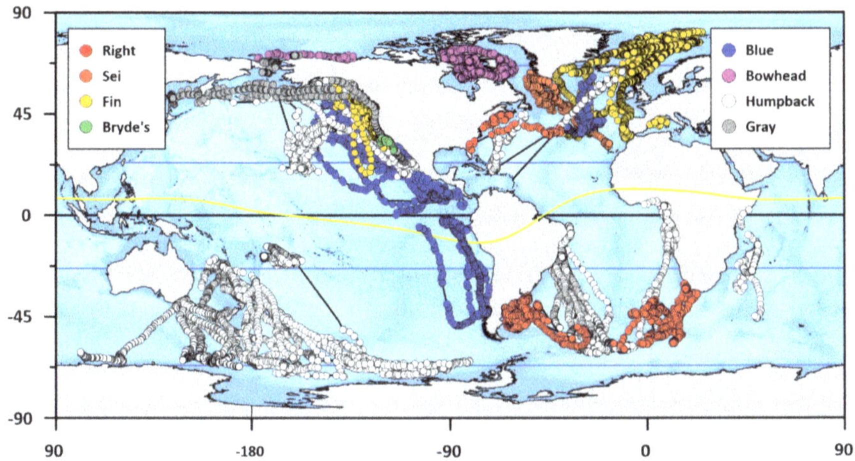

Fig. 4.1 Global compilation of some published baleen whale satellite tracking datasets (Williamson 1998; Zerbini et al. 2006; Hauser et al. 2010; Block et al. 2011; Horton et al. 2011; Silva et al. 2013; Irvine et al. 2014; Kennedy et al. 2014; Prieto et al. 2014; Garrigue et al. 2015; Kennedy et al. 2015; Mate et al. 2015; Cerchio et al. 2016; Zerbini et al. 2016; Horton et al. 2017; Andrews-Goff et al. 2018; Hucke-Gaete et al. 2018; Riekkola et al. 2018; Fortune et al. 2020a; Horton et al. 2020; Irvine et al. 2020; Lydersen et al. 2020; Palacios and Mate 2020). Blue lines show the position of the Tropics and Polar Circles. Thicker black line shows the geographic equator. Yellow curve shows the position of the magnetic equator in 2015. Circles represent platform transmitting terminal (PTT) whale location estimates and are colored by taxonomic group as indicated in the legends

fired into free-ranging whales and later recovered when carcasses were processed by whaleships, provided endpoints for the movements of marked whales that showed evidence of some very long-distance migratory movements (Brown 1954). This generalized and overly simplistic pattern of migration includes annual movements between foraging areas in polar or subpolar regions and breeding/calving habitats in tropical environments (Dawbin 1966; Norris 1967). For most baleen whales, the annual migration to higher-latitude environments occurs in spring and summer for feeding, while return migrations to lower-latitude environments typically occur in autumn and winter for breeding and calving.

This generalized pattern of baleen whale migration has variation within it. For example, some species, including blue whales and bowhead whales, likely feed throughout their annual range, whereas humpback whales and gray whales concentrate their feeding in specific higher-latitude areas at specific times of year. Migrations associated with mating and calving are similarly complex. While bowhead whales calve in the spring at higher latitudes in frigid waters (Nerini et al. 1984), gray, humpback, and right whales calve during winter in warmer and shallower waters of coastal lower-latitude environments (Lockyer 1984). Although there is an overarching pattern of baleen whale migratory behavior, it is not universal. Documenting and explaining the diversity in migratory behavior remain important goals of baleen whale research.

Another goal of current research on baleen whales is understanding the drivers and motivations of the long-distance migratory behaviors exhibited by most species. A number of hypotheses have been posed during recent decades, including (1) calf thermoregulation (Norris 1967, but see Sumich 2021); (2) energy optimization (Brodie 1975); (3) calf survivorship in calm water (Whitehead and Moore 1982); (4) vestigial behavior (Evans 1987); (5) resource tracking (Payne 1994); (6) predator avoidance (Corkeron and Connor 1999; Connor and Corkeron 2001); (7) breeding and calving in more equable environments (Clapham 2001); and (8) skin maintenance (Pitman et al. 2020). When considering this prolonged scientific debate, and the evidence both for and against the various hypotheses, one thing becomes clear: It is unreasonable to expect one hypothesis to accurately explain the diversity of baleen whale migratory behaviors.

For example, the recognition of non-migratory behaviors in some humpback (e.g., Straley 1990), fin (Geijer et al. 2016), Bryde's (Best 2001) and Omura's whales (Cerchio and Yamada 2018) suggests that traditional models are too simplistic (Geijer et al. 2016). Perhaps some species migrate with resultant benefits of reduced risk of predation and increased calf survivorship (Corkeron and Connor 1999; Connor and Corkeron 2001). The role humpback whale escorts play in actively protecting migrating mother-calf pairs from killer whale (*Orcinus orca*) and white shark (*Carcharodon carcharias*) predation is compelling (Pitman et al. 2015). Yet, it is just as reasonable to suspect that others likely migrate due to potentially harmful diatomaceous coatings of the skin (Durban and Pitman 2012; Pitman et al. 2020) or some combination of factors.

Although we do not yet know why many baleen whales migrate, technological advances have enabled a rapid growth in our knowledge of when and where whales are located. In the following subsections, we summarize representative knowledge, including the spatial patterning and timing of baleen whale migrations at the species level. Population specific examples are included to demonstrate the diversity of baleen whale migratory behaviors.

4.2.1 *Bowhead Whale (*Balaena mysticetus*)*

The migratory patterns of bowhead whales, the only baleen whale endemic to Arctic waters, vary among populations and are influenced by seasonal patterns of sea ice cover (Ferguson et al. 2010). Sea of Okhotsk (off Eastern Russia) bowhead whales are believed to be resident to that region and may occupy different subregions at different times of the year without undertaking long migrations (Shpak and Paramonov 2018). However, other bowhead whale populations maintain a characteristic meridional movement pattern from more southern environments into more northern environments in late spring and summer, followed by a reverse migration in autumn (Fig. 4.2).

Bowhead whales in the Western Arctic (the Bering-Chukchi-Beaufort population, Chap. 12) overwinter in the Northern Bering Sea (Alaska, USA) and migrate

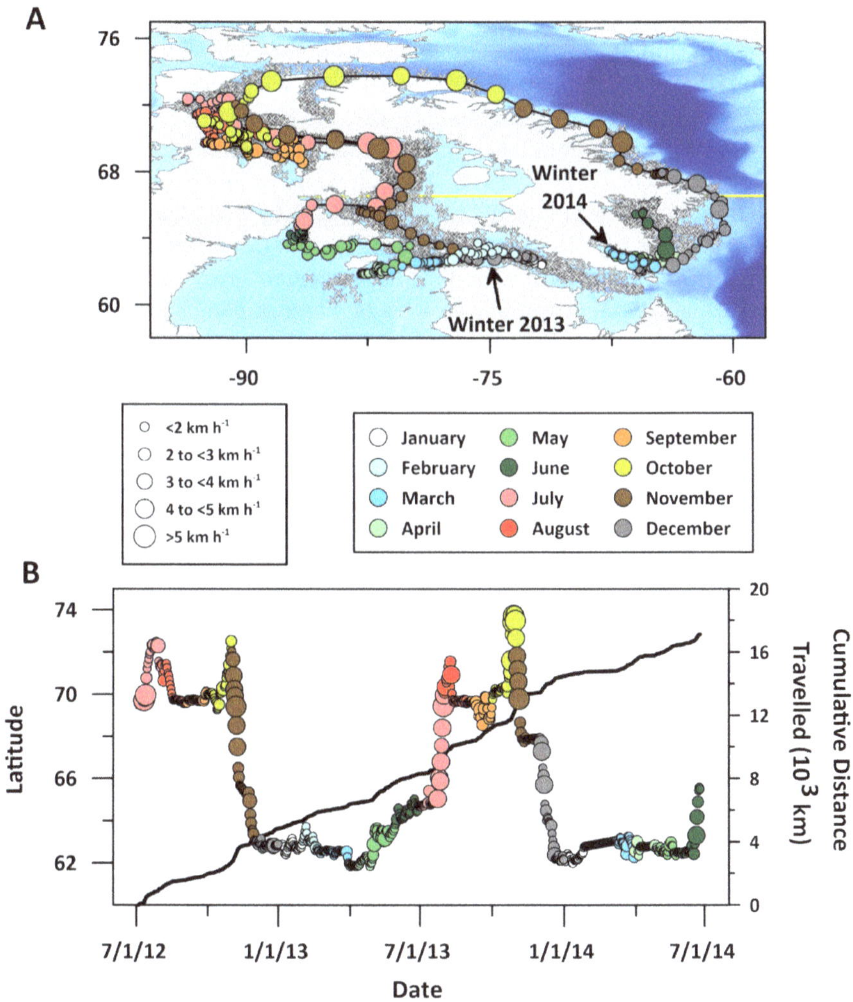

Fig. 4.2 Map (**A**) and time-series plot (**B**) of the multi-annual migrations of a bowhead whale, PTT#114495 (Fortune et al. 2020a), over a 723 day period starting in July 2013 and ending in June 2014. Panel **A** presents a plate carrée projection of PTT#114495's satellite-monitored migrations near Baffin Island, Eastern Canadian Arctic (circles; size represents swimming velocity and color represents month, as indicated in the legends). Gray crosses show every tenth Argos location estimates for 59 different bowhead whales tracked by Fortune and others (2020a). Blue-scale marine base map depicts bathymetry, and the Arctic Circle is represented by the yellow line. Panel **B** presents Julian calendar timing of PTT #114495's movements with respect to latitude (symbols as in **A**) and cumulative distance traveled (black line plot)

northward through the Chukchi Sea during spring into summer feeding grounds in the Canadian Beaufort Sea before returning westward toward Chukotka (Russia) in the late fall and then south into the Bering Sea in early winter. The spring migration occurs in stages, with whales of different age, sex, and reproductive status migrating at slightly different times. For instance, mothers with calves tend to be the last to move north in the spring (George et al. 2004). This migratory pattern has been confirmed by multiple lines of evidence, including satellite tracking (e.g., Quakenbush et al. 2013; Citta et al. 2015, 2018), passive acoustic studies (Moore et al. 1989, 2010; Clark et al. 2015), photographic identification and aerial surveys (e.g., Rugh et al. 2008; Clarke et al. 2016), geographic analysis of whale ship log books (e.g., Smith et al. 2012) and indigenous knowledge (e.g., Noongwook et al. 2007). Likewise, bowhead whales from the Eastern Canada-Western Greenland population exhibit similar north–south migratory patterns (Heide-Jørgensen et al. 2006, 2012, Fortune et al. 2020a; Fig. 4.2). In contrast, the Spitsbergen population of bowhead whales appears to have an opposite migratory pattern, reflecting the diversity of migratory behaviors in different populations. Spitsbergen bowheads tend to disperse to the south and east during spring and summer, while returning northward into the ice in autumn and winter (Lydersen et al. 2012; Kovacs et al. 2020).

Bowhead whales are believed to feed throughout their range, with breeding and calving occurring in spring (Nerini et al. 1984; Fortune et al. 2020b, c) rather than winter. Thus, bowhead whale migrations appear to be strongly associated with the spatiotemporal availability of food and the distribution of sea ice (Kovacs et al. 2020; Citta et al. 2021; Heide-Jørgensen et al. 2021). Bowhead whale migratory swimming velocities typically range between 3 and 6 km h^{-1}, over one-way migratory distances of several hundred kilometers to more than 3000 km migrations that can last as long as 9 weeks.

Although much is known about bowhead whale migrations, the application of passive acoustic technologies has continued to expand our knowledge. For example, a recorder placed in northern Fram Strait beginning in 2008 led to the discovery of a wintering, and likely breeding, ground for the critically endangered Spitsbergen population of bowhead whales (Stafford et al. 2012, 2018). This is a region that is remote, covered in sea ice most of the year and largely inaccessible as a consequence. The discovery of this hotspot has since led to an interdisciplinary program combining helicopter-based aerial surveys and satellite tagging that has yielded information on a population estimate and seasonal movements of these bowhead whales (Vacquié-Garcia et al. 2017; Kovacs et al. 2020).

4.2.2 *Right Whale (*Eubalaena australis; E. glacialis; E. japonica*)*

North Atlantic (*Eubalaena glacialis*), North Pacific (*E. japonica*), and Southern right whale (*E. australis*) movements are diverse. Some right whale populations adhere to the generalized baleen whale migratory pattern of predominantly north–south (i.e., meridional) movements between lower-latitude calving areas and higher-latitude feeding areas (e.g., Clapham et al. 2004; Munger et al. 2008; Mate et al. 2011). In contrast, some North Atlantic right whales follow a condition-dependent partial migration within neritic ecosystems off North America (Gowan et al. 2019).

In the Southwest South Atlantic, some southern right whales depart shallow water calving grounds and swim north rather than south (Zerbini et al. 2018). Yet, other southern right whales depart the continental shelf in predominantly zonal (i.e., east–west) long-distance migrations (e.g., Zerbini et al. 2018; Mackay et al. 2020), showing a remarkable diversity in movement behaviors. North Atlantic right whale migrations reinforce this complexity. In this species, mating and calving occur in different marine environments at different times of the year, suggesting a likely spatiotemporal division within a right whale population depending on reproductive status (Cole et al. 2013).

Intensive long-term visual surveys (e.g., Charlton et al. 2019), including photographic identification, passive acoustic monitoring studies (e.g., Davis et al. 2017), genetic studies (e.g., Carroll et al. 2020), and satellite tagging research (Zerbini et al. 2018; Mackay et al. 2020) have all made important contributions to unraveling the movements and migratory behaviors of endangered right whales. For example, passive acoustic monitoring identified critical habitat for the North Pacific right whale (Mellinger et al. 2004; Munger et al. 2008; Širović et al. 2015) and, as with the Spitsbergen bowhead, has directed broader research programs in photo and genetic identification of individual whales (Wade et al. 2011). Similarly, satellite tracking has demonstrated that right whales can migrate several thousand kilometers over a period of weeks while maintaining overall migratory swimming speeds between 3.5 and 6.5 km h^{-1} (Mate et al. 1997, 2011). The timing of many right whale migrations, revealed by both satellite tagging and passive acoustic monitoring, follows the general seasonal pattern of equatorward movement in autumn and poleward movement in spring (Mate et al. 2011; Davis et al. 2017).

4.2.3 *Pygmy Right Whale (***Caperea marginata***)*

Extremely little is known about the biology and movement ecology of pygmy right whales. They are believed to be a temperate to circumpolar Southern Hemisphere species capable of performing seasonal migrations between oceanic and neritic environments (Kemper 2009), not unlike their larger namesake.

4.2.4 *Blue Whale (***Balaenoptera musculus***)*

Many blue whales adhere to the stereotypical baleen whale pattern of predominantly north–south, long-distance migrations between seasonal feeding and breeding grounds (Fig. 4.3). A diverse suite of methods, including satellite tagging (e.g., Mate et al. 1999; Hucke-Gaete et al. 2018; Abrahms et al. 2019a, b), passive acoustics (Stafford et al. 1999a; b; Stafford 2003; Buchan et al. 2014; Oestreich et al. 2020), photographic identification studies (e.g., Calambokidis et al. 1990; Sears et al. 1990) as well as geographic analysis of whaling vessel log books (e.g., Reeves et al. 2004), has produced similar information on migratory patterns.

The locations of, and movements between, blue whale seasonal habitats vary regionally. Antarctic blue whales (*B. m. intermedia*) typically summer near the Antarctic continent, south of 60 S, followed by migrations to productive temperate waters during the winter. However, migratory destinations are generally not well known (Branch et al. 2007). Year-round occurrence of Antarctic blue whales in high-latitude environments (Širović et al. 2009; Thomisch et al. 2016) suggests that some blue whales may not migrate.

In contrast, pygmy blue whales (*B. m. brevicauda*) typically do not migrate to Antarctic waters in the summer, but in some regions migrate between temperate and tropical habitats (Double et al. 2014). Satellite tagging of pygmy blue whales off Southwest Australia suggests that their movement behaviors, including long-distance migrations, are not unlike those of other blue whales (Möller et al. 2020).

Chilean blue whales (*B. m.* subsp.) are known to migrate from summering habitats off the coast of Chile toward lower-latitude habitats in the Equatorial Pacific Ocean, close to the Galapagos Islands (Buchan et al. 2015; Hucke-Gaete et al. 2018). In the North Pacific, blue whale feeding grounds tend to be in the middle to higher latitudes, often less than 250 km from the coast, where upwelling can fuel seasonal blooms in primary production (e.g., Burtenshaw et al. 2004; Branch et al. 2007, Abrahms et al. 2019a).

Blue whale breeding and calving areas are believed to be in tropical marine environments, including offshore biological hotspots such as the Costa Rica Dome (Cromwell 1958) and the Galapagos Islands (Palacios 1999; Hucke-Gaete et al. 2018). Blue whale migratory swimming velocities range between ~2 and ~7 km h^{-1}, with some of the longest migrations exceeding 6000 km traveled over >6 weeks of essentially continuous directional swimming (Hucke-Gaete et al. 2018) during autumn and spring (Fig. 4.2).

Passive acoustic monitoring has been particularly instrumental in understanding seasonal changes in the global distribution of blue whales and from this, inferring migratory pathways and timing (Chap. 9). Blue whales produce sequences of sounds (songs) that are stereotyped and remain relatively unchanged over decades. Because their songs are often distinct, they have been used to define 'acoustic populations' of blue whales and their geographic ranges. For example, in the Eastern North Pacific, the complementary seasonal acoustic detection of blue whale songs has shown that Eastern North Pacific blue whales range from the Gulf of Alaska to the Equator, while Southeastern Pacific blue whales move northward from Chilean Patagonia to the Equator (Stafford et al. 1999a, b; Stafford 2003; Buchan et al. 2014). Similarly, Antarctic blue whale song detections suggest that these whales migrate northward from the ice edge to the tropics in three ocean basins (Stafford et al. 2004; Samaran et al. 2019).

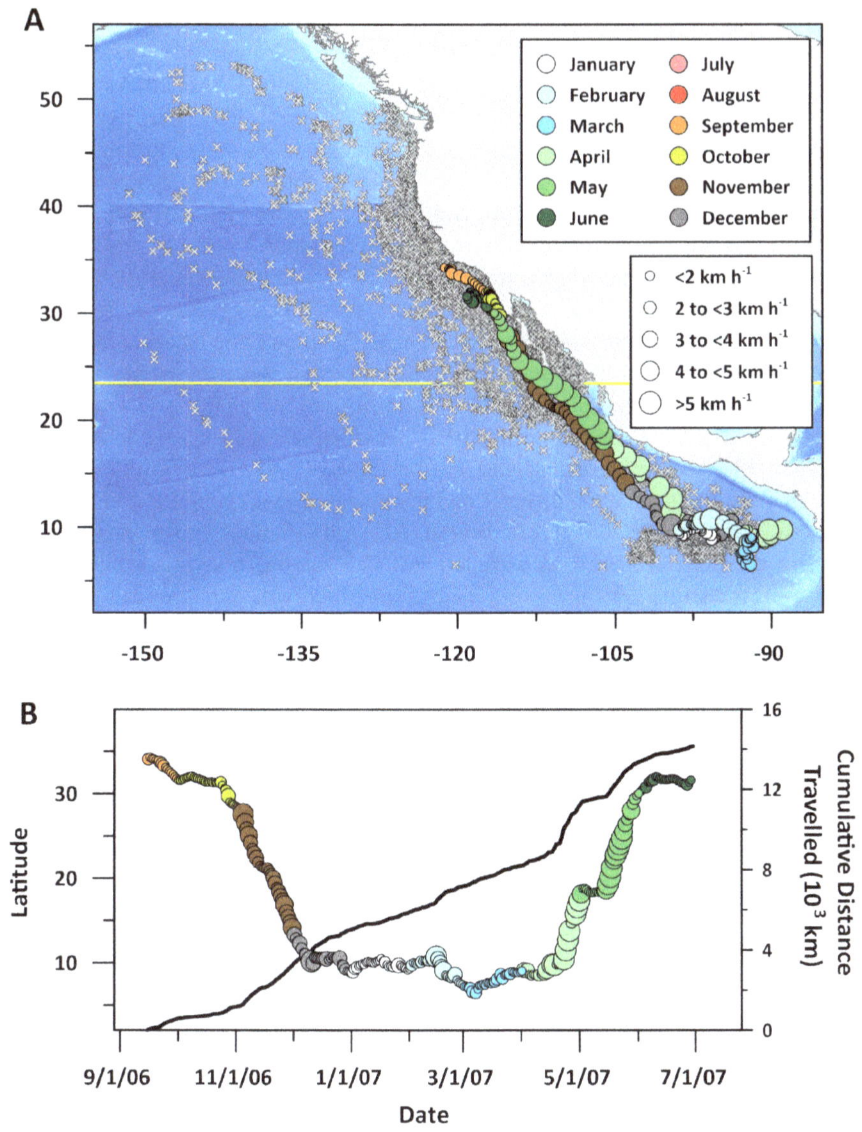

Fig. 4.3 Southward (October–December 2006) and northward (April–May 2007) return migration of satellite tracked blue whale, PTT#2083 (Irvine et al. 2014). Panel **A** presents a plate carrée projection of PTT#2083's satellite-monitored migration across the Northeast Pacific Ocean (circles; size represents swimming velocity and color represents month, as indicated in the legends). Gray crosses show Argos location estimates for 139 different blue whales. Blue-scale marine base map depicts bathymetry, and the Tropic of Cancer is represented by the yellow line. Panel **B** presents Julian calendar timing of PTT #2083's movements with respect to latitude (symbols as in **A**) and cumulative distance traveled (black line plot). An animation of these data can be viewed at: http://animove.org/wp-content/uploads/2019/04/Daniel_Palacios_animate_moveVis.html

Importantly, passive acoustic monitoring data have informed the management and conservation of blue whales globally (Cummings and Thompson 1971; Stafford et al. 2001, 2011; Stafford 2003; Mellinger and Clark 2003; Nieukirk et al. 2004; Širović et al. 2009; Buchan et al. 2015, 2020, Balcazar et al. 2015; Leroy et al. 2016; Shabangu et al. 2019; Torterotot et al. 2020). We now know that blue whales in the tropical Eastern Pacific Ocean and Indian Ocean might belong to sympatric populations of blue whales from different hemispheres (Stafford et al. 1999b; Samaran et al. 2013). Further, acoustic data have been used to look back in time to estimate historical catches of blue whales and assign them to populations based on acoustic data (Monnahan et al. 2014).

Whalers' observations of harvested animals with intermediate characteristics between blue and fin whales led to the discovery that the two species hybridize (Cocks 1887a, b; Berubé and Aguilar 1998), despite not being phylogenetically close (Arnason et al. 2018; Crossman et al. 2016). A particularly well-known living hybrid known as 'Flue' has been seen at least 19 times since 2004 in the North Pacific, both in its summering range off California and in its wintering range in the southwestern Gulf of California, Mexico (Jefferson et al. 2021). Its summer movements between southern and central California were studied with satellite tracking, while genetic analyses revealed its sex (male) and parentage (a blue whale mother and fin whale father; Jefferson et al. 2021).

Hybridization in blue and fin whales appears to be largely unidirectional (male fin x female blue), and while it is an understudied phenomenon in baleen whales, it has been documented in multiple ocean basins, suggesting that its frequency might be underestimated (Pampoulie et al. 2021). If the acoustic features of hybrids are reliably intermediate between the parents, passive acoustic monitoring and tracking might offer a tool to study the frequency and behavioral ecology of living hybrids (e.g., Watkins et al. 2004). For example, it has been hypothesized that the 'Watkins' whale' (also known as the '52-Hz whale') may have been a bluefin hybrid based on the time frequency characteristics of its song, which are intermediate between the two species in the North Pacific (Watkins et al. 2004; Stafford et al. 2007).

*4.2.5 Bryde's Whale (***Balaenoptera edeni***)*

Very little is known about Bryde's whale migratory behaviors. However, at the coarsest level, Bryde's whales are perhaps best considered as a contrast to bowheads: Bryde's whales are only known to inhabit tropical and temperate environments (Kato and Perrin 2009). If bowheads have an affinity for cold, Bryde's whales have an affinity for warmth.

Although they appear to inhabit warm water environments year-round, some Bryde's whales perform directional long-distance migrations. For example, satellite tagging data revealed that a Bryde's whale, tagged off Southern California, swam at least 600 km to the south-southeast (~152° bearing) parallel to the Baja California coast over a period of about one week in mid-October 2015 (Fig. 4.1; Lagerquist et al. 2017; Mate et al. 2018). Unfortunately, the tag ceased transmissions while the whale was still swimming southward at a typical baleen whale migratory swimming speed of ~4 km h^{-1} (Lagerquist et al. 2017; Mate et al. 2018). As abbreviated as it is, this track fits the general pattern of migrating toward the equator during autumn that is characteristic of rorquals.

Satellite tagging of two Bryde's whales in the offshore Western North Pacific also fits the general pattern, demonstrating that >2000 km equatorward migrations can occur as early as mid-summer (Murase et al. 2016). However, other Western North Pacific Bryde's whales appear to prefer spending their summers in the coastal waters off Southern Japan (Kishiro 2018). As one of the lesser understood and least studied rorquals, much remains to be learned about Bryde's whales, including the recently described balaenopterid, Rice's whale (*Balaenoptera ricei*; Rosel et al. 2021), a resident population in the Northern Gulf of Mexico that was formerly considered Bryde's whales (Corkeron et al. 2017).

There have been relatively few passive acoustic studies of Bryde's whales, and most of these have focused on characterizing the acoustic repertoire of the species (Oleson et al. 2003; Heimlich et al. 2005; Figueiredo and Simao 2014). Not one of these multiple studies was temporally or spatially extensive enough to identify potential migratory patterns, but the geographic variation in Bryde's whale signals (Širović et al. 2013) may be useful in identifying migratory patterns given more data. There is, however, evidence of seasonality in detections of Bryde's whale sounds (Kerosky et al. 2012), which could indicate that Bryde's whales are only vocal seasonally or that they migrate out of the range of recorders. Time difference of arrival acoustic monitoring methods has enabled the quantification of Bryde's whale swimming speeds and directions at the local scale off Kauai, HI (Helble et al. 2016).

*4.2.6 Fin Whale (***Balaenoptera physalus***)*

Fin whales are similar to blue whales in their adherence to the overall pattern of predominantly north–south migrations between seasonal habitats. However, year-round presence in some feeding grounds has been documented in some regions (Moore et al. 2006; Stafford et al. 2007), and populations in semi-enclosed and relatively warm waters demonstrate limited seasonal movements (Geijer et al. 2016; Panigada et al. 2017; Jiménez-Lopez et al. 2019).

Results from satellite tagging in the North Atlantic, North Pacific and South Pacific Oceans (Silva et al. 2013; Scales et al. 2017; Sepúlveda et al. 2018; Irvine et al. 2020), passive acoustic monitoring (e.g., Aulich et al. 2019; Davis et al. 2020); photographic identification (e.g., Gauffier et al. 2018), and analysis of whale ship records (Mizroch et al. 2009) generally support predominantly north–south migration. Feeding locations are not well known, as fin whales in both the Atlantic Ocean and Pacific Ocean exhibit movement behaviors consistent with area-restricted search (ARS) for prey in both temperate (Silva et al. 2013; Sepúlveda et al. 2018; Jiménez López et al.s 2019) and circumpolar environments (Silva et al. 2013; Irvine et al.

2020). However, seasonal migrations can be inferred given that discovery marking and harvests of fin whales in summer months in the northernmost Pacific regions were concentrated in a number of hotspots and that catches declined dramatically by October (Mizroch et al. 2009).

Less is known about fin whale breeding/calving ground locations compared to feeding locations. A global synthesis of post-whaling era fin whale distribution data, including visual surveys and acoustic monitoring, suggests an equatorial hiatus (ca. 20° N to 20° S) in fin whale distribution (Edwards et al. 2015). However, open ocean sightings demonstrate that fin whales have been known to congregate in the tropical South Pacific Ocean, notably thousands of kilometers from shore, in late autumn, presumably for reproductive purposes (Acevedo et al. 2012), and the recent sighting of a solitary fin whale off the coast of Colombia (ca. 10°N) is another exception to the equatorial hiatus hypothesis (Acosta et al. 2020). More localized acoustic tracking of fin whales (e.g., Soule and Wilcock 2013; Wiggins and Hildebrand 2020) has helped reveal whale movements, facilitating conservation and management in these areas.

The observation of young fin whales entering the North Atlantic via the Strait of Gibraltar (Gauffier et al. 2018) suggests that calving of migratory Northeast North Atlantic (i.e., Castellote et al. 2010) and resident whales (Geijer et al. 2016) occur in the 35° N to 45° N latitude Mediterranean Sea. Although more data are needed, satellite tagging of six migrating whales in the North Atlantic Ocean revealed that fin whales are fast, exhibiting an overall migratory swimming speed of nearly 8 km h^{-1} (Silva et al. 2013). Clarke (2004) suggests that the distribution and movement behaviors of pygmy fin whales in the Southeast Pacific are similar to those of pygmy blue whales.

*4.2.7 Minke Whale (*Balaenoptera acutorostrata; B. bonaerensis*)*

Both common (*B. acutorostrata*) and Antarctic (*B. bonaerensis*) minke whales follow the general pattern of baleen whale migration exemplified by humpback whales and other balaenopterids. Satellite tracks (e.g., Vikingsson and Heide-Jørgensen, 2015; Lee et al. 2017), passive acoustics (e.g., Risch et al. 2014; Shabangu et al. 2020), discovery marks (Horwood 1990), and photo-identification (e.g., Gill and Fairbainrns 1995) results suggest that minke whales in both hemispheres perform long-distance return migrations between circumpolar feeding areas and temperate to tropical breeding/calving areas. The dwarf minke whale is an unnamed subspecies to the common minke whale, and their seasonal occurrence in low-latitudes (Arnold et al. 1987; Zerbini et al. 1996) suggests the movement behaviors of dwarf and common minkes are likely similar.

Although minke whales are more difficult to study than many other baleen whales, in part due to their elusive behaviors and pelagic habitat (Risch et al. 2019), it is clear that minke whales are capable of ~7000 km one-way migrations from Antarctic to equatorial waters (Lee et al. 2017), with migratory swimming speeds ranging between 4.6 and 7.3 km h^{-1}. Equatorward migrations tend to occur in autumn, while poleward migrations tend to occur in spring (Risch et al. 2014; Vikingsson and Heide-Jørgensen 2015; Lee et al. 2017). Although the duration of seasonal residency and locations of minke whales in warmer water environments is one of the lesser-known aspects of their migratory pattern, the common occurrence of cookie-cutter shark (*Isistius brasiliensis*) bites and the commensal pseudostalked barnacle (*Xenobalanus globicipitis*) on minke whales provides compelling evidence of seasonal residence in warmer water environments (Towers et al. 2013). In a novel demonstration of the incredible nuances present in long-distance baleen whale movement behaviors, genetic studies have demonstrated that Antarctic minke whales have migrated to the Arctic (Glover et al. 2010), and that the two species have successfully crossbred and back-crossbred (Glover et al. 2013).

*4.2.8 Omura's Whale (***Balaenoptera omurai***)*

Very little is known about Omura's whales, which were only recently described as a species (Wada et al. 2003, Chap. 15). Research in Madagascar suggests Indian Ocean Omura's whales do not undertake seasonal long-distance migrations (Cerchio and Yamada 2018). In stark contrast to most other baleen whales, Omura's whales seem to both feed and breed/calve in tropical to temperate waters of the Indian and Western Pacific Oceans. However, the recent discovery of Omura's whales in the central equatorial Atlantic (Moreira et al. 2020) demonstrates that there remains much to be learned about this cryptic balaenopterid.

*4.2.9 Sei Whale (***Balaenoptera borealis***)*

The migratory behavior of sei whales is more like that of fin whales, despite sei whales being more closely related to Bryde's whales (Rosel and Wilcox 2014) and blue whales (Árnason et al. 2018) than fin whales. As a predominantly oceanic species, sei whales are inherently difficult to study. However, results from satellite tagging (Olsen et al. 2009; Prieto et al. 2014), passive acoustic (e.g., Nieukirk et al. 2020), and photo-identification studies (Weir et al. 2020) and analysis of whaling records (Horwood 1987; Gregr et al. 2000) provide a strong foundation for our current understanding of sei whales. Sei whale migrations connect lower-latitude calving areas to circumpolar feeding areas (Horwood 2009).

Like fin whales, sei whales are fast swimmers, with average migratory swimming speeds of ~ 7.4 km h^{-1} (Prieto et al. 2014). Passive acoustic monitoring in Fram Strait (ca. 79° N), in the polar North Atlantic near Svalbard, recorded sei whale vocalizations as early as May and as late as October, with annual peaks in the summer (Nieukirk et al. 2020). These data suggest that poleward migrations occur in spring, and equatorward migrations occur in autumn over periods of 4–8 weeks, as supported by satellite tracking (e.g., Prieto et al. 2014) and analysis of whaling records (e.g., Gregr et al. 2000).

4.2.9.1 Gray Whale (*Eschrichtius robustus*)

The migratory movements of Eastern North Pacific gray whales are relatively well studied. This large body of knowledge derives from the fact that this population of gray whales migrates in a north–south migratory corridor close to shore along the west coast of North America (Lagerquist et al. 2019), facilitating data collection and monitoring. Visual and acoustic surveys have confirmed that Eastern North Pacific gray whales migrate between calving/breeding grounds in Baja California Sur and feeding grounds in the Northeast Pacific (Lagerquist et al. 2019), and Chukchi Seas (Fig. 4.4). Gray whales depart their northern feeding grounds in autumn, pass Granite Canyon, California, from late December through early February, and arrive in their Baja California breeding areas a week or two later (Rugh et al. 2001). The onset of the northward migration peaks in early March (Swartz et al. 2007), with most whales migrating past Vancouver Island in the second-half of March (Burnham and Duffus 2020). Gray whales are known to arrive off Sivuqaq (St. Lawrence Island, ca. 63° N) in May and June (Pike 1962).

Although based on relatively small datasets, satellite tagging data demonstrate that migrating gray whales are relatively fast swimmers, maintaining northward and southward average speeds of 5.5–6.5 km h^{-1} (Mate and Urban-Ramirez 2003; Mate et al. 2015). Importantly, gray whale satellite tracking across the Sea of Okhotsk, Bering Sea, and Gulf of Alaska (Mate et al. 2015) adds two unexpected insights to our understanding of gray whale migratory behavior: (1) rather than solely relying on nearshore orientation cues, gray whales are also exceptional open ocean navigators as they cross the North Pacific Ocean when moving from feeding habitats off the Russian Far East toward the North American coast (Fig. 4.4; Mate et al. 2015); and (2) the status of Western North Pacific gray whales, a population once thought to be extinct, remains ambiguous (Mate et al. 2015).

4.2.9.2 Humpback Whale (*Megaptera novaeangliae*)

Humpback whale annual movements between tropical calving/breeding grounds and mid-latitude to high-latitude feeding areas, often with high levels of feeding and breeding area site fidelity (e.g., Calambokidis et al. 2001; Horton et al. 2017), represent a quintessential baleen whale behavior. Humpback whale migrations, some of the longest distance mammalian migrations (Rasmussen et al. 2007; Robbins et al. 2011), connect productive higher-latitude feeding ecosystems to nutrient-deficient tropical areas. These migrations help cycle limiting nutrients like nitrogen and iron via excretion (e.g., feces, urine, sloughed skin) and vertical mixing of the water column caused by their surface and diving behaviors (e.g., Doughty et al. 2016).

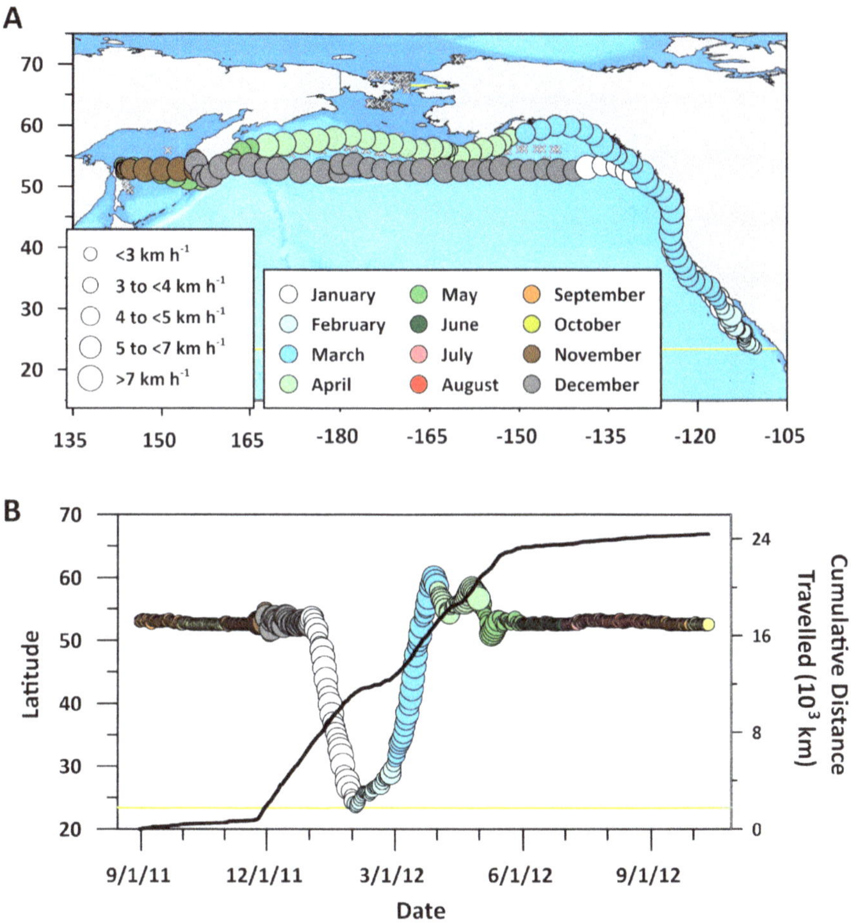

Fig. 4.4 Eastward (November–December, 2011), then southward (January, 2012), and northward (February–March, 2012), then westward (April–May, 2012) return migration of satellite tracked gray whale, PTT#1388 (Mate et al. 2015). Panel A presents a plate carrée projection of PTT#1388's satellite-monitored migration across the North Pacific Ocean (circles; size represents swimming velocity and color represents month, as indicated in the legends). Gray crosses show Argos location estimates for 12 different gray whales. Blue-scale marine base map depicts bathymetry, and the Tropic of Cancer is represented by the yellow line. Panel B presents Julian calendar timing of PTT #1388's movements with respect to latitude (symbols as in A) and cumulative distance travelled (black line plot)

Similar to gray whales, humpback whale migratory behaviors are relatively well documented. Poleward migrations typically occur during spring, and the return equatorward migrations tend to happen in late autumn (Fig. 4.5). Humpback whale migratory swimming speeds appear to range between 3 and 8 km h^{-1} across vast expanses of open ocean or parallel to coast lines, with some whales maintaining slower or faster swimming speeds for the duration of their ~6-week-long migrations (e.g., Horton

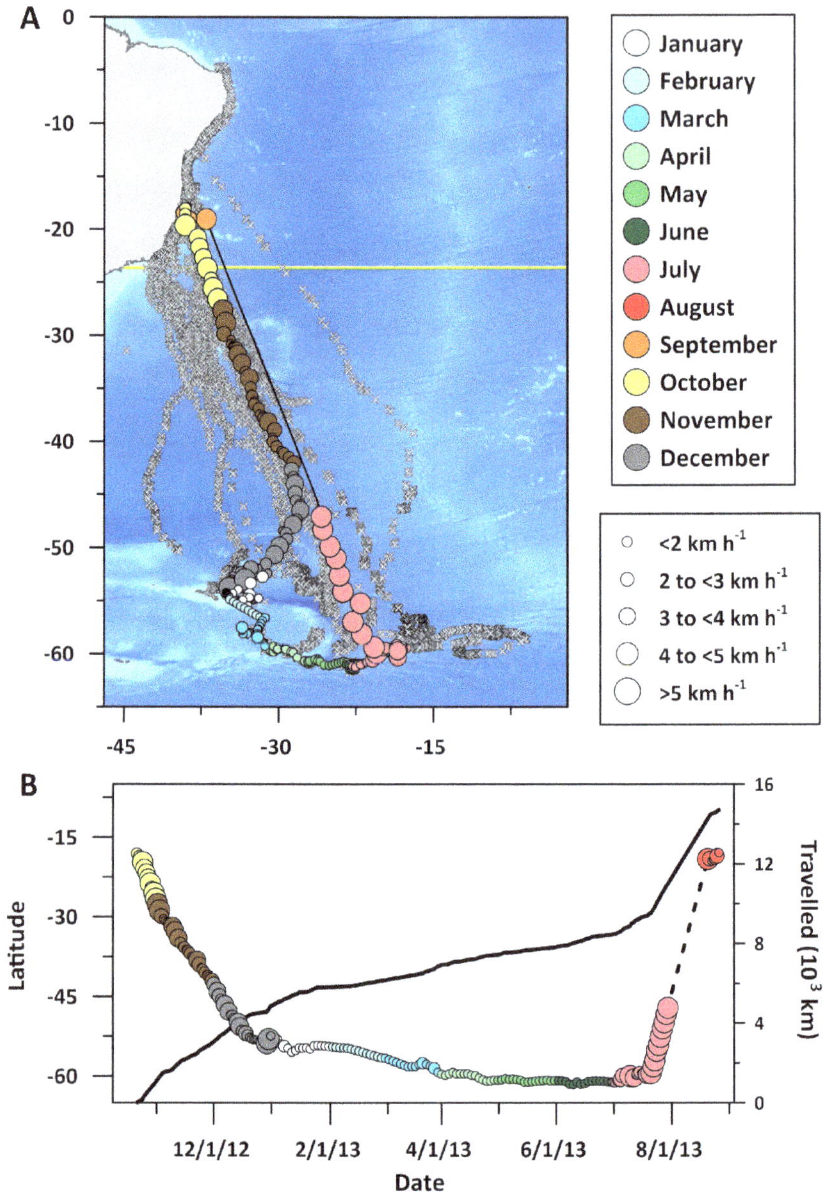

Fig. 4.5 Southward (October–December 2012) and northward (July–August 2013) return migration of satellite tracked humpback whale, PTT#121189 (Horton et al. 2020; this study). Panel **A** presents a plate carrée projection of PTT#121189's satellite-monitored migration across the Southwest Atlantic Ocean (circles; size represents swimming velocity and color represents month, as indicated in the legends). Gray crosses show every third Argos location estimates for 137 different humpback whales. Blue-scale marine base map depicts bathymetry, and the Tropic of Capricorn is represented by the yellow line. Panel **B** presents Julian calendar timing of PTT #121189's movements with respect to latitude (symbols as in **A**) and cumulative distance traveled (black line plot)

et al. 2020). Humpbacks are capable of maintaining remarkably directional (i.e., low variability in heading) movements during migration or migratory segments (Horton et al. 2011), and large proportions of regional populations appear to maintain migratory route fidelity to open-ocean corridors for 50 years or more (Horton et al. 2020). Interestingly, different humpback whale populations cross both the geographic and magnetic equators (Fig. 4.1; Rasmussen et al. 2007) and one or more ocean basins while migrating (Forestell 2007; Stevick et al. 2011, 2016; Félix et al. 2020).

A number of insights into humpback whale migrations demonstrate both agreement with the simplified general migratory pattern of many baleen whales and novel, but poorly understood, behaviors. Detailed analyses of humpback whales killed during the whaling era and long-term photographic identification results provide a strong basis of knowledge for this species (e.g., Chittleborough 1958; Craig et al. 2003). The timing of humpback whale migrations appears to be correlated with age, sex, and reproductive status (Chittleborough 1958; Craig et al. 2003). Juveniles and females without calves arrive on lower-latitude breeding/calving grounds earlier than adult males and females with calves; juveniles and females without calves also depart breeding/calving grounds earlier (Craig et al. 2003). The timing of humpback whale migrations is likely related to the different energy requirements of whales in different stages of development or reproductive status (Craig et al. 2003), and as extreme examples of capital breeders, humpback whales appear to balance the energy trade-offs of feeding, migrating, and breeding by optimizing swimming velocity during migration and the amount of time spent resting versus traveling (Braithewaite et al. 2015).

However, the total amount of time individual humpbacks spend migrating relative to residing in breeding/calving/feeding areas remains largely unknown. The small number of satellite tags ($n = 9$) that have transmitted for longer than 160 days reveals that poleward migrations can be as short as 30 days and as long as 60 days. The solitary humpback tracked across a complete annual migratory cycle (PTT 121189; Fig. 4.5), a male Western South Atlantic humpback, spent ~60 days migrating poleward during mid-late spring, ~210 days in feeding areas during summer and autumn, and ~30 days migrating equatorward during summer. Although humpback whales typically follow the general pattern of seasonal annual migration between higher-latitude feeding grounds and lower-latitude breeding and calving grounds (excluding the Arabian Sea population; Pomilla et al. 2014), there appears to be a high level of variability in the spatiotemporal dynamics of humpback whale migrations.

Technology-driven knowledge of humpback whale migratory behaviors has led to a number of distinctive conservation impacts. For example, satellite tracking of humpback whales in the Eastern Pacific (Guzman et al. 2013) led to the creation of a 'Whale Avoidance Strategy' for year-round vessel management in the Gulf of Panama, where more than 17,000 commercial ships pass between the Atlantic and Pacific each year. The spatially and temporally dynamic strategy was specifically designed to reduce, by as much as 93%, the potential for collision-induced whale trauma in one of the world's busiest shipping lanes (Guzman et al. 2013). Similarly, in central Polynesia, satellite tracking of humpback whales (Hauser et al. 2010; Horton et al. 2017) helped inform the identification and management of special, unique marine areas, or SUMAs, within the Cook Islands Marine Park—Marae Moana, which at 1.98 million km^2 is currently the world's largest marine protection area. By revealing when and where individual whales are located, satellite tracking can provide precise spatial and temporal information that is fundamental to the development of successful conservation strategies, marine protection, and sustainable management (e.g., de Castro et al. 2014; Panigada et al. 2017).

4.3 Part II: Biological Impacts of Technology

Technology has played a crucial role in advancing our understanding of baleen whale migratory behavior. For example, satellite tagging has allowed researchers to identify, with relatively high precision and accuracy, when and where whales are located. The resulting latitude, longitude, date, and time data have helped identify previously unknown whale habitats, migration routes, and movement attributes, including speed, direction, and minimum distance traveled. When combined with oceanographic, environmental, and geophysical datasets and models, deeper insights can be made regarding why certain whales do the remarkable things they do.

Passive acoustic monitoring has also had a major impact. For example, migratory pathways, including changes in the phenology of baleen whales, have been inferred by treating widely spaced hydrophones as 'turnstiles' by which the population passes at different times of the year. In the Western North Atlantic, an acoustic study determined that five different species of baleen whale (blue, fin, sei, humpback, and right) shifted their distributions northward over a decade, and these shifts likely reflected climate-related changes in the phenology of their disparate prey species (Davis et al. 2017, 2020). In this same region, some of these same acoustic datasets were used to document the movements of minke whales from low-latitude breeding grounds to higher-latitude feeding grounds and how the migration corridor varied based on time of year (Risch et al. 2014). Below, we briefly summarize two specific topics, whale navigation and hotspots, where technological advances were fundamental to elevating our knowledge and understanding of baleen whale migration.

4.3.1 Insights into Whale Navigation

Although the specific mechanisms by which baleen whales navigate remain unknown, a number of hypotheses have been proposed. In the mid-1980s, Kirschvink et al. (1986) hypothesized that cetacean orientation and navigation are informed by the topography of the Earth's magnetic field based on significant correlations between whale stranding sites and geologically induced local magnetic minima. Likewise, studies of satellite-tagged humpback whales found that animals have the ability to maintain highly directional headings during long-distance migrations (Mate et al. 1998; Horton et al. 2011). The potential linkage between cetacean navigation and the temporal dynamics of the Earth's magnetic field has been proposed as a causal contributor to whale stranding events (Vanselow et al. 2018; Vanselow 2020). A more recent analysis, however, did not find any statistical associations between globally distributed whale stranding datasets and magnetic storms documented in the instrumental record (Pulkkinen et al. 2020). Four key points emerge: (1) The influence of the Earth's magnetic field on cetacean orientation and navigation remains ambiguous; (2) deeper understanding is developed through the integration of seemingly disparate sub-disciplines (e.g., biology, geology, physics, and astronomy); (3) the mechanisms and mechanics of baleen whale navigation remain unknown; and (4) technological advances, including high-resolution mapping of Earth's magnetic and gravitation fields, facilitate both abductive and inductive approaches to elucidating how whales most likely navigate.

The interdisciplinary application of recent technological innovations has also enabled new interpretations of baleen whale navigation. For example, Horton et al. (2017) found that temporally and spatially dynamic gravitational cues were stronger predictors of humpback whale migrations than magnetic cues. In this example, technological advances created an unprecedented opportunity to explore the geophysical context of biological behaviors. Specifically, NOAA's satellite-derived Enhanced Magnetic Model (Maus 2010), a more accurate tool for quantifying the combined bedrock and main field magnetic variables at a given place and time, and the satellite-derived World Gravity Map (Balmino et al. 2012) were combined with humpback whale satellite tracking data in an empirical analysis of the geophysical cues whales experienced at the individual level. Such biogeophysical approaches are relatively new and represent an important integrated approach to solving the enduring mystery of how whales navigate.

4.3.2 Identifying Whale Hotspots

Anticipating when and where whales will be located is one of the biggest challenges of cetacean ecology. Most whales are difficult to study due to their reclusive behavior and large, and often highly inaccessible, oceanic home ranges. Passive acoustics represents an important, noninvasive technological advance that has started to advance our understanding of baleen whale hotspots. For example, long-term acoustic monitoring in the Juan Fernandez Archipelago, Chile, revealed that blue, fin and/or minke whales are present in the area during most weeks of the year (Buchan et al. 2020). Acoustic monitoring has also been a key tool in documenting the return of endangered whales to the vicinity of South Georgia Island, a pre-whaling hotspot of cetacean biodiversity (Jackson et al. 2018; Calderan et al. 2020).

Satellite tracking has proven to be a remarkably valuable tool for understanding baleen whale migrations and describing previously unknown movement patterns and whale hotspots. For example, humpback whale satellite tagging studies in the North Pacific and the Southern Hemisphere identified migratory corridors and important offshore feeding habitats (Zerbini et al. 2006, 2011; Andrews-Goff et al. 2018; Riekkola et al. 2018; Horton et al. 2020). Satellite tracking has also revealed novel diving behaviors and an association with seamounts in remote low and middle-latitude migration routes (Mate et al. 2007; Garrigue et al. 2015; Derville et al. 2019). Intensive and prolonged whale satellite tracking in the Bering, Chukchi, and Beaufort Seas helped identify a number of bowhead whale hotspots (i.e., core-use areas; Citta et al. 2015). Similarly, another prolonged satellite tracking campaign off the west coast of the USA helped not only identify the areas of highest use for feeding blue whales, but also the extent of their overlap with busy shipping lanes (Irvine et al. 2014). Robust datasets like these create powerful opportunities to anticipate and assess the impacts of the current period of rapid environmental, oceanographic, and climatic change on the health and behavior of baleen whales.

Integrated biological, oceanographic, climatic, and geophysical approaches are already underway (e.g., Abrahms et al. 2019b, Rockwood et al. 2020). Along the California coast, blue and humpback whale and krill survey data, collected as part of the Applied California Current Ecosystem Studies (ACCESS) program, were integrated with a variety of ocean variables and climate indices to produce a rigorous data-driven predictive model of when and where whales might occur (Rockwood et al. 2020). In the shipping lanes of Boston, passive acoustic monitoring has recorded North Atlantic right whales since 2009 (https://portal.nrwbuoys.org/ab/dash/), and another initiative called WhaleWatch 2.0 (https://coastwatch.pfeg.noaa.gov/projects/whalewatch2/about_whalewatch2.html) generates near-real-time online predictions of blue whale occurrence as a decision support tool to minimize ship strike risk (Abrahms et al. 2019b). Further, incorporation of stakeholder input into potential risk scenarios has proved fruitful in generating more informed management measures (Redfern et al. 2019).

4.4 A Look to the Horizon

Understanding baleen whale migration is strongly aligned to the United Nation's Decade of Ocean Science for Sustainable Development (2021–2030). The Intergovernmental Oceanographic Commission (IOC) states that, 'The main purpose of the International Decade of Ocean Science for Sustainable Development would be therefore to seek and communicate scientific solutions to the problems,' (IOC-UNESCO 2017). According to the IOC, specific goals of the 'Decade' include: (1) reversing ocean degradation; (2) filling identified knowledge gaps; (3) promoting good ocean stewardship; (4) mapping and quantifying the state of the ocean; (5) establishing sustained ocean observation; (6) strengthening interdisciplinary ocean research; (7) making technology, modeling tools, and training more accessible; (8) enhancing trust, cooperation, coordination, information exchange, communication, and education. Advancing our understanding of how and why baleen whales migrate, within the context of rapidly changing climate and ocean health, would support these long-term goals. Indeed, the themes of the future are not unlike the themes of the past, as espoused by Lythberg and Ngata's (2019) whale–human synergy.

Kaitiakitanga/guardianship, mātakitaki/observing, mātauranga/knowledge, and kotahitanga/unity will remain pillars of the relationship that connects humans, oceans, and whales, no matter what the future holds. With respect to baleen whale migration, the scientific community is perhaps at a turning point where its attention shifts from describing who, what, where, and when, to answering the questions of why and how. For example: Why are some migratory whale populations, including North Atlantic right whales and South Pacific humpback whales (i.e., Paikea; Lythberg and Ngata 2019), not recovering as quickly as others? How will migratory whales respond to changing seascapes and continued ocean development?

Realizing the complementary goals of sustainable migratory baleen whale populations and healthy oceans requires both a willingness to learn and a commitment to change. We suggest that looking to the past, while acting in the present, will help us navigate toward a more sustainable future for both humans and whales. Or, as another whakataukī more elegantly puts it:

> Kia whakatōmuri te haere whakamua.
>
> I walk backwards into the future with my eyes fixed on my past.
>
> (Rameka 2016).

References

Abrahms B, Hazen EL, Aikens EO, Savoca MS, Goldbogen JA, Bograd SJ, Jacox MG, Irvine LM, Palacios DM, Mate BR (2019a) Memory and resource tracking drive blue whale migrations. Proc Natl Acad Sci 116(12):5582–5587

Abrahms B, Welch H, Brodie S, Jacox MG, Becker EA, Bograd SJ, Irvine LM, Palacios DM, Mate BR, Hazen EL (2019b) Dynamic ensemble models to predict distributions and anthropogenic risk exposure for highly mobile species. Divers Distrib 25(8):1182–1193

Acevedo J, O'Grady M, Wallis B (2012) Sighting of the fin whale in the Eastern Subtropical South Pacific: Potential breeding ground? Rev Biol Mar Oceanogr 47(3):559–563

Acosta NB, do Amaral KB, Barrera DC, Curtidor NF, Moreno-Sotomayor JH, Campo HM, Montoya AA, Muñoz KG (2020) First Record of a Fin Whale (*Balaenoptera physalus*) in the Rosario and San Bernardo Corals National Natural Park, Colombian Caribbean. Bull Marine Coastal Res 49:259–266

Andrews-Goff V, Bestley S, Gales N, Laverick S, Paton D, Polanowski A, Schmitt N, Double M (2018) Humpback whale migrations to Antarctic summer foraging grounds through the southwest Pacific Ocean. Sci Rep 8(1):1–14

Árnason Ú, Lammers F, Kumar V, Nilsson MA, Janke A (2018) Whole-genome sequencing of the blue whale and other rorquals finds signatures for introgressive gene flow. Sci Adv 4(4):eaap9873

Arnold P, Marsh H, Heinsohn G (1987) The occurrence of two forms of minke whales in east Australian waters with a description of external characters and skeleton of the diminutive or dwarf form. Sci Rep Whales Res Inst Tokyo 38:1–46

Aulich MG, McCauley RD, Saunders BJ, Parsons MJ (2019) Fin whale (*Balaenoptera physalus*) migration in Australian waters using passive acoustic monitoring. Sci Rep 9(1):1–12

Balcazar NE, Tripovich JS, Klinck H, Nieukirk SL, Mellinger DK, Dziak RP, Rogers TL (2015) Calls reveal population structure of blue whales across the southeast Indian Ocean and the southwest Pacific Ocean. J Mammal 96(6):1184–1193

Balmino G, Vales N, Bonvalot S, Briais A (2012) Spherical harmonic modelling to ultra-high degree of Bouguer and isostatic anomalies. J Geodesy 86(7):499–520
Berube M, Aguilar A (1998) A new hybrid between a blue whale, Balaenoptera musculus, and a fin whale, B. physalus. Frequency and implication of hybridization. Mar Mamm Sci 14:82–98
Best P (2001) Distribution and population separation of Bryde's whale *Balaenoptera edeni* off southern Africa. Mar Ecol Prog Ser 220:277–289
Block BA, Jonsen ID, Jorgensen SJ, Winship AJ, Shaffer SA, Bograd SJ, Hazen EL, Foley DG, Breed G, Harrison A-L (2011) Tracking apex marine predator movements in a dynamic ocean. Nature 475(7354):86–90
Braithwaite JE, Meeuwig JJ, Hipsey MR (2015) Optimal migration energetics of humpback whales and the implications of disturbance. Conserv Physiol 3(1):cov001
Branch TA, Stafford K, Palacios D, Allison C, Bannister J, Burton C, Cabrera E, Carlson C, Galletti Vernazzani B, Gill P (2007) Past and present distribution, densities and movements of blue whales *Balaenoptera musculus* in the Southern Hemisphere and northern Indian Ocean. Mammal Rev 37(2):116–175
Brodie PF (1975) Cetacean energetics, an overview of intraspecific size variation. Ecology 56(1):152–161
Brown SG (1954) Dispersal in blue and fin whales. Discov Rep 26:357–384
Buchan SJ, Hucke-Gaete R, Rendell L, Stafford KM (2014) A new song recorded from blue whales in the Corcovado Gulf, Southern Chile, and an acoustic link to the Eastern Tropical Pacific. Endangered Species Res 23(3):241–252
Buchan SJ, Stafford KM, Hucke-Gaete R (2015) Seasonal occurrence of southeast Pacific blue whale songs in southern Chile and the eastern tropical Pacific. Mar Mamm Sci 31(2):440–458
Buchan SJ, Balcazar-Cabrera N, Stafford KM (2020) Seasonal acoustic presence of blue, fin, and minke whales off the Juan Fernández Archipelago, Chile (2007–2016). Mar Biodivers 50(5):1–10
Burnham RE, Duffus DA (2020) Maternal behaviors of gray whales (*Eschrichtius robustus*) on a summer foraging site. Mar Mamm Sci 36(4):1212–1230
Burtenshaw JC, Oleson EM, Hildebrand JA, McDonald MA, Andrew RK, Howe BM, Mercer JA (2004) Acoustic and satellite remote sensing of blue whale seasonality and habitat in the Northeast Pacific. Deep Sea Res Part II 51(10–11):967–986
Calambokidis J, Steiger G, Cubbage J, Balcomb° K, Ewald C, Kruse S, Wells R, Sears R (1990) Sightings and Movements of Blue Whales off Central California. Rep Int Whal Commn (Special Issue 12):343–348
Calambokidis J, Steiger GH, Straley JM, Herman LM, Cerchio S, Salden DR, Jorge UR, Jacobsen JK, Ziegesar OV, Balcomb KC (2001) Movements and population structure of humpback whales in the North Pacific. Mar Mamm Sci 17(4):769–794
Calderan SV, Black A, Branch TA, Collins MA, Kelly N, Leaper R, Lurcock S, Miller BS, Moore M, Olson PA (2020) South Georgia blue whales five decades after the end of whaling. Endangered Spec Res 43:359–373
Carroll EL, Ott PH, McMillan LF, Galletti Vernazzani B, Neveceralova P, Vermeulen E, Gaggiotti OE, Andriolo A, Baker CS, Bamford C (2020) Genetic Diversity and Connectivity of Southern Right Whales (*Eubalaena australis*) Found in the Brazil and Chile-Peru Wintering Grounds and the South Georgia (Islas Georgias del Sur) Feeding Ground. J Hered 111(3):263–276
Castellote M, Clark CW, Lammers MO (2010) Population identity and migration movements of fin whales (*Balaenoptera physalus*) in the Mediterranean Sea and Strait of Gibraltar. Paper SC/62/SD2, Scientific Committee of International Whaling Commission, Agadir, Morocco
Cerchio S, Yamada TK (2018) Omura's whale: *Balaenoptera omurai*. In: Encyclopedia of marine mammals. Elsevier, pp 656–659
Cerchio S, Trudelle L, Zerbini AN, Charrassin J-B, Geyer Y, Mayer FX, Andrianarivelo N, Jung J-L, Adam O, Rosenbaum HC (2016) Satellite telemetry of humpback whales off Madagascar reveals insights on breeding behavior and long-range movements within the southwest Indian Ocean. Mar Ecol Prog Ser 562:193–209

Charlton C, Ward R, McCauley RD, Brownell RL Jr, Salgado Kent C, Burnell S (2019) Southern right whale (*Eubalaena australis*), seasonal abundance and distribution at Head of Bight, South Australia. Aquat Conserv Mar Freshwat Ecosyst 29(4):576–588

Chittleborough R (1958) The breeding cycle of the female humpback whale, *Megaptera nodosa* (Bonnaterre). Mar Freshw Res 9(1):1–18

Citta JJ, Quakenbush LT, Okkonen SR, Druckenmiller ML, Maslowski W, Clement-Kinney J, George JC, Brower H, Small RJ, Ashjian CJ (2015) Ecological characteristics of core-use areas used by Bering–Chukchi–Beaufort (BCB) bowhead whales, 2006–2012. Prog Oceanogr 136:201–222

Citta JJ, Lowry LF, Quakenbush LT, Kelly BP, Fischbach AS, London JM, Jay CV, Frost KJ, Crowe GOC, Crawford JA (2018) A multi-species synthesis of satellite telemetry data in the Pacific Arctic (1987–2015): overlap of marine mammal distributions and core use areas. Deep Sea Res Part II 152:132–153

Citta J, Quakenbush L, George J (2021) Distribution and behavior of Bering-Chukchi-Beaufort bowhead whales as inferred by telemetry. In: The bowhead whale. Elsevier, pp 31–56

Clapham P (2001) Why do baleen whales migrate? A response to Corkeron and Connor. Mar Mamm Sci 17(2):432–436

Clapham PJ, Good C, Quinn SE, Reeves RR, Scarff JE, Brownell Jr RL (2004) Distribution of North Pacific right whales (*Eubalaena japonica*) as shown by 19th and 20th century whaling catch and sighting records. Publications, Agencies and Staff of the U.S. Department of Commerce 95

Clark CW, Berchok CL, Blackwell SB, Hannay DE, Jones J, Ponirakis D, Stafford KM (2015) A year in the acoustic world of bowhead whales in the Bering, Chukchi and Beaufort seas. Prog Oceanogr 136:223–240

Clarke JT, Kennedy AS, Ferguson MC (2016) Bowhead and gray whale distributions, sighting rates, and habitat associations in the eastern Chukchi Sea, summer and fall 2009–15, with a retrospective comparison to 1982–91. Arctic 69(4):359–377

Clarke R (2004) Pygmy fin whales. Mar Mamm Sci 20(2):329–334

Cocks AH (1887a) The fin whale fishery of 1886 on the Lapland coast. Bull United States Fish Commission 7:109–110

Cocks AH (1887b) The fin whale fishery of 1886 on the Lapland coast. Zoologist 11(Ser. 3):207–222

Cole TV, Hamilton P, Henry AG, Duley P, Pace RM III, White BN, Frasier T (2013) Evidence of a North Atlantic right whale *Eubalaena glacialis* mating ground. Endangered Species Res 21(1):55–64

Connor RC, Corkeron PJ (2001) Predation past and present: killer whales and baleen whale migration. Mar Mamm Sci 17(2):436–439

Corkeron PJ, Connor RC (1999) Why do baleen whales migrate? 1. Mar Mamm Sci 15(4):1228–1245

Corkeron P, Reeves R, Rosel P (2017) *Balaenoptera edeni* (Gulf of Mexico subpopulation). IUCN Red List of Threatened Species

Craig A, Gabriele C, Herman L, Pack A (2003) Migratory timing of humpback whales (*Megaptera novaeangliae*) in the central North Pacific varies with age, sex and reproductive status. Behaviour 140(8–9):981–1001

Cromwell T (1958) Thermocline topography, horizontal currents and "ridging" in the eastern tropical Pacific. Inter-Am Trop Tuna Commission Bull 3(3):133–164

Crossman CA, Taylor EB, Barrett Lennard LG (2016) Hybridization in the Cetacea: widespread occurrence and associated morphological, behavioral and ecological factors. Ecol Evol 6:1293–1303

Cummings WC, Thompson PO (1971) Underwater sounds from the blue whale, *Balaenoptera musculus*. J Acoust Soc Am 50(4B):1193–1198

Davis GE, Baumgartner MF, Bonnell JM, Bell J, Berchok C, Thornton JB, Brault S, Buchanan G, Charif RA, Cholewiak D, Clark CW, Corkeron P, Delarue J, Dudzinski K, Hatch L, Hildebrand JA, Hodge L, Klinck H, Kraus S, Martin B, Mellinger DK, Moors-Murphy H, Nieukirk S, Nowacek DP, Parks S, Read AJ, Rice AN, Risch D, Širović A, Soldevilla M, Stafford KM, Stanistreet

JE, Summers E, Todd S, Warde A, Van Parijs SM (2017) Long-term passive acoustic recordings track the changing distribution of North Atlantic right whales (*Eubalaena glacialis*) from 2004 to 2014. Sci Rep 7(1):1–12

Davis GE, Baumgartner MF, Corkeron PJ, Bell J, Berchok C, Bonnell JM, Bort Thornton J, Brault S, Buchanan GA, Cholewiak DM, Clark CW, Delarue J, Hatch LT, Klinck H, Kraus SD, Martin B, Mellinger DK, Moors-Murphy H, Nieukirk S, Nowacek DP, Parks SE, Parry D, Pegg N, Read AJ, Rice AN, Risch D, Scott A, Soldevilla MS, Stafford KM, Stanistreet JE, Summers E, Todd S, Van Parijs SM (2020) Exploring movement patterns and changing distributions of baleen whales in the western North Atlantic using a decade of passive acoustic data. Glob Change Biol 26:4812–4840

Dawbin WH (1966) The seasonal migratory cycle of humpback whales. Whales, Dolphins, and Porpoises University of California Press, Berkeley and Los Angeles, pp 145–170

de Castro FR, Mamede N, Danilewicz D, Geyer Y, Pizzorno JLA, Zerbini AN, Andriolo A (2014) Are marine protected areas and priority areas for conservation representative of humpback whale breeding habitats in the western South Atlantic? Biol Cons 179:106–114

Derville S, Torres LG, Dodémont R, Perard V, Garrigue C (2019) Aquat Conserv Marine Freshw Ecosyst 29(10):1697–1711. https://doi.org/10.1002/aqc.3127

Dingle H, Drake VA (2007) What is migration? Bioscience 57(2):113–121

Double MC, Andrews-Goff V, Jenner KCS, Jenner M-N, Laverick SM, Branch TA, Gales NJ (2014) Migratory movements of pygmy blue whales (*Balaenoptera musculus* brevicauda) between Australia and Indonesia as revealed by satellite telemetry. PLoS One 9(4):e93578

Doughty CE, Roman J, Faurby S, Wolf A, Haque A, Bakker ES, Malhi Y, Dunning JB, Svenning JC (2016) Global nutrient transport in a world of giants. Proc Natl Acad Sci 113(4):868–873

Durban J, Pitman R (2012) Antarctic killer whales make rapid, round-trip movements to subtropical waters: evidence for physiological maintenance migrations? Biol Let 8(2):274–277

Edwards EF, Hall C, Moore TJ, Sheredy C, Redfern JV (2015) Global distribution of fin whales *Balaenoptera physalus* in the post-whaling era (1980–2012). Mammal Rev 45(4):197–214

Evans PG (1987) Whales and dolphins. Facts on File, New York

Felix F, Abras DR, Cheeseman T, Haase B, Santos JDAF, Marcondes MCC, Southerland K, Acevedo J (2020) A new case of interoceanic movement of a Humpback Whale in the Southern Hemisphere: the El Nino Link. Aquat Mamm 46(6):578–584

Ferguson SH, Dueck L, Loseto LL, Luque SP (2010) Bowhead whale *Balaena mysticetus* seasonal selection of sea ice. Mar Ecol Prog Ser 411:285–297

Figueiredo LD, Simão SM (2014) Bryde's Whale (Balaenoptera edeni) vocalizations from southeast Brazil. Aquat Mamm 40:225–231

Forestell PH (2007) Movement of a humpback whale (*Megaptera novaeangliae*) between the Revillagigedo and Hawaiian archipelagos within a winter breeding season. Latin Am J Aquat Mammals 6(1):97–102

Fortune SM, Young BG, Ferguson SH (2020a) Age-and sex-specific movement, behaviour and habitat-use patterns of bowhead whales (*Balaena mysticetus*) in the Eastern Canadian Arctic. Polar Biol 43(11):1725–1744

Fortune SM, Ferguson SH, Trites AW, Hudson JM, Baumgartner MF (2020b) Bowhead whales use two foraging strategies in response to fine-scale differences in zooplankton vertical distribution. Sci Rep 10(1):1–18

Fortune SM, Ferguson SH, Trites AW, LeBlanc B, LeMay V, Hudson JM, Baumgartner MF (2020c) Seasonal diving and foraging behaviour of Eastern Canada-West Greenland bowhead whales. Mar Ecol Prog Ser 643:197–217

Garrigue C, Clapham PJ, Geyer Y, Kennedy AS, Zerbini AN (2015) Satellite tracking reveals novel migratory patterns and the importance of seamounts for endangered South Pacific humpback whales. R Soc Open Sci 2(11):150489

Gauffier P, Verborgh P, Giménez J, Esteban R, Sierra JMS, de Stephanis R (2018) Contemporary migration of fin whales through the Strait of Gibraltar. Mar Ecol Prog Ser 588:215–228

Geijer CK, Notarbartolo di Sciara G, Panigada S (2016) Mysticete migration revisited: are Mediterranean fin whales an anomaly? Mammal Rev 46(4):284–296

George J, Zeh J, Suydam R, Clark C (2004) Abundance and population trend (1978–2001) of western Arctic bowhead whales surveyed near Barrow, Alaska. Mar Mamm Sci 20:755–773

Gill A, Fairbainrns R (1995) Photo-identification of the minke whale *Balaenoptera acutorostrata* off the Isle of Mull, Scotland. In: Developments in marine biology, vol 4. Elsevier, pp 129–132

Glover KA, Kanda N, Haug T, Pastene LA, Øien N, Goto M, Seliussen BB, Skaug HJ (2010) Migration of Antarctic minke whales to the Arctic. PLoS One 5 (12):e15197

Glover KA, Kanda N, Haug T, Pastene LA, Øien N, Seliussen BB, Sørvik AG, Skaug HJ (2013) Hybrids between common and Antarctic minke whales are fertile and can back-cross. BMC Genet 14(1):1–11

Gowan TA, Ortega-Ortiz JG, Hostetler JA, Hamilton PK, Knowlton AR, Jackson KA, George RC, Taylor CR, Naessig PJ (2019) Temporal and demographic variation in partial migration of the North Atlantic right whale. Sci Rep 9(1):1–11

Gregr EJ, Nichol L, Ford JK, Ellis G, Trites AW (2000) Migration and population structure of Northeastern Pacific whales off coastal British Columbia: an analysis of commercial whaling records from 1908–1967. Mar Mamm Sci 16(4):699–727

Guzman HM, Gomez CG, Guevara CA, Kleivane L (2013) Potential vessel collisions with Southern Hemisphere humpback whales wintering off Pacific Panama. Mar Mamm Sci 29(4):629–642

Hauser N, Zerbini AN, Geyer Y, Heide-Jørgensen M-P, Clapham P (2010) Movements of satellite-monitored humpback whales, *Megaptera novaeangliae*, from the Cook Islands. Mar Mamm Sci 26(3):679–685

Heide-Jørgensen M, Laidre K, Jensen M, Dueck L, Postma L (2006) Dissolving stock discreteness with satellite tracking: bowhead whales in Baffin Bay. Mar Mamm Sci 22(1):34–45

Heide-Jørgensen MP, Hansen R, Shpak O (2021) Distribution, migrations, and ecology of the Atlantic and the Okhotsk Sea Populations. In: The bowhead whale. Elsevier, pp 57–75

Heide-Jørgensen MP, Laidre KL, Quakenbush LT, Citta JJ (2012) The northwest passage opens for bowhead whales. Biol Let 8(2):270–273

Heimlich SL, Mellinger DK, Nieukirk SL, Fox CG (2005) Types, distribution, and seasonal occurrence of sounds attributed to Bryde's whales (Balaenoptera edeni) recorded in the eastern tropical Pacific, 1999–2001. J Acoust Soc Am 118:1830

Helble TA, Henderson EE, Ierley GR, Martin SW (2016) Swim track kinematics and calling behavior attributed to Bryde's whales on the Navy's Pacific Missile Range Facility. J Acoust Soc Am 140:4170–4177

Horton TW, Hauser N, Zerbini AN, Francis MP, Domeier ML, Andriolo A, Costa DP, Robinson PW, Duffy CA, Nasby-Lucas N (2017) Route fidelity during marine megafauna migration. Front Mar Sci 4:422

Horton TW, Holdaway RN, Zerbini AN, Hauser N, Garrigue C, Andriolo A, Clapham PJ (2011) Straight as an arrow: humpback whales swim constant course tracks during long-distance migration. Biol Let 7(5):674–679

Horton TW, Zerbini AN, Andriolo A, Danilewicz D, Sucunza F (2020) Multi-decadal humpback whale migratory route fidelity despite oceanographic and geomagnetic change. Front Mar Sci 7:414

Horwood J (1987) The Sei whale: population biology, ecology & management. Routledge, 375 pp

Horwood JW (1990) Biology and exploitation of the minke whale. CRC Press, 238 pp

Horwood J (2009) Sei whale: *Balaenoptera borealis*. In: Encyclopedia of marine mammals. Elsevier, pp 1001–1003

Hucke-Gaete R, Bedrinana-Romano L, Viddi FA, Ruiz JE, Torres-Florez JP, Zerbini AN (2018) From Chilean Patagonia to Galapagos, Ecuador: novel insights on blue whale migratory pathways along the Eastern South Pacific. PeerJ 6:e4695

IOC-UNESCO (2017) International (UN) decade of ocean science for sustainable development "Towards the ocean we need for the future we want." ICO/INF-1341. Paris, 23 pp

Irvine LM, Mate BR, Winsor MH, Palacios DM, Bograd SJ, Costa DP, Bailey H (2014) Spatial and temporal occurrence of blue whales off the US West Coast, with implications for management. PLoS One 9(7):e102959

Irvine LM, Follett TM, Winsor MH, Mate BR, Palacios DM (2020) Data from: study. Blue and fin whales Southern California 2014–2015-Argos data

Jackson JA, Carroll EL, Kennedy A, Leaper R, Calderan S, Leslie MS, Andriolo A, Stepien E, Zerbini AN, Miller BS (2018) Sightings and acoustic records of right whales collected in South Georgia (Islas Georgias del Sur) waters January-February 2018. Paper SC/67B/SH/20, Scientific Committee of International Whaling Commission

Jefferson TA, Palacios DM, Calambokidis J, Baker CS, Hayslip CE, Jones PA, Lagerquist BA, Jørgensen M, Schulman-Janiger A (2021) Sightings and satellite tracking of a blue/fin whale hybrid in its wintering and summering ranges in the eastern North Pacific. Adv Oceanogr Mar Biol 2(4):1–9

Jimenez Lopez ME, Palacios DM, Jaramillo Legorreta A, Urbán R J, Mate BR (2019) Fin whale movements in the Gulf of California, Mexico, from satellite telemetry. PLoS One 14(1):e0209324

Kato H, Perrin WF (2009) Bryde's whales: Balaenoptera edeni/brydei. In: Encyclopedia of marine mammals. Elsevier, pp 158–163

Kellogg R (1929) What is known of the migrations of some of the whalebone whales. Annual Reports of the Smithsonian Institute for 1928, Washington, DC, p 46

Kemper CM (2009) Pygmy right whale: *Caperea marginata*. In Encyclopedia of marine mammals. Elsevier, pp 939–941

Kennedy AS, Zerbini AN, Vásquez OV, Gandilhon N, Clapham PJ, Adam O (2014) Local and migratory movements of humpback whales (*Megaptera novaeangliae*) satellite-tracked in the North Atlantic Ocean. Can J Zool 92(1):9–18

Kennedy A, Zerbini A, Rone B, Clapham P (2015) Location-only satellite telemetry data for gray whales in the Bering and Chukchi Sea collected by Alaska Fisheries Science Center, National Marine Mammal Laboratory from 2012-08-25 to 2013-10-27 (NCEI Accession 0139361). Version 1.1. NOAA National Centers for Environmental Information. Dataset

Kerosky SM, Širović A, Roche LK, Baumann-Pickering S, Wiggins SM, Hildebrand JA (2012) Bryde's whale seasonal range expansion and increasing presence in the Southern California Bight from 2000 to 2010. Deep Sea Res Part I 65:125–132

Kirschvink JL, Dizon AE, Westphal JA (1986) Evidence from strandings for geomagnetic sensitivity in cetaceans. J Exp Biol 120(1):1–24

Kishiro T (2018) Satellite tracking of coastal Bryde's whales *Balaenoptera edeni* along the southwest coast of Japan. Cetacean Popul Stud 1(1):3–13

Kovacs KM, Lydersen C, Vacquiè-Garcia J, Shpak O, Glazov D, Heide-Jørgensen MP (2020) The endangered Spitsbergen bowhead whales' secrets revealed after hundreds of years in hiding. Biol Let 16(6):20200148

Lagerquist BA, Palacios DM, Irvine LM, Hayslip CE, Mate BR (2017) Movements and habitat preferences of a Bryde's whale (*Balaenoptera edeni*) satellite-tagged in southern California. In: Program Book, 22nd Biennial Conference on the Biology of Marine Mammals, Society for Marine Mammalogy, Halifax, NS, Canada

Lagerquist BA, Palacios DM, Winsor MH, Irvine LM, Follett TM, Mate BR (2019) Feeding home ranges of Pacific Coast Feeding Group gray whales. J Wildl Manag 83(4):925–937

Lavery TJ, Roudnew B, Seymour J, Mitchell JG, Smetacek V, Nicol S (2014) Whales sustain fisheries: Blue whales stimulate primary production in the Southern Ocean. Mar Mamm Sci 30:888–904

Lee JF, Friedlaender AS, Oliver MJ, DeLiberty TL (2017) Behavior of satellite-tracked Antarctic minke whales (*Balaenoptera bonaerensis*) in relation to environmental factors around the western Antarctic Peninsula. Anim Biotelemetry 5(1):1–12

Leroy EC, Samaran F, Bonnel J, Royer JY (2016) Seasonal and diel vocalization patterns of Antarctic blue whale (*Balaenoptera musculus intermedia*) in the Southern Indian Ocean: a multi-year and multi-site study. PLoS One 11 (11):e0163587

Lockyer C (1984) Review of baleen whale (Mysticeti) reproduction and implications for management. Rep Int Whal Commn (Special Issue 6):27–50
Lydersen C, Freitas C, Wiig Ø, Bachmann L, Heide-Jørgensen MP, Swift R, Kovacs KM (2012) Lost highway not forgotten: satellite tracking of a bowhead whale (*Balaena mysticetus*) from the critically endangered Spitsbergen stock. Arctic 65(1):76–86
Lydersen C, Vacquié-Garcia J, Heide-Jørgensen MP, Øien N, Guinet C, Kovacs KM (2020) Autumn movements of fin whales (*Balaenoptera physalus*) from Svalbard, Norway, revealed by satellite tracking. Sci Rep 10(1):1–3
Lythberg B, Ngata W (2019) Te Aitanga a Hauiti and Paikea. RCC Perspect 5:105–112
Mackay AI, Bailleul F, Carroll EL, Andrews-Goff V, Baker CS, Bannister J, Boren L, Carlyon K, Donnelly DM, Double M (2020) Satellite derived offshore migratory movements of southern right whales (*Eubalaena australis*) from Australian and New Zealand wintering grounds. PLoS One 15(5):e0231577
Mackintosh NA (1972) Biology of the populations of large whales. Sci Progress Oxford 60:449–464
Mackintosh NA, Wheeler JFG (1929) Southern blue and fin whales. Discov Rep 1:257–540
Mate BR, Urban-Ramirez J (2003) A note on the route and speed of a gray whale on its northern migration from Mexico to central California, tracked by satellite-monitored radio tag. J Cetac Res Manage 5(2):155–157
Mate BR, Nieukirk SL, Kraus SD (1997) Satellite-monitored movements of the northern right whale. J Wildl Manag 61(4):1393–1405
Mate BR, Gisiner R, Mobley J (1998) Local and migratory movements of Hawaiian humpback whales tracked by satellite telemetry. Can J Zool 76(5):863–868
Mate BR, Lagerquist BA, Calambokidis J (1999) Movements of North Pacific blue whales during the feeding season off southern California and their southern fall migration. Mar Mamm Sci 15(4):1246–1257
Mate B, Mesecar R, Lagerquist B (2007) The evolution of satellite-monitored radio tags for large whales: one laboratory's experience. Deep Sea Res Part II: Topical Stud Oceanogr 54(3–4):224–247. https://doi.org/10.1016/j.dsr2.2006.11.021
Mate B, Best PB, Lagerquist B, Winsor M (2011) Coastal, offshore, and migratory movements of South African right whales revealed by satellite telemetry. Mar Mamm Sci 27(3):455–476
Mate BR, Ilyashenko VY, Bradford AL, Vertyankin VV, Tsidulko GA, Rozhnov VV, Irvine LM (2015) Critically endangered western gray whales migrate to the eastern North Pacific. Biol Let 11(4):20150071
Mate BR, Palacios DM, Baker CS, Lagerquist BA, Irvine LM, Follett T, Steel D, Hayslip CE, Winsor MH (2018) Baleen whale tagging in support of marine mammal monitoring across multiple navy training areas covering the years 2014, 2015, 2016, and 2017. Final Report. Prepared for Commander, U.S. Pacific Fleet. Submitted to Naval Facilities Engineering Command Southwest under Contract No. N62470-15-8006-17F4016 issued to HDR, Inc., San Diego, California. October 2018
Maus S (2010) An ellipsoidal harmonic representation of Earth's lithospheric magnetic field to degree and order 720. Geochemist Geophys Geosyst 11(6)
Mellinger D, Clark C (2003) Blue whale (*Balaenoptera musculus*) sounds from the North Atlantic. J Acoust Soc Am 114:1108–1119
Mellinger DK, Stafford KM, Moore SE, Munger L, Fox CG (2004) Detection of North Pacific right whale (*Eubalaena japonica*) calls in the Gulf of Alaska. Mar Mamm Sci 20(4):872–879
Mizroch SA, Rice DW, Zwiefelhofer D, Waite J, Perryman WL (2009) Distribution and movements of fin whales in the North Pacific Ocean. Mammal Rev 39(3):193–227
Monnahan CC, Branch TA, Stafford KM, Ivashchenko YV, Oleson EM (2014) Estimating historical eastern North Pacific blue whale catches using spatial calling patterns. PLoS One 9(6):e98974
Moore SE, Bennett JC, Ljungblad D (1989) Use of passive acoustics in conjunction with aerial surveys to monitor the fall bowhead whale (*Balaena mysticetus*) migration. Rep Int Whal Commn 39:291–295

Moore SE, Stafford KM, Mellinger DK, Hildebrand JA (2006) Listening for large whales in offshore waters of Alaska. Bioscience 56:49–55

Moore SE, Stafford KM, Munger LM (2010) Acoustic and visual surveys for bowhead whales in the western Beaufort and far northeastern Chukchi seas. Deep Sea Res Part II 57(1–2):153–157

Moreira S, Weksler M, Sousa-Lima R, Maia M, Sukhovich A, Royer J-Y, Marcondes M, Cechio S (2020) Occurrence of Omura's whale, *Balaenoptera omurai* (Cetacea: Balaenopteridae), in the Equatorial Atlantic Ocean based on passive acoustic monitoring. J Mammal 101(6):1727–1735

Munger LM, Wiggins SM, Moore SE, Hildebrand JA (2008) North Pacific right whale (*Eubalaena japonica*) seasonal and diel calling patterns from long-term acoustic recordings in the southeastern Bering Sea, 2000–2006. Mar Mamm Sci 24(4):795–814

Murase H, Tamura T, Otani S, Nishiwaki S (2016) Satellite tracking of Bryde's whales *Balaenoptera edeni* in the offshore western North Pacific in summer 2006 and 2008. Fish Sci 82(1):35–45

Möller LM, Attard CR, Bilgmann K, Andrews-Goff V, Jonsen I, Paton D, Double MC (2020) Movements and behaviour of blue whales satellite tagged in an Australian upwelling system. Sci Rep 10(1):1–19

Nerini MK, Braham HW, Marquette WM, Rugh DJ (1984) Life history of the bowhead whale, *Balaena mysticetus* (Mammalia: Cetacea). J Zool 204(4):443–468

Nieukirk SL, Stafford KM, Mellinger DK, Dziak RP, Fox CG (2004) Low-frequency whale and seismic airgun sounds recorded in the mid-Atlantic Ocean. J Acoust Soc Am 115(4):1832–1843

Nieukirk SL, Mellinger DK, Dziak RP, Matsumoto H, Klinck H (2020) Multi-year occurrence of sei whale calls in North Atlantic polar waters. J Acoust Soc Am 147(3):1842–1850

Noongwook G, Savoonga NVo, Gambell NVo, Huntington HP, George JC (2007) Traditional knowledge of the bowhead whale (*Balaena mysticetus*) around St. Lawrence Island, Alaska. Arctic 60(1):47–54

Norris K (1967) Some observations on the migration and orientation of marine mammals. Some observations on the migration and orientation of marine mammals. In: Storm RM (ed) Animal orientation and migration. Oregon State University Press, Corvallis, pp 101–125

Oestreich WK, Fahlbusch JA, Cade DE, Calambokidis J, Margolina T, Joseph J, Friedlaender AS, McKenna MF, Stimpert AK, Southall BL, Goldbogen JA (2020) Animal-borne metrics enable acoustic detection of blue whale migration. Curr Biol 30(23):4773–4779

Oleson E, Barlow J, Gordon J, Rankin S, Hildebrand J (2003) Low frequency calls of Bryde's whales. Mar Mamm Sci 19:407–419

Olsen E, Budgell WP, Head E, Kleivane L, Nottestad L, Prieto R, Silva MA, Skov H, Víkingsson GA, Waring G (2009) First satellite-tracked long-distance movement of a sei whale (*Balaenoptera borealis*) in the North Atlantic. Aquat Mamm 35(3):313

Palacios DM (1999) Blue whale (*Balaenoptera musculus*) occurrence off the Galápagos Islands, 1978–1995. J Cetac Res Manage 1(1):41–51

Palacios D, Mate BR (2020) Blue and Fin Whale movements from tags deployed in California (2014–2017) and in Oregon (2018). Research Workspace. https://doi.org/10.24431/rw1k463. Version: 10.24431_rw1k463_2020_7_28_05439. https://search.dataone.org/view

Pampoulie C, Gilson D, Ólafsdóttir G, Chosson V, Halldórsson SD et al (2021) Evidence of unidirectional hybridization and second-generation adult hybrid between the two largest animals on Earth, the fin and blue whales. Evol Appl 14:314–321

Panigada S, Donovan GP, Druon J-N, Lauriano G, Pierantonio N, Pirotta E, Zanardelli M, Zerbini AN, di Sciara GN (2017) Satellite tagging of Mediterranean fin whales: working towards the identification of critical habitats and the focussing of mitigation measures. Sci Rep 7(1):1–12

Payne R (1994) Among whales. Nat Hist 103(1):40–47

Pershing AJ, Christensen LB, Record NR, Sherwood GD, Stetson PB (2010) The impact of whaling on the ocean carbon cycle: why bigger was better. PloS one 5 (8):e12444

Pike GC (1962) Migration and feeding of the gray whale (*Eschrichtius gibbosus*). J Fisher Board Canada 19(5):815–838

Pitman RL, Totterdell JA, Fearnbach H, Ballance LT, Durban JW, Kemps H (2015) Whale killers: prevalence and ecological implications of killer whale predation on humpback whale calves off Western Australia. Mar Mamm Sci 31(2):629–657

Pitman RL, Durban JW, Joyce T, Fearnbach H, Panigada S, Lauriano G (2020) Skin in the game: epidermal molt as a driver of long-distance migration in whales. Mar Mamm Sci 36(2):565–594

Pomilla C, Amaral AR, Collins T, Minton G, Findlay K, Leslie MS, Ponnampalam L, Baldwin R, Rosenbaum H (2014) The world's most isolated and distinct whale population? Humpback whales of the Arabian Sea. PloS One 3;9(12):e114162

Prieto R, Silva MA, Waring GT, Gonçalves JM (2014) Sei whale movements and behaviour in the North Atlantic inferred from satellite telemetry. Endangered Species Res 26(2):103–113

Pulkkinen A, Moore K, Zellar R, Uritskaya O, Karaköylü E, Uritsky V, Reeb D (2020) Statistical Analysis of the Possible Association Between Geomagnetic Storms and Cetacean Mass Strandings. J Geophys Res Biogeosci 125(10):e2019JG005441

Quakenbush L, Small R, Citta J (2013) Satellite tracking of bowhead whales: movements and analysis from 2006 to 2012. US Dept of the Interior, Bureau of Ocean Energy Management, Alaska Outer Continental Shelf Region, Anchorage, AK OCS Study BOEM 1110:60

Rameka L (2016) Kia whakatōmuri te haere whakamua: 'I walk backwards into the future with my eyes fixed on my past.' Contemp Issues Early Child 17(4):387–398

Rasmussen K, Palacios DM, Calambokidis J, Saborío MT, Dalla Rosa L, Secchi ER, Steiger GH, Allen JM, Stone GS (2007) Southern Hemisphere humpback whales wintering off Central America: insights from water temperature into the longest mammalian migration. Biol Let 3(3):302–305

Redfern JV, Moore TJ, Becker EA, Calambokidis J, Hastings SP, Irvine LM, Mate BR, Palacios DM (2019) Evaluating stakeholder-derived strategies to reduce the risk of ships striking whales. Divers Distrib 25(10):1575–1585

Reeves RR, Smith TD, Josephson EA, Clapham PJ, Woolmer G (2004) Historical observations of humpback and blue whales in the North Atlantic Ocean: Clues to migratory routes and possibly additional feeding grounds. Mar Mamm Sci 20(4):774–786

Riekkola L, Zerbini AN, Andrews O, Andrews-Goff V, Baker CS, Chandler D, Childerhouse S, Clapham P, Dodémont R, Donnelly D, Friedlaender A (2018) Application of a multi-disciplinary approach to reveal population structure and Southern Ocean feeding grounds of humpback whales. Ecol Ind 89:455–465

Risch D, Castellote M, Clark CW, Davis GE, Dugan PJ, Hodge LE, Kumar A, Lucke K, Mellinger DK, Nieukirk SL (2014) Seasonal migrations of North Atlantic minke whales: novel insights from large-scale passive acoustic monitoring networks. Mov Ecol 2(1):1–17

Risch D, Norris T, Curnock M, Friedlaender A (2019) Common and Antarctic minke whales: conservation status and future research directions. Front Mar Sci 6:247

Robbins J, Dalla Rosa L, Allen JM, Mattila DK, Secchi ER, Friedlaender AS, Stevick PT, Nowacek DP, Steel D (2011) Return movement of a humpback whale between the Antarctic Peninsula and American Samoa: a seasonal migration record. Endangered Species Res 13(2):117–121

Rockwood RC, Elliott ML, Saenz B, Nur N, Jahncke J (2020) Modeling predator and prey hotspots: Management implications of baleen whale co-occurrence with krill in Central California. PLoS One 15(7):e0235603

Roman J, McCarthy JJ (2010) The whale pump: marine mammals enhance primary productivity in a coastal basin. PLoS One 5 (10):e13255

Rosel PE, Wilcox LA (2014) Genetic evidence reveals a unique lineage of Bryde's whales in the northern Gulf of Mexico. Endangered Species Res 25(1):19–34

Rosel PE, Wilcox LA, Yamada TK, Mullin KD (2021) A new species of baleen whale (Balaenoptera) from the Gulf of Mexico, with a review of its geographic distribution. *Marine Mammal Science* 1–34. https://doi.org/10.1111/mms.12776

Rugh DJ, Shelden K, Schulman-Janiger A (2001) Timing of the gray whale southbound migration. J Cetac Res Manage 3(1):31–40

Rugh DJ, Koski WR, George JC, Zeh JE (2008) 195 Interyear re-identifications of bowhead whales during their spring migration past Point Barrow, Alaska, 1984–1994. J Cetacean Res Manag 10(3):195–200
Sakakibara C (2017) People of the Whales: climate change and cultural resilience among Inupiat of Arctic Alaska. Geogr Rev 107(1):159–184
Samaran F, Stafford KM, Branch TA, Gedamke J, Royer J-Y, Dziak RP, Guinet C (2013) Seasonal and geographic variation of southern blue whale subspecies in the Indian Ocean. PLoS One 8(8):e71561
Samaran F, Berne A, Leroy EC, Moreira S, Stafford KM, Maia M, Royer JY (2019) Antarctic blue whales (*Balaenoptera musculus intermedia*) recorded at the Equator in the Atlantic Ocean. Mar Mamm Sci 35(2):641–648
Scales KL, Schorr GS, Hazen EL, Bograd SJ, Miller PI, Andrews RD, Zerbini AN, Falcone EA (2017) Should I stay or should I go? Modelling year-round habitat suitability and drivers of residency for fin whales in the California current. Divers Distrib 23(10):1204–1215
Sears R, Williamson JM, Wenzel FW, Bérubé M, Gendron D, Jones P (1990) Photographic identification of the blue whale (Balaenoptera musculus) in the Gulf of St. Lawrence, Canada. Paper SC/A88/ID23. Scientific Committee of International Whaling Commission
Sepúlveda M, Pérez-Álvarez MJ, Santos-Carvallo M, Pavez G, Olavarría C, Moraga R, Zerbini AN (2018) From whaling to whale watching: Identifying fin whale critical foraging habitats off the Chilean coast. Aquat Conserv Mar Freshwat Ecosyst 28:821–829
Shabangu FW, Findlay KP, Yemane D, Stafford KM, Van den Berg M, Blows B, Andrew RK (2019) Seasonal occurrence and diel calling behaviour of Antarctic blue whales and fin whales in relation to environmental conditions off the west coast of South Africa. J Mar Syst 190:25–39
Shabangu FW, Findlay K, Stafford KM (2020) Seasonal acoustic occurrence, diel-vocalizing patterns and bioduck call-type composition of Antarctic minke whales off the west coast of South Africa and the Maud Rise, Antarctica. Marine Mammal Sci 36(2):658–675
Shpak O, Paramonov AY (2018) The bowhead whale, *Balaena mysticetus* Linnaeus, 1758, in the Western Sea of Okhotsk (2009–2016): distribution pattern, behavior, and threats. Russ J Mar Biol 44(3):210–218
Silva MA, Prieto R, Jonsen I, Baumgartner MF, Santos RS (2013) North Atlantic blue and fin whales suspend their spring migration to forage in middle latitudes: building up energy reserves for the journey? PLoS One 8(10):e76507
Širović A, Hildebrand JA, Wiggins SM, Thiele D (2009) Blue and fin whale acoustic presence around Antarctica during 2003 and 2004. Mar Mamm Sci 25(1):125–136
Širović A, Bassett HR, Johnson SC, Wiggins SM, Hildebrand JA (2013) Bryde's whale calls recorded in the Gulf of Mexico. Mar Mamm Sci 30:399–409
Širović A, Johnson SC, Roche LK, Varga LM, Wiggins SM, Hildebrand JA (2015) North Pacific right whales (*Eubalaena japonica*) recorded in the northeastern Pacific Ocean in 2013. Mar Mamm Sci 31(2):800–807
Smith TD, Reeves RR, Josephson EA, Lund JN (2012) Spatial and seasonal distribution of American whaling and whales in the age of sail. PLoS One 7(4):e34905
Soule DC, Wilcock WSD (2013) Fin whale tracks recorded by a seismic network on the Juan de Fuca Ridge, Northeast Pacific Ocean. J Acoust Soc Am 133:1751–1761
Stafford KM (2003) Two types of blue whale calls recorded in the Gulf of Alaska. Mar Mamm Sci 19(4):682–693
Stafford KM, Nieukirk SL, Fox CG (1999a) An acoustic link between blue whales in the eastern tropical Pacific and the northeast Pacific. Mar Mamm Sci 15(4):1258–1268
Stafford KM, Nieukirk SL, Fox CG (1999b) Low-frequency whale sounds recorded on hydrophones moored in the eastern tropical Pacific. J Acoust Soc Am 106(6):3687–3698
Stafford KM, Nieukirk SL, Fox CG (2001) Geographic and seasonal variation of blue whale calls in the North Pacific. J Cetac Res Manage 3(1):65–76

Stafford KM, Bohnenstiehl DR, Tolstoy M, Chapp E, Mellinger DK, Moore SE (2004) Antarctic-type blue whale calls recorded at low latitudes in the Indian and eastern Pacific Oceans. Deep Sea Res Part I 51(10):1337–1346

Stafford KM, Mellinger DK, Moore SE, Fox CG (2007) Seasonal variability and detection range modeling of baleen whale calls in the Gulf of Alaska, 1999–2002. J Acoust Soc Am 122:3378–3390

Stafford KM, Chapp E, Bohnenstiel DR, Tolstoy M (2011) Seasonal detection of three types of "pygmy" blue whale calls in the Indian Ocean. Mar Mamm Sci 27(4):828–840

Stafford KM, Moore SE, Berchok CL, Wiig Ø, Lydersen C, Hansen E, Kalmbach D, Kovacs KM (2012) Spitsbergen's endangered bowhead whales sing through the polar night. Endangered Species Res 18(2):95–103

Stafford K, Lydersen C, Wiig Ø, Kovacs KM (2018) Extreme diversity in the songs of Spitsbergen's bowhead whales. Biol Let 14(4):20180056

Stevick PT, Neves MC, Johansen F, Engel MH, Allen J, Marcondes MC, Carlson C (2011) A quarter of a world away: female humpback whale moves 10 000 km between breeding areas. Biol Let 7(2):299–302

Stevick PT, Berrow SD, Bérubé M, Bouveret L, Broms F, Jann B, Kennedy A, Suárez PL, Meunier M, Ryan C (2016) There and back again: multiple and return exchange of humpback whales between breeding habitats separated by an ocean basin. J Mar Biol Assoc UK 96(4):885

Straley JM (1990) Fall and winter occurrence of humpback whales (*Megaptera novaeangliae*) in southeastern Alaska. Rep Int Whal Commn (Special Issue 12):319–323

Sumich JL (2021) Why Baja? A bioenergetic model for comparing metabolic rates and thermoregulatory costs of gray whale calves (*Eschrichtius robustus*). Marine Mammal Sci 1–18 https://doi.org/10.1111/mms.12778

Swartz SL, Urbán J, González S, Gómez-Gallardo A, Troyo B, Nájera M, Rojas-Bracho L (2007) Preliminary Report of the 2007 gray whale studies at Laguna San Ignacio, BCS, Mexico. Paper SC/59/BRG39. Scientific Committee of International Whaling Commission

Thomisch K, Boebel O, Clark CW, Hagen W, Spiesecke S, Zitterbart DP, Van Opzeeland I (2016) Spatio-temporal patterns in acoustic presence and distribution of Antarctic blue whales *Balaenoptera musculus* intermedia in the Weddell Sea. Endangered Species Res 30:239–253

Torterotot M, Samaran F, Stafford KM, Royer J-Y (2020) Distribution of blue whale populations in the Southern Indian Ocean based on a decade of acoustic monitoring. Deep Sea Res Part II: Top Stud Oceanogr 179:104874

Towers JR, McMillan CJ, Malleson M, Hildering J, Ford JK, Ellis GM (2013) Seasonal movements and ecological markers as evidence for migration of common minke whales photo-identified in the eastern North Pacific. J Cetac Res Manage 13(3):221–229

Vacquié-Garcia J, Lydersen C, Marques TA, Aars J, Ahonen H, Skern-Mauritzen M, Øien N, Kovacs KM (2017) Late summer distribution and abundance of ice-associated whales in the Norwegian High Arctic. Endangered Species Res 32:59–70

Vanselow KH (2020) Where are Solar storm-induced whale strandings more likely to occur? Int J Astrobiol 19(5):413–417

Vanselow KH, Jacobsen S, Hall C, Garthe S (2018) Solar storms may trigger sperm whale strandings: explanation approaches for multiple strandings in the North Sea in 2016. Int J Astrobiol 17(4):336–344

Vikingsson GA, Heide-Jørgensen MP (2015) First indications of autumn migration routes and destination of common minke whales tracked by satellite in the North Atlantic during 2001–2011. Mar Mamm Sci 31(1):376–385

Wada S, Oishi M, Yamada TK (2003) A newly discovered species of living baleen whale. Nature 426(6964):278–281

Wade PR, Kennedy A, LeDuc R, Barlow J, Carretta J, Shelden K, Perryman W, Pitman R, Robertson K, Rone B (2011) The world's smallest whale population? Biol Let 7(1):83–85

Watkins WA, Daher MA, George JE, Rodriguez D (2004) Twelve years of tracking 52-Hz whale calls from a unique source in the North Pacific. Deep-Sea Res 51:1889–1901

Weir CR, Oms G, Baracho-Neto CG, Wedekin LL, Daura-Jorge FG (2020) Migratory movement of a sei whale (Balaenoptera borealis) between Brazil and the Falkland Islands (Malvinas). Mar Mamm Sci 36(3):1050–1057

Whitehead H, Moore MJ (1982) Distribution and movements of West Indian humpback whales in winter. Can J Zool 60(9):2203–2211

Wiggins SM, Hildebrand JA (2020) Fin whale 40-Hz calling behavior studied with an acoustic tracking array. Mar Mamm Sci 36(3):964–971

Williamson JM (1998) WhaleNet–interactive education and research utilizing advanced technologies. Mar Technol Soc J 32(1):106

Zerbini AN, Secchi ER, Siciliano S, Simöes-Lopes PC (1996) The dwarf form of the minke whale, *Balaenoptera acutorostrata* Lacépède, 1804, in Brazil. Rep Int Whal Commn 46:333–340

Zerbini AN, Andriolo A, Heide-Jørgensen MP, Pizzorno JL, Maia YG, VanBlaricom GR, DeMaster DP, Simões-Lopes PC, Moreira S, Bethlem C (2006) Satellite-monitored movements of humpback whales Megaptera novaeangliae in the Southwest Atlantic Ocean. Mar Ecol Prog Ser 313:295–304

Zerbini, AN, Andriolo, A, Heide-Jørgensen, MP, Moreira, SC, Pizzorno, JL, Maia, YG, Bethlem, C, VanBlaricom, GR and DeMaster, DP (2011) Migration and summer destinations of humpback whales (*Megaptera novaeangliae*) in the western South Atlantic Ocean. J Cetacean Res Manag (Special Issue 3):113–118

Zerbini AN, Rosenbaum H, Mendez MA, Sucunza F, Andriolo A, Harris GU, Clapham PJ, Sironi M, Uhart MA, Ajó AF (2016) Tracking southern right whales through the southwest Atlantic: an update on movements, migratory routes and feeding grounds. Paper SC/66b/BRG/26, Scientific Committee of International Whaling Commission

Zerbini AN, Fernandez Ajos A, Andriolo A, Clapham PJ, Crespo E, Gonzalez R (2018) Satellite tracking of Southern right whales (*Eubalaena australis*) from Golfo San Matias, Rio Negro Province, Argentina. Paper SC/67b/CMP/17. Scientific Committee of International Whaling Commission

Chapter 5
Feeding Strategies of Baleen Whales Through a Behavioral Ecology and Evolutionary Lens

Ari S. Friedlaender

Abstract Baleen whales are some of the largest animals to have lived on the planet. They have evolved a feeding mechanism that allows for large amounts of prey-laden water to be processed, which in turn promotes extraordinary energetic gains and helps promote the evolution of gigantic body size. To contextualize the feeding strategies that baleen whales exhibit, it is imperative to first understand the evolution of baleen and filter feeding, how and why it developed, and how it has provided the foundation for the behaviors that are observed today across a range of species in varying habitats. Our first glimpses into baleen whale underwater foraging behavior initially came from a combination of at-sea visual observations and thorough examination of the anatomy and functional morphology of feeding structures in fossil and extant species. More recently, biologging tools have opened new doors to quantify and measure the biomechanics, timing, rates, organization, performance, and energetics of feeding behavior. Most importantly, this emergent technology has allowed researchers to understand how baleen whale behavior is influenced by and relates to their environment and prey. These insights are fundamentally changing our understanding of the roles of baleen whales in ocean ecosystems and provide valuable information for mitigating the negative impacts that human activity has on the health of baleen whales. The continued development of technologies provide greater opportunity to learn how we can use these mammalian giants as sentinels of ocean health.

Keywords Filter feeding · Kinematics · Biologging · Prey behavior · Foraging

5.1 Evolution of Baleen and Baleen Whales

Fifty million years ago during the Eocene, ancestral species of whales began transitioning from terrestrial to aquatic habitats, in part to take advantage of abundant resources in the ocean. Over the next several million years, cetaceans evolved and refined the suite of morphological, physiological, and anatomical adaptations that

A. S. Friedlaender (✉)
Ocean Sciences Department, University of California, Santa Cruz, CA, USA
e-mail: ari.friedlaender@ucsc.edu

C.W. Clark and E. C. Garland (eds.), *Ethology and Behavioral Ecology of Mysticetes*, Ethology and Behavioral Ecology of Marine Mammals,
https://doi.org/10.1007/978-3-030-98449-6_5

would make them successful foragers in a three-dimensional and heterogeneous marine environment. Sometime during the late Oligocene (~30 mya), mysticete whales began to lose their dentition in favor of a specialized keratinized feeding structure called baleen (Tsai and Fordyce 2015). Symmetrical racks of these baleen plates, sometimes numbering in the 100 s, extend from the upper jaw and form a mesh network of hair-like strands that act to trap small prey inside a large mouth and buccal cavity. This emergence of baleen in place of teeth enabled animals to capture numerous prey in one mouthful and effectively sieve water through small gaps in overlapping teeth, leaving behind the captured prey. The evolution of baleen occurred coincident to changes in oceanographic conditions that promoted sustained and consistent high levels of primary productivity that led to similarly consistent and aggregated regions of phyto and zooplankton; this allowed cetaceans access to an enormous and largely unexploited biomass of small-bodied prey in the oceans that could successfully be filtered out of the water to make feeding extremely profitable (Slater et al. 2017a, b). Over time, different forms of baleen adapted to take advantage of a variety of types of prey and feeding conditions. For example, gray whales (*Eschrichtius robustus*) evolved relatively short and coarse baleen that is used to forage on larger crustaceans that are often buried in substrate, whereas northern right whales (*Eubalena glacialis*) have adapted very long and fine baleen that is adept at filtering out extremely small zooplankton such as copepods.

Throughout evolutionary time, large-bodied suspension feeders have arisen on several occasions in diverse clades and have maintained a 100-million-year dynasty as the largest animals in the ocean (Friedman et al. 2010). These include sharks, rays, fishes, and baleen whales. The largest of these animals—the baleen whales—represent a unique group of animals, and the mechanisms that promote and maintain gigantism have only recently been evaluated. Coincident to the evolution of baleen was the development of the Antarctic Circumpolar Current and other large-scale oceanographic processes that promoted seasonal, stable, persistent, and large-scale ocean productivity. Increased diversity among mysticetes initially occurred later, toward the end of the Miocene (~10 mya), with the onset of coastal upwelling systems. From the late Pliocene onwards (~3.5 mya), the world's oceans also experienced increasingly intensified wind-driven upwelling that likely contributed to delphinid as well as further mysticete diversification (Slater et al. 2017a, b).

Gigantism in mysticete whales did not occur until the Late Pliocene concurrent to a shift in oceanographic conditions and concomitant changes in the intensity and distribution of primary productivity (Goldbogen and Madsen 2018; Marx and Uhen 2010; Slater et al. 2010). It seems that high prey densities, rather than simply high prey abundance, increase the energetic efficiency of bulk filter feeding events and that extended foraging bouts on ephemeral patches are the defining characteristics of baleen whale life history (Goldbogen et al. 2017b). This large size leads to lower mass-specific metabolic rates to buffer against seasonal resource availability, lower costs of transport (using less energy to swim) to allow for long-distance migrations to and from feeding locations, and highly efficient feeding performance (Goldbogen 2018; Goldbogen et al. 2019). While all baleen whales are characterized by their capacity to engulf large amounts of small-bodied prey, present-day mysticetes employ three different filter feeding strategies that are all facilitated through slight differences in morphology (Goldbogen et al. 2017a, b; Goldbogen 2010b). These include suction feeding (gray whale), continuous ram feeding (right whale, *Eubalena* sp. and bowhead whales, *Balaena mysticetus*), and lunge feeding (rorquals).

5.1.1 Suction Feeding

Intermittent suction feeding is a feeding strategy that is primarily employed by gray whales. The baleen in gray whales is thick, coarse, and relatively short, which serves to withstand the rigors of repeated contact with the seafloor and sediments. While the diet of gray whales is largely benthic invertebrates, they also feed on swarming shrimp and krill via gulping or skim feeding (Norris et al. 1983). Gray whales and right whales lack the large and elastic ventral groove blubber commonly found in rorqual whales that allows for large volumes of water to be processed in one gulp. Rather, gray whales use well-developed gular muscles and their tongue to create suction in their mouth cavities that vacuums up prey-laden sediment (Johnston and Berta 2011). This feeding style typically involves the whales rolling onto their side and contacting the benthos with the (largely right) side of their mouth at slow speeds and then episodically sucking up sediment that is then filtered out by their baleen (Woodward and Winn 2006). This results in what have been called feeding pits, and long series of these occur on the seafloor in gray whale foraging locations. Gray whales often feed in shallow, intertidal areas where conspicuous mud plumes and depressions in the sea floor are left in their wake during feeding (Oliver and Kvitek 1984).

5.1.2 Ram Feeding

A ram-filter feeding strategy—characterized by slow swimming speeds and similar to grazing on small prey—is employed by right and bowhead whales. This mode is analogous in nature to other large marine filter-feeders, such as basking sharks and manta rays. The baleen in right and bowhead whales is extremely long and consists of very fine-strands, suitable for the small copepods and euphausiids that are the main prey for these whales. Their mouths include a very high and arched rostrum with a trough between the baleen and the lower lip that promotes continuous water flow through the baleen and out of the mouth (see Potvin and Werth 2017 for details). Feeding generally occurs at depths less than 100 m and is often at or directly below the surface. While foraging dives can last for up to 20 min, the whales typically feed continuously for 2–3 min at a time (Baumgartner and Mate 2003; Parks et al. 2011; van der Hoop et al. 2019). Even though these feeding bouts occur at slow speeds, given the size of the mouth and the drag associated with filtering water, the energetic cost of swimming while feeding can be up to five times greater than swimming while non-feeding (Goldbogen et al. 2012).

5.1.3 *Lunge Feeding*

Rorquals employ a high-speed energetically costly yet remarkably efficient intermittent gulp feeding strategy called lunge feeding. An enlarged skull, increased gape angle, hyper-extensible ventral groove blubber, and invertible tongue are among the morphological adaptations that allow rorquals to engulf a volume of water on each lunge feeding event that can well exceed the mass of the animal itself (Goldbogen 2010a; b; Goldbogen et al. 2010, 2007). These whales target dense aggregations of euphausiids, copepods, and schooling fish and feed by accelerating rapidly and opening their mouths—often times while rotating around the body axis. Acceleration drops precipitously as water fills the extended buccal cavity and the tongue is pushed back and everted. Once the mouth is closed, the tongue and muscles in the ventral groove blubber act to push the prey-laden water out through baleen plates. Across species, rorqual baleen plates are shorter and more coarse than in right and bowhead whales, but less so than in gray whales. Blue whales (*Balaenoptera musculus*), which are exclusively krill feeders, have the finest baleen in the rorqual group, whereas minke (*Balaenoptera acutorostrata* and *Balaenoptera bonaerensis*) and humpback whales (*Megaptera novaeangliae*) have coarser baleen that suits a more varied and more often piscivorous diet.

5.2 Baleen Whale Foraging Ecology

Baleen whales are large and wide-ranging animals that have evolved behaviors to maximize energetic gains relative to the costs associated with locating and consuming prey. One of the evolutionary benefits of large body size is that it provides animals with a buffer that allows individuals to sustain themselves on their own energy stores if/when prey are heterogeneously distributed, i.e., patchy in both space and time. The movement patterns exhibited by baleen whales are tightly coupled to those of their prey at annual, seasonal, and daily scales (Cade et al. 2020; Curtice et al. 2015; Friedlaender et al. 2006; Nowacek et al. 2011a; Weinstein and Friedlaender 2017). At the broadest scale, this has manifested in a seasonal migratory pattern, whereby baleen whales give birth and breed in tropical regions where they generally do not feed. The advent of georeferencing satellite tag technology has provided a wealth of information on habitat use and animal migration at these greater spatial scales (Heide-Jorgensen et al. 2003; Irvine et al. 2014; Kennedy et al. 2014; Mate et al. 2007; Rosenbaum et al. 2014).

Baleen whales migrate 1000 s of kms each year to temperate and polar regions that seasonally have high densities of prey due to conditions that favor primary and secondary productivity (e.g., high nutrients and light, stratified water column). Based on ecological theory that animals forage in a manner that maximizes energy gain versus expenditure (optimal foraging theory) and that a threshold exists below which a prey patch no longer provides those energetic gains (marginal value theorem) (Charnov 1976), the behavior of whales in the environment will be influenced by the distribution and quality of these prey patches (Abrahms et al. 2019). As a result, whales exhibit nested hierarchical scales of movement and foraging behaviors that reflect the spatial and temporal scales of individual prey patches (Hazen et al. 2009). As prey is not uniformly distributed, the first order of movements will be confined to the area that contains high numbers of individual prey patches (Kirchner et al. 2018). There may then be smaller-scale oceanographic features such as eddies or upwelling centers that either entrain planktonic prey or provide resources for more nektonic prey (Cotte et al. 2009). Within these finer-scale areas, prey are yet again distributed in a number of patches that vary in quality (e.g., density, biomass, depth, and shape) (Cade et al. 2020).

Consider humpback whales that feed in the waters surrounding the Antarctic Peninsula. At the broadest scale, these whales inhabit waters on the continental shelf in areas that largely mirror the broad-scale distribution of their prey, Antarctic krill (*Euphausia superba*) (Curtice et al. 2015). Throughout the summer months, whales exhibit combinations of movement patterns that can be differentiated by the persistence of or lack of swimming directionality. These behavioral states of movement are often used in state-space modeling approaches to quantify broad behavioral patterns and likelihoods of remaining in or switching among behavioral states (Jonsen et al. 2013, 2005). The first broad behavioral state considered is transiting, which is defined as directed movement from one location to another over time. Depending on the sampling rate of the data being collected, time steps can be minutes, hours, or even days. Thus, in seeking out periods of transiting in tag records, researchers look for autocorrelation or similarity in the heading of an animal from one time step to the next.

Area-restricted search (ARS) behavior is the opposite of transiting and is the process by which whales remain in a given area by constant changes in heading or turning that limit how far an animal moves from a central location (unlike transiting which is unidirectional movement). Thus, ARS is defined by high rates of turning over a period of time. While feeding on a prey patch, a whale exhibits ARS behavior to maintain its proximity to that identified area.

Whales that are socializing or resting also exhibit ARS. While the amount of time an animal spends in a given behavioral state can be informative about the habits of an animal, the frequency or likelihood of transitioning from one state to another can provide additional information about how the environment may influence an animal's behavior. For example, if an animal spends 50% of its time in ARS and 50% of its time transiting but has a very low probability of state switching, this may indicate very large prey patches that are spread out in space. At the same percentages of time, if the probability of state switching is very high, it would be consistent with a hypothesis that the environment contains many smaller prey patches that are relatively close together, and the whale spends an equal amount of time in these patches and transits between them regularly. This technical approach to interpreting positional data from tags can help to elucidate nested scales of movement and provide insight into the prey scape of the environment in which a whale lives.

Several studies have used data derived from long-term tags. These instruments, containing a battery, transmitter, and antenna cased in a metal housing, are deployed into the skin and blubber layer and remain anchored for extended periods of weeks to months (Mate et al. 2000). Transmitters are programmed to generate georeferenced positions via satellite telemetry at regular intervals to provide data on the amount of time that individual whales spend in different behavioral states on foraging grounds, and how they change within and across seasons. In the Antarctic ecosystem, these types of data have shown that over the course of the feeding season, humpback whales increase the amount of time spent in area-restricted search, decrease the amount of time spent transiting between patches, and occupy smaller areas later in the season (Weinstein et al. 2017; Weinstein and Friedlaender 2017). With this information, one of the unresolved questions in this system is whether the whales are responding to changes in their physical environment (e.g., changes in sea ice condition) or to changes by the prey themselves or a combination of the two. Note that these behavioral states (i.e., foraging, transiting, and resting) are generally inferred from tag data. For data from satellite tags that supply positional information only, the inference likely has lower accuracy than behavior that is derived from more robust motion, audio, and video sensors. Interpreting these data into more refined and quantitative behaviors will be discussed below.

In the Antarctic, sea ice develops throughout the late fall through winter, and as it encroaches, it limits the amount of available open water foraging habitat for humpback whales. If the distribution and movement patterns of the whales were simply affected by the increase in sea ice and less available habitat, there is no reason to believe that the amount of time spent in each behavioral state would change (i.e., it would simply happen in a smaller area). However, the amount of time that whales spend in ARS increases and the amount of time transiting decreases, suggesting that krill are changing their behavior and coalescing into larger aggregations that are moving into new areas. Indeed, previous work evaluating the occurrence of krill has shown that during summer months krill are broadly distributed over the continental shelf in discrete and dispersed patches. But in fall and winter, krill move inshore into coastal bays and straits and often form massive, high-density patches containing over two million tons of krill (Espinasse et al. 2012; Nowacek et al. 2011b). Thus, by measuring and understanding physical and biological processes of a marine ecosystem and linking these to direct information on movement patterns of baleen whales, we can gain fundamental insights into how whale foraging patterns are differentially affected by physical changes to their foraging habitat versus biological changes in prey behavior.

As there are a select number of times and places where conditions are favorable enough to allow for baleen whales to successfully forage, different species exhibit spatial and temporal overlap. How foraging whales behave to minimize the potential for competition over a limited resource provides a glimpse into the feeding strategies and behavior of individual species. In California, blue and humpback whales likely facilitate sympatry by partitioning their foraging in three ways. Blue whales feed almost exclusively on krill that aggregate in dense, deep layers along shelf breaks for a defined period during the summer. In contrast, humpback whales feed more broadly on a variety of prey (including small schooling fish and euphausiids) over longer periods of time during the summer and distribute themselves across shallow and deep-water areas (Calambokidis et al. 2014; Croll et al. 2005; Fossette et al. 2017; Irvine et al. 2014; Rockwood et al. 2020). It is precisely these differences in behavior that allow for species to coexist and also invite a more detailed and precise study of their underwater behavior.

While the broad-scale foraging behavior of baleen whales can be inferred from their positional data, to understand feeding behaviors and how whales interact with their prey requires intimate knowledge of their subsurface behavior. Over the past 20–30 years, the development of short-term dive-recording tags has revolutionized the information that can be collected from baleen whales, which in turn has informed our understandings of dive patterns, feeding behaviors, energetic costs, and ecological roles (Goldbogen et al. 2017a; Goldbogen et al. 2013). More recently, the advent of high-resolution motion sensors and video-recording tags has enabled researchers to study the fine-scale foraging behaviors of baleen whales and help to provide context to the patterns and behaviors that are measured by these sensors (Fig. 5.1). While motion-sensing tags allow researchers the ability to describe the myriad feeding strategies that baleen whales employ, greater inferences regarding these strategies likely come from concurrent information on the available prey field. Knowledge about prey behavior and availability can be used to determine variation in feeding at the scale of a single dive or over greater spatiotemporal scales.

5.2.1 *Bubble Net Feeding in Humpback Whales*

One of the most unique feeding strategies that humpback whales employ is bubble net feeding. This choreographed behavior can be performed by a single animal or by groups of over 15 whales. Owing in part to their long pectoral flippers and large flukes that allow humpback whales to maneuver gracefully in the water column, an individual humpback or a group of humpbacks encircle a patch of prey by producing either a stream, curtain, or pulses of bubbles that corral prey and allow the whales to feed on dense aggregations. As this feeding strategy culminates with whales actively breaking the surface (often times with mouths agape), it has been well documented and observed passively from research platforms since it was first described in the 1970s (Jurasz and Jurasz 1979). The behavior appears to have developed independently in geographically isolated stocks of humpbacks in the northwestern Atlantic, northeastern Pacific, and southeastern Pacific Oceans. Because of the conspicuousness of this foraging behavior, researchers have been able to identify individual whales that consistently bubble net feed and use individualized feeding strategies (Fig. 5.2) (Ware et al. 2006; Wiley et al. 2011). Since humpback whales are social feeders, but also have fluid social associations during the feeding season, it has also

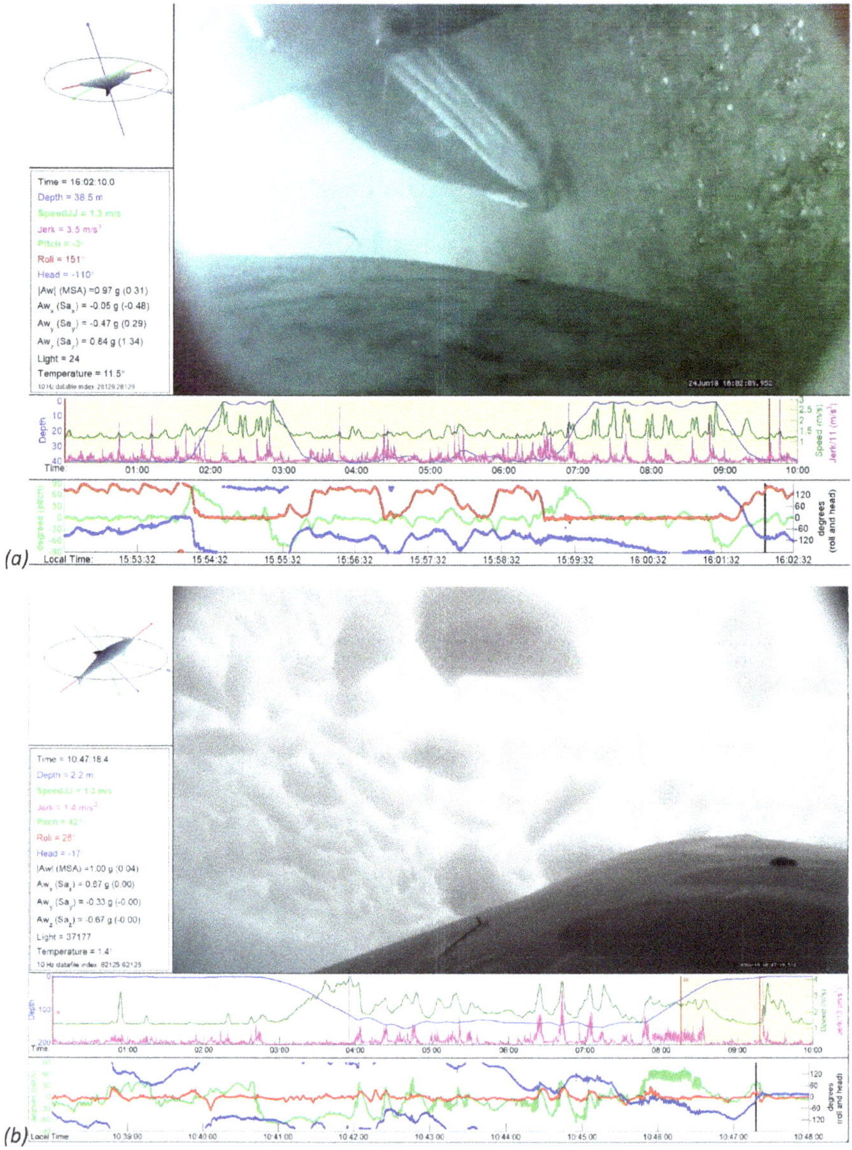

Fig. 5.1 Advent of motion-sensing and video-recording tags has provided researchers with novel ways of quantifying cetacean behavior concurrently with gaining new insights on the environmental context surrounding behavior. **a** Taken from a tag on the right side of a humpback whale in the Stellwagen Bank National Marine Sanctuary shows a second humpback whale, mouth agape, feeding on sand lance directly at the seafloor. Numerous sand lance can be seen carpeting the seafloor and also in the water column surrounding the whales. **b** Taken from a tag deployed on an Antarctic minke whale, shows the tagged whale approaching the surface that is covered entirely by small pieces of brash ice. **c** Shows a California Sea Lion scavenging on anchovy that were left over after a feeding event by a humpback whale in Monterey Bay, California

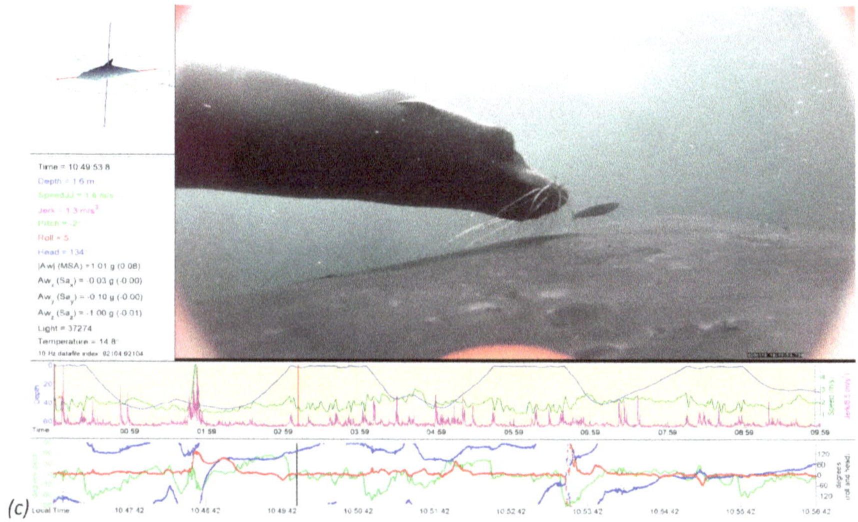

Fig. 5.1 (continued)

been concluded that information regarding feeding strategies can be learned and disseminated very rapidly throughout a feeding population (Chap. 8).

Motion-sensing tags have been fundamental to our understanding of the subsurface nature of bubble net feeding behavior, whether individuals learn and maintain distinct roles in feeding groups, and if individual whales vary in how they produce bubble nets. Tagging studies have demonstrated that whales that create bubble nets do so in individualized and stereotyped ways (Allen et al. 2016; Wiley et al. 2011). Whereas one whale may produce a single ring of bubbles, another may make two spiraling rings, and yet another may surface after a single ring and then create a second independent ring combined with a slapping of the flukes at the surface between rings. This individuality appears to be maintained across feeding events and feeding partners. It appears that humpback whales employ bubble net feeding as a strategy for single prey types within a region: herring (*Clupea pallasii*) in Alaska, sand lance (*Ammodytes* sp.) in Massachusetts, krill (*Euphausia* spp.) in Antarctica.

Concurrent consideration of the types of prey humpback whales are targeting has shown that the number of whales that participate in bubble net feeding groups seems to correlate with the size of the prey. In the Antarctic where small and less mobile Antarctic krill are targeted, foraging groups are small—ranging from 1 to 4 whales. Off Massachusetts where medium sized sand lance is targeted, groups range from 1 to 8 whales, while in Alaska when whales target larger and more mobile herring, groups can contain more than 15 individuals. Because the amount of coordination required for larger groups of individuals to successfully work in concert to capture increasingly mobile prey is high, humpback whales in Alaska often use an audible feeding call that is believed to be produced by the whale that is creating the bubble

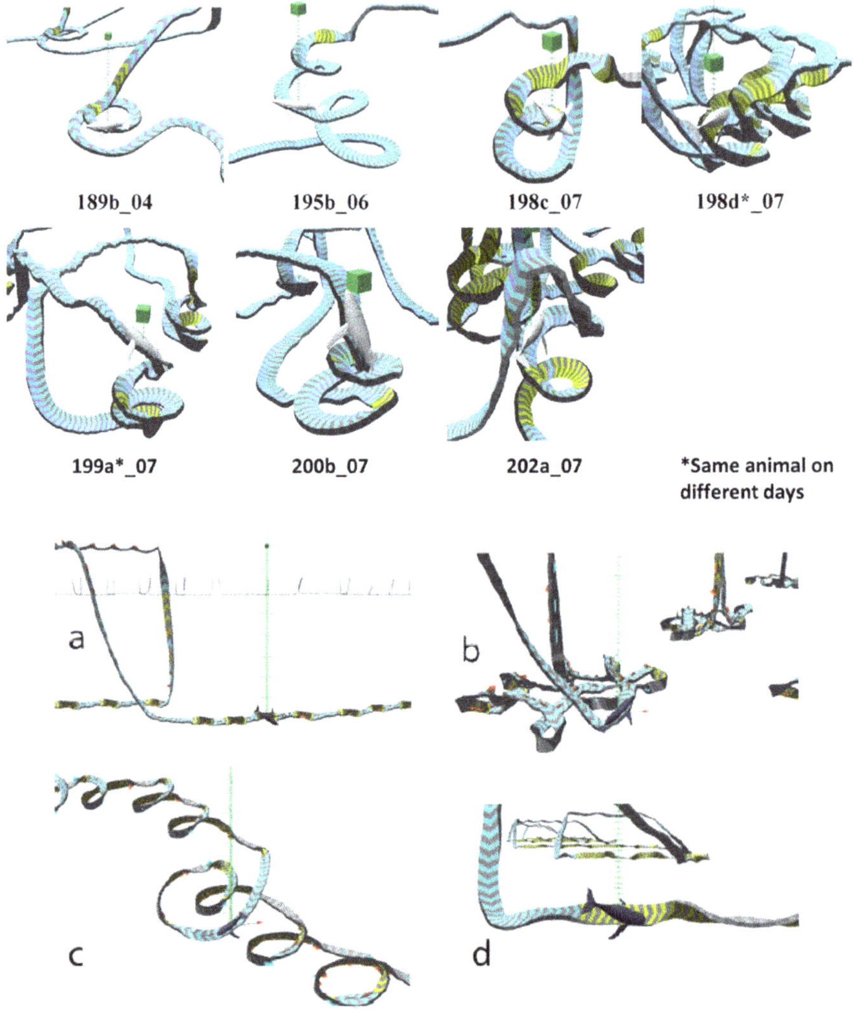

Fig. 5.2 Trackplot animations of motion-sensing tag data taken from Ware et al. (2011) showing bubble net and bottom feeding strategies of humpback whales tagged in the Stellwagen Bank National Marine Sanctuary. The top two panels show unique individualized bubble net feeding strategies that include different forms and numbers of loops and spirals. The bottom panels similarly show a myriad of approaches that humpback whales have adapted for feeding along the bottom. In these approaches, some whales feed in a linear fashion along the seafloor, rolling on to their sides at regular intervals, whereas other whales feed in more circular motions or even complex star-shaped patterns. From Wiley et al. (2011) and Ware et al. (2013)

net and may serve a dual purpose of frightening the prey as well as coordinating movements among individual whales (Chap. 10).

While the knowledge of the underwater maneuvers that whales perform to create bubble nets has been gleaned from observations and tag data, why and when they do so requires detailed information on prey behavior. Combining tag deployments with detailed measurements of prey patches has revealed that bubble net feeding only occurs during daylight hours and that whales target prey that is vertically oriented (rather than horizontally) in the water column, and at depths less than 40 m (Friedlaender et al. 2009; Hazen et al. 2009). Sand lance generally aggregate in near-surface schools during daylight, and often descend to the seafloor and burrow in substrate at night. In response to this diurnal change in prey behavior, humpback whales change their feeding behavior accordingly and switch from bubble net feeding during the day to yet another feeding strategy along the bottom at night. More recently, humpback whales have been anecdotally shown to switch between surface and bottom feeding within the same feeding bouts, depending on how prey are distributed.

Due to observations of abraded skin and mandibles of humpback whales in the western North Atlantic, it was suggested that they employ a feeding strategy to come in direct contact with the bottom and feed on prey in the substrate (Hain et al. 1995). Combined with knowledge of the behavior of sand lance, a compelling, albeit circumstantial, case was made that humpback whales feed along the bottom. Eventually, tag data revealed that humpback whales not only feed along the bottom, but they do so utilizing an array of individualized feeding strategies to target sand lance along the sea floor (similar to the individuality in bubble net feeding behavior observed in this same population) (Fig. 5.2). Some animals roll laterally on to their side repeatedly in a linear fashion multiple times in a single dive, whereas others perform tight spirals along the sea floor (Ware et al. 2013). More recently, by using motion-sensing tags that also contain concurrent video-recording capabilities, researchers have found remarkable evidence that humpback whales feed along the sea floor in coordinated groups with defined roles among individuals, similar to individual behavioral roles as observed in bubble netting groups. With concurrent tags on multiple animals, it is now possible to quantify how individuals interact with one another to optimize prey consumption and to determine how stable the groups are in terms of individual membership over time. Maintaining foraging group cohesion when switching feeding strategies—along with plasticity in foraging strategies in response to changes in prey behavior—is remarkable among baleen whales and is only discernable from the combined approach of motion-sensing tags and concurrent environmental sampling at the same scales over which whales are feeding. Video data also provide an environmental context to determine the behavior of sand lance that whales take advantage of during this feeding strategy.

5.3 Lateralization of Feeding Behavior in Baleen Whales

Baleen whales appear to balance energetic costs of foraging versus benefits gained from prey. This balance is not unique to cetaceans, but baleen whale foraging strategies may represent an upper limit given their enormous body sizes and that they consequently have the highest energetic requirements of any animal. Thus, there are selective advantages for individuals or species that develop adaptive strategies to maximize energetic efficiency through consistently repeated behaviors.

One manner for animals to increase efficiency is to develop a mechanism for performing repeated tasks, i.e., "practice makes perfect." Among all classes of vertebrates that have been studied, from amphibians to reptiles to mammals, a right-side bias exists in repeated behaviors that are visually mediated, allowing for greater precision and effectiveness (MacNeilage 2013). Visually, such a bias is derived from the contralateral connection between the left hemisphere of the vertebrate brain and the right optic nerve. In humans and primates, this is manifested as right-handedness. Among cetaceans, a similar right-side bias has been demonstrated across a number of species, generally related to feeding behavior. For example, gray whales have been shown to forage predominantly on their right sides (e.g., Woodward and Winn 2006), and bottlenose dolphins exhibit foraging strategies with a right-side bias.

Among humpback whales, a right-side bias occurs in bottom feeding, with whales preferentially rolling onto their right sides when feeding at or near the seafloor (Ware et al. 2013). Similarly, blue whales show a right-side bias in their rolling behavior associated with deep-water and surface feeding. In fact, the general kinematic movements that baleen whales perform to feed includes acceleration toward a patch of prey and then rolling laterally 90° while opening the mouth. The ubiquity of this pattern likely confers whales the greatest opportunity to maximize prey capture. However, in one case, blue whales have a left-side bias when rolling during feeding, and it is the only known example of a left-sided bias in foraging (Fig. 5.3). When targeting shallow and small prey patches that require visually mediated precision for successful feeding, blue whales generally approach from below, countershading the prey patch against the surface. As they approach from below at a steep pitch angle and because their eyes are on the sides of their heads, whales roll to their left side and continue this process through a complete 360° rotation as they feed. Presumably, an adaptive advantage of this behavior is that it keeps a whale's right eye in constant visual contact of the small prey patch. Thus, while the whales are rotating to the left, contrary to all previously described lateralized behavior, the whales are doing so to maintain a right-side/eye bias for maintaining a visual connection to their prey that enables them to more accurately and successfully target these small, and discrete patches (Friedlaender et al. 2017).

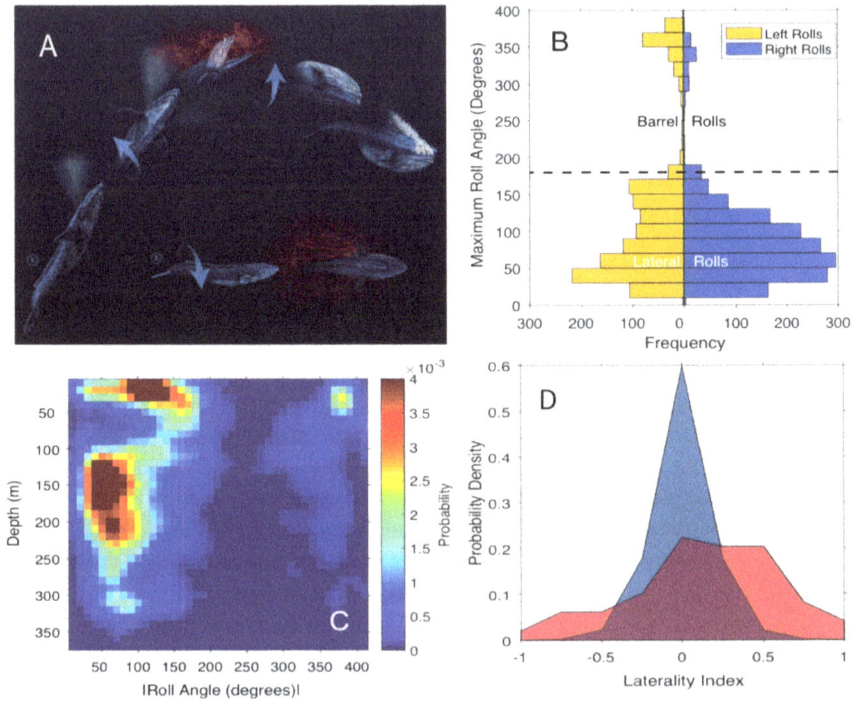

Fig. 5.3 This graphic shows the lateralized feeding strategies of blue whales tagged off the California coast. **a** Describes a unique left-side bias that blue whales have when targeting shallow and small prey patches from below. Highlighted here is that the whale maintains visual contact with its right eye (presumed to be responsible for increased accuracy of performance in visually mediated tasks) to the approaching prey patch. **b** Provides information that blue whales perform lateral rolls more frequently to the right side than the left side but that 360° rolls are more left-side dominant. This links with the heat map in **c** indicating the amount of roll that animals typically perform based on the depth of prey. In this image, the vast majority of rolls that are mostly lateral in nature occur throughout the water column, but 360° rolls take place at shallow depths. **d** Compares a null model (blue) in which a population of animals displays no laterality or preference for performing a task more from one side than the other. The distribution of data in red is from tagged blue whales showing that while many individuals do not display a lateralization in the side to which they roll during feeding events, there are a small number of individuals that display a left-side preference for all feeding rolls (indicated by a negative laterality index), and a larger proportion that show a strong preference for rolling to the right side during feeding events. Taken from Friedlaender et al. (2017)

5.4 Balancing Energy Gain with Oxygen Use

That blue whales perform such acrobatic underwater maneuvers is presently only known from tag data. When and why such feeding strategies exist can again be answered by incorporating information on prey, as well as physiological parameters. As air-breathing, diving predators, baleen whales balance costs associated with breath-holding and oxygen use versus energetic gains from feeding. Because lunge feeding by baleen whales is such an energetically costly activity, dive times are limited (Croll et al. 2001; Goldbogen et al. 2007, 2012). The amount of time required to process a single mouthful of water scales with body size; blue whales require over a minute to process a single mouthful, whereas minke whales, the smallest baleen whale, require less than ten seconds. Thus, in terms of number of feeding events per unit time, feeding rates scale inversely with body size. However, because the feeding apparatus also scales with body size, the engulfment capacity, in terms of water filtered, scales positively with body size (J. A. Goldbogen et al. 2012). While a minke whale consumes a mouthful of its prey five times more frequently than a blue whale, in the same amount of time, a blue whale filters prey from 16 times more water volume, leading to vast differences between the energetic gains and feeding efficiencies of the two species. Blue whales feeding on shallow, low density (<100 krill/m^3) prey patches perform more acrobatic and more energy-intensive maneuvers (i.e., rolling), but when feeding on deep, higher density (>400 krill/m^3) prey patches, they engage in less energy-intensive maneuvers (Goldbogen et al. 2011; Hazen et al. 2015). The whales are balancing competing needs of energy gain and oxygen use, and blue whales prioritize oxygen conservation over energy (caloric) gains when feeding on more shallow, less profitable prey patches; they maximize energy gains at the expense of oxygen when feeding on deeper, denser prey patches.

Thus, for the largest animal on the planet, baleen whales are constrained in their opportunities to gain energy primarily when they feed on deep and dense prey layers. Such a strategy can only persist if prey routinely and predictably aggregate under these conditions and if the whales are able to locate prey patches by utilizing some memory of the places where and times when conditions exist that promote efficient feeding to maximize energy gain (Abrahms et al. 2019; Fossette et al. 2017; Oestreich et al. 2020). Blue whales and baleen whales in general have been extremely successful at locating ephemeral and heterogeneously distributed prey and are not indiscriminate grazers that wander ocean basins and occasionally run into profitable prey patches. Rather, these animals likely use a combination of sensory capabilities (e.g., sight, sound, and touch) and cultural memory to detect prey over a range of spatial scales and venture to the places that over time routinely offer the best availability of food (Torres et al. 2016). While all animals have evolved successful feeding strategies, the suite of anatomical, physiological, and behavioral adaptations that baleen whales (and specifically blue whales) employ across multiple spatial and temporal scales allow them to optimize energy gain in a way that has led to them being the largest animals to have ever existed.

5.5 Looking Toward the Future

Baleen whales likely evolved and became enormous in response to environmental conditions that promoted predictable and abundant food in the world's oceans. Observations of whales from the surface have formed the foundation upon which we understand how whales behave and interact both with one another and their environment.

Detailed examination of beach cast whales has also provided greater knowledge of the anatomical, morphological, and physiological adaptations that whales have evolved for efficient and effective bulk filter feeding. However, the recent development of motion sensors and the miniaturization of electronics has revolutionized our ability to observe and quantify the underwater behavior of whales and provide insight and context to when, why, and how whales feed.

One obstacle that remains for researchers is the ability to collect high-resolution, continuous data on animal movement and behavior over spatial scales that are relevant to whales and their environment. For example, current motion-sensing and suction cup tag technology is limited to collecting data and remaining on a whale for up to several days due to battery capacity and the limits of passive suction. As long-lived species that feed over a period of months over vast areas of the ocean, 24 or 48 h may not be a relevant time period over which to determine, for example, consumption rates or energy use. In polar environments where the sun may not set for a long period of time, in areas where tidal cycles are not predictably diurnal, or where features such as sea ice may be present for months at a time, we require tools that can continuously measure the fine-scale behaviors of animals over longer time periods. It is with information such as this that we can better understand the patterns and rhythms at time and spatial scales that are more appropriately matched to the actual sensory scales of the whales.

Implicit to this valiant quest for understanding is similarly detailed knowledge of the environment which whales inhabit. While we know that the foraging behavior of baleen whales is intimately tied to that of their prey, we still lack fundamental knowledge on how the behavior and aggregation of prey affect whales and vice versa. If autonomous underwater vehicles that can carry oceanographic sensors including echosounders and water profilers for weeks to months could communicate with and stay proximate to whales equipped with long-duration biologgers, we could fundamentally change our ability to understand whale behavior. To be successful, this framework would require critical and creative thinking from scientists, engineers, mathematicians, and others—and a collaborative spirit. Many of the threats that marine mammals face are directly linked to how human activities diminish ocean ecosystem health and function. To understand and mitigate these threats require greater information on the needs and behaviors of baleen whales, something that can only be achieved through a commitment to the development of new tools and creative ways to study whales underwater.

Acknowledgements I thank my friends, colleagues, mentors, and students with whom I have shared innumerable conversations about science, marine mammals, and our natural world. You have all inspired me and taught me so much more than I ever thought I could learn. I am grateful for your advice and wisdom, your ideas and thoughts, and most of all your support. My thanks to the funders and granting agencies that have supported the numerous field projects that are represented throughout this chapter. And to Caroline, thank you for giving me the shoulders to stand on.

References

Abrahms B, Hazen EL, Aikens EO, Savoca MS, Goldbogen JA, Bograd SJ, Mate BR et al. (2019) Memory and resource tracking drive blue whale migrations. J Proc Natl Acad Sci 201819031. https://doi.org/10.1073/pnas.1819031116

Allen AN, Goldbogen JA, Friedlaender AS, Calambokidis J (2016) Development of an automated method of detecting stereotyped feeding events in multisensor data from tagged rorqual whales. Ecol Evol 6(20):7522–7535. https://doi.org/10.1002/ece3.2386

Baumgartner MF, Mate BR (2003) Summertime foraging ecology of North Atlantic right whales. Mar Ecol-Prog Ser 264:123–135. https://doi.org/10.3354/meps264123

Cade D, Seakamela M, Findlay K, Fahlbusch J, Kahane-Rapport S, Oestreich W, Calambokidis J et al (2020) The Patchy distribution of Krill within temporally transient large swarms drives collective exploitation by Super Groups of Rorqual Whales. Paper presented at the Ocean Sciences Meeting 2020

Calambokidis J, Munger LM, Soldevilla MS, Ferguson MC, Havron AM, Camachots DL, Hildebrand JA (2014) Seasonal distribution and abundance of cetaceans off Southern California estimated from CalCOFI cruise data from 2004 to 2008. Fish Bull 112(2–3):198–220. https://doi.org/10.7755/fb.112.2-3.7

Charnov EL (1976) Optimal foraging, the marginal value theorem. Theor Popul Biol 9(2):129–136

Cotte C, Guinet C, Taupier-Letage I, Mate B, Petiau E (2009) Scale-dependent habitat use by a large free-ranging predator, the Mediterranean fin whale. Deep-Sea Res Part I-Oceanogr Res Pap 56(5):801–811. https://doi.org/10.1016/j.dsr.2008.12.008

Croll DA, Acevedo-Gutiérrez A, Tershy BR, Urbán-Ramírez J (2001) The diving behavior of blue and fin whales: is dive duration shorter than expected based on oxygen stores? Compar Biochem Physiol Part A Mol Integr Physiol 129(4):797–809

Croll DA, Marinovic B, Benson S, Chavez FP, Black N, Ternullo R, Tershy BR (2005) From wind to whales: trophic links in a coastal upwelling system. Mar Ecol Prog Ser 289:117–130. https://doi.org/10.3354/meps289117

Curtice C, Johnston DW, Ducklow H, Gales N, Halpin PN, Friedlaender AS (2015) Modeling the spatial and temporal dynamics of foraging movements of humpback whales (Megaptera novaeangliae) in the Western Antarctic Peninsula. Mov Ecol 3(1):13. https://doi.org/10.1186/s40462-015-0041-x

Espinasse B, Zhou M, Zhu YW, Hazen EL, Friedlaender AS, Nowacek DP, Carlotti F et al (2012) Austral fall-winter transition of mesozooplankton assemblages and krill aggregations in an embayment west of the Antarctic Peninsula. Mar Ecol Prog Ser **452**:63–80. https://doi.org/10.3354/meps09626

Fossette S, Abrahms B, Hazen EL, Bograd SJ, Zilliacus KM, Calambokidis J, Croll DA et al (2017) Resource partitioning facilitates coexistence in sympatric cetaceans in the California Current. Ecol Evol 7(21):9085–9097. https://doi.org/10.1002/ece3.3409

Friedlaender AS, Halpin PN, Qian SS, Lawson GL, Wiebe PH, Thiele D, Read AJ (2006) Whale distribution in relation to prey abundance and oceanographic processes in shelf waters of the Western Antarctic Peninsula. Mar Ecol Prog Ser 317:297–310. https://doi.org/10.3354/meps317297

Friedlaender AS, Hazen EL, Nowacek DP, Halpin PN, Ware C, Weinrich MT, Wiley D et al (2009) Diel changes in humpback whale Megaptera novaeangliae feeding behavior in response to sand lance Ammodytes spp. behavior and distribution. Mar Ecol Prog Ser 395:91–100. Retrieved from <Go to ISI>://WOS:000273383900007

Friedman M, Shimada K, Martin LD, Everhart MJ, Liston J, Maltese A, Triebold M (2010) 100-million-year dynasty of giant planktivorous bony fishes in the mesozoic seas. Science 327(5968):990–993. https://doi.org/10.1126/science.1184743

Friedlaender AS, Herbert-Read JE, Hazen EL, Cade DE, Calambokidis J, Southall BL, Goldbogen JA et al (2017) Context-dependent lateralized feeding strategies in blue whales. Curr Biol 27(22):R1206–R1208. Retrieved from <Go to ISI>://WOS:000415815800005

Goldbogen JA (2010a) Mechanics and energetics of rorqual lunge feeding (Ph.D.). University of British Columbia, Vancouver, BC

Goldbogen JA (2010b) The ultimate mouthful: lunge feeding in rorqual whales. Am Sci 98(2):124–131. Retrieved from <Go to ISI>://WOS:000274615200021

Goldbogen JA (2018) Physiological constraints on marine mammal body size. Proc Natl Acad Sci USA 115(16):3995–3997. https://doi.org/10.1073/pnas.1804077115

Goldbogen J, Madsen P (2018) The evolution of foraging capacity and gigantism in cetaceans. J Exp Biol 221(11):jeb166033

Goldbogen JA, Pyenson ND, Shadwick RE (2007) Integrating morphology, behavior and phylogeny to determine the mechanics and energetics of rorqual lunge-feeding. J Morphol 268(12):1079–1079. Retrieved from <Go to ISI>://WOS:000251266000182

Goldbogen JA, Calambokidis J, Oleson EM, Potvin J, Schorr G, Shadwick RE (2010) Big heads, big gulps and high drag: mechanics and energetics of rorqual lunge feeding. Integr Compar Biol 50:E62–E62. Retrieved from <Go to ISI>://WOS:000280297000248

Goldbogen JA, Calambokidis J, Oleson E, Potvin J, Pyenson ND, Schorr G, Shadwick RE (2011) Mechanics, hydrodynamics and energetics of blue whale lunge feeding: efficiency dependence on krill density. J Exp Biol 214(Pt 1):131–146. https://doi.org/10.1242/jeb.048157

Goldbogen JA, Calambokidis J, Croll DA, McKenna MF, Oleson E, Potvin J, Tershy BR et al (2012) Scaling of lunge-feeding performance in rorqual whales: mass-specific energy expenditure increases with body size and progressively limits diving capacity. Funct Ecol 26(1):216–226. https://doi.org/10.1111/j.1365-2435.2011.01905.x

Goldbogen JA, Friedlaender AS, Calambokidis J, McKenna MF, Simon M, Nowacek DP (2013) Integrative approaches to the study of baleen whale diving behavior, feeding performance, and foraging ecology. Bioscience 63:90–100

Goldbogen J, Cade D, Boersma A, Calambokidis J, Kahane-Rapport S, Segre P, Friedlaender A et al (2017a). Using digital tags with integrated video and inertial sensors to study moving morphology and associated function in large aquatic vertebrates. Anatom Rec 300(11):1935–1941

Goldbogen J, Cade D, Calambokidis J, Friedlaender A, Potvin J, Segre P, Werth A (2017b) How baleen whales feed: the biomechanics of engulfment and filtration. Ann Rev Mar Sci 9:367–386

Goldbogen JA, Cade DE, Wisniewska DM, Potvin J, Segre PS, Savoca MS, Pyenson ND (2019) Why whales are big but not bigger: physiological drivers and ecological limits in the age of ocean giants. Science 366(6471):1367–1372. https://doi.org/10.1126/science.aax9044

Hain JHW, Ellis SL, Kenney RD, Clapham PJ, Gray BK, Weinrich MT, Babb IG (1995) Apparent bottom feeding by humpback whales on Stellwagen Bank. Mar Mamm Sci 11(4):464–479

Hazen EL, Friedlaender AS, Goldbogen JA (2015) Blue whales (*Balaenoptera musculus*) optimize foraging efficiency by balancing oxygen use and energy gain as a function of prey density. Sci Adv 1(9):e1500469. https://doi.org/10.1126/sciadv.1500469

Hazen EL, Friedlaender AS, Thompson MA, Ware CR, Weinrich MT, Halpin PN, Wiley DN (2009) Fine-scale prey aggregations and foraging ecology of humpback whales Megaptera novaeangliae. Mar Ecol-Prog Ser 395:75–89. https://doi.org/10.3354/meps08108

Heide-Jorgensen MP, Laidre KL, Wiig O, Jensen MV, Dueck L, Maiers LD, Hobbs RC et al (2003) From Greenland to Canada in ten days: tracks of bowhead whales, Balaena mysticetus, across Baffin Bay. Arctic 56(1):21–31. Retrieved from <Go to ISI>://WOS:000182434600003

Irvine LM, Mate BR, Winsor MH, Palacios DM, Bograd SJ, Costa DP, Bailey H (2014). Spatial and temporal occurrence of blue whales off the U.S. West Coast, with implications for management. PLoS One 9(7):e102959. https://doi.org/10.1371/journal.pone.0102959

Johnston C, Berta A (2011) Comparative anatomy and evolutionary history of suction feeding in cetaceans. Mar Mamm Sci 27(3):493–513. https://doi.org/10.1111/j.1748-7692.2010.00420.x

Jonsen ID, Flenming JM, Myers RA (2005) Robust state-space modeling of animal movement data. Ecology 86(11):2874–2880. Retrieved from <Go to ISI>://000233419600004

Jonsen ID, Basson M, Bestley S, Bravington MV, Patterson TA, Pedersen MW, Wotherspoon SJ et al (2013) State-space models for bio-loggers: a methodological road map. Deep Sea Res Part II: Top Stud Oceanogr 88–89:34—46. https://doi.org/10.1016/j.dsr2.2012.07.008

Jurasz CM, Jurasz VP (1979) Feeding modes of the humpback whale, *Megaptera novaeangliae*, in southeast Alaska. Sci Rep Whales Res Inst 31:69–83
Kennedy AS, Zerbini AN, Vasquez OV, Gandilhon N, Clapham PJ, Adam O (2014) Local and migratory movements of humpback whales (Megaptera novaeangliae) satellite-tracked in the North Atlantic Ocean. Can J Zool 92(1):9–18. https://doi.org/10.1139/cjz-2013-0161
Kirchner T, Wiley DN, Hazen EL, Parks SE, Torres LG, Friedlaender AS (2018) Hierarchical foraging movement of humpback whales relative to the structure of their prey. Mar Ecol Prog Ser 607:237–250. https://doi.org/10.3354/meps12789
MacNeilage PF (2013) Vertebrate whole-body-action asymmetries and the evolution of right handedness: a comparison between humans and marine mammals. Dev Psychobiol 55(6):577–587
Marx FG, Uhen MD (2010) Climate, critters, and cetaceans: cenozoic drivers of the evolution of modern whales. Science 327(5968):993–996. https://doi.org/10.1126/science.1185581
Mate BR, Krutzikowsky GK, Winsor MH (2000) Satellite-monitored movements of radio-tagged bowhead whales in the Beaufort and Chukchi seas during the late-summer feeding season and fall migration. Can J Zool 78(7):1168–1181. https://doi.org/10.1139/cjz-78-7-1168
Mate B, Mesecar R, Lagerquist B (2007) The evolution of satellite-monitored radio tags for large whales: one laboratory's experience. Deep Sea Res Part II 54(3–4):224–247. https://doi.org/10.1016/j.dsr2.2006.11.021
Nowacek DP, Friedlaender AS, Halpin PN, Hazen EL, Johnston DW, Read AJ, Zhu Y et al (2011a) Super-aggregations of krill and humpback whales in Wilhelmina Bay, Antarctic Peninsula. PLoS One 6(4):e19173. https://doi.org/10.1371/journal.pone.0019173
Nowacek DP, Friedlaender AS, Halpin PN, Hazen EL, Johnston DW, Read AJ, Zhu Y et al (2011b) Super-aggregations of Krill and Humpback Whales in Wilhelmina Bay, Antarctic Peninsula. Plos One 6(4). e1917310.1371/journal.pone.0019173
Norris KS, Villa-Ramirez B, Nichols G, Würsig B, Miller K (1983) Lagoon entrance and other aggregations of gray whales (*Eschrichtius robustus*). pp. 259–293. In: Payne R (ed) Communication and behavior of whales. Westview Press, Boulder
Oestreich WK, Fahlbusch JA, Cade DE, Calambokidis J, Margolina T, Joseph J, Ryan JP et al (2020) Animal-borne metrics enable acoustic detection of blue whale migration. Curr Biol. https://doi.org/10.1016/j.cub.2020.08.105
Oliver JS, Kvitek RG (1984) Side-scane sonar redords and diver observations of the gray whale feeding grounds. Biol Bull 176:264–269
Parks SE, Warren JD, Stamieszkin K, Mayo CA, Wiley D (2011) Dangerous dining: surface foraging of North Atlantic right whales increases risk of vessel collisions. Biol Lett. https://doi.org/10.1098/rsbl.2011.0578
Potvin J, Werth AJ (2017) Oral cavity hydrodynamics and drag production in Balaenid whale suspension feeding. PloS one 12(4):e0175220
Rockwood RC, Elliott ML, Saenz B, Nur N, Jahncke J (2020) Modeling predator and prey hotspots: Management implications of baleen whale co-occurrence with krill in Central California. PLoS ONE 15(7):e0235603. https://doi.org/10.1371/journal.pone.0235603
Rosenbaum HC, Maxwell SM, Kershaw F, Mate B (2014) Long-range movement of humpback whales and their overlap with anthropogenic activity in the South Atlantic Ocean. Conserv Biol 28(2):604–615. https://doi.org/10.1111/cobi.12225
Slater GJ, Price SA, Santini F, Alfaro ME (2010) Diversity versus disparity and the radiation of modern cetaceans. Proc R Soc b Biol Sci 277(1697):3097–3104. https://doi.org/10.1098/rspb.2010.0408
Slater GJ, Goldbogen JA, Pyenson ND (2017a) Independent evolution of baleen whale gigantism linked to Plio-Pleistocene ocean dynamics. Proc R Soc B Biol Sci 284(1855). https://doi.org/10.1098/rspb.2017.0546
Slater GJ, Goldbogen JA, Pyenson ND (2017b) Independent evolution of baleen whale gigantism linked to Plio-Pleistocene ocean dynamics. Proc R Soc B 284(1855):20170546

Torres LG, Orben RA, Tolkova I, Thompson DR (2016) Animal movement analysis through residence in space and time. Peer J Preprints 4:e2480v1. https://doi.org/10.7287/peerj.preprints.2480v1

Tsai CH, Fordyce RE (2015) The earliest gulp-feeding mysticete (Cetacea: Mysticeti) from the oligocene of New Zealand. J Mamm Evol 22(4):535–560. https://doi.org/10.1007/s10914-015-9290-0

van der Hoop JM, Nousek-McGregor AE, Nowacek DP, Parks SE, Tyack P, Madsen PT (2019) Foraging rates of ram-filtering North Atlantic right whales. Funct Ecol 33(7):1290–1306. https://doi.org/10.1111/1365-2435.13357

Ware C, Arsenault R, Plumlee M (2006) Visualizing the underwater behavior of humpback whales. IEEE Comput Graphics Appl 26(4):14–18. https://doi.org/10.1109/mcg.2006.93

Ware C, Friedlaender AS, Nowacek DP (2011) Shallow and deep lunge feeding of humpback whales in fjords of the West Antarctic Peninsula. Marine Mammal Sci 27(3):587–605

Ware C, Wiley D, Friedlaender AS, Weinrich M, Hazen EL, Bocconcelli A, Abernathy K et al (2013) Bottom side-roll feeding by humpback whales (Megaptera novaeangliae) in the southern Gulf of Maine, U.S.A. Mar Mammal Sci. https://doi.org/10.1111/mms.12053

Weinstein BG, Double M, Gales N, Johnston DW, Friedlaender AS (2017) Identifying overlap between humpback whale foraging grounds and the Antarctic krill fishery. Biol Conserv 210, Part A:184–191. https://doi.org/10.1016/j.biocon.2017.04.014

Weinstein BG, Friedlaender AS (2017) Dynamic foraging of a top predator in a seasonal polar marine environment. Oecologia. https://doi.org/10.1007/s00442-017-3949-6

Wiley D, Friedlaender A, Weinrich M, Bocconcelli A, Cholewiak D, Thompson M, Ware C (2011) Underwater components of humpback whale bubble-net feeding behaviour. Behaviour 148(5):575–602. https://doi.org/10.1163/000579511x570893

Woodward BL, Winn JP (2006) Apparent lateralized behavior in gray whales feeding off the central British Columbia coast. Mar Mammal Sci 22(1):64–73. Retrieved from <Go to ISI>://WOS:000234942300006

Chapter 6
Viewing the Lives of Whales Through a Molecular Lens

Emma L. Carroll and Ellen C. Garland

Abstract Genetics and genomics tools are providing unprecedented insights into the hidden social lives of baleen whales, from parentage to diet to defining population segments. Genetic information reveals simple yet critical information about individual whales, from their sex to their age to their kin. Integrating genetic and genomic information with other data such as photo-identification records, stable isotope data on foraging ecology, and hormone studies on reproductive status can provide detailed pictures of individuals and populations. Here we show through published examples how such datasets can be used to provide information on mating systems and assess population recovery. Genetic tools such as DNA diet studies have been instrumental at uncovering new aspects of whale ecology, such as identifying previously unknown prey that are key to the persistence of non-migratory baleen whale populations. Finally, we discuss how genetics, acoustics, and culture interact in whale populations, and how these data are being used in complementary ways to identify units to conserve.

Keywords Molecular ecology · Genomics · DNA · Stable isotope · Population structure · Song · Whaling

6.1 Putting the Molecular in Marine Mammal Ecology

Despite the challenges inherent in studying often migratory, elusive, or endangered marine mammals, cetaceans are among the best studied wild animals when it comes to molecular ecology (Cammen et al. 2016). Part of this is by necessity; many behaviors that can be observed directly in terrestrial species, such as mating and foraging, are often difficult or impossible to directly study in the parvorder Mysticeti, or baleen

---

E. L. Carroll (✉)
School of Biological Sciences, University of Auckland - Waipapa Taumata Rau, Auckland, New Zealand

E. C. Garland
Sea Mammal Research Unit, School of Biology, University of St. Andrews, St. Andrews, UK

C.W. Clark and E. C. Garland (eds.), *Ethology and Behavioral Ecology of Mysticetes*, Ethology and Behavioral Ecology of Marine Mammals,
https://doi.org/10.1007/978-3-030-98449-6_6

whales. Genetic studies offer an alternative way to assess important characteristics of individuals and populations. Several populations of Mysticetes are now the subject of long-term genetic monitoring programs, providing a wealth of data on genetic population structure, diversity, and population demographic parameters (Fig. 6.1). As with acoustics, combining molecular methods with classical behavioral observations, such as photo-identification, as well as other tools such as stable isotope analysis, allows us to build more complete and complex pictures of the lives of whales. For

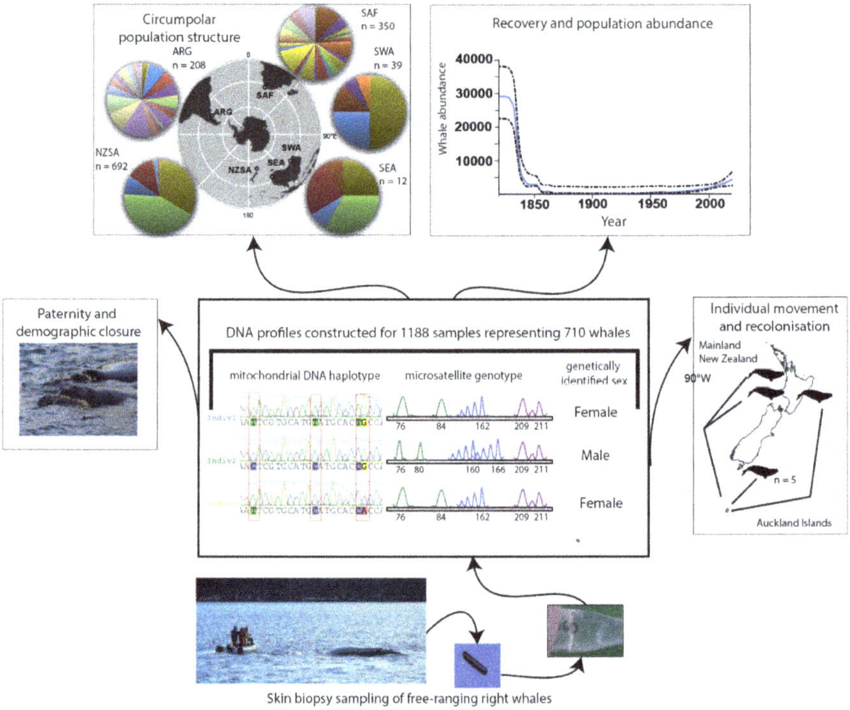

Fig. 6.1 Genetic monitoring of New Zealand southern right whales. This population has been the subject of genetic monitoring surveys to the New Zealand sub-Antarctic Auckland Islands in the austral winters of 1995–1998 and 2006–2009 (Patenaude et al. 1998; Carroll et al. 2013). This work begins with the collection of small skin biopsy samples. High quality DNA is extracted from these samples and used to build DNA profiles comprised of the mtDNA control region haplotype, multilocus microsatellite genotype and genetically identified sex of individual animals. These DNA profiles have been used to assess individual movement and recolonization, including movements of eight genetically identified individuals between the Auckland Islands and Mainland New Zealand (Carroll et al. 2014); estimate population recovery, growth rate and abundance for the New Zealand population (Carroll et al. 2013; Jackson et al. 2016); investigate circumpolar population structure by comparison to data from other wintering grounds, such as mtDNA haplotype frequency data shown for the New Zealand sub-Antarctic (NZSA), Southeast Australia (SEA), Southwest Australia (SWA), South African (SAF) and Argentine (ARG) wintering grounds (Carroll et al. 2019b); and investigate paternity and mating systems (Carroll et al. 2012)

example, combining genetic data and stable isotope data allows us to test hypotheses on migratory culture (Chap. 8).

Genetic studies rely on the collection of tissue, blow or scat samples from wild animals, and the type of samples collected from cetaceans typically depends on the accessibility of the population and the question under consideration. Dedicated field projects can collect small samples of skin and blubber from whales at sea using remote biopsy systems (Lambertsen 1987; Krützen et al. 2002). Such skin samples produce high-quality DNA for many different molecular applications, but the skin can also be used for stable isotope analyses, while the blubber can be used for hormone and pollutant analyses. Sloughed skin samples can also be collected after animals breach or from animals using rough pads on the end of poles (termed "scrubbies") (Harlin et al. 1999; Valsecchi et al. 2002), but these can produce a variable quality and quantity of DNA which is not suitable for every application. Additionally, samples from beach cast or entangled animals have also proven critical for molecular ecology applications of some species (Dalebout et al. 2005; Oremus et al. 2009).

As with sample collection, the type of genetic marker used depends on the question at hand. In this section, we discuss different classes of genetic markers and the characteristics which make them useful in the study of marine mammal ethology. First, a few words on terminology (Table 6.1). A genetic locus (plural loci) is the position on a chromosome: This can refer to a single base pair or multiple base pairs (bp) of DNA. Loci can have a known functional purpose, typically involved in coding or gene expression such as a gene, or loci can be in a non-coding region of the genome. Genetic variation within loci can be functional, in that the variation could impact gene expression, or "neutral", in that it does not. This means that you can have neutral variation in coding regions, as not all mutations will impact gene coding. Different variants at haploid loci, such as loci on the mitochondrial genome, are termed haplotypes, and different variations of diploid loci are termed alleles; the total variation at a locus is termed its genetic polymorphism or variation.

The mitochondrial genome is found in the cellular organelle called the mitochondrion (plural mitochondria), which is involved in cellular energy generation. Due to this critical role, the mitochondrion is often found in multiple copies per cell, compared with just a single copy of the nuclear genome per cell. This high copy number has made mitochondrial DNA a reliable marker to work with even with low-quality DNA templates. The whole mitochondrial genome is about 16,000 bp in cetaceans and comprises 37 genes critical in its energy production role. However, most cetacean studies investigate variation at the control region or d-loop. This non-coding region can accumulate more neutral variation than the genes in the mitochondrion, providing more haplotypes, or higher polymorphism, that is useful for discriminating between populations. Mitochondrial DNA (mtDNA) also shows a maternal inheritance pattern that has been useful as cetaceans display behaviors that are passed from mother to offspring, or that live in groups in which the females are related (termed matrilineal groups: Chap. 8).

The nuclear genome or nuclear DNA markers are inherited from both parents, and different variants of these are used in molecular ecology studies. The most commonly applied nuclear DNA markers are microsatellite loci and single-nucleotide

Table 6.1 List of commonly used genetic terms and their definitions, sourced from Waples et al. (2018)

Term	Definition
Alleles	One of two or more alternative forms of a gene or nucleotide sequence at a given locus
Effective population size, N_e	The size of an "ideal" population that would experience the same rate of genetic drift as the population in question. In an ideal population (also called a Wright–Fisher population), generations are discrete, mating is random, and every individual has an equal probability of contributing genes to the next generation. In most species, including cetaceans, N_e is typically smaller than the number of individuals in the census population *N*, as the assumptions are not met
Functional variation	Variation in DNA code that impacts gene expression and can be subject to natural selection
Genetic drift	Random change in allele frequencies from one generation to the next. Drift is expected to have a greater effect as the effective population size of the population decreases
Genetic locus	A stretch of DNA at a particular place on a particular chromosome; often used to refer to a gene
Haplotypes	The combination of alleles at loci that are found on a single chromosome or DNA molecule and thus tend to be inherited together. In cetaceans, haplotype typically refers to the mitochondrial DNA sequence held by an individual
Hybridization	Mating between individuals from two genetically distinct populations or species
Microsatellite locus	Genetic marker comprised of short DNA sequence units that are repeated multiple times (e.g., ATATATATAT), generally bi-parentally inherited on the nuclear genome
Mitochondrial genome	Full complement of DNA found in the mitochondrion
Neutral variation	Variation in DNA code that does not impact gene expression and is not influenced by natural selection
Nuclear DNA	DNA found in a cell's nucleus. In cetaceans, autosomal nuclear DNA is bi-parentally inherited, such that an individual's genotype at a given locus contains one allele inherited from its mother and another allele inherited from its father. Nuclear DNA also includes DNA found on sex chromosomes

(continued)

Table 6.1 (continued)

Term	Definition
Population	A group of individuals that co-occur in space and time and freely interbreed; synonyms include "subpopulation" and "stock"; the latter can also refer to units of management convenience that do not imply interbreeding
Single-nucleotide polymorphisms (SNPs)	DNA sequence variation that occurs when a single nucleotide (A, T, C, or G) differs at a specific site among individuals or within an individual (for diploid markers)

polymorphisms (SNPs). Microsatellite loci are also called variable number tandem repeats, short tandem repeats, or simple sequence repeats. These names stem from the fact that microsatellites comprise short motifs of DNA, typically 2–6 bp long, that are repeated multiple times. For example, in the sequence *TAGATAGATAGA*, the motif *TAGA* is repeated three times. This structure of repeated elements helps explain why microsatellites have high polymorphism; it is easy for the DNA to "slip" during replication and either delete or insert new repeats. It also explains why microsatellites are defined by sizes rather than their sequence; we measure the number of repeated elements by the length of the allele. For example, in one allele the motif *TAGA* is repeated three times, *TAGATAGATAGA*, to give a length of 12 base pairs, but in another allele, *TAGATAGATAGATAGATAGA*, it is repeated five times, to give a length of 20 base pairs. Two alleles at each locus (one from each parent) define an individual's genotype, and individuals are typically scored at many loci to produce a multilocus genotype. The high polymorphism and relatively inexpensive manner of inferring alleles by length have led to the widespread adoption of microsatellites in cetacean molecular ecology studies over the past 30 years.

Increasingly, SNPs are being used in molecular ecology studies in place of microsatellite markers. SNPs typically only have two alleles per locus (but theoretically up to four), and so this comparatively low polymorphism compared with microsatellites means that more SNPs are needed to provide the same amount of genetic information. Luckily, next-generation sequencing platforms allow us to genotype individuals at a large number of SNPs simultaneously, and the lowering costs of sequencing have led to an increase in SNP-based studies. If each locus can be thought of as an independent sample of the genome, increasing your sample size from 10–20 microsatellite loci to 20,000 SNPs is indicative of the increase in power these new technologies provide. There are several "reduced representation" genomic approaches that make it cost-effective to consistently obtain large numbers of markers across samples and populations. As the methodology is continuing to advance, and is beyond the scope of this chapter, please see recent review papers for more information (Carroll et al. 2018). These advances combined with the relative ease of genotyping, analyzing, and modeling SNP data due to their digital/binary nature mean that the popularity of SNPs is likely to increase (Cammen et al. 2016; Carroll et al. 2018).

Combined with genetically identified sex and mtDNA haplotype, multilocus genotypes from either microsatellites or SNPs can give each whale a unique "DNA profile" that can be used to infer identity and relatedness. Such data can be used over time to contribute to the long-term monitoring of a population (Schwartz et al. 2007). For example, genetic monitoring of New Zealand southern right whales (*Eubalaena australis*), Tohorā nō Aotearoa in the Māori language, has been used to estimate demographic parameters such as population abundance and growth rates to inform management (Fig. 6.1) (Carroll et al. 2013, 2011). The DNA profiles from New Zealand southern right whales have also been used to infer mating systems, distribution, and population structure (Carroll et al. 2012, 2015, 2019b).

As well as being used to estimate and track changes in demographic parameters, genetic monitoring can be used to measure changes in genetic diversity. This is often accomplished with an estimate of "effective population size". This refers to the size of an ideal population that would experience genetic drift, or the random change in allele frequencies over time, at the same rate as the actual population under study. By "ideal", we mean a population that meets the following conditions: has a constant size, equal sex ratio and random mating over generations; and these generations do not overlap (for more see Frankham 1995). Clearly, whale populations rarely meet these conditions and tend to have an effective population size much lower than the census population size. When generations overlap, as they do in whales and dolphins, the effective population size becomes an estimate of the effective number of breeders in the population (Waples 2006).

As effective population size is a genetic measure that does not readily relate to the census population, it can be difficult to use it to compare diversity across species. However, it is useful for tracking changes within species or populations over time. Long-term genetic monitoring work on the Māui dolphin (*Cephalorhynchus hectori maui*), a dolphin subspecies endemic to the North Island of New Zealand, has documented declines in both the effective and census population sizes over the past few decades. The decline in effective population size has had a time lag relative to census population size, as genetic diversity decreases on the scale of generations (Baker et al. 2013a, 2016). Specifically, while effective population size (69 dolphins, 95% confidence limits (CL): 40–168) was similar to census population size (69 dolphins, 95% CL: 38–125) in 2007, by 2016 effective population size (34, 95% CL: 24–51) had halved while census population size had stayed comparable (63, 95% CL: 57–75) (Baker et al. 2016). Declining genetic diversity can be a signal of trouble ahead, as low genetic diversity is correlated with issues such as inbreeding and outbreeding depression, loss of adaptive potential, increased risk of extinction and other issues (for a recent review see Leroy et al. 2018). Therefore, monitoring effective as well as census population size is recommended in conservation biology (Schwartz et al. 2007; Frankham et al. 2014).

With these basic ideas now understood, we will explore how Mysticete ethology has been investigated using genetic and genomic methods, often in combination with other tools, with a focus on parentage and mating systems, hybridization, diet, and defining populations. Finally, we consider how genetic tools have been used to assess the impact of whaling, and how similar work could be done using other methodologies.

6.2 Parentage and Mating Systems

Mysticetes typically have a seasonal breeding cycle. After summering in productive high-latitude foraging grounds, baleen whales generally migrate to winter calving and breeding grounds. The details of Mysticete breeding cycles vary between species, but in all cases females invest heavily in reproduction. Typically, females gestate for an extended period, typically around 12 months, with lactation thought to be 6–12 months (Lockyer 1984; Best 1994; Burnell 2001). Depending on the species, females might become pregnant while nursing their calf (*e.g.*, minke whales, *Balaenoptera acutorostrata*) or require a year or two rest period to regain condition before conceiving again (*e.g.*, bowhead, *Balaena mysticetus* and right whales, *Eubalaena* spp.). The latter lowers the proportion of females that are receptive in a given breeding season, leading to a skew in the number of reproductively available females compared with the number of reproductively available males (operational sex ratio). As the number of males exceeds the number of reproductively available females, how female mate choice and male competition operates is an important ecological and evolutionary question. This is particularly interesting in Mysticetes; as marine mammals that mate in the water (Bartholomew 1970), typical polygynous breeding systems cannot be established (Emlen and Oring 1977). For example, there are no obvious resources that can be defended by males on Mysticete wintering grounds, such as foraging patches, and females themselves cannot be easily defended as they are widely dispersed on wintering grounds, are either the same size or larger than males and movement in three dimensions limits guarding. Indeed, females in several baleen whale species have been shown to have offspring sired by multiple males, suggesting a polygynandrous rather than polygynous mating system (Clapham and Palsbøll 1997; Frasier et al. 2007; Lang et al. 2009). Thus, Mysticetes must employ tactics other than resource defense to maximize reproductive success, such as sperm competition in southern right whales and perhaps lekking in humpback whales (*Megaptera novaeangliae*) (Ginsberg and Huck 1989; Clapham 1996; Frasier et al. 2007).

Genetic analyses have shone light on the social and mating behavior of right whales. These species have the largest testis size of any mammal species (Brownell and Ralls 1986; Frasier et al. 2007), likely facilitating sperm competition. Right whales aggregate in socializing and calving grounds during winter, during which whales form surface active groups (SAGS) (Kraus and Hatch 2001; Best et al. 2003). Females in these courtship groups vocalize to attract males, leading them to jostle for prime position in the SAG, and multiple copulations occur over the course of minutes to hours. Interestingly, a study in South Africa showed that female southern right whales (*E. australis*) that are the center of SAGs were juvenile or primiparous, so that the mating behavior was highly unlikely to result in conception (Best et al. 2003). This study, which used a combination of genetics to identify the sex of whales in SAGs and photo-identification (photo-ID) to track which females had calves the following year, challenged the idea that coastal wintering grounds were also breeding grounds.

It also highlights how genetic and photo-ID datasets can provide more information on Mysticete ethology when integrated into long-term individual records.

Molecular tools give us the ability to directly confirm maternities, often inferred by the close association of cow and calf, and identify paternities, which are otherwise virtually impossible to identify in baleen whales. Parentage assignment has typically been done using microsatellite markers in marine mammals using the principle of Mendelian inheritance; at a given locus, an offspring inherits one allele from its mother and one from its father. When government permitting allows, it is much more powerful to collect biopsy samples from both the mother and the calf while the calf is still dependent on the mother. In this case, the maternal alleles are easily identified. By the principle of exclusion, we can infer the father's allele in the offspring. For example, using our repeat motif of *TAGA* above, the offspring has a genotype of three and five repeat alleles and the mother has two copies (homozygous) for the three repeat allele. This means that the father and offspring must share the five repeat allele. By creating multilocus genotypes for mother, offspring, and candidate fathers, we can identify males that could be potential fathers. By identifying potential paternities, we can start investigating whether males have equal reproductive success, or if there is a skew whereby males with certain phenotypic (*e.g.*, acoustic display) or genetic characteristics (*e.g.*, higher polymorphism) have higher than expected number of offspring.

Paternity analyses have been applied to both southern and North Atlantic right whales (*E. australis* and *E. glacialis*, respectively). In New Zealand, work conducted on 34 cow–calf pairs and 314 candidate males suggested that mating occurs in the New Zealand population, but did not find evidence of significant reproductive skew (Carroll et al. 2012). In contrast, paternity analyses in North Atlantic right whales found highly skewed reproductive success among males, based on 87 cow–calf pairs and 151 males genotyped at 28 microsatellite loci (Frasier et al. 2007). The study showed that there was an excess of males not assigned any paternities and an excess of males assigned multiple paternities. Furthermore, males were not successful fathers until ~15 years of age, suggesting that age, size, and/or experience matters. Perhaps this explains the focus of SAGs on juvenile females in South Africa; males are gaining experience or learning techniques to increase future reproductive success (Kraus et al. 2001). Alternatively, older males could have higher reproductive success due to testes development (Frasier et al. 2007). Steroid hormone analyses to detect pregnancy are now possible using both baleen plates (Hunt et al. 2018) and blubber samples (Vu et al. 2015; Pallin et al. 2018). The integration of testosterone and genetic analyses in future work looks likely to shed further light on the mating systems of Mysticetes.

Sperm competition is also believed to be important in bowhead and gray whales (*Eschrichtius robustus*), as these species also have high testes-to-body weight ratios (Brownell and Ralls 1986). Therefore, it is not surprising that a reproductive skew was detected among males in the western North Pacific gray whale population (Lang et al. 2009). This study assigned paternities to approximately one-half of the offspring in 57 cow–calf pairs from a pool of 42 candidate males; this represented 83% of whales identified on the Sakhalin feeding ground at the time. The skew detected was significant, as the proportion of males assigned paternities was lower than predicted by simulations of random mating. Using a standardized measure of reproductive variance, the skew was similar between the North Atlantic right whales and western gray whales, suggesting that sperm competition can generate variance in reproductive success in both species (Lang et al. 2009).

A milder skew in reproductive success was determined in the Mexico Pacific population of humpback whales (Cerchio et al. 2005). In a study of 125 cow–calf pairs and 297 potential fathers that used 13 microsatellite loci, 62 paternities assigned to 51 males were made under relaxed criteria. The detected skew in reproductive success was not severe; most successful males sired only one calf, and no male was assigned more than three paternities. The milder skew in the Mexico Pacific population of humpbacks compared with North Atlantic right and gray whales is consistent with differences in their mating systems. Another hypothesis is that the reproductive skew could be amplified in small populations due to either stochastic effects or the avoidance of mating with close relatives. This could potentially explain the difference in observed male reproductive skew between the small and poorly recovering North Atlantic and now strongly recovering New Zealand and other southern right whale populations (Carroll et al. 2013; Pace et al. 2017).

Probably the most well-studied Mysticete mating system, that of the humpback whale, is characterized by males singing highly stereotyped songs that may provide information to both females and other males (Chap. 11). How this display relates directly to mate choice and reproductive success has not yet been directly studied using parentage analyses. However, combining genetic parentage assessments with singing behavior and song from males (both successful and unsuccessful) would be a powerful way to investigate the evolutionary role of song. Additionally, a molecular age biomarker (Jarman et al. 2015) has been developed for humpback whales (Polanowski et al. 2014) that allows the age of individuals to be determined using a skin biopsy sample. This would also provide information about whether age-related experience or physiological development contributes to reproductive success in male humpback whales.

In addition to shedding light on mating systems, parentage analyses can be used to investigate the degree of demographic isolation and size of whale populations. For example, Garrigue et al. (2004) conducted parentage analysis of the New Caledonian humpback whale population to investigate the degree of connectivity with other Oceanian (South Pacific) wintering grounds. The work was based on 16 cow–calf pairs and 133 candidate males using nine loci microsatellite genotypes. Using the sample of males as the initial capture, and paternity assignment to five calves as the recapture, the authors estimated that male abundance from this "gametic mark–recapture" was 379 (CV = 0.30). This was in close agreement with the sex-specific estimate for the New Caledonian wintering ground based on the DNA profile or "organismal" recapture using a two-sample model (382, CV = 0.22) (Garrigue et al. 2004). A similar agreement was found between the organismal and gametic mark–recapture estimates in the New Zealand southern right whale wintering ground (Carroll et al. 2012). Using the sample of males as the initial capture, and paternity assignment as the recapture, the authors estimated that male abundance from this "gametic mark–recapture" was 1001 (95% CI 542, 1469), similar to the male census estimate (Carroll et al. 2013). In both cases, the findings supported the hypothesis that whales on each of these wintering grounds were members of an autonomous population unit that is relatively closed to demographic and reproductive interchange.

Recent work has built upon these parentage studies to develop "close kin" mark–recapture methods (CKMR). This method essentially extends conventional capture–recapture models, which are predicated on the recapture of individuals, to focus on the recapture of close kin to estimate demographic parameters. The benefits of CKMR compared with conventional mark–recapture models include requiring a smaller sample size and shorter time series (Bravington et al. 2016), suggesting that CKMR approaches hold the potential to become a key tool in marine mammal molecular ecology in the future.

6.3 Diet

Our understanding of diet in baleen whales has come from a tool kit of different approaches, including stomach content analyses from stranded, commercially killed and bycaught whales; stable isotope analysis of skin, bone, and baleen; and increasingly, molecular methods. Molecular studies use DNA-based methods to identify prey species in whale scat. Initially, techniques included differentiating among prey species using variation in the length of gene regions amplified by PCR. This approach was used to identify which krill species among eight possible candidates a pygmy blue whale (*Balaenoptera musculus brevicauda*) was feeding on in the waters off Victoria, Australia (Jarman et al. 2002).

Today, most molecular diet studies are based on DNA metabarcoding, which identifies multiple taxa from a single environmental sample by combining DNA-sequence-based methods for species identification with high-throughput sequencing (Epp et al. 2012). This provides a powerful method for linking organisms across food webs through DNA-based diet analysis (Pompanon et al. 2012). To date, the method has not been applied widely to study baleen whale diet. This is partly due to the difficulty in collecting whale scat and because the foraging grounds of many baleen whales are in inaccessible high-latitude and offshore regions. Those studies that have been carried out have focused on populations that do not migrate and have sought to understand how these large mammals survive in one region year-round. One such study investigated the diet of Northern Indian Ocean blue whales (*Balaenoptera musculus*), which typically inhabit low-latitude waters around Somalia, the Seychelles, Maldives, and Sri Lanka (de Vos et al. 2018). The most common dietary components found in the 34 whale scat samples were Dendrobranchiata (94% of samples), specifically Sergestid shrimp, Euphausiacea (56%) and Amphipoda (37%). The Sergestid shrimp are locally abundant in Sri Lanka and appear to be a unique prey item for Northern Indian Ocean blue whales compared with the Antarctic blue whale in the Southern Ocean and pygmy blue whales off Australia, both of which feed primarily on euphausiids.

Metabarcoding was also employed to understand how the diet of Bryde's whales (*Balaenoptera brydei*) in the Hauraki Gulf, New Zealand, changes in the context of seasonal variations in their potential prey (Carroll et al. 2019a). Whale scat and plankton community samples were taken over the course of 12 months (Fig. 6.2) to investigate how seasonal changes in prey availability might affect whale diet. Multilocus metabarcoding found significant changes in the plankton community diversity and composition linked with warm and cool water seasons. Despite this, data from 16 whale scats showed that Bryde's whales preferred the same types of prey items year-round, including krill-like crustaceans, copepods, salps, and ray-finned fishes.

Fig. 6.2 Steps in the collection of Bryde's whale scat: (1) Find Bryde's whales, which are often foraging (top left); (2) identify defecation in the water based on discoloration (pink color in this case, middle left); (3) use custom built net to scoop up the scat (bottom left) and bring into the boat (top right); and (4) rinse out net contents with ethanol into a sterile container for storage. Photo credit: Rochelle Constantine

These two examples show how molecular methods can shed light on unique or unsuspected behaviors, as both the salps and Sergestid shrimp were unexpected prey items. Identification of prey and trophic linkages of cetaceans provides important insights into their place in the food web and has implications for conservation management (Chap. 5).

6.4 Defining Units to Conserve

The definition and assessment of appropriate management units are critically important: In the past, this process was fatally flawed and led to the over-hunting and collapse of most great whale populations (Clapham et al. 2008). Today, lessons have been learned, and this process now underpins conservation (Waples et al. 2018). It is now recognized by the International Whaling Commission that unsustainable impacts on a population could lead to local extirpation, and an irrevocable loss of biodiversity (Waples et al. 2018), or to the lack of recovery of a population in a time frame relevant for management (i.e., decades) (Clapham et al. 2008). Due to their migratory traditions, acoustic repertoire, and other behavioral traits, cetaceans provide some of the clearest examples of the importance of combining genetics with ethology to guide the delineation of stocks and populations.

With this perspective in mind, we can now assess the degree of isolation and differentiation between groups and populations in different ways; ecologically, using factors such as spatial distribution, niche, and movement of animals; genetically, using estimates of connectivity; demographically, using estimates of recruitment and mortality; and finally by assessing cultural and behavioral variation, including acoustic displays (Wade and Angliss 1997; Crandall et al. 2000; Brakes et al. 2019). In baleen whales, the unit of management is often termed the "stock" or "population"; defined as a group of individuals that co-occur in space and time and freely interbreed (following Waples and Gaggiotti 2006; Table 6.1). Differentiation among populations occurs along a continuum (Fig. 6.3): At one end is "panmixia" whereby every individual has an equal chance of mating with any other individual; and at the other end is complete isolation, whereby only individuals within distinct populations mate. Despite the capacity to swim thousands of kilometers, most whale populations sit in between these two extremes, as there is a surprising degree of genetic population structure observed in humpback whales (Baker et al. 1993; Jackson et al. 2014; Kershaw et al. 2017), southern right whales (Patenaude et al. 2007; Carroll et al. 2019b), fin whales, *Balaenoptera physalus* (Bérubé et al. 1998; Archer et al. 2013), blue whales (Attard et al. 2010; Torres-Florez et al. 2014; Leduc et al. 2017; Barlow et al. 2018), and gray whales (Lang et al. 2014; Brüniche-Olsen et al. 2018). Such patterns of genetic diversity are shaped by the adaptive challenges and demographic histories of populations and species (Jobling 2012), oceanographic and climatic factors such as the distribution of suitable prey (*e.g.,* Pastene et al. 2007; Bowen et al. 2016), and habitat resources (*e.g.,* Fontaine et al. 2010). However, behavior and social organization (Baker et al. 1993; Palsboll et al. 1995; Whitehead 1998) are

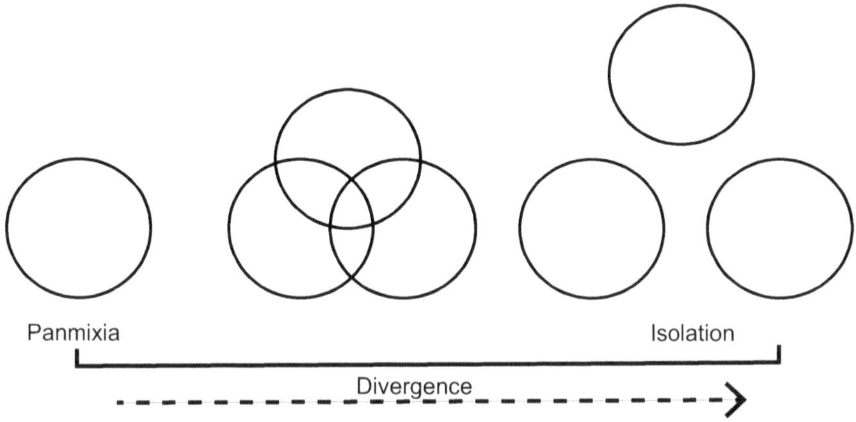

Fig. 6.3 Schema showing the different levels of connectivity of populations. Modified and copied with permission (Waples and Gaggiotti 2006)

often key drivers of genetic patterns in marine mammals and have been increasingly studied in recent years.

Maternally directed learning of both breeding and feeding grounds appears to be a strong behavioral driver of genetic population structure in two particularly well-studied species—humpback whales and southern right whales (Baker et al. 1993; Kershaw et al. 2017; Carroll et al. 2019b). The approximately one-year nursing period in these species provides an opportunity for a calf to travel with its mother from its place of birth to her preferred migratory destinations, facilitating vertical transmission of migratory preferences or "migratory culture" (Chap. 8). Such behavior has an impact on an evolutionary timescale; it is likely leading to structuring of maternal lineages or mtDNA haplotypes across wintering grounds in both species, and among humpback whale foraging grounds in the North Atlantic (Baker et al. 1993, 2013b; Kershaw et al. 2017; Carroll et al. 2019b). In southern right whales, there is also a small but significant level of differentiation in nuclear DNA markers among whales on wintering grounds, as well as evidence of structuring of maternal lineages on feeding grounds inferred from isotope-mtDNA haplotype correlations (Valenzuela et al. 2009; Carroll et al. 2015, 2019b; Chap. 8).

On a longer timescale, the effects of philopatry and migratory traditions likely contribute to speciation. The three recognized species of right whales have distinct maternal lineages that correspond with North Atlantic, North Pacific, and Southern Ocean basins, and no maternal lineages are shared across species (Rosenbaum et al. 2000). Fixed genetic differences among these species have been attributed to their anti-tropical distributions as well as differences in maternal traditions (McGowen et al. 2020). Similarly, there has been a recent call to divide humpback whales into oceanic subspecies as there is strong mtDNA differentiation between the North Atlantic, North Pacific and Southern Oceans; in fact, only three of the 219 maternal lineages found worldwide are shared between ocean basins (Jackson et al. 2014).

As with southern right whales, there are also signals of divergence in nuclear DNA among ocean basins (Jackson et al. 2014). Interestingly, this structuring of maternal lineages, and correspondingly genetic diversity, into discrete subpopulations, has been suggested as a mechanism for the maintenance of high mtDNA diversity in humpback whales (Jackson et al. 2014).

Linking such genetic structure to migratory culture can require long-term studies that demonstrate philopatry to natal wintering and/or parental foraging grounds. This has been possible in several well-studied humpback and southern right whale populations (*e.g.,* southern right whales on their wintering ground at Península Valdés, Argentina (Valenzuela et al. 2009), humpback whales on their feeding grounds in Alaska (Pierszalowski et al. 2016), Gulf of Maine (Clapham et al., 2003), and South Africa (Barendse et al. 2013)). Maternally directed site fidelity has also been demonstrated for fin whales in the Gulf of Maine (Clapham and Seipt 1991). However, fidelity to migratory habitats has been demonstrated in a wider range of baleen whale species and populations. For example, site fidelity to foraging grounds has been demonstrated for blue whales in the Gulf of California (Costa-Urrutia et al. 2013) and off New Zealand (Barlow et al. 2018), as well as in gray whales off the US Pacific Coast (Lang et al. 2014). How migratory habitats are learned, and whether they can be learned from conspecifics (horizontal transmission) as well as from parents (vertical transmission), will be key determinants in whether site fidelity in these species contributes to genetic population structure. For example, if both behavior and genes are passed vertically, then over time there could be concomitant structuring of both.

Social structure and communication signals are also a strong driver of genetic patterns of diversity in cetaceans (Brakes et al. 2019). In baleen whales, song is one method of communication, although its precise function is still a matter of debate (Chap. 2). In blue, humpback and fin whales, song is typically highly stereotyped and stable over time in some areas (Chapters 9 and 11). This means song has been useful for identifying, for example, different species of blue whale (Chap. 9), and different populations of fin (Delarue et al. 2009; Castellote et al. 2012; Širović et al. 2017) and humpback whales (Garland et al. 2015; Darling et al. 2019). In the latter species, interpopulation differences in humpback whale song have been used to define population structure and mixing at the ocean basin scale (Garland et al. 2015; Darling et al. 2019). Such studies demonstrate that some components/features of their population structures can be assessed with acoustics, and ground-truthed with genetics where possible; an approach that may also be applicable to other baleen whale species (Mellinger and Barlow 2003; Hatch and Clark 2004; McDonald et al. 2006). However, any comparison of genetic and acoustic data needs to take account of the type of information being conveyed. For example, only male whales are known to sing, so song can provide insight into male-mediated gene flow and connectivity. As mentioned above, Garland et al. (2015) used song to resolve populations previously identified with genetics. In contrast, as discussed above, female philopatry has meant that maternally inherited mitochondrial markers have been incredibly useful to assess population structure in baleen whales. Comparing song and mtDNA could therefore provide an over-arching understanding of baleen whale population structure for both sexes.

Furthermore, the temporal and spatial resolutions of song and genetic data need to be considered. Genetic methods can detect long-standing barriers to gene flow using differentiation indices such as the F-statistic (Nei 1973). However, if two groups have only recently diverged, or have a reasonably high level of gene flow, such differentiation indices may not reflect the level of connectivity (Palsbøll et al. 2010). This is because indices like F-statistics change on a time span of generations, which can be decades in whales (Landguth et al. 2011). In such cases, song may change to reflect distinct changes in population demographics faster than genetic methods can detect those changes. However, this depends on the species. For example, humpback whale song change can be rapid (Noad et al. 2000; Garland et al. 2011) and can act as a highly sensitive indicator of current population contact and mixing, or seasonal overlap in distribution (Garland et al. 2011, 2015; Darling et al. 2019). In contrast, while it is suspected that there are several populations of Antarctic blue whale, the species has one song type found throughout its range (Chap. 9).

Below population level, baleen whale song has not proved informative thus far, although there is potential for group-level resolution in bowhead and North Pacific right whales, *E. japonica* (Crance et al. 2019). However, genomic data combined with new statistical methods for examining the spatial pattern of kin relationships can reveal distinct population segments and show the impact of recent barriers to gene flow (Escoda et al. 2019).

Finally, although there appear to be behavioral mechanisms leading to isolation between whale populations that can lead to speciation, it does not appear to be a one-way process. There is increasing genetic evidence of hybridization and introgression between baleen whale species (Attard et al. 2012; Árnason et al. 2018). We still have much to learn about the balance between processes that drive differentiation, such as philopatry and mate choice, and those that promote connectivity and, potentially, hybridization (Carroll et al. 2019b).

6.5 Impact of Whaling: Genetics and Behavior

Baleen whales have been the focus of multiple studies that have attempted to detect the signature of whaling in their genomes. Whaling caused a large decline in the abundance of whale populations, and this change in census size can also reduce the amount of genetic diversity held by a population, called a "genetic bottleneck". The severity of a genetic bottleneck is determined by its duration in terms of generations and by how small the population gets in terms of effective population size (Nei 1975). Declines in genetic diversity have primarily been detected in baleen whale species that have a long history of exploitation—long in terms of 20+ year whale generations, that is (Taylor et al. 2007). One case of this was a simulation study showing that there has been a reduced mtDNA haplotype number and diversity in southern right whales due to its centuries-long demographic bottleneck from whaling (Jackson et al. 2008).

However, the relative impact of recent events such as whaling and other important historical events can be difficult to discern. Bowhead whales, for example, appear to have fewer unique mitochondrial lineages in contemporary samples compared with historical samples, which was linked to both habitat loss during the Little Ice Age and whaling (Alter et al. 2012a; b). Any decline in diversity in nuclear genes like microsatellites is likely to be less severe than declines in maternally inherited mtDNA, given the larger effective population size of these bi-parentally inherited markers. Indeed, high levels of nuclear diversity have been found in some great whale species, supporting the hypothesis that they have had large, long-term effective population sizes prior to whaling (*e.g.,* Ruegg et al. 2013).

Increasingly, the importance of maintaining variation in social behaviors and culture is now being highlighted as a conservation priority (Brakes et al. 2019). New evidence suggests that declines in abundance lead to a concomitant decline in song diversity and complexity in Hawaiian forest birds, highlighting how culturally transmitted behaviors can be rapidly altered by anthropogenic impacts (Paxton et al. 2019), or be lost, leading to increasing extinction risks (Crates et al. 2021). However, it is hard to study the change in behavioral displays through time in cetaceans as extinct songs cannot be described and migratory traditions cannot be easily inferred if populations have been extirpated. But perhaps it is worth considering how to use a comparative framework to understand how both genetic diversity and cultural variation could have been impacted by whaling. For example, although there is no doubt the species underwent a dramatic decline in abundance, humpback whale populations still harbor very high levels of genetic diversity (Baker et al. 1993; Jackson et al. 2014), likely because the species' demographic bottleneck was typically short (1–2 generations). This species also has a broad behavioral repertoire (Chaps. 10 and 11). In contrast, right whales have had a prolonged demographic bottleneck (5–10 generations), with evidence of a decline in genetic diversity (Waldick et al. 2002; Jackson et al. 2008), so perhaps all three right whale species have also undergone a decline in cultural or behavioral variation. Such a decline would be hard to distinguish from a lack of observational data, although recent analyses have shown that the North Pacific right whale has a previously undocumented form of song (Crance et al. 2019). Future work could investigate whether song is also present in North Atlantic and southern right whale populations, and if so, how genetic diversity, migratory, or other behavioral traditions and song variation co-vary among the right whale species and populations.

6.6 Conclusion

Molecular methods have proven useful as a key part of the cetacean ethologist's toolkit to investigate diet, parentage, and mating systems. Genetic methods have been particularly useful for assessing units for management in Mysticetes. In the future, we will be able to build more complete pictures of individuals through combining DNA profiles, molecular measures of age, hormone assays of reproductive status, and assessment of acoustic repertoire. Such complex datasets will allow us to test hypotheses on the drivers of mate choice and reproductive success and to provide complementary spatial, demographic, and temporal information for conservation and management.

Acknowledgements ELC is supported by a Rutherford Discovery Fellowship from the Royal Society of New Zealand Te Apārangi. ECG is supported by a Royal Society University Research Fellowship. Thanks to C. Scott Baker and Philippa Brakes for providing comments and reviewing an earlier draft of the chapter.

References

Alter SE, Rosenbaum HC, Postma LD et al (2012a) Gene flow on ice: The role of sea ice and whaling in shaping holarctic genetic diversity and population differentiation in bowhead whales (*Balaena mysticetus*). Ecol Evol 2:2895–2911. https://doi.org/10.1002/ece3.397

Alter SE, Newsome SD, Palumbi SR (2012b) Pre-whaling genetic diversity and population ecology in eastern pacific gray whales: Insights from ancient DNA and stable isotopes. PLoS ONE 7:1–12. https://doi.org/10.1371/journal.pone.0035039

Archer FI, Morin PA, Hancock-Hanser BL et al (2013) Mitogenomic phylogenetics of fin whales (*Balaenoptera physalus* spp.): genetic evidence for revision of subspecies. PLoS One 8. https://doi.org/10.1371/journal.pone.0063396

Árnason Ú, Lammers F, Kumar V et al (2018) Whole-genome sequencing of the blue whale and other rorquals finds signatures for introgressive gene flow. Sci Adv 4:eaap9873

Attard CRM, Beheregaray LB, Jenner C et al (2010) Genetic diversity and structure of blue whales (*Balaenoptera musculus*) in Australian feeding aggregations. Conserv Genet 11:2437–2441. https://doi.org/10.1007/s10592-010-0121-9

Attard CRM, Beheregaray LB, Jenner KCS et al (2012) Hybridization of southern hemisphere blue whale subspecies and a sympatric area off Antarctica: impacts of whaling or climate change? Mol Ecol 21:5715–5727

Baker CS, Perry A, Bannister JL et al (1993) Abundant mitochondrial DNA variation and world-wide population structure in humpback whales. Proc Natl Acad Sci 90:8239–8243

Baker CS, Steel DJ, Calambokidis J et al (2013a) Strong maternal fidelity and natal philopatry shape genetic structure in North Pacific humpback whales. Mar Ecol Prog Ser 494:291–306. https://doi.org/10.3354/meps10508

Baker CS, Hamner RM, Cooke J et al (2013b) Low abundance and probable decline of the critically endangered Māui's dolphin estimated by genotype capture-recapture. Anim Conserv 16:224–233. https://doi.org/10.1111/j.1469-1795.2012.00590.x

Baker CS, Hamner RM, Oremus M et al (2016) Estimating the abundance and effective population size of Māui's dolphins using microsatellite genotypes in 2010–11, with retrospective matching to 2001–07. Dep Conserv Auckl 1–44

Barendse J, Best P, Carvalho I, Pomilla C (2013) Mother knows best: occurrence and associations of resighted humpback whales suggest maternally derived fidelity to a Southern Hemisphere coastal feeding ground. PLoS One 8. https://doi.org/10.1371/journal.pone.0081238

Barlow DR, Torres LG, Hodge KB et al (2018) Documentation of a New Zealand blue whale population based on multiple lines of evidence. Endanger Species Res 36:27–40. https://doi.org/10.3354/esr00891

Bartholomew G (1970) A model for the evolution of pinniped polygyny. Evolution 24:546–559

Bérubé M, Aguilar A, Dendanto D et al (1998) Population genetic structure of North Atlantic, Mediterranean Sea and Sea of Cortez fin whales, *Balaenoptera physalus* (Linnaeus 1758): analysis of mitochondrial and nuclear loci. Mol Ecol 7:585–599

Best P (1994) Seasonality of reproduction and the length of gestation in southern right whales *Eubalaena australis*. J Zool 232:175–189

Best P, Schaeff C, Reeb D, Palsbøll PJ (2003) Composition and possible function of social groupings of southern right whales in South African waters. Behaviour 140:1469–1494

Bowen BW, Gaither MR, Dibattista JD et al (2016) Comparative phylogeography of the ocean planet. Proc Natl Acad Sci USA 113:7962–7969

Brakes P, Dall SRX, Aplin LM et al (2019) Animal cultures matter for conservation. Science 363:1032–1034. https://doi.org/10.1126/science.aaw3557

Bravington MV, Skaug HJ, Anderson EC (2016) Close-Kin Mark-recapture. Stat Sci 31:259–274. https://doi.org/10.1214/16-STS552

Brownell R, Ralls K (1986) Potential for sperm competition in baleen whales. Rep Int Whal Comm Special Is:97–112

Brüniche-Olsen A, Urban RJ, Vertyankin VV et al (2018) Genetic data reveal mixed-stock aggregations of gray whales in the North Pacific Ocean. Biol Lett 14:1–4. https://doi.org/10.1098/rsbl.2018.0399

Burnell SR (2001) Aspects of the reproductive biology, movements and site fidelity of right whales off Australia. J Cetacean Res Manag Spec Issue 2:89–102

Cammen KM, Andrews KR, Carroll EL et al (2016) Genomic methods take the plunge: Recent advances in high-throughput sequencing of marine mammals. J Hered 107:481–495. https://doi.org/10.1093/jhered/esw044

Carroll EL, Patenaude NJ, Childerhouse SJ et al (2011) Abundance of the New Zealand subantarctic southern right whale population estimated from photo-identification and genotype mark-recapture. Mar Biol 158. https://doi.org/10.1007/s00227-011-1757-9

Carroll EL, Childerhouse SJ, Christie M et al (2012) Paternity assignment and demographic closure in the New Zealand southern right whale. Mol Ecol 21. https://doi.org/10.1111/j.1365-294X.2012.05676.x

Carroll EL, Childerhouse SJ, Fewster R et al (2013) Accounting for female reproductive cycles in a superpopulation capture—recapture framework. Ecol Appl 23:1677–1690. https://doi.org/10.1890/12-1657.1

Carroll EL, Rayment WJ, Alexander AM et al (2014) Reestablishment of former wintering grounds by New Zealand southern right whales. Mar Mammal Sci 30:206–220. https://doi.org/10.1111/mms.12031

Carroll EL, Baker CS, Watson M et al (2015) Cultural traditions across a migratory network shape the genetic structure of southern right whales around Australia and New Zealand. Sci Rep 5:16182

Carroll EL, Bruford MW, Dewoody JA et al (2018) Genetic and genomic monitoring with minimally invasive sampling methods. Evol Appl. https://doi.org/10.1111/eva.12600.10.1111/eva.12600

Carroll EL, Gallego R, Sewell MA et al (2019a) Multi-locus DNA metabarcoding of zooplankton communities and scat reveal trophic interactions of a generalist predator. Sci Rep 9:281. https://doi.org/10.1038/s41598-018-36478-x

Carroll EL, Alderman R, Bannister JL et al (2019b) Incorporating non-equilibrium dynamics into demographic history inferences of a migratory marine species. Heredity (edinb) 122:53–68. https://doi.org/10.1038/s41437-018-0077-y

Castellote M, Clark CW, Lammers MO (2012) Fin whale (*Baleanoptera physalus*) population identity in the western Mediterranean Sea. Mar Mammal Sci 28:325–344

Cerchio S, Jacobsen J, Cholewiak D et al (2005) Paternity in humpback whales, *Megaptera novaeangliae*: assessing polygyny and skew in male reproductive success. Anim Behav 70:267–277

Clapham PJ (1996) The social and reproductive biology of humpback whales: an ecological perspective. Mamm Rev 26:27–49

Clapham PJ, Seipt IE (1991) Resightings of independent fin whales, *Balaenoptera physalus*, on maternal summer ranges. J Mammal 72:788–790
Clapham PJ, Palsbøll PJ (1997) Molecular analysis of paternity shows promiscuous mating in female humpback whales (*Megaptera novaeangliae*, Borowski). Proc R Soc Biol Sci 264:95–98. https://doi.org/10.1098/rspb.1997.0014
Clapham PJ, Barlow J, Bessinger M et al (2003) Abundance and demographic parameters of humpback whales from the Gulf of Maine, and stock definition relative to the Scotian Shelf. J Cetacean Res Manag 5:13–22
Clapham PJ, Aguilar A, Hatch LT (2008) Determining spatial and temporal scales for management of cetaceans: lessons from whaling. Mar Mammal Sci 24:183–201
Costa-Urrutia P, Sanvito S, Victoria-Cota N et al (2013) Fine-Scale population structure of blue whale wintering aggregations in the Gulf of California. PLoS ONE 8:1–10. https://doi.org/10.1371/journal.pone.0058315
Crance JL, Berchok CL, Wright DL et al (2019) Song production by the North Pacific right whale, *Eubalaena japonica*. J Acoust Soc Am 145:3467–3479. https://doi.org/10.1121/1.5111338
Crandall K, Bininda-Emonds ORP, Mace GM, Wayne RK (2000) Considering evolutionary processes in conservation biology. Trends Ecol Evol 15:290–295
Crates R, Langmore N, Ranjard L et al (2021) Loss of vocal culture and fitness costs in a critically endangered songbird. Proc R Soc B 288:20210225. https://doi.org/10.1098/rspb.2021.0225
Dalebout M, Robertson K, Frantzis A et al (2005) Worldwide structure of mtDNA diversity among Cuvier's beaked whales (*Ziphius cavirostris*): implications for threatened populations. Mol Ecol 14:3353–3371
Darling JD, Acebes JMV, Frey O et al (2019) Convergence and divergence of songs suggests ongoing, but annually variable, mixing of humpback whale populations throughout the North Pacific. Sci Rep 9:7002. https://doi.org/10.1038/s41598-019-42233-7
de Vos A, Faux CE, Marthick J et al (2018) New determination of prey and parasite species for Northern Indian Ocean blue whales. Front Mar Sci 5. https://doi.org/10.3389/fmars.2018.00104
Delarue J, Todd SK, Van PSM et al (2009) Implications for stock structure assessment Geographic variation in Northwest Atlantic fin whale (*Balaenoptera physalus*) song: implications for stock structure. J Acoust Soc Am 125:1774–1782. https://doi.org/10.1121/1.3068454
Emlen S, Oring L (1977) Ecology, sexual selection and the evolution of mating systems. Science 197:215–223
Epp LS, Boessenkool S, Bellemain EP et al (2012) New environmental metabarcodes for analysing soil DNA: potential for studying past and present ecosystems. Mol Ecol 21:1821–1833. https://doi.org/10.1111/j.1365-294X.2012.05537.x
Escoda L, Fernández-González Á, Castresana J (2019) Quantitative analysis of connectivity in populations of a semi-aquatic mammal using kinship categories and network assortativity. Mol Ecol Resour 19:310–326. https://doi.org/10.1111/1755-0998.12967
Fontaine MC, Tolley KA, Michaux JR et al (2010) Genetic and historic evidence for climate-driven population fragmentation in a top cetacean predator: the harbour porpoises in European water. Proc Biol Sci 277:2829–2837. https://doi.org/10.1098/rspb.2010.0412
Frankham R (1995) Inbreeding and extinction: a threshold effect. Conserv Biol 9:792–799
Frankham R, Bradshaw CJ, Brook BW (2014) Genetics in conservation management: revised recommendations for the 50/500 rules, Red List criteria and population viability analyses. Biol Conserv 170:56–63. https://doi.org/10.1016/j.biocon.2013.12.036
Frasier TR, Hamilton P, Brown MW et al (2007) Patterns of male reproductive success in a highly promiscuous whale species: the endangered North Atlantic right whale. Mol Ecol 16:5277–5293
Garland EC, Goldizen AW, Rekdahl ML et al (2011) Dynamic Horizontal Cultural Transmission of Humpback Whale Song at the Ocean Basin Scale. Curr Biol 21:687–691. https://doi.org/10.1016/j.cub.2011.03.019
Garland EC, Goldizen AW, Lilley MS et al (2015) Population structure of humpback whales in the western and central South Pacific Ocean as determined by vocal exchange among populations. Conserv Biol 29:1198–1207. https://doi.org/10.1111/cobi.12492

Garrigue C, Dodemont R, Steel D, Baker CS (2004) Organismal and 'gametic' capture-recapture using microsatellite genotyping confirm low abundance and reproductive autonomy of humpback whales on the wintering grounds of New Caledonia. Mar Ecol Prog Ser 274:251–262

Ginsberg JR, Huck UW (1989) Sperm competition in mammals. Trends Ecol Evol 4:74–79. https://doi.org/10.1080/14647270410001699054

Harlin A, Wursig B, Baker CS (1999) Skin swabbing for genetic analysis: application to dusky dolphins (*Lagenorhynchus obscurus*). Mar Mammal Sci 14:409–425

Hatch LT, Clark CW (2004) Acoustic differentiation between fin whales in both the North Atlantic and North Pacific Oceans, and integration with genetic estimates of divergence

Hunt KE, Lysiak NSJ, Matthews CJD et al (2018) Multi-year patterns in testosterone, cortisol and corticosterone in baleen from adult males of three whale species. Conserv Physiol 6:1–16. https://doi.org/10.1093/conphys/coy049

Jackson JA, Patenaude NJ, Carroll EL, Baker CS (2008) How few whales were there after whaling? Inference from contemporary mtDNA diversity. Mol Ecol 17:236–251. https://doi.org/10.1111/j.1365-294X.2007.03497.x

Jackson JA, Steel DJ, Beerli P et al (2014) Global diversity and oceanic divergence of humpback whales (*Megaptera novaeangliae*). Proc R Soc B Biol Sci 281:20133222

Jackson JA, Carroll EL, Smith TD et al (2016) An integrated approach to historical population assessment of the great whales : case of the New Zealand southern right whale. R Soc Open Sci 3:150669

Jarman SN, Gales NJ, Tierney M, Gill PC, Elliott NG (2002) A DNA based method for identification of krill species and its application to analysing the diet of marine vertebrate predators. Mol Ecol 11(12):2679–2690. https://doi.org/10.1046/j.1365-294X.2002.01641.x

Jarman SN, Polanowski AM, Faux CE et al (2015) Molecular biomarkers for chronological age in animal ecology. Mol Ecol 24:4826–4847. https://doi.org/10.1111/mec.13357

Jobling MA (2012) The impact of recent events on human genetic diversity. Philos Trans R Soc B 367:793–799

Kershaw F, Carvalho I, Loo J et al (2017) Multiple processes drive genetic structure of humpback whale (*Megaptera novaeangliae*) populations across spatial scales. Mol Ecol 26:977–994. https://doi.org/10.1111/mec.13943

Kraus SD, Hatch LT (2001) Mating strategies in the North Atlantic right whale (*Eubalaena glacialis*). J Cetacean Res Manag Spec Issue 2:237–244

Kraus SD, Hamilton P, Kenney R et al (2001) Reproductive parameters of the North Atlantic right whale. J Cetacean Res Manag Spec Issue 2:231–236

Krützen M, Barré L, Möller L et al (2002) A biopsy system for small cetaceans; darting success and wound healing in *Tursiops* spp. Mar Mammal Sci 18:863–878

Lambertsen R (1987) A biopsy system for large whales and its use for cytogenetics. J Mammal 68:443–445

Landguth E, Cushman S, Schwartz MK et al (2011) Quantifying the time lag to detect barriers in landscape genetics. Mol Ecol 19:4179–4191

Lang A,Weller D, LeDuc RG et al (2009) Delineating patterns of male reproductive success in the western gray whale (Eschrichtius robustus) population. Report SC/62/BRG10 submitted to the Scientific Committee of the International Whaling Commission, Cambridge, UK

Lang A, Calambokidis J, Scordino J et al (2014) Assessment of genetic structure among eastern North Pacific gray whales on their feeding grounds. Mar Mammal Sci 30:1473–1493. https://doi.org/10.1111/mms.12129

Leduc RG, Archer FI, Lang A et al (2017) Genetic variation in blue whales in the eastern pacific: implication for taxonomy and use of common wintering grounds. Mol Ecol 26:740–751. https://doi.org/10.1111/mec.13940

Leroy G, Carroll EL, Bruford MW et al (2018) Next-generation metrics for monitoring genetic erosion within populations of conservation concern. Evol Appl 11:1066–1083. https://doi.org/10.1111/eva.12564

Lockyer C (1984) Review of baleen whale (Mysticeti) reproduction and implications for management. Rep Int Whal Comm Spec Issue 6:27–50
McDonald M, Mesnick S, Hildebrand JA (2006) Biogeographic characterisation of blue whale song worldwide: using song to identify populations. J Cetacean Res Manag 8:55–65
McGowen MR, Tsagkogeorga G, Álvarez-Carretero S et al (2020) Phylogenomic Resolution of the Cetacean Tree of Life Using Target Sequence Capture. Syst Biol 69:479–501. https://doi.org/10.1093/sysbio/syz068
Mellinger D, Barlow J (2003) Future directions for acoustic marine mammal surveys: Stock assessment and habitat use (NOAA OAR Special Report, NOAA/PMEL Contribution No. 2557). La Jolla, CA
Nei M (1973) Analysis of gene diversity in subdivided populations. Proceeding Natl Acad Sci USA 70:3321–3323. https://doi.org/10.1073/pnas.70.12.3321
Nei M (1975) Molecular population genetics and evolution. North-Holland, Amsterdam
Noad M, Cato DH, Bryden MM et al (2000) Cultural revolution in whale songs. Nature 408:537
Oremus M, Gales R, Dalebout ML et al (2009) Worldwide mitochondrial DNA diversity and phylogeography of pilot whales (*Globicephala* spp.). Biol J Linn Soc 98:729–744
Pace RM, Corkeron PJ, Kraus SD (2017) State-space mark-recapture estimates reveal a recent decline in abundance of North Atlantic right whales. Ecol Evol 8730–8741. https://doi.org/10.1002/ece3.3406
Pallin LJ, Baker CS, Steel D et al (2018) High pregnancy rates in humpback whales (*Megaptera novaeangliae*) around the western antarctic peninsula, evidence of a rapidly growing population. R Soc Open Sci 5:180017. https://doi.org/10.1098/rsos.180017
Palsbøll PJ, Clapham PJ, Mattila DK et al (1995) Distribution of mtDNA haplotypes in North Atlantic humpback whales: The influence of behaviour on population structure. Mar Ecol Prog Ser 116:1–10. https://doi.org/10.3354/meps116001
Palsbøll PJ, Peery MZ, Bérubé M (2010) Detecting populations in the "ambiguous" zone: kinship-based estimation of population structure and genetic divergence. Mol Ecol Resour 10:797–805
Pastene LA, Goto M, Kanda N et al (2007) Radiation and speciation of pelagic organisms during periods of global warming: the case of the common minke whale, *Balaenoptera acutorostrata*. Mol Ecol 16:1481–1495. https://doi.org/10.1111/j.1365-294X.2007.03244.x
Patenaude NJ, Baker CS, Gales N (1998) Observations of southern right whales on New Zealand's subantarctic wintering grounds. Mar Mammal Sci 14:350–355
Patenaude NJ, Portway V, Schaeff C et al (2007) Mitochondrial DNA diversity and population structure among southern right whales (*Eubalaena australis*). J Hered 98:147–157
Paxton KL, Sebastián-González E, Hite JM et al (2019) Loss of cultural song diversity and the convergence of songs in a declining Hawaiian forest bird community. R Soc Open Sci 6:190719. https://doi.org/10.1098/rsos.190719
Pierszalowski SP, Gabriele CM, Steel DJ et al (2016) Local recruitment of humpback whales in Glacier Bay and Icy Strait, Alaska, over 30 years. Endanger Species Res 31:177–189. https://doi.org/10.3354/esr00761
Polanowski AM, Robbins J, Chandler D, Jarman SN (2014) Epigenetic estimation of age in humpback whales. Mol Ecol Resour 14:976–987. https://doi.org/10.1111/1755-0998.12247
Pompanon F, Deagle BE, Symondson WOC, Brown DS, Jarman SN, Taberlet P (2012) Who is eating what: diet assessment using next generation sequencing. Mol Ecol 21(8):1931–1950. John Wiley & Sons, Ltd. https://doi.org/10.1111/j.1365-294X.2011.05403.x
Rosenbaum H, Brown MW, Schaeff C et al (2000) Worldwide differentiation of Eubalaena: questioning the number of right whale species. Mol Ecol 9:1793–1802
Ruegg K, Rosenbaum HC, Anderson EC et al (2013) Long-term population size of the North Atlantic humpback whale within the context of worldwide population structure. Conserv Genet 14:103–114. https://doi.org/10.1007/s10592-012-0432-0
Schwartz MK, Luikart G, Waples RS (2007) Genetic monitoring as a promising tool for conservation and management. Trends Ecol Evol 22:25–33. https://doi.org/10.1016/j.tree.2006.08.009

Širović A, Oleson EM, Buccowich J et al (2017) Fin whale song variability in southern California and the Gulf of California. Sci Rep 7:10126. https://doi.org/10.1038/s41598-017-09979-4
Taylor BL, Chivers SJ, Larese J, Perrin WF (2007) Generation length and percent mature estimates for IUCN assessments of cetaceans. Adm Rep LJ-07-01 24p. 10.1.1.530.4789
Torres-Florez JP, Hucke-Gaete R, Leduc R et al (2014) Blue whale population structure along the eastern South Pacific Ocean: evidence of more than one population. Mol Ecol 23:5998–6010. https://doi.org/10.1111/mec.12990
Valenzuela LO, Sironi M, Rowntree VJ, Seger J (2009) Isotopic and genetic evidence for culturally inherited site fidelity to feeding grounds in southern right whales (*Eubalaena australis*). Mol Ecol 18:782–791. https://doi.org/10.1111/j.1365-294X.2008.04069.x
Valsecchi E, Glockner-Ferrari D, Ferrari MJ, Amos W (2002) Molecular analysis of the efficiency of slough skin sampling in whale population genetics. Mol Ecol 7:1419–1422
Vu ET, Clark C, Catelani K et al (2015) Seasonal blubber testosterone concentrations of male humpback whales (*Megaptera novaeangliae*). Mar Mammal Sci 31:1258–1264. https://doi.org/10.1111/mms.12191
Wade P, Angliss R (1997) Guidelines for assessing marine mammal stocks: Report of the GAMMS workshop April 3–5, 1996, Seattle, WA. U.S. Department of Commerce, NOAA Technical Memoradum NMFS-OPR-12, 93 pp
Waldick RC, Kraus SD, Brown MW, White BN (2002) Evaluating the effects of historic bottleneck events: an assessment of microsatellite variability in the endangered North Atlantic right whale. Mol Ecol 11:2241–2249
Waples RS (2006) A bias correction for estimates of effective population size based on linkage disequilibrium at unlinked gene loci. Conserv Genet 7:167–184
Waples RS, Gaggiotti O (2006) What is a population? An empirical evaluation of some genetic methods for identifying the number of gene pools and their degree of connectivity. Mol Ecol 15:1419–1439. https://doi.org/10.1111/j.1365-294X.2006.02890.x
Waples RS, Hoelzel RA, Gaggiotti OE et al (2018) Guidelines for genetic data analysis. J Cetacean Res Manag 18:33–80
Whitehead H (1998) Cultural selection and genetic diversity in matrilineal whales. Science 282:1708–1711

Chapter 7
Social Organization of Baleen Whales

Peter L. Tyack

Abstract Baleen whales are specialized to find and exploit prey that form dense patches seasonally within broad ocean areas. Most populations have an annual reproductive cycle that separates their breeding season from their feeding season, often with long-distance migrations between feeding and breeding areas. The longest bond we know of for most baleen whale species is the mother-calf bond. Calves suckle for five to seven months in most species and wean by the next feeding season. The species whose social behavior has been most studied is the humpback whale, enabled by their distribution, which often is near shore, and their individually distinctive natural markings. Some humpback whales feeding on mobile prey form stable groups where each individual learns specific roles to perform coordinated group foraging. Aside from these groups, which may last for many years, most baleen whale groups are reported to be fluid with few strong associations between individuals other than mother and calf. However, most researchers define whale groups in terms of the number of whales that are close enough to be sighted within a certain distance of one another. Sound propagates so well underwater that whale sounds can be heard at distances of tens to hundreds of kilometers away. This means that whales may be able to maintain contact over much greater ranges than are usually assessed by human observers—they may form long-range "heards" in addition to shorter range "herds." The social organization of whales during the breeding season is structured in part by songs—acoustic reproductive advertisement displays. The potential scale of "heards" is indicated by the ability of scientists to track one singing blue whale for 43 days as it swam > 1700 km. Scientists will start to develop a fuller understanding of the social organization of baleen whales when they apply methods that can make observations and test hypotheses over the temporal and spatial scales at which baleen whales move and communicate.

Keywords Baleen whale · Social organization · Annual cycle · Feeding · Migration · Breeding and calving · Song · Effective range of communication

P. L. Tyack (✉)
School of Biology, University of St. Andrews, St. Andrews, UK
e-mail: plt@st-andrews.ac.uk

C.W. Clark and E. C. Garland (eds.), *Ethology and Behavioral Ecology of Mysticetes*, Ethology and Behavioral Ecology of Marine Mammals,
https://doi.org/10.1007/978-3-030-98449-6_7

7.1 Introduction

7.1.1 Annual Cycle of Baleen Whales

Most baleen whale species have an annual cycle of a winter breeding and calving season in low latitudes, migration to summer feeding areas in higher latitudes, and another migration back to the winter breeding and calving areas (Fig. 7.1). The selection pressures operating on social behavior likely differ during these different phases of the annual cycle, leading to the potential for differences in social organization in different seasons. Most baleen whales feed little if at all during the breeding and calving season, so selection focuses on maternal care of the young and on the mating system. The migration seasons select for efficient and safe movements over thousands of kilometers, but also may involve feeding, care of young, and behavior associated with the mating system. The primary feeding adaptation of baleen whales involves their ability to engulf large aggregations of prey in one mouthful, through ram filtration in balaenids (Simon et al. 2009), benthic and epi-benthic suction feeding in gray whales (Nerini 1984), or gulping in balaenopterids (Goldbogen et al. 2006). Baleen

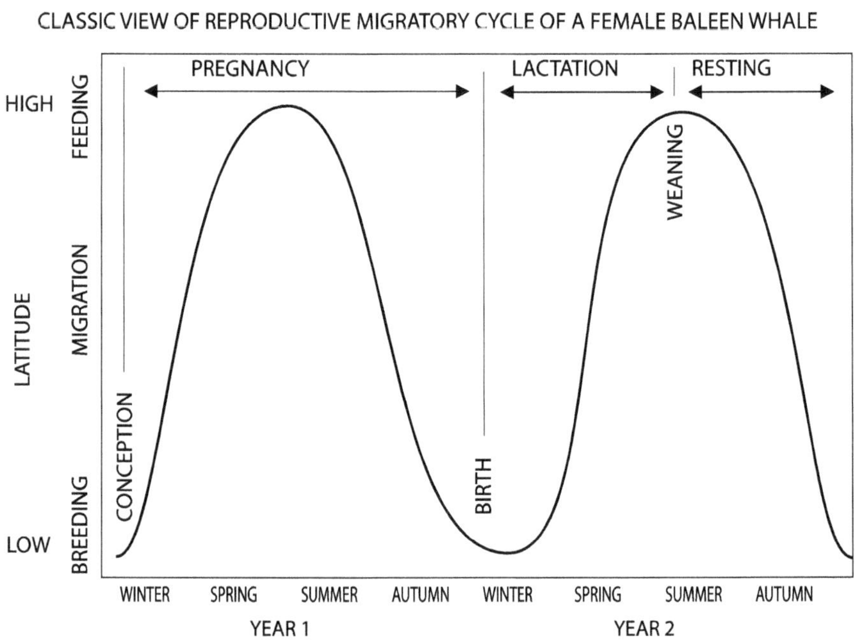

Fig. 7.1 Schematic timeline for the reproductive and migratory cycles of female baleen whales (adapted from Lockyer 1984). Some minke (*Balaenoptera acutorostrata*) and humpback whales (*Megaptera novaeangliae*) may conceive soon after birth, with no resting period, but balaenid whales have multi-year resting periods (adapted from Lockyer 1984). Some resident populations of baleen whale do not migrate.

whales have a variety of social adaptations to improve foraging efficiency, and social organization during the feeding season is driven primarily by foraging ecology. The abundance of dense patches of prey during summer in high latitudes is thought to represent the primary factor for migrating to summer feeding areas (Chap. 4).

The reasons for the migration of baleen whales to lower latitude breeding areas are less well understood (Corkeron and Connor 1999). Several hypotheses focus on energetics. Brodie (1975) argued for a metabolic advantage for whales swimming from high-latitude waters that are cold in winter to warmer lower latitude waters. However, most modeling studies suggest that baleen whales have a low energetic cost for maintaining body temperature in cold water. Payne (1995) argued that foraging opportunities might be better in lower latitudes than in higher latitudes during winter. Southern right whales (*Eubalaena australis*) feed on copepods during the breeding season on the Península Valdés, Argentina, calving ground (D'Agostino et al. 2016), and Norris et al. (1977) observed gray whales (*Eschrichtius robustus*) engaged in behavior that looked like foraging near their Baja California, Mexico, breeding grounds. However, the whaling literature conclusively shows for the most heavily whaled populations that foraging provides a minor contribution to energy reserves outside of the feeding season when fat reserves are laid down, as lipid stores decrease during the migration and breeding seasons (e.g., Chittleborough 1965 for humpback whales (*Megaptera novaeangliae*). Therefore, it seems unlikely that rare foraging opportunities drive selection for the fall migration to the breeding and calving grounds. The bowhead whale (*Balaena mysticetus*) has an unusual high-latitude range and may be an exception to this pattern. Many bowhead whales in the North Pacific spend the winter in the Bering Sea, which is a summer foraging habitat for gray whales. Stable isotope analyses suggest that they acquire much of their energy through foraging in fall and winter (Schell et al. 1989); foraging in spring is also important for bowheads in some regions of the North Atlantic (Heide-Jørgensen et al. 2013).

Most of the other hypotheses suggested for migration to low-latitude breeding grounds focus on neonate calves. Dingle (1996) argued that calves may face difficulties thermoregulating in cold high-latitude waters, but Watts et al. (1993) and Kanwisher and Sundnes (1966:408) argue that baleen whale calves "are theoretically insulated sufficiently for the coldest seas." Another hypothesis for migration to low latitudes focuses on safety for young calves. Whitehead and Moore (1982) point out that female humpback whales with calves prefer calm waters inshore of coral reefs off the Dominican Republic, an observation that is replicated in Madagascar by Ersts and Rosenbaum (2003) who report that humpback mothers with calves prefer shallow waters near shore. Parks (Chap. 13) summarizes evidence that right whale mothers with calves also prefer shallow coastal waters. This preference for shallow inshore waters may reduce the odds of detection by predators, such as a shark or killer whale (*Orcinus orca*), which are likely to come from offshore, and may reduce the predator's abilities to maneuver and attack from below (Ford and Reeves 2008). However, some baleen whale species that migrate to lower latitudes during the breeding season do not congregate in inshore waters. For example, some blue (*Balaenoptera musculus*) and fin whale (*Balaenoptera physalus*) populations are thought to disperse in offshore waters during the breeding season. Corkeron and Connor (1999) argue that reducing the risk of killer whale predation may be a prime driver for autumn migrations to lower latitudes for both inshore and offshore species. They cite data that the densities of killer whales are much lower in lower latitudes, and they suggest that pregnant female whales may migrate to the lower latitudes to reduce the risk of predation on their neonate calves.

Ford and Reeves (2008) agree with Corkeron and Connor (1999) that killer whale predation has been an important selection pressure for baleen whales. They propose that the inshore vs. offshore strategies for calving grounds may stem from two categories of baleen whale responses to killer whale attacks. They review evidence that five species of baleen whale, the southern and North Atlantic right whales (*Eubalaena glacialis*), bowhead whale (*Balaena mysticetus*), gray whale, and humpback whale, fight killer whales to protect themselves and their calves. By contrast, they argue that six balaenopterid species have a flight strategy. These flight species can sustain prolonged flight reactions at speeds of 20–40 km/h, but show little evidence of defense if overtaken. Ford and Reeves (2008) suggest that it is the fight species that seek shallow coastal refuges for their newborn calves. They argue that the flight species disperse into pelagic habitats where they can escape in all directions. They suggest that for this flight strategy to work in a calving area, young calves in the flight species must also be able to sustain prolonged rapid swimming.

The multi-year breeding cycle for most female baleen whales means that many adult females are not pregnant at the end of each feeding season. It is not known how many non-pregnant females of any baleen whale species migrate to the calving grounds for breeding. If they are not going to give birth to a calf, then the advantages of low-latitude calving grounds may not be relevant for deciding about migratory destinations. This raises the possibility that there might be breeding areas separate from the calving areas. In right whales, which tend to take several years after having a calf to conceive the next one, females tend not to be sighted on the calving grounds the year before they give birth, suggesting that they mate elsewhere (Chap. 13). If not enough years have passed for a female right whale to become pregnant again, she may choose winter habitats that are neither breeding nor calving areas. Here, the priority may be to continue feeding to reduce the time it takes to build up enough energy stores to reproduce again. In some other baleen whale species, the rate of post-partum ovulation is high enough (Lockyer 1984; Chittleborough 1958) to select for adult males to follow pregnant females for mating opportunities (Corkeron and Connor 1999). If males form a mating assemblage around the areas where females calve, non-pregnant females may have to migrate to the same area for mating. However, the reproductive advertisement displays of fin and humpback whales have been recorded during every month of the year on some feeding grounds (Vu et al. 2012; Morano et al 2012), suggesting a broader distribution of mating opportunities for females, some of whom may not migrate for mating.

The classic view of the migratory cycle of baleen whales (Fig. 7.1) stems from observations from the whaling industry, which focused on finding times and places when whales were concentrated in areas where they could efficiently be hunted. However, there is much more variability in baleen whale annual movements than suggested by this classic view. Most populations of humpback whales have an annual migratory cycle, but there is a population that resides all months of the year in the Arabian Sea (Mikhalev 1997). Similarly, genetically distinct populations of fin whales are mostly resident year-round in the Mediterranean and in other ocean regions, with seasonal movements to good feeding habitats (Geijer et al. 2016). As Geijer et al. (2016) emphasize, methods such as tagging, genetics, photo-id, and acoustic monitoring can identify the full range of migratory patterns in baleen whales. Understanding how these vary with varying ecological conditions can help sharpen our understanding of how and when baleen whales choose to reside or move, which will broaden our understanding of their social organization.

7.1.2 Mother-Calf Bond

The longest social bond known in most baleen whale species is the mother-calf bond. Mothers are thought to nurse their calf for five to seven months or more in most species (Oftedal 1997). This means that the calf will accompany its mother on the migration from the calving area to the feeding area (Fig. 7.1). The calf must be ready for migration after about three months on the calving ground. This requires an intense period of lactation during which a blue whale mother may produce 4000 MJ/d (Oftedal 1997) and a southern right whale calf may grow in volume by 0.08 m^3/day, while the mother shrinks by 0.13 m^3/day (Christiansen et al. 2018). In some species, the calf may be able to start feeding on prey during this first summer, and in most species, the calves appear to learn how to migrate to their feeding ground from their mother on this first migration (e.g., for humpbacks: Clapham and Mayo 1987; Barendse et al. 2013) and even specific regions within a feeding ground (Weinrich 1998). Studies of social learning define this kind of transmission of information from mother to offspring as vertical transmission, in contrast to horizontal transmission, which is defined as learning from peers (Chap. 8).

In some baleen whale species, calves may associate with the mother for up to a year. Valsecchi et al. (2002) analyzed genetic relationships of humpback whales migrating together and found mothers traveling with offspring estimated to be 11–12 months old, and Baker et al. (1987) sighted yearling humpback whales with their mothers on the Hawaiian breeding grounds. Taber and Thomas (1984) studied southern right whales on their calving ground and found that mothers returned in late winter with their yearling calves who remained with the mother for the first two to six weeks after return. Consistent with the idea that mothers put a high priority on protecting their offspring, Taber and Thomas (1984) found that young right whale calves spent nearly 90% of their time within ¼ body length of the mother, and the mothers were responsible for maintaining this proximity. By contrast, after a year of age, it is the yearlings who are primarily responsible for maintaining proximity. Even though mothers and calves spend most of their time close together, they do separate, with infants departing more often than approaching and mothers approaching more often than departing. This pattern reverses for yearlings (Taber and Thomas 1984).

When mother and calf are within ¼ body length, they can probably sense one another's proximity by sight or touch. Visibility is so limited in seawater, however, that they may lose visual contact during separations of more than a body length. The ability to reliably regain contact is critical for the lactating mother and dependent calf. Acoustic communication plays an important role in maintaining contact when whales separate, and contact calls have been identified for right whales and humpback whales. While making acoustic recordings, Clark (1983) observed the behavior of southern right whales, and he reports that in about half of the cases when right whales made an upsweep identified as a contact call, another whale called back and both whales exchanged upcalls as they approached one another, stopping calling once they had joined. Videsen et al. (2017) and Nielsen et al. (2019) attached acoustic recording tags on lactating female humpback and southern right whales, respectively, and showed that mothers and calves are more likely to call during active dives which are likely to be associated with separations. Contact calls are used by both species in a variety of settings, including adult male right whales seeking contact with adult females (Parks and Tyack 2005) and including humpbacks on the feeding grounds (Wild and Gabriele 2014). However, killer whale predators can detect the presence of whales by eavesdropping on their calls. Videsen et al. (2017) and Nielsen et al. (2019) report that the source levels of calls between mothers and calves are quite low, yielding an effective range of communication of 200 m or less. They argue that while mothers and calves must call to maintain contact during separations, they adopt a strategy of acoustic crypsis to prevent detection by more distant threats.

7.2 Feeding

During the feeding season, the social organization of baleen whales has been studied directly for species where individuals have distinctive natural markings and forage near shorelines accessible to field workers. Field workers typically describe whales as being in the same group if they are sighted within a few body lengths of one another with coordinated behavior. During the feeding season, most humpback whales are reported as alone or in pairs, with fluid fission/fusion groups seldom lasting for more than hours to days (e.g., Whitehead 1983; Clapham 1993; Ramp et al. 2010). One of the reasons suggested for the fluidity of groupings on a feeding ground is the high variability in the size of prey patches that may lead to variability in the number of whales that can be supported by each patch. The size of humpback groups feeding off Newfoundland is proportional to the size of the prey patch (Whitehead 1983). However, there is some evidence for social bonds in foraging baleen whales that are stable over several years. Some of the best evidence stems from humpback whales during the feeding season. Ramp et al. (2010) studied association patterns of humpback whales feeding in the Gulf of St Lawrence over nine years and report that some pairs of non-lactating females had stable associations that lasted for up to six feeding seasons. They hypothesized that whales that learn how to forage cooperatively may maintain stable associations of individuals who have learned to coordinate their social foraging. More detail on coordinated behavior in social foraging groups is available for humpback whales that feed on mobile euphausiid and herring prey in Southeast Alaska. D'Vincent et al. (1985) provide detailed observations of this social foraging specialization. They observed a group of eight humpback whales vertical lunge feeding in Southeast Alaska that not only was stable over three days, but "always surfaced simultaneously in an identical formation" and "each whale maintained a constant physical orientation and spatial relationship within the group" [p. 42]. Some of the vertical lunge feeding involved whales exhaling underwater to form bubble nets that are thought to constrain and/or concentrate prey (Sharpe and Dill 1997). Stereotyped vocalizations appeared to initiate this cooperative foraging and to synchronize simultaneous surfacing in the lunge-feeding group (D'Vincent et al. 1985). Sharpe (2001) studied patterns of association of humpbacks foraging in Southeast Alaska for five feeding seasons. Most of these whales had weak associations, but 25 of the 257 whales formed a stable association with one or more of the other 24 whales, which he called a core community. The sex of 19 of these 25 whales was determined genetically: 11 were male and eight were female. Sharpe (2001) found no relationship between the coefficient of association and genetic relatedness of these whales. Pierszalowski (2014) tested the genetic relationships of humpback whales in large groups on these foraging grounds and found "that feeding in large stable groups did not typically occur among closely related individuals" [p. 105]. Calves born to mothers in the core community were not found to feed preferentially with their mothers in following years. When members of this core community were foraging in large groups, these groups usually included at least one whale that was not part of the community. In such cases, non-community whales associated with a core group for an average of 1.9 days, but most of these visitors were only observed feeding with a community group on one day.

When the core group of humpback whales studied by Sharpe (2001) was lunge feeding, specific whales produced the vocalizations used to coordinate foraging, individual whales dove in the same order, and individuals maintained a specific position within the group. These observations led Sharpe (2001) to suggest that this form of group foraging may require individuals to learn the timing and path they must swim to maintain the spatial configuration of the group during lunging, as well as roles such as bubble blower, vocalizer, and herder. The structure of these groups suggests that social learning of these roles involves horizontal transmission of information. The actual dynamics of learning have not been documented in this case, but they were observed for a novel feeding behavior in the Gulf of Maine. One whale was observed in 1980 lobtailing just before bubble net feeding. The high numbers of researchers observing whales in the Gulf of Maine made it possible to track the spread of this lobtail feeding behavior over 27 years. Allen et al. (2013) analyzed when individual whales started lobtail feeding coupled with data on the amount of time different individuals associated with one another to test different models of social transmission of the behavior. These models showed that whales who associated with lobtail feeders were more likely to start lobtail feeding, providing strong evidence for social transmission. Having a mother who engaged in lobtail feeding did not increase the chances of the calf lobtail feeding. These results show that horizontal transmission, not vertical transmission, is most likely responsible for the acquisition of lobtail feeding. As in the case of social foraging in Southeast Alaska, horizontal social transmission may speed the ability of whales to learn behaviors that are tuned to local conditions (Chap. 8).

Outside of these cooperative groups, baleen whales often congregate as they feed on patches of prey, but the groupings are fluid with few strong associations between individuals other than mother and calf. It is not well understood how baleen whales find dynamic patches of concentrated prey as they search over large distances. It is possible that individual whales each find good prey patches on their own, with congregations resulting from several animals independently finding the same patch. Alternatively, Payne and Webb (1971) hypothesized that if a whale finds a patch of food that is too large for it to consume alone, it could call to broadcast the location to other whales nearby. If nearby whales reciprocated, Payne and Webb hypothesized that this could improve the foraging efficiency of the local population. Several species of whale do in fact call when feeding on a large patch (e.g., fin whales; Croll et al. 1992) or when feeding in groups (e.g., blue whales; Oleson et al. 2007). Watkins and Schevill (1979) used an airplane to observe fin whales approaching a large group of feeding whales, making a beeline from distances as far as 10 km away, and starting to feed immediately as soon as they joined the group. This hypothesis that whales broadcast the location of prey would change our understanding of social organization of baleen whales on the feeding grounds, but it has not been tested systematically.

7.3 Migration

The migrations of many baleen whales cover thousands of kilometers and may take considerable energy and a significant proportion of the annual cycle. Migratory behavior has been best studied for species that migrate along a populated coast where whales can be visually observed, such as the migration of gray whales off the west coast of North America or of humpback whales along the east coast of Australia (Chap. 4). It has long been observed that different age/sex classes may migrate at different times. For example, in most species, pregnant females are the first to arrive at the summer feeding ground and the last to leave, probably because of the high energetic demand of pregnancy and lactation, while lactating females with calves are late to arrive, perhaps, slowed down by the calves (Lockyer 1984). Many species tend to migrate in small groups rather than alone, but Valsecchi et al. (2002) found no indication that whales swimming in the same group tend to be more related, except for groups containing a mother-calf or mother-yearling pair. This suggests that most groups on migration are transitory and not based on kinship. Migrating bowhead whales have been recorded exchanging calls as they navigate around ice, with whales closer to the obstacle potentially signaling the route to those behind (Würsig and Clark 1993). This indicates acoustic coordination among whales that would not be judged to be in the same group by most field workers based on visual observation.

At the end of the breeding season, when whales start to migrate to their feeding grounds, there is not a sudden complete stop to breeding behavior. An example comes from humpback whales that have been observed closely during their migration off the east coast of Australia. Humpback song has commonly been recorded during this migration as well as during the breeding season, and many of the interactions observed between migrating singers and other migrating humpback whales were the same as those observed between singers and other animals on a breeding ground (Smith et al. 2008). Data from whaling off the east coast of Australia showed a peak of ovulation during the winter breeding season in August, but some females ovulated up to three times in one year, which may extend some ovulation into the migration season (Chittleborough 1965). Male humpbacks have a broad reproductive season with spermatozoa present from June to October in the Southern Hemisphere (Chittleborough 1965), and song has been recorded outside of the classical breeding season (Vu et al. 2012; Garland et al. 2013). These observations suggest that at least for some baleen whale species, mating may take place outside of the classic breeding season.

Similarly, as whales migrate from the breeding grounds to the foraging grounds, they may take advantage of foraging opportunities on the way. For example, a bloom of phytoplankton starts at around 35 ° North latitude in the North Atlantic in December and January and spreads northward through the spring (Visser et al. 2011). The timing of this bloom and its northward spread varies from year to year. Visser et al. (2011) studied blue, fin, humpback, and sei (*Balaenoptera borealis*) whales sighted off the Azores at 38° N over a four-year period. The whales were observed feeding on krill in the area, and the timing of whales sighted off the Azores tracked the onset of the spring bloom with time lags of 13–16 weeks. These observations suggest that whales time their migration to take advantage of opportunities to forage on zooplankton prey species whose density is linked to the timing of the phytoplankton bloom. Silva et al. (2013) tagged fin and blue whales off the Azores and documented traveling behavior consistent with migration as well as whales remaining in a restricted area, a behavior consistent with foraging.

7.4 Breeding and Calving

The duration of pregnancy in most baleen whales is about 12 months, so the breeding season is also the season when calves are born. As discussed above, one of the primary hypotheses for the baleen whale migration to lower latitudes during the breeding season is to provide a safer place for adult females to give birth and to care for their neonates without disturbance. Ford and Reeves (2008) distinguish two baleen whale anti-predator strategies. They review evidence that several baleen whale species, including balaenid whales, as well as the gray whale and the humpback whale, physically defend calves against killer whales. In most of these "fight" species, females with neonates seek out calm inshore waters at lower latitudes than the foraging grounds. For example, gray whale mothers and calves tend to stay far inshore in lagoons on the Mexican breeding ground, with mating behavior concentrated near lagoon mouths (Norris et al. 1977). South Atlantic right whale females with calves concentrate in shallow inshore waters (Best 1990). The bowhead whale forages in high-latitude Arctic waters and spends the winter breeding season in the lower latitudes of the shallow Bering Sea (Braham et al. 1980) and about 65° N in the Davis Strait (Reeves et al. 1983). Rather than seeking shelter inshore, they appear to use pack ice as a refuge. Several studies of the habitat frequented by different humpback groups during the breeding season found that mothers with calves preferred shallow inshore waters compared to habitats preferred by other groups (Martins et al. 2001; Ersts and Rosenbaum 2003; Félix and Botero-Acosta 2011). Other balaenopterid whale species show a flight strategy involving rapid swimming away from killer whales. These other balaenopterid whale species, such as blue, fin, minke (*Balaenoptera acutorostrata*), and sei whales, appear to disperse in temperate pelagic waters, while Bryde's whales (*Balaenoptera edeni*) rarely venture outside of tropical pelagic waters.

Balaenopterid species that depend upon flight reactions are not known to rely on social defense against predation, while species that defend themselves against predators will group tightly if under attack (Ford and Reeves 2008). However, even for fight species, mothers with calves do not appear to form large groups to provide a social defense when not under attack from predators. For example, very few humpback groups are sighted with more than one neonate calf in the group (e.g., Martins et al. 2001; Félix and Botero-Acosta 2011), and between-group distances are greater for humpback groups with a calf than those without (Smultea 1994). Females with young calves may only spend a few months in the calm protected waters, a relatively short time for the calves to grow enough for efficient migration. Humpback whale calves were observed to spend about 20% of the time in a suckling position, while females are often resting (Videsen et al. 2017). Their vocalizations are so faint as to be detectable only out to a range of ~ 100 m. These observations are consistent with the hypothesis that mothers with calves are moving to a refuge location where threats are less likely and relying on a strategy of crypsis to maximize undisturbed time for resting and suckling.

Predators can pose a lethal threat for neonates, but adult males also may disrupt normal behavior of mother-calf groups. Elwen and Best (2004) report that neonate southern right whales were less likely to strand in areas dominated by mother-calf pairs than in those where adults without calves are more common. The segregation of mother-calf groups inshore of other humpback groups on the breeding ground may represent a strategy on the part of mothers to avoid disturbance by males as well as to avoid predators. A mother and calf humpback are seldom sighted with another female on the breeding ground, but may be sighted with one or more males. An increase in activity is often observed when a mother and calf are in a group with more than one adult humpback male, suggesting that the mother and calf expend more energy in these groups (Sullivan and Cartwright 2009). The group is usually less active when a single adult male escorts a mother and calf, suggesting that females may also have a strategy of selecting or accepting one male as an escort to reduce harassment from other males (Mesnick 1997). This is not a long-term association, however, as females are seen associated with different males throughout the breeding season, and males

Fig. 7.2 Structure of surface-active groups (adapted from Tyack and Whitehead 1983). NA = nuclear animal, PE = principal escort, SE = secondary escort

associate with different females (Baker and Herman 1984). Clapham and Palsbøll (1997) used genetic analyses to demonstrate that females mate with different males to father their sequence of calves across different years.

Groups of more than two adult humpback whales on the breeding ground often have a clear structure in which one whale, the principal escort (PE), swims close to a nuclear animal (NA) who is the focal point of the group (Fig. 7.2). Secondary escorts may challenge the principal escort by trying to swim between the NA and PE; these challenges often involve agonistic displays and physical contact between the escorts (Tyack and Whitehead 1983; Baker and Herman 1984). Clapham et al. (1992) determined the sex of whales in 21 competitive groups. None of the groups contained more than one adult female, but 7/21 contained only males. All of the whales that challenged the PE were male, 23/24 PEs were male, and of the 22 whales identified as the NA, 17 were female, and five were male. Clapham et al. (1992) argue that these results suggest that some of these groups likely involve males competing for dominance, but that most of the groups involved a PE guarding a NA female to limit other males gaining access to the female. Clapham et al. (1992) also observed pairs of males joining or leaving competitive groups together with no aggression toward one another. They and Darling et al. (2006) argue that some male humpbacks may form coalitions to improve their ability to compete in these competitive groups. We do not know how much choice the female has in choosing either a single male escort, or principal escort in a competitive group, but the groups of males competing for access to the female represent male-male competition that may reduce the ability of the female to choose a mate over the short term.

Groups with competitive behavior similar to that observed in humpbacks have been observed in other baleen whale species. Norris et al. (1977) report observing mother-calf pairs of gray whales in groups in the Mexican breeding lagoons that were involved in high-speed chases in which animals thought to be males were engaged

Fig. 7.3 Surface-active group of right whales (from Roman et al. 2016)

in violent competition. During one such chase, Norris et al. observed a young calf struggling to keep up with three adults. The next day a calf was found partially stranded, suggesting that these chases may involve greater risks than just energy loss. Detailed observations of competitive groups are not extensive for right whales during the breeding season, but groups with surface-active behavior (surface-active groups or SAGs) that appear similar to those observed in humpbacks have been observed in right whales during the feeding season in the Bay of Fundy (Fig. 7.3; Kraus and Hatch 2001; Chap. 1, Fig. 1.6). While some SAGs are all-male or all-female, most involve a central female or multiple females surrounded by males (Parks and Tyack 2005). Parks and Tyack (2005) used an array of hydrophones to identify which individual whale produced some of the calls in SAGs. This evidence suggests that one type of call, referred to as a scream, is produced in extended sequences by a female in the group. Right whales produce an intense impulsive sound called the gunshot; when produced in stereotyped sequences by males, it may function as an advertisement to attract females and/or a threat to other males (Parks et al. 2005). Playback of gunshot sounds recorded from a lone North Atlantic right whale male elicited no approaches from North Atlantic right whales, but males did approach playbacks of calls recorded from SAGs (Parks 2003). Similar responses were observed from Southern right whales during playbacks of Southern right whale sounds recorded from active groups (Clark and Clark, 1980). The upcalls that are thought to function as contact calls are also heard in SAGs. Data from an array of hydrophones show that some males produce gunshots when entering a group or when searching for a female, perhaps, to announce their presence to the female or other members of the group.

Brownell and Ralls (1986) argue that sperm competition may play an important role in the mating system of right, bowhead, and gray whales. Right whales have the largest testes as a proportion of body mass of any baleen whale, a trait that has been linked to sperm competition in mammals. Right whale males are less aggressive in right whale SAGs compared to males in humpback or gray whale SAGs, and multiple right whale males have been observed to mate with a female in surface-active groups. Kraus and Hatch (2001) observed right whale males joining SAGs from distances as far as 8 km. The observations that females produce screams and that males approach playback of calls from SAGs are consistent with the hypothesis that female right whales may call to attract males and incite male-male competition expressed in part as competition for access to the female and in part as sperm competition.

The most detailed observations of right whale SAGs come from the feeding season. Mating has been observed in these SAGs, but Kraus and Hatch (2001) point out that in the Northern Hemisphere most conceptions should occur from October through December, a period with few observations of northern right whales. Therefore, many of the females calling to males in SAGs are probably not receptive during the feeding season. If mating does not lead to conception in the feeding season SAGs, what is their function? Kraus and Hatch (2001) suggest two possibilities:

1. Females may keep track of the performance of different males, make a choice, and mate with the selected male at a later time.
2. Females may require repeated practice to learn how to assess and select males in SAGs, and it may be lower risk to assess them during the feeding season before the stakes become high during ovulation.

The presence of calves and juveniles in these SAGs also suggests that SAGs may provide opportunities for play or practice, and that they may function to reinforce social bonds (Parks et al. 2007).

A glaring knowledge gap in understanding the mating system of baleen whales stems from our lack of direct data on mating. In spite of extensive observer effort on the breeding grounds, observation of copulation has not been reported for humpback whales. Sexual activity has been observed in surface-active groups of right whales, but females involved in mating in these groups are very rarely seen with calves the next year, and the timing of births suggests that most mating occurs outside of the summer season in which most observations of right whale SAGs have taken place. For species that disperse into pelagic seas, we know even less about where mating takes place. The problems described above suggest that better observations of mating behavior are unlikely for many species, and even where they are available, they may not correctly identify the mating event that led to pregnancy. Genetic methods such as those used by Clapham and Palsbøll (1997) are more likely to provide the necessary data, especially when linked to observational data (e.g., Cerchio 2003).

7.5 Reproductive Advertisement Displays and Song

In many species of baleen whale, the mating strategies of males include producing reproductive advertisement displays that are often referred to as songs. The functions of reproductive advertisements in animals have been discussed for more than

150 years, with Darwin (1859, 1871) arguing that they evolve by sexual selection. Songs can advertise a challenge to other males within earshot, with intra-sexual selection operating between males. Females may also use song to select a male for mating, with inter-sexual selection operating between the sexes. Males may sing for either or both of these functions. The acoustic features of many animal songs include complex sequences of notes that are repeated. We all have experiences listening to the songs of birds, and scientists have written about bird song for more than 300 years, but it was just 50 years ago that Payne and McVay (1971) published a scientific paper describing that whales sing as well. Perhaps, the best-known whale song is the song of the humpback whale. The frequency range and tempo of humpback song are similar to those of human music, as is its complex repeated phrase structure. It is so accessible to our ears that in 1970, a vinyl recording of humpback song sold more than 100,000 copies. See Chap. 11 for more information on the acoustic structure of humpback song and Chaps. 9–15 for detailed species-specific song or call repertoire information.

Evidence that whale songs are associated with reproduction comes from studies discussed in Chaps. 9–15 showing that songs tend to be produced during the breeding season and that singing whales whose sex has been determined have been male. The humpback is the whale species for which we have the most detailed observations of interactions between singing whales and other whales, as is discussed in Chap. 11. These observations provide evidence that song functions in male-male competition and also for female choice of a mate. Darling et al. (2006) used visual observations along with listening underwater to report that 80% of singers on the Hawaiian breeding grounds were lone whales, 12% were escorting a mother and calf, and 8% accompanied an adult without a calf. The most common interactions observed by Darling et al. (2006) involved singers joining lone males; the other 11% involved singers stopping singing and approaching and joining groups of males competing directly for access to a female. Smith et al. (2008) studied interactions of humpbacks singing during the migration along the east coast of Australia and were able to distinguish whether a singer joined other whales or whether the other whales joined the singer. They report that singers were more likely to join mother-calf groups than any other kind of group, and they sang for a longer time when with mother-calf groups than with other groups. By contrast, the whales that joined with singers tended to be lone whales (those whose sex could be determined were males); the singer usually responded by stopping, and the interactions were briefer than when a singer joined with a mother-calf pair. There are rarer reports of female humpbacks approaching and joining a singer. Medrano et al. (1994) report two cases when a mother and calf or a female with one other adult joined a singer with no signs of aggressive behavior, and Tyack (1981) reports a case of a mother and calf approaching a singer who stopped singing and also approached the mother and calf. After they joined, behaviors were observed that have been associated with courtship and mating in baleen whales. The observations of Smith et al. (2008) are that 27 of 63 observed associations were with lone adults (10 of which were males), 20 of 63 observed associations of singing whales were with mother-calf pairs, and the observations of females approaching singers all suggest that both males and females are audiences for the song and that song mediates both intra-sexual and inter-sexual interactions.

Research on interactions between songbirds during the breeding season is much more detailed and more extensive than that on whales, which can provide a useful comparative perspective. Females of some songbird species prefer mates with a familiar local dialect, and males may have difficulties establishing a territory if they have a foreign dialect. The idea that song dialects play a role in reproductive isolation of different populations led to studies (reviewed in Slabbekoorn and Smith 2002) of how well variation in song dialects corresponded with genetic boundaries. Some of these studies report a match between acoustic and genetic variation, but many others report that song dialects do not match well-defined genetic differences. One important reason for so much variation is that many species of songbird learn their song. If a male can disperse to a new dialect zone and learn the new dialect, this will break the connection between any natal genotype and his new local song phenotype. Learning takes place over much shorter time scales than genetic differentiation of a population, so dialects may be established through social learning well before there is any genetic differentiation.

The way in which the songs of some baleen whale species change provides evidence that they are also learned. The strongest evidence comes from the way most of the humpback whales within a region sing similar songs at one time, with the song evolving over time (Payne et al. 1983; Payne and Payne 1985), and the way whales in one area can rapidly adopt song elements from another area (Noad et al. 2000; Garland et al. 2011). Chapter 12 describes how bowhead whales also sing complex songs that change over time, but with much more variability than humpbacks in song from different whales within an area at any one time. Chapter 9 describes that along with a basic pattern of stability in song units of blue whales within a region, and of stability in how these units are sequenced, there also have been long-term changes in the frequency of some tonal units and also of timing of units in some areas. All of these observations are hard to explain unless these whales learn acoustic features of their song from hearing the song of other whales within earshot.

The great interest in conservation of baleen whale populations that were decimated by commercial whaling led many marine bioacousticians to explore the potential for using song dialects in whales as a proxy for biological population structure, as defined by genetic differences. As with the earlier work on bird song, the data do not suggest a strong match between genetic definitions of populations and song dialect boundaries. Multiple song dialects have been recorded in one population of Southeast Pacific blue whales (Buchan et al. 2015). As described in Chap. 9, multiple populations sing one song type among Antarctic blue whales and another song type is shared among North Atlantic blue whale populations. As with songbirds that learn their song, the pace of change of learned song is so much faster than changes in the genetic structure of populations that there are many reasons to expect that song dialects may not map perfectly onto the genetic structure of populations. The sounds of blue and fin whale song may carry hundreds of kilometers, so whales may hear one another's song over distances well beyond what we normally think of as sympatric. Song features that are audible across long ranges may be learned across ocean basins that contain several populations. Some higher frequency acoustic features of humpback song are not audible across ranges of more than tens of kilometers, but migratory patterns may also support cultural diffusion across greater ranges than genetic diffusion. For example, if humpback whales in the South Pacific hear songs when populations overlap on the feeding ground and then return to breed on their natal breeding ground, then song may be shared across populations if one singer picks up song from another population (Garland et al. 2013).

This review suggests that song dialects cannot be used as a reliable indicator of population structure in whales, just as they cannot in most songbird populations. It would, therefore, be problematic to manage protection of biological populations of whales based upon song dialects. Modern genetic methods enable measuring actual biological population structure using tissue that can be collected by non-invasive or minimally invasive methods. Song dialects may play a role in establishing geographical strategies for sampling whale genetics, even if they are not sufficient to define populations. Brakes et al. (2019) argue that animal cultures are worthy of conservation as well. This is the setting in which learned song dialects may prove most useful for conservation biology.

As described above, many studies of birdsong have focused on the role song might play as a reproductive isolating mechanism, testing whether song dialect boundaries matched population boundaries. Another hypothesis has focused on the ways sound propagation and ambient noise may provide selective pressures on the acoustic structure of songs, with different acoustic features propagating better in different habitats. Date and Lemon (1993) proposed that the acoustic features of habitats may select for the formation of song dialects, and Slabbekoorn and Smith (2002) proposed that if different habitats select for different traits within a species, that song can act as a marker for females to select a male best suited for that habitat. Boncoraglio and Saino (2007) performed a meta-analysis of 26 birdsong studies and found weak support for this acoustic adaptation hypothesis, but other factors such as energetic costs of singing and risk of eavesdropping by predators and parasites are likely also important. In the next section, we consider how the acoustic structure of whale songs may be affected by the ecological and acoustic setting in which they are produced.

7.6 The Ecological Setting in Which Songs of Different Whale Species Are Produced Has Selected for Their Acoustic Structure

Propagation of sound in the ocean is driven by different factors than the terrestrial habitat features, such as dense vegetation, that are thought to affect bird song, but there is a clear pattern where the acoustic features of songs of different baleen whale species appear to be selected to propagate well over the ranges required for effective communication (Clark and Ellison 2004). As discussed in the introduction, Ford and Reeves (2008) review evidence that species such as humpback and bowhead whales, that fight to protect their young, tend to congregate during the breeding season in high density in areas where they can find refuge from predators. By contrast, many populations of flight species, such as blue and fin whales, disperse into temperate and subtropical pelagic waters during the winter breeding season. The difference in expected range from singer to potential receivers between these two settings may have influenced the evolution of acoustic properties of the songs of these species.

Sound can propagate very efficiently in the deep ocean environment where blue and fin whales tend to sing during the breeding season. Their songs are relatively intense, with source levels of about 180 dB re 1 μPa m and have most of their energy well below 100 Hz. At frequencies this low, there is very little absorption of sound energy as it passes through seawater. The fundamental frequencies of finback and blue whale song also tend to center in a 10–30 Hz band at which Curtis et al. (1999) report particularly low levels of ambient noise deep in the North Pacific. The combination of high source level, low propagation loss, and low ambient noise means that fin and blue whale songs can be detected at long ranges. The effective range for detecting fin whale song in the deep ocean has been estimated using acoustic propagation models to be 400–1000 km (Spiesberger and Fristrup 1990). Using the US Navy Integrated Undersea Surveillance System (SOSUS), which is optimized for long-range detection of low-frequency sounds in open ocean, Clark and Gagnon (2002) report being able to detect, localize, and track a singing fin or blue whale at ranges of nearly 3000 km. Blue whale songs have been localized in the Antarctic at ranges of 200 km using recorders placed at depths that are less well suited for long-range detection than those of the SOSUS (Širović et al. 2007).

Humpback whales tend to select shallow coastal waters for their breeding grounds, and the Bering Sea breeding grounds of bowhead whales are quite shallow with an average depth of about 100 m. In shallow waters, whales cannot increase their range of communication simply by lowering the frequency of song (Mercado and Frazer 1999). In shallow water, there is a cutoff frequency below which sound will not propagate well; Mercado and Frazer (1999) estimate this frequency to range between 1–100 Hz for the water depths where humpbacks usually sing, suggesting selection for higher frequencies. Clark and Ellison (2004) suggest that in these shallow waters, the frequencies from 100–400 Hz offer a low noise window for communication. Cholewiak et al. (2018) modeled the effective range of communication for songs of fin and humpback whales in the relatively shallow waters of Stellwagen Bank in Cape Cod Bay. Even though the average source level of finback whales was 180 dB re 1 μPa m and that of humpback song was 170 dB re 1 μPa m, the higher frequency humpback song had a modeled effective range of 23 km versus 15 km for fin whale song under normal ambient noise in Cape Cod Bay.

Tervo et al. (2012) studied the songs of bowhead whales recorded in March in Disko Bay on the west coast of Greenland, which has an average depth of 200 m. The mean source level of bowhead song was 185 dB re 1 μPa m, which is within the range reported for fin and blue whales (Širović et al. 2007). Thus, the main difference for estimating the effective range of bowhead songs stems from the higher frequencies (mean frequency of 444 Hz) and broader bandwidth (284 Hz) of the bowhead song compared to fin whale song, whose main element ranges from 17 to 25 Hz. During the March recording period, ice cover in Disko Bay reduced ambient noise levels to 65 dB re 1 μPa over the frequency band of bowhead song and 51 dB re 1 μPa over the much smaller frequency band of finback song.

Using information on sound propagation in ice-covered Arctic seas, Tervo et al. (2012) estimate an effective range in this quiet environment of 130 km for bowhead song and about 3500 km for finback song. These are likely overestimates because of the very low noise levels used and because the local sound propagation conditions would not hold for 3500 km, which is greater than any distance in the Arctic Ocean, but the main point is that the songs of bowhead whales have about 4% the range of that of fin whales. High-frequency (> 1 kHz) components of bowhead song, which are likely important features of the complex song repertoire, would have effective ranges well below 40 km, similar to that estimated for humpback song by Cholewiak et al. (2018).

Fin and blue whales are thought to disperse into low-latitude seas during the breeding season and must rely upon the long effective ranges of their songs to structure mating interactions, while humpback and bowhead whales congregate in breeding grounds where effective ranges of several tens of km are likely sufficient to reach intended receivers. Whale researchers sometimes use the words "breeding grounds" because this was the term whalers used for areas where whales were concentrated enough for whaling to be effective. Whale researchers have continued to return to many of these areas as they are good for studying whales as well, but we must remember that we do not know where many whale species breed, especially for species that do not congregate inshore.

If fin and blue whale songs were selected for long-range propagation, this may have affected other features of the song such as their more stable, simple repeated elements of the song, which may make it easier to detect at low signal-to-noise ratios. By contrast, whales listening to singing humpback and bowhead whales are often close enough to hear song well above the noise, and this may enable selection for more variable and complex songs.

7.7 Future Directions

7.7.1 Do Whales Form Herds or "Heards" (Payne and Webb 1971)?

The discussion about effective range of communication raises questions about how to define a group of whales and interactions between whales. Most of the studies on associations between individual whales define groups in terms of distance between animals and/or animals coordinating their activities (Mann 2000). The appropriate distance to define whether animals are interacting depends upon the activity. For a male and female to mate or for a calf to suckle from its mother, they must be touching one another. The coordinated foraging described for humpback groups above requires animals being within a body length or so of one another. This focus on close contact led Taber and Thomas (1984) to compare the amount of time mother and calf right whales spent more than ¼ body length away (16%) versus two body lengths away (1.5%). One would reach different conclusions about the association patterns of mother and calf depending upon which of these close distance measures one used to define a mother-calf group. As discussed above, calves may temporarily leave their mother, swimming tens to hundreds of meters away. But, the calf is dependent on the mother, and when separated beyond visible range, the mother and/or calf can use calls to reunite. In keeping with their cryptic strategy, right and humpback whale mothers and calves vocalize little outside of separations, and they tend to produce faint calls only audible at ranges of < 200 m (Nielsen et al. 2019; Videsen et al. 2017). These calls are so faint that they would seldom be audible to biologists recording them from a boat nearby; recording these whispered vocalizations reliably requires use of acoustic recording tags attached to the mother or calf. The distances between group members maintaining contact using whispered calls should be close enough that observers following whales visually from a vessel should be able to see all members of the group. By contrast, non-lactating adult female right whales on the calving grounds tend to produce higher amplitude calls such as upcalls (Parks et al. 2019). Munger et al. (2011) suggest that upcalls of North Pacific right whales are detectable at ranges of up to 100 km in the Bering Sea, appropriate for allowing widely dispersed whales to maintain contact. Payne (1995) used aerial surveys to find that right whales in the calving grounds of Peninsula Valdés, Argentina, formed loose herds about 37 km in diameter. These loose aggregations, which are within the effective range calculated for right whale high-amplitude calls, seem to define a higher level and larger scale of grouping than is typically used for baleen whales. Perhaps another important definition of being associated is being within range of such a contact call. The discussion of effective range of song also shows that whales may monitor one another at distances of tens to hundreds of kilometers. Tyack and Whitehead (1983) present an interaction in which a humpback whale singing in Hawaiian waters, stopped singing, accelerated, and made a beeline to a surface-active group of whales that was 9 km away. It took 40 min for the ex-singer swimming at high speed to join the surface-active group and understanding the interaction required following this individual whale for over four hours. This ability to notice that the singer was swimming toward a distant group and to pinpoint the location of whales was possible for boat-based observers following individual whale groups working with shore-based observers on a hill using a theodolite to pinpoint the locations of all of the whale surfacings, but 9 km is on the outer limits of the observation range of this method, which is limited to daylight hours, so it cannot track interactions for more than a day. The interaction took so long over such a large spatial scale that researchers only could understand it after it was mapped out after the fact. Field researchers need to be aware that methods are not available to study the range at which a female fin whale, for example, may assess the songs of male fin whales tens or even hundreds of km away. When I first heard of such long-range communication in whales, I recalled a quote from Henry David Thoreau "We are in great haste to construct a magnetic telegraph from Maine to Texas; but Maine and Texas, it may be, have nothing important to communicate" (Thoreau 1854, p. 58). But tag data and acoustic tracks show that fin and blue whales can swim > 100 km in a day (Ray et al. 1978; Clark et al. 2019), so a female might assess dispersed males over these ranges when selecting a mate and then swim to the selected whale. We cannot fully understand the social organization of baleen whales until we can follow such interactions, which may take days to unfold over hundreds of kilometers. Analyzing such social communication interactions will require methods specifically designed to observe interactions that most likely operate over scales of time and space that are much larger than our terrestrial intuitions. Developing these methods will be an important challenge for the next generation of researchers studying the outer limits of social organization in baleen whales.

We have similar limitations for studying social coordination of foraging behavior in baleen whales. Some baleen whales vocalize frequently on the foraging grounds, and these sounds are detectable at ranges of 5 km or more (Cholewiak et al. 2018). These ranges may not be as large as those for songs in the deep ocean, but they are well beyond the typical sighting range of vessel-based observers. As discussed in the foraging section, Watkins and Schevill (1979) used a small airplane to observe whales foraging over two days on large schools of small fish associated with a dense patch of plankton about 5 km long. More than 20 finback whales and one humpback whale were observed feeding on the fish, while two right whales and a sei whale foraged on the plankton. No other whales were sighted after a search within 20 km on the first day, but two groups of three and six fin whales were sighted on the second day about 7 and 10 km away from the large group of foraging whales. These two groups were swimming rapidly, headed directly toward the foraging whales. The group of three whales began feeding as soon as they arrived at the fish schools. These observations suggest that the groups of three and six whales were alerted by calls of the feeding finbacks as to the foraging opportunity 10 km away. Is it possible that whales over hundreds of km^2 might be able to monitor when whales find large patches of prey as one dispersed feeding group (as suggested by Payne and Webb 1971)?

An important adaptation of baleen whales involves their ability to cover large areas of ocean to find dense patches of evanescent prey (Goldbogen et al. 2019). We do not know how whales find these patches. Payne and Webb (1971) suggested that when whales do find a patch larger than could be consumed by their group during the period when the patch is dense enough for filter feeding, there may be relatively low cost for advertising the location of a difficult-to-find patch in settings that would select for reciprocity. The group size of feeding humpback whales has been shown to be proportional to the size of the prey patch (Whitehead 1983), but how with their fluid fission/fusion groupings do whales aggregate into appropriately sized groups at a patch? The methods marine mammologists have used to study foraging in the past are not well suited to studying these scales, but passive acoustic monitoring of calling behavior coupled with observations of the prey field and of foraging and traveling whales could address the hypothesis that baleen whales may broadcast the location of foraging opportunities and other whales may use these signals to find large evanescent patches of prey.

Baleen whales do not have the specialized high-frequency echolocation systems that evolved in toothed whales. However, it has been suggested that they might use lower frequency sound to listen for echoes from their own sounds to locate landmarks during migration as discussed by Clark and Ellison (2004) or dense patches of some kinds of prey during the foraging season. There have also been suggestions that two or more whales might use sound together to locate prey schools. Weston (1967) and Diachok (2000) report that when sound of about 1 kHz in frequency propagates underwater through schools of fish with air-filled swim bladders, sound is attenuated much more than when it passes through seawater. Tyack (1997) reviews evidence that whales separated by one to tens of km might be able to detect schools of fish with air-filled swim bladders by detecting attenuation of vocalizations when a fish school comes between caller and receiver. Gong et al. (2010) show how a receiver at some distance from a ~ 1 kHz sound source can detect echoes from fish schools to a range of 100 km. This suggests that a whale might be able to detect fish schools by listening for echoes from the sounds of other whales. If whales in groups use sounds in these ways to find prey patches, future testing of these hypotheses would change our understanding of whale sociality and foraging ecology on the feeding grounds.

7.7.2 Do Baleen Whales Have Long-Term Individual-Specific Social Relationships Beyond Mother-Calf?

In my 1986 review of social behavior in cetaceans, I argued that, except for the mother-calf bond, individual-specific relationships appeared to be the exception to the norm in baleen whales (Tyack 1986). Certainly, if groupings of whales are defined by being sighted together on a feeding ground, during migration, or on a breeding ground, then groups appear to be very transient, with few stable associations between individuals outside of special cases such as the humpbacks in Alaska and the Gulf of St Lawrence that feed in coordinated groups. However, animals such as bottlenose dolphins (*Tursiops truncatus*; Quintana-Rizzo 2006) and chimpanzees (*Pan troglodytes*; Lehmann and Boesch 2004) that maintain stable and strong individual-specific social relationships lasting many years may live in a fission–fusion society in which groupings are fluid and do not last long. There is strong evidence that individuals of these species can recognize one another using several sensory modalities, and individual recognition is common among mammals. Therefore, observation of fluid groups by itself does not demonstrate lack of individual-specific social relationships. Furthermore, the discussion above on range of communication raises the possibility that our definitions of groups, often defined as sightings of individuals within a few body lengths, may not match the ranges over which whales are able to maintain their social bonds.

Since 1986, a series of papers has suggested the potential for longer term, individual-specific social relationships in baleen whales. Considering that adult female right whales call males into competitive mating groups during the feeding season when females are not ovulating, Kraus and Hatch (2001) suggested that females may keep track of the performance of different males, make a choice, and mate with the selected male at a later time. This would require the ability of the female to identify the same male during the feeding and breeding seasons. Clapham (1993) studied associations of humpback whales on the Gulf of Maine feeding ground. He found that adult male–female pairs were more common among groups than expected by chance, and he also surmised that perhaps "males establish bonds with many females in the summer with a possible payoff on the breeding grounds during the winter" [p. 142]. Clapham et al. (1992) and Brown and Corkeron (1995) determined the sex of humpback whales competing in groups during the breeding season or migration, respectively. Most of these groups fit the pattern described by Tyack and Whitehead (1983) of males competing for access to a female, but some of these groups of competing males did not include a female. Clapham et al. (1992) and Brown and Corkeron (1995) suggest that these all-male competitive groups may represent interactions in which males are sorting out their dominance status. Brown and Corkeron (1995) point out that this kind of dominance sorting requires whales to recognize one another as individuals and to remember the outcome of previous encounters with each individual. Darling and Bèrubè (2001) and Smith et al. (2008) determined that many of the lone whales which join singers are males. Darling and Bèrubè (2001) argue that these male-male interactions are also involved in dominance sorting and "that these animals know each other and continually reestablish their relationships, with some facilitation by the song" [p. 581]. Brown and Corkeron (1995) conclude: "If male humpback whales are establishing dominance hierarchies, associations between males may not be as ephemeral as described previously" [p. 175]. If they use song to track known individuals and if song can be detected to ranges of tens of km in humpbacks and farther in other species, then we would need studies designed to track associations over many tens of kilometers, much greater ranges than have previously been conducted. Carefully designed playback studies would be required to test whether a whale can identify other known individuals by their song.

The evidence for these individual-specific relationships is not very strong for baleen whales, nor do we know whether baleen whales can identify individuals acoustically using distinctive call features, visually based on natural markings, and/or by sensing distinctive combinations of odorants. However, as with most mammals, it is clear that mothers can keep in reliable contact with their young, and baleen whale mothers maintain a stable association with their young calf through migrations over thousands of kilometers and for periods of a year or more in some cases. We know right whales use a variety of calls including upcalls to maintain contact during short separations. McCordic et al. (2016) analyzed upcalls of 14 right whales of known age and sex and showed that these contact calls carry information about the identity and age/sex class of the caller. Classification based on simple acoustic features of the calls was correct 73% of the time for individual identification and 86% of the time for age/sex class. Given this demonstration, it does not seem far-fetched to hypothesize that adult whales might similarly recognize other adults. The complexity, stereotypy, and stability of bowhead song (Stafford et al. 2018) suggests the potential that a whale could recognize individuals by their songs. The way in which the songs of humpback whales change over time seems to make individual identification difficult, but some acoustic features of the song may carry individual-specific cues. The key missing test is playback experiments to test for individual recognition. Once a vocalization is identified as potentially used for individual identification, a paired playback design such as used by Sayigh et al. (1999) for common bottlenose dolphins would suffice. The key is to find two pairs of animals where A has a stronger bond with B than to D and C has a stronger bond with D than to B. Upon playback of the calls of B and D to whales A and C, the prediction would be that A would respond to B, and that C would respond to D with responses appropriate to the bond. Clark et al. (2006) used a similar design to show that Magellanic penguin (*Spheniscus magellanicus*) females respond more strongly to calls of their mate than to those of a neighbor or a stranger, but they show no difference in response to neighbors and strangers. A common playback design for songbirds tests whether a subject responds less strongly to the song of a neighbor that is played from the neighbor's location than when it is played from a territory far from the neighbor's usual location. Birds that pass this test demonstrate that they recognize the neighbor's song and associate it with a particular location (Falls and Brooks 1975).

7.7.3 Conclusions

Since the days of whaling under sail, humans have learned that some baleen whale populations live on the scale of ocean basins. Most have an annual migratory cycle where they feed in areas thousands of km away from where they breed, and they can communicate acoustically over ranges of tens to hundreds of kilometers. However, it is only in the past few decades that human researchers are starting to appreciate that the methods we use to understand their behavior and social organization must match the temporal and spatial scales on which these whales operate rather than the scales that we as terrestrial primates apply to our own communication and social behavior.

References

Allen J, Weinrich M, Hoppitt W, Rendell L (2013) Network-based diffusion analysis reveals cultural transmission of lobtail feeding in humpback whales. Science 340(6131):485–488

Baker CS, Herman LM (1984) Aggressive behavior between humpback whales (*Megaptera novaeangliae*) wintering in Hawaiian waters. Can J Zool 62(10):1922–1937

Baker CS, Perry A, Herman LM (1987) Reproductive histories of female humpback whales *Megaptera novaeangliae* in the North Pacific. Mar Ecol Prog Ser 41:103–114

Barendse J, Best PB, Carvalho I, Pomilla C (2013) Mother knows best: occurrence and associations of resighted humpback whales suggest maternally derived fidelity to a Southern Hemisphere coastal feeding ground. PloS One 8(12):e81238

Best PB (1990) Trends in the inshore right whale population off South Africa, 1969–1987. Mar Mamm Sci 6(2):93–108

Boncoraglio G, Saino N (2007) Habitat structure and the evolution of bird song: a meta-analysis of the evidence for the acoustic adaptation hypothesis. Funct Ecol 21(1):134–142
Braham HW, Fraker MA, Krogman BD (1980) Spring migration of the Western Arctic population of bowhead whales. Mar Fish Rev 42(9):36–46
Brakes P, Dall SR, Aplin LM, Bearhop S, Carroll EL, Ciucci P, Fishlock V, Ford JK, Garland EC, Keith SA, McGregor PK, et al. (2019) Animal cultures matter for conservation. Science 363(6431):1032–1034
Brodie PF (1975) Cetacean energetics, an overview of intraspecific size variation. Ecology 56:152–161
Brown M, Corkeron P (1995) Pod characteristics of migrating humpback whales (*Megaptera novaeangliae*) off the east Australian coast. Behaviour 132(3–4):163–179
Brownell RL, Ralls K (1986) Potential for sperm competition in baleen whales. Reports of the International Whaling Commission 8(Special Issue):97–112
Buchan SJ, Stafford KM, Hucke-Gaete R (2015). Seasonal occurrence of southeast Pacific blue whale songs in southern Chile and the eastern tropical Pacific. Marine Mammal Science, 31(2):440–458
Cerchio S (2003) Paternity, polygyny and alternative mating tactics in humpback whales *(Megaptera novaeangliae*). Doctoral dissertation, University of Michigan
Chittleborough RG (1958) The breeding cycle of the female humpback whale, *Megaptera nodosa* (Bonnaterre). Aust J Mar Freshw Res 9(1):1–18
Chittleborough R (1965) Dynamics of two populations of the humpback whale, *Megaptera novaeangliae* (Borowski). Aust J Mar Freshw Res 16:33–128. https://doi.org/10.1071/MF9650033
Cholewiak D, Clark CW, Ponirakis D, Frankel A, Hatch LT, Risch D, Stanistreet JE, Thompson M, Vu E, van Parijs SM (2018) Communicating amidst the noise: modeling the aggregate influence of ambient and vessel noise on baleen whale communication space in a national marine sanctuary. Endangered Species Res 36:59–75
Christiansen F, Vivier F, Charlton C, Ward R, Amerson A, Burnell S, Bejder L (2018) Maternal body size and condition determine calf growth rates in southern right whales. Mar Ecol Prog Ser 592:267–281
Clapham PJ (1993) Social organization of humpback whales on a North Atlantic feeding ground. Symp Zool Soc Lond 66:131–145
Clapham PJ, Mayo CA (1987) Reproduction and recruitment of individually identified humpback whales, *Megaptera novaeangliae*, observed in Massachusetts Bay, 1979–1985. Can J Zool 65(12):2853–2863
Clapham PJ, Palsbøll PJ (1997) Molecular analysis of paternity shows promiscuous mating in female humpback whales (*Megaptera novaeangliae*, Borowski). Proc R Soc Lond. Ser B: Biol Sci 264(1378):95–98
Clapham PJ, Palsbøll PJ, Mattila DK, Vasquez O (1992) Composition and dynamics of humpback whale competitive groups in the West Indies. Behaviour 122(3–4):182–194
Clark CW (1983) Acoustic communication and behavior of the southern right whale (*Eubalaena australis*). In: Payne R (ed) Communication and behavior of whales. Westview Press, Boulder CO, pp 163–198
Clark CW, Clark JM (1980) Sound playback experiments with southern right whales (*Eubalaena australis*). Science 207:663–665
Clark CW, Gagnon GJ (2002) Low-frequency vocal behaviors of baleen whales in the North Atlantic: Insights from Integrated Undersea Surveillance System detections, localization, and tracing from 1992 to 1997. US Navy J Underwater Acoust 52(3):609–640
Clark CW, Clapham PJ (2004) Acoustic monitoring on a humpback whale (*Megaptera novaeangliae*) feeding ground shows continual singing into late spring. Proc R Soc Lond. Ser B: Biol Sci 271(1543):1051–7
Clark CW, Ellison WT (2004) Potential use of low-frequency sounds by baleen whales for probing the environment: evidence from models and empirical measurements. In: Thomas JA, Moss CF,

Vater M (eds) Advances in the study of echolocation in bats and dolphins. University of Chicago Press, Chicago, IL, pp 564–589
Clark JA, Boersma PD, Olmsted DM (2006) Name that tune: call discrimination and individual recognition in Magellanic penguins. Anim Behav 72(5):1141–1148
Clark CW, Gagnon GJ, Frankel AS (2019) Fin whale singing decreases with increased swimming speed. R Soc Open Sci 6(6):180525
Corkeron PJ, Connor RC (1999) Why do baleen whales migrate? Mar Mamm Sci 15(4):1228–1245
Croll DA, Clark CW, Acevedo A, Tershy B, Flores S, Gedamke J, Urban J (2002) Only male fin whales sing loud songs. Nature 417(6891):809–809
Curtis KR, Howe BM, Mercer JA (1999) Low-frequency ambient sound in the North Pacific: long time series observations. J Acoust Soc Am 106(6):3189–3200
D'Agostino VC, Hoffmeyer MS, Degrati M (2016) Faecal analysis of southern right whales (*Eubalaena australis*) in Península Valdés calving ground, Argentina: *Calanus australis*, a key prey species. J Mar Biol Assoc UK 96(4):859
Darling JD, Bérubé M (2001) Interactions of singing humpback whales with other males. Mar Mamm Sci 17(3):570–584
Darling JD, Nicklin CP, Jones ME (2006) Humpback whale songs: Do they organize males during the breeding season? Behaviour 143(9):1051–1101
Darwin C (1859) The origin of species by means of natural selection, or the preservation of favoured races in the struggle for life. John Murray, London
Darwin C (1871) The descent of man, and selection in relation to sex. John Murray, London
Date EM, Lemon RE (1993) Sound transmission: a basis for dialects in birdsong? Behaviour 124(3–4):291–312
Diachok O (2000) Absorption spectroscopy: a new approach to estimation of biomass. Fish Res 47(2–3):231–244
Dingle H (1996) Migration. The biology of life on the move. Oxford University Press, New York
D'Vincent CG, Nilson RM, Hanna RE (1985) Vocalization and coordinated feeding behavior of the humpback whale in southeastern Alaska. Sci Rep Whales Res Inst Tokyo 36:41–47
Elwen SH, Best PB (2004) Female southern right whales *Eubalaena australis*: are there reproductive benefits associated with their coastal distribution off South Africa? Mar Ecol Prog Ser 269:289–295
Ersts PJ, Rosenbaum HC (2003) Habitat preference reflects social organization of humpback whales (*Megaptera novaeangliae*) on a wintering ground. J Zool 260(4):337–345
Falls JB, Brooks RJ (1975) Individual recognition by song in white-throated sparrows. II. Effects of location. Can J Zool 53(10):1412–1420
Félix F, Botero-Acosta N (2011) Distribution and behaviour of humpback whale mother–calf pairs during the breeding season off Ecuador. Mar Ecol Prog Ser 426:277–287
Ford JK, Reeves RR (2008) Fight or flight: antipredator strategies of baleen whales. Mammal Rev 38(1):50–86
Garland EC, Goldizen AW, Rekdahl ML, Constantine R, Garrigue C, Hauser ND, Poole MM, Robbins J, Noad MJ (2011) Dynamic horizontal cultural transmission of humpback whale song at the ocean basin scale. Curr Biol 21(8):687–691. https://doi.org/10.1016/j.cub.2011.03.019
Garland EC, Gedamke J, Rekdahl ML, Noad MJ, Garrigue C, Gales N (2013) Humpback whale song on the Southern Ocean feeding grounds: implications for cultural transmission. PloS One 8(11):e79422
Geijer CK, Notarbartolo di Sciara G, Panigada S (2016) Mysticete migration revisited: are Mediterranean fin whales an anomaly? Mammal Rev 46(4):284–296
Goldbogen JA, Calambokidis J, Shadwick RE, Oleson EM, McDonald MA, Hildebrand JA (2006) Kinematics of foraging dives and lunge-feeding in fin whales. J Exp Biol 209(7):1231–1244
Goldbogen JA, Cade DE, Wisniewska DM, Potvin J, Segre PS, Savoca MS, Hazen EL, Czapanskiy MF, Kahane-Rapport SR, DeRuiter SL, Gero S et al. (2019) Why whales are big but not bigger: physiological drivers and ecological limits in the age of ocean giants. Science 366(6471):1367–1372

Gong Z, Andrews M, Jagannathan S, Patel R, Jech JM, Makris NC, Ratilal P (2010) Low-frequency target strength and abundance of shoaling Atlantic herring (*Clupea harengus*) in the Gulf of Maine during the Ocean Acoustic Waveguide Remote Sensing 2006 Experiment. J Acoust Soc Am 127(1):104–123

Heide-Jørgensen MP, Laidre KL, Nielsen NH, Hansen RG, Røstad A (2013) Winter and spring diving behavior of bowhead whales relative to prey. Anim Biotelemetry 1(1):1–4

Kanwisher J, Sundnes G (1966) Thermal regulation in cetaceans. In: Norris KS (ed) Whales, dolphins and porpoises. University of California Press, Berkeley CA, pp 397–409

Kraus SD, Hatch JJ (2001) Mating strategies in the North Atlantic right whale (*Eubalaena glacialis*). J Cetacean Res Manage 2(Special Issue):237–244

Lehmann J, Boesch C (2004) To fission or to fusion: effects of community size on wild chimpanzee (*Pan troglodytes verus*) social organisation. Behav Ecol Sociobiol 56(3):207–216

Lockyer C (1984) Review of baleen whale (Mysticeti) reproduction and implications for management. Rep Int Whaling Comm (Special Issue 6):27–50

Mann J (2000) Unraveling the dynamics of social life. In: Mann J, Connor RC, Tyack PL, Whitehead H (eds) Cetacean societies: field studies of dolphins and whales. University of Chicago Press, Chicago, IL, pp 45–64

Martins CCA, Morete ME, Coitinho MHE, Freitas AC, Secchi ER, Kinas PG (2001) Aspects of habitat use patterns of humpback whales in the Abrolhos Bank, Brazil, breeding ground. Mem Queensland Museum 47(2):563–570

McCordic JA, Root-Gutteridge H, Cusano DA, Denes SL, Parks SE (2016) Calls of North Atlantic right whales *Eubalaena glacialis* contain information on individual identity and age class. Endangered Species Res 30:157–169

Medrano L, Salinas M, Salas I, Guevara PLD, Aguayo A, Jacobsen J, Baker CS (1994) Sex identification of humpback whales, *Megaptera novaeangliae*, on the wintering grounds of the Mexican Pacific Ocean. Can J Zool 72(10):1771–1774

Mercado E III, Frazer LN (1999) Environmental constraints on sound transmission by humpback whales. J Acoust Soc Am 106(5):3004–3016

Mesnick S (1997) Sexual alliances: evidence and evolutionary implications. In: Gowaty P (ed) Feminism and evolutionary biology. Springer, New York, pp 207–260

Mikhalev YA (1997) Humpback whales *Megaptera novaeangliae* in the Arabian Sea. Mar Ecol Prog Ser 149:13–21

Morano JL, Salisbury DP, Rice AN, Conklin KL, Falk KL, Clark CW (2012) Seasonal and geographical patterns of fin whale song in the western North Atlantic Ocean. J Acoust Soc Am 132(2):1207–1212

Munger LM, Wiggins SM, Hildebrand JA (2011) North Pacific right whale up-call source levels and propagation distance on the southeastern Bering Sea shelf. J Acoust Soc Am 129(6):4047–4054

Nerini M (1984) A review of gray whale feeding ecology. In: Jones ML, Swartz SL, Leatherwood S (eds) The gray whale, Academic Press, San Diego CA, pp 423–450

Nielsen ML, Bejder L, Videsen SK, Christiansen F, Madsen PT (2019) Acoustic crypsis in southern right whale mother–calf pairs: infrequent, low-output calls to avoid predation? J Exp Biol 222(13):p.jeb190728

Noad MJ, Cato DH, Bryden MM, Jenner M-N, Jenner KCS (2000) Cultural revolution in whale songs. Nature 408:537–537. https://doi.org/10.1038/35046199

Norris KS, Goodman RM, Villa-Ramirez B, Hobbs L (1977) Behavior of California gray whale, *Eschrichtius robustus*, in southern Baja California, Mexico. Fish Bull 75(1):159–172

Oftedal OT (1997) Lactation in whales and dolphins: evidence of divergence between baleen- and toothed-species. J Mammary Gland Biol Neoplasia 2(3):205–230

Oleson EM, Calambokidis J, Burgess WC, McDonald MA, LeDuc CA, Hildebrand JA (2007) Behavioral context of call production by eastern North Pacific blue whales. Mar Ecol Prog Ser 330:269–284

Parks SE (2003) Response of North Atlantic right whales (*Eubalaena glacialis*) to playback of calls recorded from surface active groups in both the North and South Atlantic. Mar Mamm Sci 19(3):563–580

Parks SE, Tyack PL (2005) Sound production by North Atlantic right whales (*Eubalaena glacialis*) in surface active groups. J Acoust Soc Am 117(5):3297–3306

Parks SE, Hamilton PK, Kraus SD, Tyack PL (2005) The gunshot sound produced by male North Atlantic right whales (*Eubalaena glacialis*) and its potential function in reproductive advertisement. Mar Mamm Sci 21(3):458–475

Parks SE, Brown MW, Conger LA, Hamilton PK, Knowlton AR, Kraus SD, Slay CK, Tyack PL (2007) Occurrence, composition, and potential functions of North Atlantic right whale (*Eubalaena glacialis*) surface active groups. Mar Mamm Sci 23(4):868–887

Parks SE, Cusano DA, Van Parijs SM, Nowacek DP (2019) Acoustic crypsis in communication by North Atlantic right whale mother–calf pairs on the calving grounds. Biol Let 5(10):20190485

Payne K, Payne R (1985) Large scale changes over 19 years in songs of humpback whales in Bermuda. Z Tierpsychol 68:89–114

Payne K, Tyack P, Payne R (1983) Progressive changes in the songs of humpback whales (*Megaptera novaeangliae*): a detailed analysis of two seasons in Hawaii. In: Payne R (ed) Communication and behavior of whales. Westview Press, Boulder CO, pp 9–57

Payne R (1995) Among whales. Scribner, New York

Payne RS, McVay S (1971) Songs of humpback whales. Science 173(3997):585–597

Payne R, Webb D (1971) Acoustic signaling in baleen whales. Ann N Y Acad Sci 188:110–141

Pierszalowski SP (2014) The influence of local fidelity and recruitment on population dynamics and specialized foraging of humpback whales in Glacier Bay and Icy Strait, Alaska. MSc Dissertation, Oregon State University

Quintana-Rizzo E (2006) Group fission-fusion dynamics and communication in the bottlenose dolphin (*Tursiops truncatus*). PhD Dissertation, University of South Florida. http://scholarcommons.usf.edu/etd/2666

Ramp C, Hagen W, Palsbøll P, Bérubé M, Sears R (2010) Age-related multi-year associations in female humpback whales (*Megaptera novaeangliae*). Behav Ecol Sociobiol 64(10):1563–1576

Ray GC, Mitchell ED, Wartzok D, Kozicki VM, Maiefski R (1978) Radio tracking of a fin whale (*Balaenoptera physalus*). Science 202(4367):521–524

Reeves R, Mitchell E, Mansfield A, McLaughlin M (1983) Distribution and migration of the bowhead whale, *Balaena mysticetus*, in the eastern North American Arctic. Arctic 36:5–64

Roman et al. (2016) Endangered Right Whales Enhance Primary Productivity in the Bay of Fundy. PLoS ONE. 11. https://doi.org/10.1371/journal.pone.0156553

Sayigh LS, Tyack PL, Wells RS, Solow AR, Scott MD, Irvine AB (1999) Individual recognition in wild bottlenose dolphins: a field test using playback experiments. Anim Behav 57(1):41–50

Schell DM, Saupe SM, Haubenstock N (1989) Bowhead whale (*Balaena mysticetus*) growth and feeding as estimated by δ^{13} C techniques. Mar Biol 103(4):433–443

Sharpe FA (2001) Social foraging of southeast Alaskan humpback whales. PhD dissertation, Simon Fraser University

Sharpe FA, Dill LM (1997) The behavior of Pacific herring schools in response to artificial humpback whale bubbles. Can J Zool 75(5):725–730

Silva MA, Prieto R, Jonsen I, Baumgartner MF, Santos RS (2013) North Atlantic blue and fin whales suspend their spring migration to forage in middle latitudes: building up energy reserves for the journey? PloS ONE 8(10):e76507

Simon M, Johnson M, Tyack P, Madsen PT (2009) Behaviour and kinematics of continuous ram filtration in bowhead whales (*Balaena mysticetus*). Proc R Soc B: Biol Sci 276(1674):3819–3828

Širović A, Hildebrand JA, Wiggins SM (2007) Blue and fin whale call source levels and propagation range in the Southern Ocean. J Acoust Soc Am 122(2):1208–1215

Slabbekoorn H, Smith TB (2002) Bird song, ecology and speciation. Philos Trans R Soc Lond. Ser B: Biol Sci 357(1420):493–503

Smith JN, Goldizen AW, Dunlop RA, Noad MJ (2008) Songs of male humpback whales, *Megaptera novaeangliae*, are involved in intersexual interactions. Anim Behav 76(2):467–477
Smultea MA (1994) Segregation by humpback whale (*Megaptera novaeangliae*) cows with a calf in coastal habitat near the island of Hawaii. Can J Zool 72(5):805–811
Spiesberger JL, Fristrup KM (1990) Passive localization of calling animals and sensing of their acoustic environment using acoustic tomography. Am Nat 135(1):107–153
Stafford KM, Lydersen C, Wiig Ø, Kovacs KM (2018) Extreme diversity in the songs of Spitsbergen's bowhead whales. Biol Lett 14(4):p.20180056
Sullivan M, Cartwright R (2009) Associations with multiple male groups increase the energy expenditure of humpback whale (*Megaptera novaeangliae*) female and calf pairs on the breeding grounds. Behaviour 146(11):1573–1600
Taber SM, Thomas PO (1984) Mother-infant interaction and behavioral development in southern right whales, *Eubalaena australis*. Behaviour 88(1–2):42–60
Tervo OM, Christoffersen MF, Simon M, Miller LA, Jensen FH et al (2012) High source levels and small active space of high-pitched song in bowhead whales (*Balaena mysticetus*). PLoS ONE 7(12):e52072. https://doi.org/10.1371/journal.pone.0052072
Thoreau HD (1854) Walden. Houghton Mifflin, Boston (1906 edition)
Tyack P (1981) Interactions between singing Hawaiian humpback whales and conspecifics nearby. Behav Ecol Sociobiol 8(2):105–116
Tyack P (1986) Population biology, social behavior and communication in whales and dolphins. Trends Ecol Evol 1(6):144–150
Tyack PL (1997) Studying how cetaceans use sound to explore their environment. In: Owings DH, Beecher MD, Thompson NS (eds) Perspectives in ethology, vol 12. Communication. Plenum, New York, pp 251–297
Tyack PL (1998) Acoustic communication under the sea. In: Hopp SL, Owren KJ, Evans CS (eds) Animal acoustic communication. Springer, Berlin, pp 163–220
Tyack P, Whitehead H (1983) Male competition in large groups of wintering humpback whales. Behaviour 83(1–2):132–154
Valsecchi E, Hale P, Corkeron P, Amos W (2002) Social structure in migrating humpback whales (*Megaptera novaeangliae*). Mol Ecol 11(3):507–518
Videsen SK, Bejder L, Johnson M, Madsen PT (2017) High suckling rates and acoustic crypsis of humpback whale neonates maximise potential for mother–calf energy transfer. Funct Ecol 31(8):1561–1573
Visser F, Hartman KL, Pierce GJ, Valavanis VD, Huisman J (2011) Timing of migratory baleen whales at the Azores in relation to the North Atlantic spring bloom. Mar Ecol Prog Ser 440:267–279
Vu ET, Risch D, Clark CW, Gaylord S, Hatch LT, Thompson MA, Wiley DN, van Parijs SM (2012) Humpback whale song occurs extensively on feeding grounds in the western North Atlantic Ocean. Aquat Biol 14(2):175–183
Watkins WA, Schevill WE (1979) Aerial observation of feeding behavior in four baleen whales: *Eubalaena glacialis*, *Balaenoptera borealis*, *Megaptera novaeangliae*, and *Balaenoptera physalus*. J Mammal 60(1):155–163
Watts P, Hansen S, Lavigne DM (1993) Models of heat loss by marine mammals: thermoregulation below the zone of irrelevance. J Theor Biol 163(4):505–525
Weinrich MT (1998) Early experience in habitat choice by humpback whales (*Megaptera novaeangliae*). J Mammal 79:163–170
Weston DE (1967) Sound propagation in the presence of bladder fish. In: Albers VM (ed), Underwater Acoustics. Plenum, New York, vol 2, pp 55–58
Whitehead H (1983) Structure and stability of humpback whale groups off Newfoundland. Can J Zool 61(6):1391–1397
Whitehead H, Moore MJ (1982) Distribution and movements of West Indian humpback whales in winter. Can J Zool 60(9):2203–2211

Wild LA, Gabriele CM (2014) Putative contact calls made by humpback whales (*Megaptera novaeangliae*) in Southeastern Alaska. Can Acoust 42(4):23–32

Würsig B, Clark C (1993) Behavior. The bowhead whale. The Society for Marine Mammalogy, Lawrence, KS, pp 157–199

Chapter 8
Culture and Social Learning in Baleen Whales

Ellen C. Garland and Emma L. Carroll

Abstract Culture, the sharing of behaviors or information within a community acquired through some form of social learning from conspecifics, represents a "second inheritance system". This assertion, while still controversial, is a clear indication that culture and the study of social learning in animals is no longer a taboo subject. Some of the strongest evidence for culture in animals has come from the study of cetaceans; while the focus has typically been on the odontocetes (mainly sperm whales, killer whales, and bottlenose dolphins), baleen whales provide important, unique, and robust evidence for cultural processes. Baleen whales undertake a myriad of behaviors across a variety of contexts. Some of these behaviors have been investigated with a cultural lens and have clearly shown maternally directed (and thus culturally transmitted) site fidelity to breeding, feeding and migratory routes, dynamic cultural transmission of song, and social transmission of novel feeding techniques. Undertaking cultural studies in large, free-ranging cetaceans requires multiyear, long-term datasets with enough detail to track changes; such datasets are rare and take decades to accumulate. However, we are now seeing a number of such datasets come to light, and the results are spectacular. Here, we first provide an overview of culture and its transmission; we then highlight some of the clearest examples of baleen whale culture to date, concluding with research considerations. Culture and its influence on the lives of cetaceans can no longer be ignored as, to paraphrase some of the pioneers in the cetacean culture field, it is now clear that culture rules their [cetaceans'] lives.

Keywords Animal culture · Social learning · Cultural processes · Song · Migration · Feeding · Vocal learning · Isotopes

---

E. C. Garland (✉)
Centre for Social Learning & Cognitive Evolution, and Sea Mammal Research Unit, School of Biology, University of St. Andrews, St. Andrews, Fife KY16 8LB, UK
e-mail: ecg5@st-andrews.ac.uk

E. L. Carroll
School of Biological Sciences, University of Auckland, Waipapa Taumata Rau, Auckland 1010, New Zealand

C.W. Clark and E. C. Garland (eds.), *Ethology and Behavioral Ecology of Mysticetes*, Ethology and Behavioral Ecology of Marine Mammals,
https://doi.org/10.1007/978-3-030-98449-6_8

8.1 Culture and Social Learning

To be human is to be cultural. Cultural traditions have shaped human societies (Ramsey 2013); this concept has led to heated debate about the existence of culture in animals (see Laland and Janik 2006; Whiten et al. 2017; Whiten 2021). Since evidence for social learning of behavioral variants and their subsequent social transmission have come to light in species from fruit flies (*Drosophila melanogaster*; Danchin et al. 2018) to southern right whales (*Eubalaena australis*; Valenzuela et al. 2009), such arguments are diminishing. Culture and its transmission have been extensively investigated in our nearest cousins, the great apes, in particular chimpanzees (*Pan troglodytes*; Whiten et al. 1999, 2007). A wide body of research illustrates the broad-scale patchwork of behavioral variants across populations and controlled experiments demonstrate the social learning and transmission of behaviors (including tool use; Whiten et al. 2007, 2017; Whiten 2017). These have set the stage for understanding animal culture.

The term "culture" is defined in a broad sense as "shared behavior or information within a community acquired through some form of social learning from conspecifics" (Rendell and Whitehead 2001; Fragaszy and Perry 2003). The term "culture" can at times be used interchangeably in the peer-review literature with the term "tradition" (Whiten 2017). The important step when asserting that a behavior is cultural is the transmission process (i.e., social learning) and subsequent diffusion of the behavior from the innovator through a population. Animals are able to innovate or invent behaviors de novo (e.g., exploit a new resource), but it is the learning of this innovation *from another* individual and the diffusion, spread, sharing, or transmission of the behavior through a population that is of interest here. This is what is meant by culture.

Influences such as genetics or ecology contribute to behavior, and these factors may explain behavioral differences between and among populations. To establish that a behavior is culturally transmitted, the so-called exclusion method involves an examination of the plausibility of other influences or confounding factors. This method has received both praise and criticism (Laland and Janik 2006; Krützen et al. 2007). For example, in the late 1990s, variation in chimpanzee behavior across Africa was described. After exclusion of ecological explanations, 39 different behavioral patterns were reported as cultural (Whiten et al. 1999). The method of exclusion is useful to create a broad-brush examination that should be refined in future analyses. A general criticism of this method is that behaviors do not develop in isolation but as an interaction between environmental experience and a genetic blue-print or predisposition (Laland and Janik 2006). The animal culture field is now moving beyond the method of exclusion toward approaches evaluating the magnitude of each potential causal factor on the development of a specific behavior (Laland and Janik 2006; Allen et al. 2013; Whitehead et al. 2019).

A potentially confounding issue when designating behaviors as having a cultural basis is the pathway of transmission. Vertical transmission is from parent to offspring (e.g., mother to calf), oblique transmission is from a non-parent model to the next generation (e.g., juvenile learning the song of an unrelated adult male), and horizontal transmission is transmission within a cohort (e.g., learning a new feeding technique from your peers) (Whitehead et al. 2019). The direction of information flow is important, especially when attempting to distinguish vertical cultural transmission from underlying genetic explanations. For example, in Shark Bay , Western Australia, a subset of the bottlenose dolphin (*Tursiops* sp.) population forages for food using marine sponges (Krutzen et al. 2005; Sargeant and Mann 2009). "Sponging" is habitat specific and is vertically transmitted from mother to calf (Krutzen et al. 2005; Kopps et al. 2014). Sponging was initially identified in a single matriline (Krutzen et al. 2005), raising concerns of the genetic influence on the purported cultural behavior. However, a second sponging matriline has been identified in Shark Bay indicating this is a more widespread foraging tactic (Kopps et al. 2014) and most likely culturally transmitted.

Here, we highlight some of the studies that have provided evidence for cultural processes in baleen whales. These include studies of migration, foraging, acoustic communication, and breeding behaviors. Cultural processes can be investigated using approaches such as molecular genetics, isotopes, acoustics, and telemetry tags. We do not cover all of the different social learning mechanisms or delve deeply into discussions of conformity (see Laland 2008). The goal of this chapter is to provide readers with the biological concept and framework of culture, and perhaps inspire readers to take a look into interpretations of their own data through a cultural lens.

8.2 Culture in Cetaceans

The past two decades have seen an explosion in cetacean culture studies. From an initial synthesis by Rendell and Whitehead (2001) through to recent work with a diverse number of species and behavioral contexts [reviewed in Whitehead and Rendell (2015)], a clear conclusion has emerged: Culture is important to cetaceans. Four genera have been the focus of most cultural studies: the sperm whale (*Physeter macrocephalus*), killer whale (*Orcinus orca*), bottlenose dolphin and humpback whale (*Megaptera novaeangliae*). Vocal clans, ecotypes, sponging and song are part of these species, respectively. These are examples of cultures that have shaped communities and species. We urge readers to explore studies on this subject, some of which are presented by species in Book 1 of this series (Würsig 2019). Here, we note two examples of odontocete culture, before focusing on baleen whales.

Killer whales can be divided into ecotypes strongly associated with diet (foraging specialization; Ford et al. 1998; Foote et al. 2016). Pods are composed of a number of related matrilines, and each pod has a pod-specific acoustic repertoire (dialect; Ford 1991). Subtle acoustic call differences are shared among members of a pod, and some call types are also shared among pods within an area, creating "acoustic clans" (Ford 1991; Miller et al. 2000; Yurk et al. 2002). These clan "dialects" undergo cultural drift and evolve (Filatova and Miller 2015); call changes can be horizontally transmitted among matrilines (Deecke et al. 2000); and calves learn their repertoires from their mothers (i.e., vertical transmission; Filatova et al. 2015). Multiple cultural processes are therefore at work. Dialects provide vocal markers not only of pods, but also of ecotypes. Like dialects, feeding specializations are vertically transmitted within the matriline; such conditions may promote gene–culture coevolution and potentially, speciation (see Foote et al. 2016; Ford 2019). Sperm whales also provide an excellent example of gene–culture coevolution, where feeding specializations and vocal dialects interact and are culturally transmitted (see Cantor et al. 2019).

Bottlenose dolphins produce patterns of frequency-modulated sounds called "signature whistles" that encode individual identity (Caldwell and Caldwell 1965; Janik et al. 2006). Bottlenose dolphins copy one another's signature whistles as a means of addressing specific social companions, in effect addressing a conspecific by using their "name" (Janik 2000; King and Janik 2013). This ability is rare in non-human animals (King and Janik 2013). Bottlenose dolphins are also able to label objects with novel learned whistle patterns (Richards et al. 1984). Thus, vocal learning abilities of bottlenose dolphins are exceptional among mammals and hint at underlying complex cognitive abilities (Janik 2009).

Tool-using culture has been documented in a subset of wild bottlenose dolphins in Shark Bay, West Coast Australia (Krutzen et al. 2005; Sargeant and Mann 2009; Connor et al. 2019; Mann 2019; Wild et al. 2019a). Individuals place marine sponges over their rostra to probe the sediment for fish in deep-water channels (Mann and Sargeant 2003; Krutzen et al. 2005). Sponging behavior is passed vertically from mother to calf (Krutzen et al. 2005, 2014), although mainly female calves acquire sponging behavior, as male offspring of sponging mothers tend to not show sponging in later life (Mann and Sargeant 2003; Krutzen et al. 2005). Initially, a single matriline was discovered to sponge (Krutzen et al. 2005); but subsequently, a second sponging matriline was identified, indicating that sponging is not a unique foraging tactic (Kopps et al. 2014), and unlikely to be confounded by genetic inheritance factors (Laland and Janik 2006). The importance of social learning on the diffusion of sponging was more recently investigated using network-based diffusion analysis, which accounts for ecological and genetic factors (Wild et al. 2019a). Results support previous findings that sponging is vertically socially transmitted from mother to (primarily female) offspring (Wild et al. 2019a). The behavioral proclivity of sponging therefore represents a clear example of tool-use culture in a marine mammal (Krutzen et al. 2005; Krützen et al. 2014).

8.3 Evidence for Culture in Baleen Whales

We focus here on two species that provide case studies for culture in baleen whales: southern right whales and humpback whales. But, we acknowledge that there are many other potential examples of cetacean culture yet to be discovered.

8.3.1 Migratory Culture in Southern Right Whales

As with other baleen whales, southern right whales move between offshore high latitude summer foraging grounds and sheltered coastal wintering grounds. Long-term studies that used photographs of natural markings (photo-ID) and unique genetic profiles to identify individual whales have shown that females demonstrate long-term fidelity to their wintering grounds (Bannister 2001; Rowntree et al. 2001; Carroll et al. 2016). Females calve in these preferred wintering grounds and migrate with their calves to preferred foraging grounds. These mother behaviors provide a way for offspring to learn their mothers' migratory traditions in their first year of life. The cultural transmission of migratory preferences, or migratory culture, has profound implications for genetic structure and connectivity across the species' migratory network. Such behavior is believed to contribute to significant genetic differentiation between wintering grounds, particularly evident in maternally inherited mitochondrial DNA (mtDNA; Patenaude et al. 2007; Carroll et al. 2015, 2018). Such a pattern is found in other baleen whales, demonstrating how molecular methods allow us to understand the behavioral lives of whales (Chap. 6).

Correlations between mtDNA, a proxy for maternal lineage or tradition, and stable isotope data, a proxy for foraging grounds, have yielded interesting results. The first study was conducted using 131 southern right whales from the Argentine wintering ground, primarily including females with calves (Valenzuela et al. 2009). The study showed there was a non-random association between maternal lineage, as indicated by mtDNA version or haplotype, and foraging ground, as indicated by stable isotope data (Fig. 8.1). Whales with the same mtDNA haplotype were more likely than expected by chance to have more similar isotope profiles. This work was confirmed and expanded based on 78 whales sampled on the Australian wintering ground (Carroll et al. 2015). Whales that shared mtDNA haplotype and were more genetically related were also more likely to have more similar isotope profiles. The simplest explanation for this finding is maternally directed learning of foraging grounds, such that related animals are foraging in isotopically similar locations.

Migratory culture also plays a role in the recovery of southern right whales following the decline of whaling and could be a key determinant of their response to climate change. Once numbering perhaps 100,000 throughout the Southern Ocean, the species declined to possibly less than 400 individuals around 1920 (Jackson et al. 2008). Today, the species shows patchy recovery: For example, few whales occur around mainland New Zealand and the east Australian wintering grounds, compared with the strongly recovering populations in the New Zealand sub-Antarctic Islands and in southwest Australia (Carroll et al. 2011, 2015). It is likely that when whales that inhabited a region were extirpated, the memory of that area as a good migratory destination was also lost. This loss of "cultural memory", exacerbated by the loss of adjacent populations and low abundance, means it is unlikely that previously inhabited areas will be recolonized on a timeframe relevant to management (i.e., decades; Clapham et al. 2008).

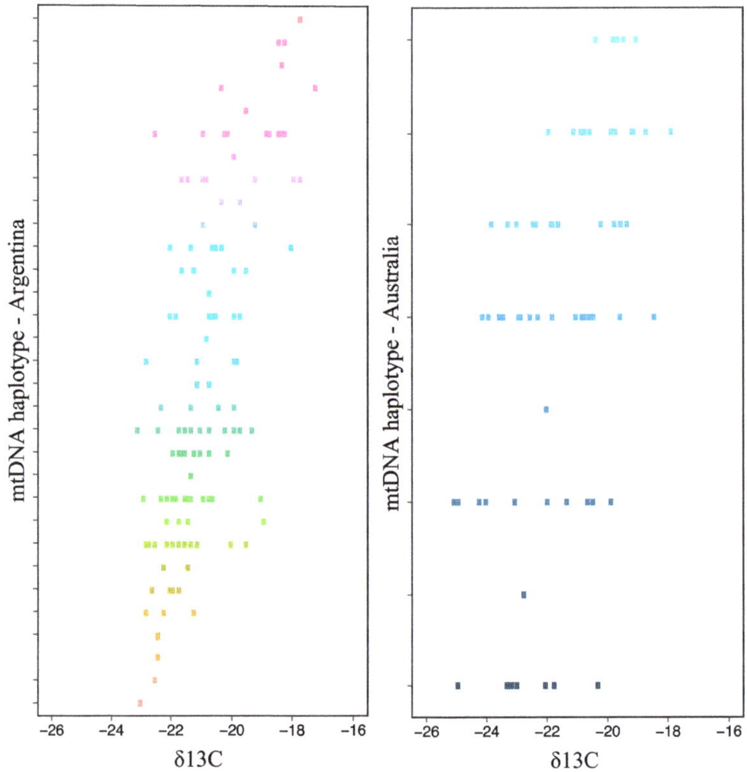

Fig. 8.1 Empirical evidence for southern right whale migratory culture: correlation between foraging ground, as shown by δ13C value, and maternal lineage, as indicated by mtDNA version or haplotype for southern right whales sampled on wintering grounds. Data are plotted independently for the Argentine wintering ground, with each square representing an individual whale [left panel, data from Valenzuela et al. (2009)] and the Australian wintering ground [right panel, data from Carroll et al. (2015)]

Southern right whale recovery also depends on their ability to adjust to changes in prey distribution. The vertical transmission of migratory culture has likely been beneficial on long time scales as it provides useful information on foraging and nursery habitats in an often visually featureless and vast sea (Whitehead 2010). Indeed, there is evidence that the recovery of southern right whales has been very dependent on environmental conditions at their feeding grounds. The reproductive success of southern right whales on their wintering grounds in Argentina and Brazil has been correlated with oceanographic conditions and krill availability at their South Georgia foraging ground, respectively (Leaper et al. 2006; Seyboth et al. 2016). However, in an era of climate change, these cultural traditions can become detrimental if they are too fixed to respond to rapid resource shifts due to climate change or other anthropogenic activities (Keith and Bull 2017). Such problems are called ecological or evolutionary traps, entailing a behavior that originally increased fitness (e.g., fidelity to rich feeding ground) but then became a hindrance in the face of rapid environmental change (e.g., changes in distribution of food resources due to climate variation and/or anthropogenic activities; Schlaepfer et al. 2002; Keith and Bull 2017). Encouragingly, there is evidence that at least some baleen whale species can behaviorally adapt to such shifts. For example, humpback whales in the Gulf of St. Lawrence, Canada, have altered the timing of their migratory journeys to coincide with earlier prey availability (Ramp et al. 2015). However, humpback whales may be uniquely plastic in their foraging behaviors (see next section), and we do not know if southern right whale culture is too conservative to cope with shifting prey resources under rapid climate change conditions.

8.3.2 Feeding Traditions and Song Culture in Humpback Whales

Humpback whales have multiple independently evolving cultural traditions within a population. These include the social learning of feeding tactics (Allen et al. 2013), maternally directed site fidelity to breeding grounds, feeding grounds and migratory routes (Baker et al. 1990), and the evolution and revolution of song displays (Payne and Payne 1985; Noad et al. 2000; Garland et al. 2011). We focus on two different behavioral contexts—song culture and feeding tactics—as migratory culture has been covered in the previous section and may be considered broadly similar between the two genera.

In 1980, a humpback whale in the Gulf of Maine population on the east coast of North America performed an innovative modification to their feeding technique, termed "lobtail feeding" (Allen et al. 2013). Humpback whales commonly forage by "bubble-net feeding", where a bubble stream is produced below and around a school of prey followed by lunging through the bubbles (Chap. 5). The lobtail feeding technique is a modification of the original bubble-net feeding behavioral sequence by adding a series of tail slaps to start the sequence (Weinrich et al. 1992). Researchers observed an increase in the feeding technique over a 27-year period, suggesting social transmission of the innovation through the population. However, its emergence coincided with the crash of one prey species, herring (*Clupea harengus*), and a switch in prey to sand lance (*Ammodytes americanus*), suggesting an ecological factor (Allen et al. 2013).

The spread of a behavioral innovation can be investigated by apportioning the relative influence of multiple factors (i.e., ecological, social, and genetic) using network-based diffusion analysis (NBDA). Allen et al. (2013) employed NBDA to analyze the spread of the lobtail feeding innovation. Results strongly indicated that social transmission was responsible for the spread as support for models including social transmission were six to 23 orders of magnitude greater than for models without social transmission (Allen et al. 2013). While there was clearly an ecological driver to initiate the innovation (i.e., change in prey abundance resulting in a switch in prey), the *spread* of the innovation through the population was overwhelmingly driven by social transmission. Modeling approaches such as NBDA assist researchers in examining multiple drivers for a behavior while not removing the behavior out of the environmental context. The NBDA method has drawbacks; for example, it includes the need for large (e.g., hundreds to thousands of sightings) datasets to have the power to tease out relative effects of different drivers. Furthermore, application of NBDA to population-wide, stable cultural traditions may be limited in situations where social transmission could have historically occurred between populations.

Let us now investigate song culture and dynamics. Male humpback whales sing a long, complex, stereotyped, and hierarchically structured vocal display termed "song" (Payne and McVay 1971; Herman and Tavolga 1980), which functions in sexual selection to attract a mate and/or mediate male–male interactions (Herman 2017). Most males within a population sing a similar song at any time; that is, they sing songs which share similar themes, phrases, and units, as well as the same arrangement of song components. Thus, there is strong cultural conformity to the current arrangement (version) of the song display in male humpback whales (Payne et al. 1983; Payne and Payne 1985). Songs also evolve through time (Payne and Payne 1985); males within the population incorporate changes into their own song to maintain the observed cultural conformity. Further information on general humpback whale song can be found in Chap. 11, while humpback whale social communication is explored in Chap. 10.

Over the last two decades, some striking song dynamics have been discovered in the South Pacific. Noad et al. (2000) discovered that over a two-year period, song from the west Australian population replaced the existing song of the east Australian population. This rapid and complete change was termed by the authors a "song revolution" to distinguish it from the common and traditional perspective of a gradual "song evolution" process. More recent studies have shown that song transmission from the west into the east Australian population is relatively frequent and represents the transfer of multiple song types between the Indian and South Pacific Ocean basins (Rekdahl 2012; Allen et al. 2018).

Payne and Guinee (1983) hypothesized three different mechanisms to allow songs to be shared among populations. First, song sharing could occur on shared feeding grounds and/or on shared or partially shared migratory routes. Second, males may share song by visiting more than one wintering ground in consecutive years. And finally, song sharing may occur by males visiting more than one wintering ground within a breeding season. Since the Australian continent separates east and west Australian breeding grounds, song transmission between the two breeding grounds was suggested to occur either through males switching breeding grounds between seasons and/or through males from the different breeding grounds occurring on the same feeding ground in the same season (Noad et al. 2000). Models of humpback whale song revolutions test which of these scenarios are most likely to initiate a song revolution (Lamoni 2018). To successfully mirror the west–east transmission pattern with song revolution data, the model required a song memory to ensure that singers did not revert back to their old song. This suggests that a cognitive capacity is required to concurrently remember the previous seasons' songs and the current song, similar to songbirds (McGregor and Avery 1986). Such capacity is likely essential to not only remember previous songs, but also to rapidly learn the ever-changing, complex and culturally driven song display.

The introduction of a substantial amount of novel material (i.e., when a new song type is introduced) requires rapid learning to reproduce the observed pattern. Garland et al. (2017) investigated how humpback whales learn a new song type by examining instances of song hybridization, where we recorded a whale thought to be in the process of learning a new song during a song revolution. Songs were segmented and then learned as whole themes, akin to how some songbirds learn song and human infants learn language in segments (Garland et al. 2017). The position in the song where a singer switched from "old" to "new" song themes was not random. Garland et al. (2017) uncovered a "*switch when similar rule*": Singers smoothly transitioned from old to new song themes at the location in each song where the similarity in unit type and arrangement was highest.

Garland et al. (2017) confirmed that song structure and syntax are important to humpback whales. To further investigate song learning, Allen et al. (2018) investigated song complexity in the east Australian population over 13 consecutive years. Song complexity increased during periods without revolutions, but decreased after revolutions, leading to oscillations in the long-term pattern of song complexity. Allen et al. (2018) suggested that relative complexity of songs in song revolutions may represent an upper limit to song learning (Allen et al. 2018), while the degree of structural complexity and syntax in the song may facilitate rapid learning of novel material (Allen et al. 2019). How memory, vocal complexity, speed of learning, song structure, and syntax interact is unknown, but the integration of these components is likely to yield intriguing results.

Garland et al. (2011) documented a striking pattern where multiple song types spread from the east Australian population eastward through the populations of the South Pacific, causing multiple song revolutions in a series of cultural waves (Fig. 8.2). Song types typically took two years to spread from east Australia, located in the western South Pacific, across to French Polynesia, in the central South Pacific, approximately 6000 km away. The new, revolutionary songs replaced the current song in each population as they spread. This pattern created a checkerboard of behavioral phenotypes at the decadal scale (Fig. 8.2). Garland et al. (2011) also traced Noad et al.'s (2000) original cultural revolution as it transited across the Pacific, representing a cultural signal that spanned two ocean basins and seven years.

Subsequent work demonstrated that song revolutions rapidly and repeatedly transit exclusively east across the South Pacific region, and song-type differences can be used to identify populations (Garland et al. 2011, 2012, 2013b, 2015, 2017; Owen et al. 2019). Why songs spread in an easterly direction remains elusive; one hypothesis is that this directionality is due to differences in population sizes across the region (Garland et al. 2011). South Pacific humpback whale song transmission remains the best example to date of repeated, population-wide, horizontal cultural transmission in which behavioral variants are passed among populations of a non-human animal. While human behavior (e.g., fashion trends) remains the best analogy for this cultural process, we suspect that another baleen whale, the bowhead whale

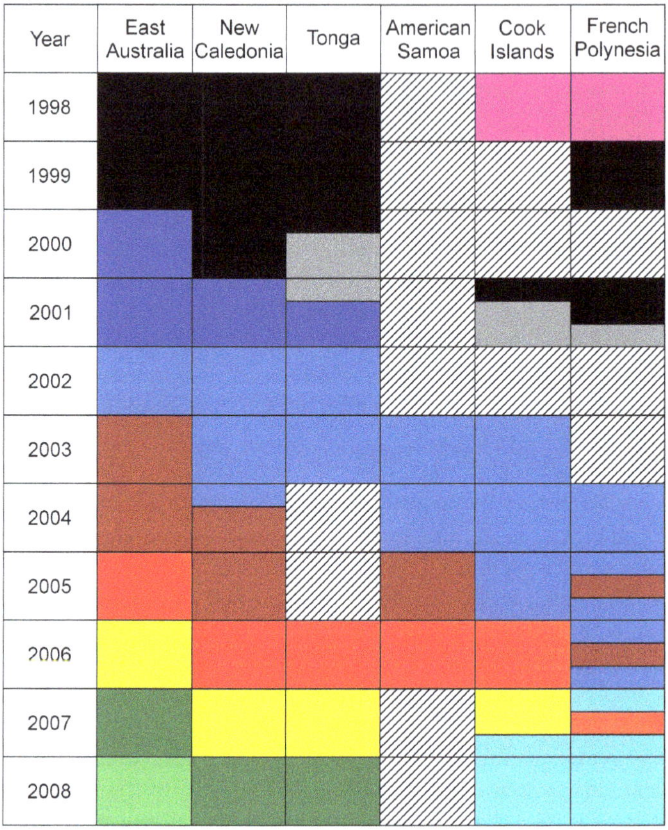

Fig. 8.2 Humpback whale song types identified in the South Pacific region from 1998 to 2008 (from Garland et al., 2011). Populations are listed from west to east across the region. Each color represents a distinct song type; song-type colors are as follows: black, gray, pink, dark blue, blue, light blue, dark red, light red, yellow, dark green, and light green. Two colors within the same year and location indicate that both song types were present. In these cases, the seasons are broken into three periods (early, middle, and late) to indicate when a new song type was recorded. Crosshatching indicates no data available. Reprinted from Current Biology, 21, Garland et al., Dynamic horizontal cultural transmission of humpback whale song at the ocean basin scale, 687–691, 2011, with permission from Elsevier

(*Balaena mysticetus*), may also be a prime candidate for future discoveries of song culture. We know significantly less about bowhead whale song than humpbacks (see Chap. 12); however, bowhead whale song is complex, changes rapidly, possibly evolves, and may be individual or group specific. These features hint that this display may have a cultural component.

8.4 Conclusions and Future Directions

By acting as a "second inheritance system" (Whiten 2017), culture provides an important and understudied aspect of animal behavior. Culture plays an important part in the lives of cetaceans (Whitehead and Rendell 2015). Here, we have presented an overview of evidence for cultural processes in baleen whales. But, why do we care and why should you? One important answer is that a significant number of cetacean populations are impacted by human activities and require conservation management and interventions. There are real population consequences arising from ignoring culture when considering outcomes (Brakes et al. 2019, 2021). As the climate changes, the ability of baleen whale populations to be behaviorally plastic and respond to change may be limited by culture. Maternally directed migratory culture in southern right whales, for example, may result in an inflexibility to change feeding areas (Keith and Bull 2017), further exacerbating a slow recovery or blocking it. Alternatively, cultural traditions may provide a buffer to environmental extremes through exploiting a different foraging niche (Gruber et al. 2019), as recently documented in Shark Bay bottlenose dolphins (Wild et al. 2019b). For example, during an extreme marine heatwave, individual bottlenose dolphins that sponge-fed were less impacted in terms of survival than non-sponge-feeders (Wild et al. 2019b). Wild et al. (2019b) suggest that spongers were buffered against the cascading effects of habitat loss following the heatwave by having access to a less severely affected foraging niche. However, long-term population viability will be impacted if conditions do not improve, as reproduction was equally impacted for both groups (Wild et al. 2019b).

The undertaking of cultural studies of free-ranging cetaceans requires multiyear and long-term datasets with enough detail to track changes. Such datasets are rare and take decades to accumulate. A number of long-term datasets are coming to fruition; we urge readers to think about what data they hold and whether re-examining these data through a cultural lens might yield interesting results. The next decade will be an exciting time for cetacean research, especially relative to considerations of culture. It appears that each time a researcher investigates their behavioral data in detail, a new example of cetacean culture is discovered. Examples of cultural traditions from cetaceans are pivotal to our understanding of culture in animals and are having an impact on the fields of animal culture and their conservation (Brakes et al. 2019, 2021).

Acknowledgements We thank Christopher Clark, Bernd Würsig, and Andrew Whiten for their comments on this chapter. We thank Luciano Valenzuela for providing raw data for Fig. 8.1. E.C.G. is supported by a Royal Society University Research Fellowship, and E.L.C. is supported by a Royal Society of New Zealand Rutherford Discovery Fellowship. This chapter stems from many extended discussions on a Friday evening commuter train between St. Andrews and Edinburgh; we therefore thank our respective favorite brands of whisky for fueling these lively discussions.

References

Allen J, Weinrich M, Hoppitt W, Rendell L (2013) Network-based diffusion analysis reveals cultural transmission of Lobtail feeding in humpback whales. Science 340:485–488. https://doi.org/10.1126/science.1231976

Allen JA, Garland EC, Dunlop RA, Noad MJ (2018) Cultural revolutions reduce complexity in the songs of humpback whales. Proc R Soc B Biol Sci 285:20182088. https://doi.org/10.1098/rspb.2018.2088

Allen JA, Garland EC, Dunlop RA, Noad MJ (2019) Network analysis reveals underlying syntactic features in a vocally learnt mammalian display, humpback whale song. Proc R Soc B Biol Sci 286:20192014. https://doi.org/10.1098/rspb.2019.2014

Baker CS, Palumbi SR, Lambertsen RH et al (1990) Influence of seasonal migration on geographic distribution of mitochondrial DNA haplotypes in humpback whales. Nature 344:238–240

Bannister JL (2001) Status of southern right whales (*Eubalaena australis*) off Australia. J Cetacean Res Manag Spec Issue 2:103–110

Brakes P, Dall SRX, Aplin LM et al (2019) Animal cultures matter for conservation. Science 363:1032–1034. https://doi.org/10.1126/science.aaw3557

Brakes P, Carroll EL, Dall SRX et al (2021) A deepening understanding of animal culture suggests lessons for conservation. Proc R Soc B Biol Sci 288:20202718. https://doi.org/10.1098/rspb.2020.2718

Caldwell MC, Caldwell DK (1965) Individualized whistle contours in bottle-nosed Dolphins (*Tursiops truncatus*). Nature 207:434–435. https://doi.org/10.1038/207434a0

Cantor M, Gero S, Whitehead H, Rendell L (2019) Sperm Whale: the largest toothed creature on earth. In: Würsig B (ed) Ethology and behavioral ecology of odontocetes. Springer, pp 261–280

Carroll E, Patenaude N, Alexander A et al (2011) Population structure and individual movement of southern right whales around New Zealand and Australia. Mar Ecol Prog Ser 432:257–268. https://doi.org/10.3354/meps09145

Carroll EL, Baker CS, Watson M et al (2015) Cultural traditions across a migratory network shape the genetic structure of southern right whales around Australia and New Zealand. Sci Rep 5:16182

Carroll EL, Fewster RM, Childerhouse SJ et al (2016) First direct evidence for natal wintering ground fidelity and estimate of juvenile survival in the New Zealand southern right whale *Eubalaena australis*. PLoS ONE. https://doi.org/10.1371/journal.pone.0146590

Carroll EL, Bruford MW, DeWoody JA et al (2018) Genetic and genomic monitoring with minimally invasive sampling methods. Evol Appl 11:1094–1119. https://doi.org/10.1111/eva.12600

Clapham P, Aguilar A, Hatch LT (2008) Determining spatial and temporal scales for management of cetaceans: lessons from whaling. Mar Mammal Sci 24:183–201

Connor RC, Sakai M, Morisaka T, Allen SJ (2019) The Indo-Pacific Bottlenose Dolphin (*Tursiops aduncus*). In: Würsig B (ed) Ethology and behavioral ecology of odontocetes. Springer, pp 345–368

Danchin E, Nöbel S, Pocheville A et al (2018) Cultural flies: conformist social learning in fruitflies predicts long-lasting mate-choice traditions. Science 362:1025–1030. https://doi.org/10.1126/science.aat1590

Deecke VB, Ford JKB, Spong P (2000) Dialect change in resident killer whales: implications for vocal learning and cultural transmission. Anim Behav 60:629–638. https://doi.org/10.1006/anbe.2000.1454

Filatova OA, Miller PJO (2015) An agent-based model of dialect evolution in killer whales. J Theor Biol 373:82–91. https://doi.org/10.1016/j.jtbi.2015.03.020

Filatova OA, Samarra FIP, Deecke VB et al (2015) Cultural evolution of killer whale calls: background, mechanisms and consequences. Behaviour 152:2001–2038. https://doi.org/10.1163/1568539X-00003317

Foote AD, Vijay N, Avila-Arcos M et al (2016) Genome-culture coevolution promotes rapid divergence of killer whale ecotypes. Nat Commun 7:11693. https://doi.org/10.1101/040295

Ford JKB (1991) Vocal traditions among resident killer whales (*Orcinus orca*) in coastal waters of British Columbia. Can J Zool 69:1454–1483

Ford JKB (2019) Killer Whales: behavior, social organization, and ecology of the Oceans' Apex Predators. In: Würsig B (ed) Ethology and behavioral ecology of odontocetes. Springer, pp 239–259

Ford JKB, Ellis GM, Barrett-Lennard LG et al (1998) Dietary specialization in two sympatric populations of killer whales (*Orcinus orca*) in coastal British Columbia and adjacent waters. Can J Zool 76:1456–1471. https://doi.org/10.1139/z98-089

Fragaszy DM, Perry S (2003) Preface. In: Fragaszy DM, Perry S (eds) The biology of traditions: models and evidence. Cambridge University Press, Cambridge, pp xiii–xvi

Garland EC, Goldizen AW, Rekdahl ML et al (2011) Dynamic horizontal cultural transmission of humpback whale song at the ocean basin scale. Curr Biol 21:687–691. https://doi.org/10.1016/j.cub.2011.03.019

Garland EC, Lilley MS, Goldizen AW et al (2012) Improved versions of the Levenshtein distance method for comparing sequence information in animals' vocalisations: tests using humpback whale song. Behaviour 149:1413–1441. https://doi.org/10.1163/1568539X-00003032

Garland EC, Gedamke J, Rekdahl ML et al (2013a) Humpback whale song on the Southern Ocean feeding grounds: implications for cultural transmission. PLoS ONE 8:e79422. https://doi.org/10.1371/journal.pone.0079422

Garland EC, Noad MJ, Goldizen AW et al (2013b) Quantifying humpback whale song sequences to understand the dynamics of song exchange at the ocean basin scale. J Acoust Soc Am 133:560–569. https://doi.org/10.1121/1.4770232

Garland EC, Goldizen AW, Lilley MS et al (2015) Population structure of humpback whales in the western and central South Pacific Ocean as determined by vocal exchange among populations. Conserv Biol 29:1198–1207. https://doi.org/10.1111/cobi.12492

Garland EC, Rendell L, Lamoni L et al (2017) Song hybridization events during revolutionary song change provide insights into cultural transmission in humpback whales. Proc Natl Acad Sci U S A 114:7822–7829. https://doi.org/10.1073/pnas.1621072114

Gruber T, Luncz L, Mörchen J et al (2019) Cultural change in animals: a flexible behavioural adaptation to human disturbance. Palgrave Commun 5:1–9. https://doi.org/10.1057/s41599-019-0271-4

Herman LM (2017) The multiple functions of male song within the humpback whale (*Megaptera novaeangliae*) mating system: review, evaluation, and synthesis. Biol Rev 92:1795–1818. https://doi.org/10.1111/brv.12309

Herman LM, Tavolga WN (1980) The communication system of cetaceans. In: Herman LM (ed) Cetacean behaviour: mechanisms and functions. Wiley, New York, pp 149–209

Jackson JA, Patenaude NJ, Carroll EL, Baker CS (2008) How few whales were there after whaling? Inference from contemporary mtDNA diversity. Mol Ecol. https://doi.org/10.1111/j.1365-294X.2007.03497.x

Janik VM (2000) Whistle matching in wild bottlenose dolphins (*Tursiops truncatus*). Science 289:1355–1357. https://doi.org/10.1126/science.289.5483.1355

Janik VM (2009) Acoustic communication in delphinids. Adv Study Behav 40:123–157. https://doi.org/10.1016/S0065-3454(09)40004-4

Janik VM, Sayigh LS, Wells RS (2006) Signature whistle shape conveys identity information to bottlenose dolphins. Proc Natl Acad Sci U S A 103:8293–8297. https://doi.org/10.1073/pnas.0509918103

Keith SA, Bull JW (2017) Animal culture impacts species' capacity to realise climate-driven range shifts. Ecography (cop) 40:296–304. https://doi.org/10.1111/ecog.02481

King SL, Janik VM (2013) Bottlenose dolphins can use learned vocal labels to address each other. Proc Natl Acad Sci U S A 110:13216–13221. https://doi.org/10.1073/pnas

Kopps AM, Krützen M, Allen SJ et al (2014) Characterizing the socially transmitted foraging tactic "sponging" by bottlenose dolphins (*Tursiops* sp.) in the western gulf of Shark Bay, Western Australia. Mar Mammal Sci 30:847–863. https://doi.org/10.1111/mms.12089

Krutzen M, Mann J, Heithaus MR et al (2005) Cultural transmission of tool use in bottlenose dolphins. Proc Natl Acad Sci U S A 102:8939–8943. https://doi.org/10.1073/pnas.0500232102

Krützen M, van Schaik C, Whiten A (2007) The animal cultures debate: response to Laland and Janik. Trends Ecol Evol 22:6. https://doi.org/10.1016/j.tree.2006.10.011

Krützen M, Sina K, MacLeod CD et al (2014) Cultural transmission of tool use by Indo-Pacific bottlenose dolphins (*Tursiops* sp.) provides access to a novel foraging niche. Proc R Soc B 281:20140374. https://doi.org/10.1098/rspb.2014.0374

Laland KN (2008) Animal cultures. Curr Biol 18:366–370. https://doi.org/10.1016/j.cub.2008.02.049

Laland KN, Janik VM (2006) The animal cultures debate. Trends Ecol Evol 21:542–547. https://doi.org/10.1016/j.tree.2006.06.005

Lamoni L (2018) The role of individual behaviour in the collective cultural evolution of humpback whale songs. PhD thesis University of St, Andrews, UK. 198 p

Leaper R, Cooke J, Trathan P et al (2006) Global climate drives southern right whale (*Eubalaena australis*) population dynamics. Biol Lett 2:289–292. https://doi.org/10.1098/rsbl.2005.0431

Mann J (2019) Maternal care and offspring development in odontocetes. In: Würsig B (ed) Ethology and behavioral ecology of odontocetes. Springer, pp 95–116

Mann J, Sargeant B (2003) Like mother, like calf: the ontogeny of foraging traditions in wild Indian Ocean bottlenose dolphins (*Tursiops* sp.). In: Fragaszy DM, Perry S (eds) The biology of traditions: models and evidence. Cambridge University Press, Cambridge, pp 236–266

McGregor PK, Avery MI (1986) The Unsung Songs of Great Tits (*Parus major*)—learning neighbors songs for discrimination. Behav Ecol Sociobiol 18:311–316

Miller PJO, Bain DE (2000) Within-pod variation in the sound production of a pod of killer whales, Orcinus orca. Anim Behav 60:617–628. https://doi.org/10.1006/anbe.2000.1503

Noad M, Cato DH, Bryden MM et al (2000) Cultural revoultion in whale songs. Nature 408:537

Owen C, Rendell L, Constantine R et al (2019) Migratory convergence facilitates cultural transmission of humpback whale song. R Soc Open Sci 6:190337. https://doi.org/10.1098/rsos.190337

Patenaude NJ, Portway VA, Schaeff CM et al (2007) Mitochondrial DNA diversity and population structure among southern right whales (*Eubalaena australis*). J Hered 98:147–157. https://doi.org/10.1093/jhered/esm005

Payne K, Payne RS (1985) Large scale changes over 19 years in songs of humpback whales in Bermuda. Z Tierpsychol 68:89–114. https://doi.org/10.1111/j.1439-0310.1985.tb00118.x

Payne K, Tyack P, Payne R (1983) Progressive changes in the songs of humpback whales (*Megaptera novaeangliae*): a detailed analysis of two seasons in Hawaii. In: Payne R (ed) Communication and behavior of Whales. Westview Press Inc., Boulder, CO, pp 9–57

Payne R, Guinee LN (1983) Humpback whale (*Megaptera novaeangliae*) songs as an indicator of 'stocks.' In: Payne R (ed) Communication and behavior of Whales. Westview Press Inc., Boulder, CO, pp 333–358

Payne RS, McVay S (1971) Songs of Humpback Whales. Science 173:585–597

Ramp C, Delarue J, Palsbøll PJ et al (2015) Adapting to a warmer ocean—seasonal shift of baleen whale movements over three decades. PLoS ONE 10:1–15. https://doi.org/10.1371/journal.pone.0121374

Ramsey G (2013) Culture in humans and other animals. Biol Philos 28:457–479. https://doi.org/10.1007/s10539-012-9347-x

Rekdahl ML (2012) Humpback whale vocal communication: use and stability of social calls and revolutions in the songs of east Australian whales. PhD thesis. The University of Queensland, Brisbane, Australia. 174 p

Rendell L, Whitehead H (2001) Culture in whales and dolphins. Behav Brain Sci 24:309–382

Richards DG, Wolz JP, Herman LM (1984) Vocal mimicry of computer-generated sounds and vocal labelling of objects by bottlenose dolphins (*Tursiops truncatus*): evidence for vocal learning. J Comp Psychol 98:10–28

Rowntree VJ, Payne R, Schell D (2001) Changing patterns of habitat use by southern right whales (*Eubalaena australis*) on their nursery ground at Península Valdés, Argentina, and in their long-range movements. J Cetacean Res Manag Spec Issue 2:133–143

Sargeant BL, Mann J (2009) From social learning to culture: Intrapopulation variation in bottlenose dolphins. In: The question of animal culture, pp 152–173

Schlaepfer MA, Runge MC, Sherman PW (2002) Ecological and evolutionary traps. Trends Ecol Evol 17:474–480. https://doi.org/10.1016/S0169-5347(02)02580-6

Seyboth E, Groch KR, Dalla Rosa L et al (2016) Southern right whale (*Eubalaena australis*) reproductive success is influenced by Krill (*Euphausia superba*) density and climate. Sci Rep 6:1–8. https://doi.org/10.1038/srep28205

Valenzuela LO, Sironi M, Rowntree VJ, Seger J (2009) Isotopic and genetic evidence for culturally inherited site fidelity to feeding grounds in southern right whales (*Eubalaena australis*). Mol Ecol 18:782–791. https://doi.org/10.1111/j.1365-294X.2008.04069.x

Weinrich MT, Schilling MR, Belt CR (1992) Evidence for acquisition of a novel feeding behaviour: lobtail feeding in humpback whales, *Megaptera novaeangliae*. Anim Behav 44:1059–1072. https://doi.org/10.1016/S0003-3472(05)80318-5

Whitehead H (2010) Conserving and managing animals that learn socially and share cultures. Learn Behav 38:329–336. https://doi.org/10.3758/LB.38.3.329

Whitehead H, Rendell L (2015) The cultural lives of whales and dolphins. The University of Chicago Press, Chicago

Whitehead H, Laland KN, Rendell L et al (2019) The reach of gene–culture coevolution in animals. Nat Commun. https://doi.org/10.1038/s41467-019-10293-y

Whiten A (2017) A second inheritance system: the extension of biology through culture. Interface Focus 7:160142

Whiten A (2021) The burgeoning reach of animal culture. Science

Whiten A, Goodall J, McGrew WC et al (1999) Cultures in chimpanzees. Nature 399:682–685

Whiten A, Spiteri A, Horner V et al (2007) Transmission of multiple traditions within and between chimpanzee groups. Curr Biol 17:1038–1043. https://doi.org/10.1016/j.cub.2007.05.031

Whiten A, Ayala FJ, Feldman MWD, Laland KN (2017) The extension of biology through culture. Proc Natl Acad Sci 114:7775–7781

Wild S, Allen SJ, Krützen M et al (2019a) Multi-network-based diffusion analysis reveals vertical cultural transmission of sponge tool use within dolphin matrilines. Biol Lett 15:20190227. https://doi.org/10.1098/rsbl.2019.0227

Wild S, Krützen M, Rankin RW et al (2019b) Long-term decline in survival and reproduction of dolphins following a marine heatwave. Curr Biol 29:R239–R240. https://doi.org/10.1016/j.cub.2019.02.047

Würsig B (2019) Ethology and behavioral ecology of odontocetes: concluding remarks. In: Würsig B (ed) Ethology and behavioral ecology of odontocetes. Springer, pp 483–494

Yurk H, Barrett-Lennard L, Ford JKB, Matkin CO (2002) Cultural transmission within maternal lineages: vocal clans in resident killer whales in southern Alaska. Anim Behav 63:1103–1119

Part II
Examples of Mysticete Acoustic Ethology

Chapter 9
The Bioacoustics of Blue Whales—Global Diversity and Behavioral Variability in a Foraging Specialist

Ana Širović and Erin M. Oleson

Abstract Blue whales are the largest animals ever to inhabit our planet. Worldwide they comprise at least four subspecies, currently recognized to make up to 11 distinct populations which share a krill-specialized foraging preference. The populations are defined by common geography, migratory behavior, and destinations but can also be delineated by unique songs that likely function as a reproductive display. Male singers in each population typically sing the same song type and share some common behavioral patterns in the use of songs and song units. Migratory behavior and characteristics of their habitat shape the differences in the structure of blue whale songs among different populations. Offshore populations tend to have simpler songs than populations occurring more coastally, possibly reflecting differing propagation environments. In areas of high sympatry, such as the Indian Ocean, the increased complexity in song structure may arise as a result of acoustic competition. Both male and female blue whales also produce other sounds, primarily including the so-called D calls, commonly heard associated with foraging behavior. These calls are spectrally similar to foraging signals of other balaenopterids, and eavesdropping may be a mechanism for locating productive areas intra- and interspecies. While some balaenopterids fast on breeding grounds and during much of the migration, blue whales forage along their migration route and in breeding areas, indicating that prey availability is an important driver of their behavior year-round. Fine-scale variability in their acoustic behaviors and relationships to their habitat and prey variability may be the keys that can help us explain the long-term drivers of blue whale population trends. We highlight research questions and possible future directions that would help further enhance our understanding of the interplay between acoustic behavior, diversity, and ecology of this species, along with understanding how some present and future threats, including climate change, may impact the populations of this charismatic species.

A. Širović (✉)
Norwegian University of Science and Technology, Trondheim, Norway
e-mail: ana.sirovic@ntnu.no

E. M. Oleson
Pacific Islands Fisheries Science Center, NOAA Fisheries, Honolulu, HI, USA

C.W. Clark and E. C. Garland (eds.), *Ethology and Behavioral Ecology of Mysticetes*,
Ethology and Behavioral Ecology of Marine Mammals,
https://doi.org/10.1007/978-3-030-98449-6_9

Keywords Blue whales · Acoustic behavior · Whale song · Population structure · Song variability

9.1 Foreward by A Širović

I had been listening to them for years, but I had never met them in real life. No, I am not talking about my favorite band; it turns out it was much easier for me to meet U2 than Antarctic blue whales. The former took about five years, the latter nearly 19.

I wrote my dissertation on Antarctic blue whales and despite spending about eight months at sea in the Southern Ocean, during that time, the closest I came to one was when a visual observer on one of my research trips reported a "like blue" sighting. However, by the time I made it up to the bridge from my acoustics corner in the lab on the main deck of the vessel, the whale was not to be seen again. I heard them plenty, though. Their calls travel hundreds of miles, but as I was always on multidisciplinary research trips with focus on something other than blue whales, we never had the time to divert potentially hundreds of miles off course to find them. At times, it was hard for my fellow passengers, scientists, and crew alike, to fathom that I am hearing blue whales maybe even daily, yet we were not seeing any, but that is the power of sound in the marine environment.

Luckily for me, it turns out you can learn a lot about blue whales by just listening to them. We have been doing that for decades now. Our understanding of their distribution and diversity would be much poorer without eavesdropping on them. And in the end, it took a dedicated eavesdropping effort for me to finally meet my first Antarctic blue, courtesy of an invitation from Dr. Brian Miller to join a voyage of the Australian Antarctic Division.

It was a gray and hazy day, the kind that dominates in the Southern Ocean despite all the clear blue skies you see in nature shows about Antarctica. We tracked down a pair of blue whales by localizing their calls and then spent a few hours in their vicinity while all the teams were collecting their data: photo ID, drone footage, more acoustic recordings, etc. When you see a blue whale, it takes time to realize just how large the animal is because you only see one small part at a time as they surface. But that surfacing goes on and on and on and on and on as the body of the largest animal on the planet slowly reveals itself in front of you until, finally, its disproportionally small dorsal fin appears. And even then, it is the sound, the deep rumble of its strong exhale that fully conveys that you are looking at a special sight.

What amazed our fellow krill biologists on board was that in addition to blue whales, our acoustic tracking also led us to large krill aggregations. And often, fin whales and humpback whales, too. It turns out that by listening to blue whales, we may be able to learn about a lot more than just the sounds that they make.

9.2 Introduction

Blue whales (*Balaenoptera musculus*) are, in many ways, the epitome of charismatic megafauna and a great example of why humans are often fascinated by such large animals. As the largest animals to ever inhabit our planet (Goldbogen and Madsen 2018) at over 30 m in length, blue whales were the primary target of the twentieth century, pre-WWII whaling industry (Branch et al. 2007). They feed very selectively on krill that are only a few centimeters long (Nemoto 1957; Kawamura 1980). Blue whales are considered a cosmopolitan species, although adaptations to local environmental and ecological factors have led to differentiation into several recognized subspecies and even more finely into multiple populations within some subspecies. These populations are defined by common geography, migratory behavior, and destinations (Torres-Florez et al. 2014; Barlow et al. 2018), as well as their unique song dialects (Fig. 9.1, McDonald et al. 2006). Here, we use the term "population" in its biological sense, while noting that we are largely basing it on observations of singing behavior, which is an easily observable behavioral trait produced by males. Furthermore, we identify and differentiate between populations based on the uniquely

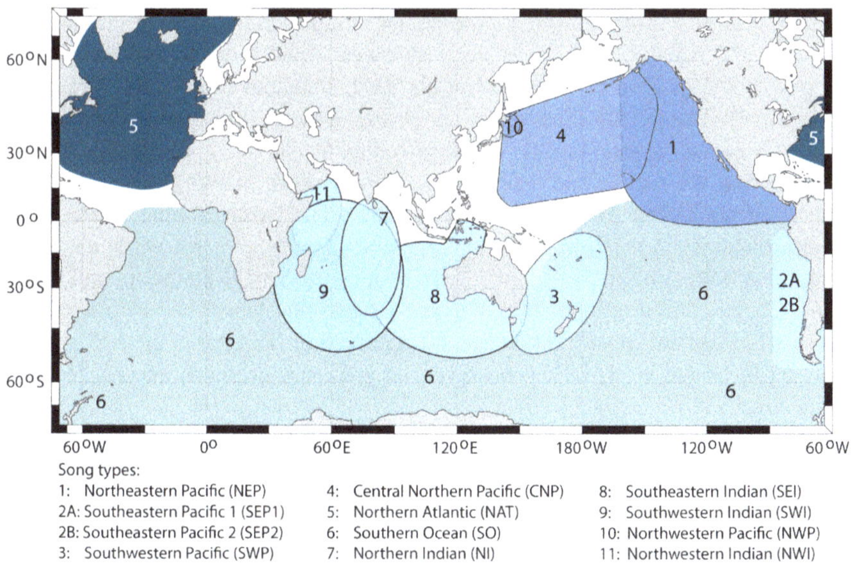

Fig. 9.1 Biogeography of blue whale song occurrence, showing exemplars of 11 confirmed or putative blue whale song types. Song types 1–9 are largely as reported by McDonald et al. (2006), with an addition of a new variant of the Southeast Pacific song type (Buchan et al. 2014). Song types 10 and 11 are two entirely new, putative blue whale song types identified since the McDonald et al. (2006) review: song type 10 off Japan in the Northwestern Pacific (McDonald et al. 2017) and song type 11 off Oman in the Northwestern Indian Ocean (Cerchio et al. 2020). Other names that have been used for these songs in publications are 2A and 2B: Chilean, 3: New Zealand, 4: North Pacific, 7: Sri Lanka, 8: Indo-Australian, 9: Madagascar

obvious features of songs produced by singers in different ocean regions (i.e., "song types"). While populations are referred to by a variety of names in the literature, for simplicity, here we use names based on the ocean basins and regions within which they are most commonly associated, although in Fig. 9.1 we provide a reference to other names that have been used in the literature (e.g., the Northern Indian population has also been referred to as Sri Lankan blue whales, or the Southwestern Pacific population is also known as New Zealand blue whales, etc.). We also refer to song types by the same ocean basin-based names (e.g., the Northeast Pacific song type). The link between acoustic characteristics and population structure makes blue whales an interesting case study for understanding behavior in baleen whales and for considering how acoustic behaviors can be used to better understand species' ecology and population dynamics. Additionally, population level studies of naturally occurring acoustic behavior are critical for understanding and thus possibly mitigating risks from threats facing blue whale populations.

Blue whale subspecies are generally subdivided based on geographic, genetic, and morphological differences, which are traditional markers of speciation in mammals. The Committee on Taxonomy of the Society for Marine Mammalogy (2020) presently recognizes four distinct blue whale subspecies, two in the Southern Hemisphere and two in the Northern Hemisphere. The Antarctic blue whale (*B. m. intermedia*), the largest in size despite its subspecies moniker, is found in the circumpolar region around Antarctica and seasonally at lower latitudes of the Southern Hemisphere (Širović et al. 2004; Stafford et al. 2004; Samaran et al. 2013; Sears and Perrin 2018). The pygmy blue whale (*B. m. brevicauda*), the smallest of the blue whale subspecies, is found in the southern Indian Ocean and southwestern Pacific Ocean during the winter and typically uses areas north of 60° S as its foraging grounds (Branch et al. 2007; Leroy et al. 2018b). The Northern blue whale, *B. m. musculus*, resides in the North Atlantic and North Pacific Oceans, while the North Indian blue whale, *B. m. indica*, primarily resides off Sri Lanka (Sears and Perrin 2018). Genetic and morphometric analyses suggest that blue whales off Chile are sufficiently different to warrant additional subdivision (Pastene et al. 2020; Leslie et al. 2020; Leduc et al. 2017); however, they are not presently recognized as a subspecies.

While blue whale subspecies and populations are distinguished by a variety of biological and ecological factors, extreme specificity in prey selection that is found in most, if not all blue whale populations worldwide, drives the ecological commonality against which many other behavioral similarities and differences can be considered. Unlike most other baleen whale species that feed on a variety of swarming crustacean and schooling fish species, blue whales feed primarily on krill (Nemoto 1957). Blue whales specifically target dense aggregations of krill (Hazen et al. 2015) and often have a preference for a particular species of krill. In the Southern Ocean, Antarctic krill, *Euphausia superba*, is the main prey item for much of marine life in the region (Kawamura 1994). In the eastern North Pacific, blue whale diet is dominated by two krill species: *Thysanoessa spinifera* and *E. pacifica* (Croll et al. 2005, 1998; Fiedler et al. 1998), with a preference for *T. spinifera* (Nickels et al. 2018), while in the North Atlantic and the western North Pacific, *T. inermis* is the primary choice

(Nemoto 1957). At lower latitudes in the Southrn Hemisphere, blue whales feed on *Nyctiphanes australis* off Australia (Gill 2002) which is also their likely prey in the waters off New Zealand (Torres 2013). Blue whales off South Africa consume *E. recurva* and *E. diomedeae,* and their primary prey off Crozet and Kerguelen Islands are *E. vallentini* (Kawamura 1980).

Blue whale social behavior has not been closely studied, but what we know points toward more commonalities than differences across populations. Blue whales generally are seen alone or in small groups (Calambokidis et al. 2007; Sears and Perrin 2018) and have been reported to form semi-stable male–female pairs during summers on their feeding grounds (Schall et al. 2019; Sears and Larsen 2002; Berchok et al. 2006). The details of their mating behavior are unknown; pairs of blue whales have not been tracked beyond feeding grounds, though the nature of these pairs suggests that they may be a part of breeding behaviors. In feeding areas, blue whales form loose aggregations (Miller et al. 2019; Lomac-MacNair and Smultea 2016), often engaging in lunge-feeding behaviors during the day (Oleson et al. 2007b; Caruso et al. 2020; Gill 2002; Calambokidis et al. 2007).

There also appears to be some consistency in the behavioral context associated with acoustic signaling. Blue whale breeding displays include songs (Edds-Walton 1997), which are population-specific (McDonald et al. 2006, Fig. 9.1) and consist of one to five highly stereotypical units that last 8–20 s and are repeated approximately every 1–2 min (Mellinger and Clark 2003; McDonald et al. 2006). These songs are produced by males (Oleson et al. 2007b; Lewis et al. 2018). Blue whales also produce non-song sounds, with the most common a variable downsweep generally referred to as a D call (Fig. 9.2). These calls are more variable in their spectral and temporal characteristics, are produced by both sexes (Oleson et al. 2007b), and do not appear to consistently vary geographically (Thode et al. 2000; Rankin et al. 2005; Schall et al. 2019; Mellinger and Clark 2003; Ljungblad et al. 1997; Berchok

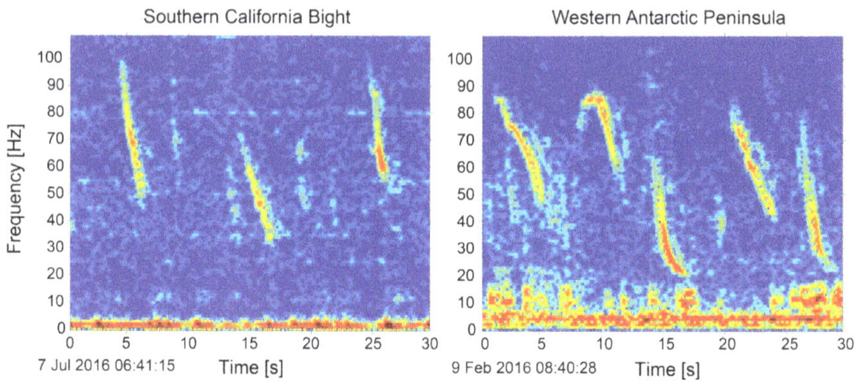

Fig. 9.2 Examples of diversity in blue whale D calls, from Northeast Pacific population recorded off Southern California (left) and the Southern Ocean population recorded off the Western Antarctic Peninsula (right) recorded in 2016. Spectrograms created with 1 Hz frequency and 0.1 s time resolutions

et al. 2006). These calls have been recorded across a broad range of social contexts, generally between or among pairs or dispersed groups (McDonald et al. 2001; Oleson et al. 2007b) during foraging (Oleson et al. 2007b) and during agonistic interactions likely related to mating (Schall et al. 2020). Both blue whale songs and D calls are generally of very high intensity (Thode et al. 2000; Širović et al. 2007; Gavrilov and McCauley 2013; Miller et al. 2021). In most regions, the peak period of song detections is different from the peak period of D call detections. On feeding grounds, D calls are most commonly detected during spring and summer, while songs are most commonly detected during late summer, fall and in some places into early winter (Clark and Gagnon 2002; Watkins et al. 2000; Stafford et al. 2001; Širović et al. 2004; Oleson et al. 2007c; Samaran et al. 2013; Paniagua-Mendoza et al. 2017). A similar temporal separation is also observed in the Benguela system, an overwintering area for Antarctic blue whales (Shabangu et al. 2019). The variable nature of D calls has made them difficult to detect reliably and efficiently using automated algorithms, resulting in relatively few studies of the presence of D calls in long-term datasets from other potential breeding areas. Blue whales have been recorded producing song units as single calls, as well as other sounds (described in Mellinger and Clark 2003; Oleson et al. 2007b; Berchok et al. 2006), although the behavioral context and prevalence of these sounds within the overall vocal repertoire has not been well-studied.

Despite common foraging specificity, as well as social and overall vocal behavior, several differences also exist among blue whale populations worldwide. The established paradigm of baleen whale migrations is that they spend their summers in high latitude, productive feeding grounds and their winters in low latitude, breeding grounds (Kellogg 1929, Chap. 4). Productive regions with high krill densities, including upwelling zones, largely overlap with the known blue whale foraging grounds (Buchan et al. 2014; Fiedler et al. 1998; Croll et al. 2005). Blue whale migrations, however, are not always simple (Chap. 4). While most populations undergo long-range migrations, one population in the Southwestern Pacific appears to be largely resident to New Zealand waters (Barlow et al. 2018). In addition, although generally migratory, there is evidence that some portion of the Antarctic blue whale population remains at high latitudes year-round (Thomisch et al. 2016; Širović et al. 2004). In the eastern North Pacific, the primary blue whale foraging grounds are actually at mid-latitudes (32°–45° N) with some part of the population extending farther north into the Gulf of Alaska (Calambokidis et al. 2009; Rice et al. 2021; Stafford 2003). And while there is little evidence of large-scale migration across the equator in most blue whale populations, the Northern Indian population, which is confined from the north by the Asian continent, performs their seasonal migrations into the southern Indian Ocean. Finally, even for populations that migrate to productive higher-latitude feeding grounds foraging along migration routes and at lower latitudes appears to be common (Mate et al. 1999; Bailey et al. 2010; Hucke-Gaete et al. 2018). Unlike other baleen whale populations, however, blue whales cannot endure a prolonged fasting period on breeding grounds and are thought to choose low latitude regions that support high densities of krill (Matteson 2010).

The variation in the extent and timing of the specific migratory behavior, as well as the preferred habitat of individual blue whale populations, is likely driven by environmental conditions, including the distribution of their primary prey (Szesciorka et al. 2020). Sustained year-round availability of krill in the waters off New Zealand (Torres 2013) may be one reason that the Southwestern Pacific population can eschew traditional migratory behavior in favor of adequate localized prey resources. Several migratory populations of blue whales have foraging grounds that occur relatively close to shore, with the Southeastern Indian blue whale population foraging in productive waters off south and west Australia, the Northeastern Pacific population foraging in coastal waters off the US west coast, and Southeastern Pacific blue whales foraging in Chilean fjords. For the most part, these areas feature high-productivity upwelled waters relatively close to the coast. In contrast, the Northern Atlantic, Southern Ocean, and Central Northern Pacific blue whale populations occur over a much larger geographic region and in deeper oceanic waters.

Bass and Clark (2003) and Clark and Ellison (2004) proposed that the character of baleen whale songs and contact calls is related to the habitat and social context in which they are produced. According to their hypothesis, tonal blue whale songs are optimized for long-range acoustic propagation across their broad oceanic habitat, as opposed to humpback (*Megaptera novaeangliae*) or bowhead whales (*Balaena mysticetus*) which have optimized their songs to take advantage of local ambient noise conditions in the more localized regions where those whales aggregate. This same concept may be extended to relating the habitat characters of common feeding grounds for each blue whale population to the complexity observed within the song of that population. We propose that differences in the structure of blue whale songs are correlated with, if not influenced by the variation in the physical environment encountered by blue whale populations worldwide. Populations with broad oceanic distribution (Central Northern Pacific, Northern Atlantic, Southern Ocean) have songs consisting entirely of simple tonal units, while populations found in more coastal environments (Northeastern Pacific, Southeastern Pacific, Southwestern Pacific) have more complex song phrases consisting of both pulsed and tonal units (Fig. 9.3; McDonald et al. 2006). Although the complex songs of pelagic pygmy blue whale populations in the Indian Ocean do not appear to follow this relationship between song structure and habitat, the greater diversity and sympatry of blue whale populations in this ocean basin may well play a role in the evolution of the complexity of their songs, within and between populations (Naugler and Ratcliffe 1994). Detailed examination of fine-scale variation in blue whale songs has only recently begun (Lewis and Širović 2018; Jolliffe et al. 2019; Carbaugh-Rutland et al. 2021). Additional study of these differences in song complexity and the communicative significance of that complexity will allow a more detailed examination into the relationships between acoustic behavior and individual population habitat preferences, migratory behavior, and other factors that potentially influenced the evolution of each subspecies and population of blue whales.

Acoustic methods have already provided a multitude of insights and new revelations into the distribution, behavior, and ecology of blue whales (e.g., McDonald et al. 2006; Oleson et al. 2007b; Širović and Hildebrand 2011; Stafford et al. 2009;

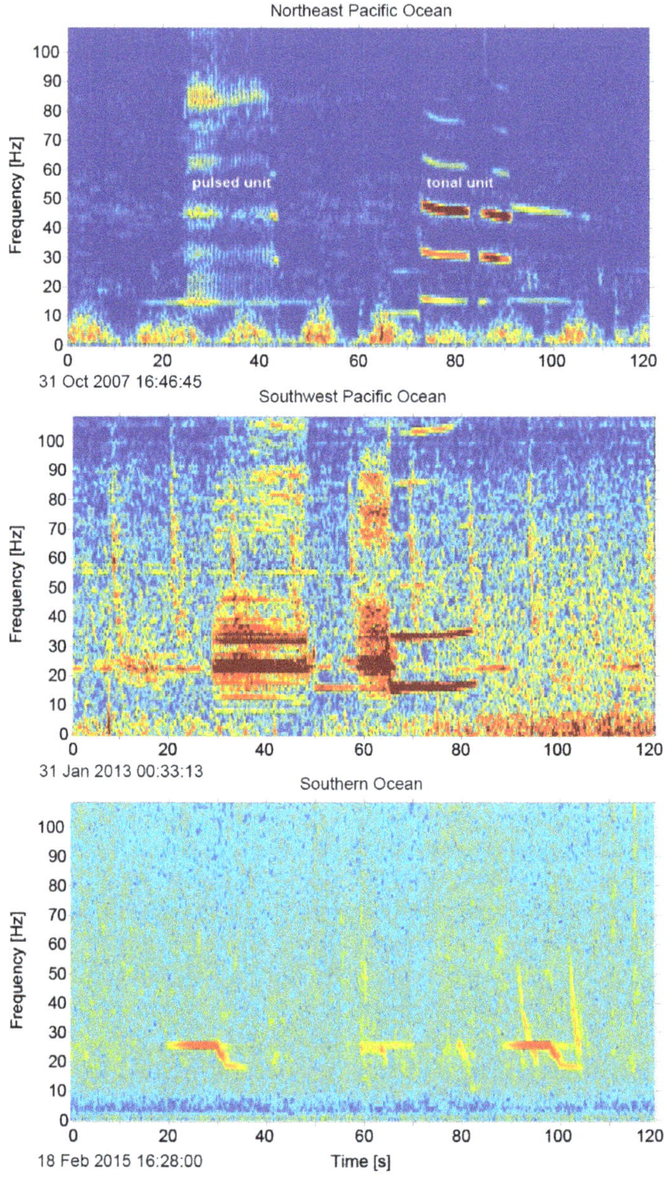

Fig. 9.3 Examples of diversity in blue whale songs, from the Northeastern Pacific Ocean population recorded in Southern California (top), Southwestern Pacific Ocean population recorded off New Zealand (middle), and the Southern Ocean population recorded off the Western Antarctic Peninsula (bottom). Spectrograms created with 1 Hz frequency and 0.1 s time resolutions. Note pulsed and tonal units marked in Northeastern Pacific song, as well as the presence of a second, lower intensity tonal unit after the marked one. Lower intensity tone of Southern Ocean song is also visible between two higher intensity song units. Full blue whale song syntax is frequently not recorded because of propagation effects; only the most intense segments are typically dominant in long-term recordings. Southwestern Pacific song recording by Miller et al. (2014a, b)

Balcazar et al. 2015; Miller et al. 2019). In this overview, we offer a synthesis of our improved understanding of the variability in blue whale song and D-calling behaviors within and across populations, primarily but not exclusively since McDonald et al. (2006) reviewed blue whale song characteristics and proposed a link to population structure. We then focus on new questions and directions for passive acoustic studies of blue whales that further our exploration into the distribution, characteristics, and fine-scale acoustic behavior and feeding ecology. We hope this will help guide us to better understand the population structure, foraging and migratory behavior, and status of this feeding specialist in varied environments worldwide.

9.3 Recent Advances in Blue Whale Acoustic Research

9.3.1 Spatial and Temporal Variability in Blue Whale Song

Although blue whale populations can be distinguished by their characteristic songs (Fig. 9.1), recent finer-scale study of the variability in blue whale songs within and across populations provides further clues regarding the behavioral plasticity in this species, as well as the role of signaling within their life history and ecology. Discussion of this variability requires some consistent terminology, similar to that defined for humpback whales (Payne and McVay 1971). When discussing blue whales, the structure of *songs,* or their *song syntax,* consists of *phrases*, which in turn are made up of individual *units*. The time between units, measured from the end of one unit to the start of the next, is the *interunit interval*, and the time between song phrases is the *interphrase interval.*

9.3.1.1 Long-Term Variation: Frequency Shifts and Variability in Song Length

Although most blue whale songs show remarkable stability in the characteristics of individual units and the assembly of those units into phrases, long-term drift in song unit features has been observed in several populations where multiple years of blue whale song recordings are available for comparison. Many blue whale populations have at least one tonal unit, and repeated measures of the frequency of peak energy (i.e., peak frequency) in these tonal units through time have revealed a decline in frequency for at least nine populations (McDonald et al. 2009; Gavrilov et al. 2012; Gavrilov et al. 2011; Leroy et al. 2018a; Malige et al. 2020; Miller et al. 2014a; Miksis-Olds et al. 2018). The two song types that do not have a tonal unit, the Southeastern Pacific songs, also show a decline in both the peak frequency and the pulse rate (i.e., repetition rate of individual pulses that make up the unit) of the highest signal-to-noise ratio (SNR) units (Malige et al. 2020). The frequency of the Northeastern Pacific song tonal unit, for example, changed from 21.9 to 15.2 Hz between 1963 and 2008 (McDonald et al. 2009). The rate of the frequency decline varies across populations, from a high of 0.8% change per year for Northeastern Pacific blue whale song to 0.3% annually for Northern Indian Ocean blue whale songs (Malige et al. 2020). Subsequent analyses of multi-year datasets from Northern Indian (Miksis-Olds et al. 2018), Southeastern Indian (Gavrilov et al. 2011), Southwestern Pacific (Miller et al. 2014a) and Southern Ocean song units (Gavrilov et al. 2012) have shown a continued rate of peak frequency decline similar to that measured for the same regions by McDonald et al. (2009). In several populations, the rate of peak frequency decline has been measured for several song units. In Southwestern Pacific blue whale song, the rate of decline of all four units appears consistent over time (Miller et al. 2014b), similar to the decline in peak frequency for two pulsed units of one Southeastern Pacific blue whale song (SEP2, Malige et al. 2020). In contrast, the rates of peak frequency decline were different for two measured song units in the Northern Indian blue whale song as well as one Southeastern Pacific song (SEP1) but still ranged between 0.3 and 0.6% per year (Miksis-Olds et al. 2018; Malige et al. 2020). The only population with a documented change in the rate of the song unit tonal frequency decline is the Northeastern Pacific, whose decline slowed from 0.8% per year through 2008 (McDonald et al. 2009), to 0.5% per year measured for 2012–2014 (Širović 2016).

There is still no broadly accepted explanation why blue whale song unit frequencies are decreasing for so many populations. McDonald et al. (2009) considered several causes for this decline across populations, and generally rejected the influence of increases in body size post-whaling, changes in ocean ambient noise or ocean chemistry, and interference from other low-frequency sound producers. Changes in diving behavior later in the season are also not supported (Lewis et al. 2018), and variability in physical propagation conditions or Doppler shifts are insufficient to explain the magnitude of the variation (Miller et al. 2014b). McDonald et al. (2009) favored a hypothesis that related cultural conformity and sexual selection for lower frequency songs, with reductions in necessary communication space as populations increase after the cessation of commercial whaling and whales are more closely spaced. Two data points challenge the proposed link between population increase and frequency decrease. The abundance of Northeastern Pacific blue whales appears to have been stable over several decades (Calambokidis and Barlow 2004; Campbell et al. 2014), even as the frequency decline has persisted. Similarly, the increase in population size for the Antarctic blue whales of about 7–8% per year (Olson et al. 2018) is much higher than predicted from the rate of frequency shift (1.3%) postulated by McDonald et al. (2009). The disconnect between population growth and frequency decline does not necessarily negate the influence of sexual selection and cultural conformity; rather, it just removes the dependence on effective communication space and allows instead for conformity driven by a population-level advantage for singing at the same frequency. A further twist in the story of the frequency shift is that the physical constraints of sound production predict the decline of tonal frequency cannot persist forever, as eventually calls will be too low in frequency and have very low source level and their emission will be limited by blue whale anatomy (Aroyan et al. 2000, Chap. 3).

In addition to the long-term inter-annual decline in tonal song unit frequency, Southern Ocean blue whale songs have an intra-annual seasonal trend; they decrease in peak frequency from March through October (Gavrilov et al. 2012), but then increase between December and March (Miller et al. 2014b). This seasonal variation in tonal song peak frequency led Miller et al. (2014b) to propose an alternative hypothesis coupling the frequency shift to sexual selection, supported by a potential link between tonal peak frequency and changes in body mass over the feeding season. They supported this suggestion by correlating historic whaling data with the timing of the present-day seasonal change in tonal frequency. Blubber thickness is a strong predictor for overall body condition and could serve as an honest indication of a male's quality (Rolland et al. 2016). A seasonal change in tonal peak frequency linked with blubber thickness could indicate a physical acoustic relationship that is beyond a whale's control, providing a proxy for a whale's quality as it is signaling to possible mates. If combined with cultural conformity (Malige et al. 2020), it may provide a plausible explanation for this ubiquitous phenomenon.

Studies into the variation in song length have suggested long-term evolution of song structure in some blue whale populations. Perhaps the most detailed such study to date has been conducted on Southeastern Indian blue whales recorded in Perth Canyon off Western Australia and the Bonney Upwelling region south of Australia (Beck 2019; Jolliffe et al. 2019). Comparison of songs in these two regions showed a difference: There was a monotonic increase in the duration of song phrases recorded in Perth Canyon over a decade (Jolliffe et al. 2019), while the interval between phrases recorded in the Bonney Upwelling region did not lengthen over that time (Beck 2019). In Southwestern Pacific blue whales, song underwent several changes: from 1964 to 1997 interunit interval decreased and then it increased from 1997 to 2013, and there was a consistent decrease in pulse rate for all units (Miller et al. 2014a). Conversely, the interunit intervals between A and B units of the Northeastern Pacific blue whale song off Southern California remained consistent over a 5–10 year interval (Lewis and Širović 2018; Oleson et al. 2007a). These different findings may illustrate greater variability in the function and context of songs seasonally and geographically.

Another feature of Southeastern Indian songs recorded in Perth Canyon was a seasonal increase in the interval between song phrases (which the authors term "inter-song interval") during the northbound migration from February to June, which reset to a lower interval in August prior to the southbound migration (Beck 2019). This finding is similar to the gradual seasonal lengthening of inter-pulse intervals (the time interval between adjacent song units) observed for fin whale (*Balaenoptera physalus*) song in the eastern North Pacific (Oleson et al. 2014; Širović et al. 2017) and a more distinct seasonal shift in inter-pulse intervals in fin whale song in the North Atlantic (Morano et al. 2012). For both species, this seasonal change coincides with an expected transition toward breeding behaviors, perhaps suggesting that the lengthening of intervals conveys information about breeding state. Contrary to measures of fin whale inter-pulse interval, the variability in inter-phrase interval for blue whale songs measured during a season off Perth Canyon is quite high, leading Beck (2019) to alternatively propose that individual whales may maintain a signature song. This signature song hypothesis, however, is challenged by several observations, including differences in the relative proportion of phrase types between the feeding areas at the Bonney Upwelling, Perth Canyon, and the migratory corridor off Cape Leeuwin (Gavrilov et al. 2011; Beck 2019).

9.3.1.2 Variability in Song Syntax

Variability in song syntax has not been as widely studied over long periods for most blue whale populations. The only record of change in units has been documented in Southwestern Pacific song where the final unit was dropped and a new beginning unit was added between 1967 and 1997 (Miller et al. 2014a). Variability in phrasing of units has been observed in two populations. Along the west coast of Australia, Southeastern Indian blue whale songs consist of six phrase variations each composed of three common units, with their relative occurrences varying spatially and temporally (Jolliffe et al. 2019). All song variants included at least one unit II, the loudest unit, and the phrases were generally distinguished based on the occurrence of different units and the inter-unit interval between them (Jolliffe et al. 2019). Gavrilov et al. (2011) suggested the greater prevalence of the high-intensity unit II in Perth Canyon may be a compensatory response to higher ambient noise levels there, enabling the whales to be heard above the background din of fellow pygmy blue whales. Observations in the South Australian Bight foraging ground revealed greater occurrence of song units in offshore areas, and higher overall sound production rates in regions with fewer whales (Garcia-Rojas et al. 2018), thereby supporting a behavioral response to local conditions as a driver of changes in song phrasing and occurrence in this population. However, this does not explain an increase in phrase use in Perth Canyon as the density of whales there increased.

The large variability in song phrasing and intervals between song bouts observed for Southeastern Indian blue whales is not evident in other well-studied blue whale populations. Only two phrase types have been reported for Northeastern Pacific blue whale song, with simple AB phrasing or repeated B unit phrasing (termed ABB in several studies) with a variable number of B units following each A unit. Closer examination of song phrasing off southern California revealed that more than half of song phrases detected at an offshore site included ABB phrases, whereas standard AB phrasing was more common at an inshore monitoring location, where the prevalence of this phrase type increased seasonally (Lewis and Širović 2018). Greater use of tonal B units may enable songs to be heard at greater distance, since detection area in waters farther from the shore is greater than inshore (Širović et al. 2015), potentially increasing the likelihood of a male's song being heard by listening females (Lewis and Širović 2018) or other males. In the waters off southern California, observations of solitary whales are more common farther from the coast than are those of pairs or larger groups (Lomac-MacNair and Smultea 2016), supporting the idea that whales may produce different song variants based on social context. There is limited variation in phrase structure evident in Northern Atlantic blue whale song (Berchok et al. 2006; Mellinger and Clark 2003; Nieukirk et al. 2004). Although syntax in the Northeastern Pacific and Northern Atlantic songs is simpler than in the Southeastern Indian song, the underlying mechanisms driving the variability may be the same. The physical constraints on acoustic propagation and the likelihood of reception by receivers drive selection for high intensity, simple song units when solitary and traveling, and more complex calls when engaged in foraging behavior and near other whales.

9.3.1.3 Variability in Individual Song Units (or Calls)

Although the characteristics of song units and syntax of songs allow identification of blue whale populations, the frequency and timing of individual units can be variable when examined in detail. Blue whales produce the same sounds typically considered song units outside of song phrases (Oleson et al. 2007a, 2007b; Berchok et al. 2006), although with some differences in frequency content and unit duration (Oleson et al. 2007a). The behavioral context of the singular occurrence of units or phrases might well be different from when they are produced in full song. Additionally, slight variations in frequency and duration characteristics of units may also convey a variety of meanings to the receiver. Broken units, or those with a silent gap within the unit, have been recorded from many blue whale populations. Berchok et al. (2006) examined whether broken units observed from Northern Atlantic blue whales in the St. Lawrence estuary could be created by different propagation effects and found that most calls with single breaks could be explained through interference generated by bottom reflections. Bottom reflections would be uncommon in deeper water environments, so this is likely not a sufficient explanation for breaks in song units generally. Rather, Berchok et al. (2006) concluded that these breaks are likely controlled by the whale and may be another way to convey different information about the environment or the individual, similar to the increased phrase variability in Southeastern Indian blue whale songs observed in some years (Beck 2019).

Detailed analyses of the frequency and duration of individual song units have been carried out for Southeastern Indian, Northeastern Pacific, Central Northern Pacific, Southeastern Pacific, Northern Atlantic, and Southern Ocean blue whale songs, generally focusing on the loudest song unit (Širović et al. 2016; Beck 2019; Berchok et al. 2006; Gavrilov et al. 2012; Miller et al. 2014a; Malige et al. 2020). Geographic variation in the frequency and duration of song units measured across distant feeding grounds in the eastern North Pacific and North Atlantic may suggest distinct regional dialects representing finer-scale structure across a broadly distributed population (Carbaugh-Rutland et al. 2021; Berchok et al. 2006).

When combined, studies showing rapid change and variability in vocal behavior, including changes in song structure and phrase composition, suggest that song production in blue whales is driven at least in part by response to local environmental conditions and social context, requiring some degree of vocal learning (Širović et al. 2016; Lewis and Širović 2018). At the same time, broad similarities in the characteristics of song units and calls may also point to genetically based predispositions, possibly even at the population level. Some aspects of the observed song variation, then, may be outside of the whale's control and instead may serve as a true measure of whale fitness and be a truthful reproductive display.

9.3.2 Links Between Songs and Genetics

If songs serve as an identification of a singer's population, the patterns of song occurrence should be correlated with patterns of phylogeography identified through genetics. This may be complicated by the varying time scales of different factors acting on populations. Songs may represent current day population biogeography, whereas genetic patterns reflect the evolutionary history of the population and are often not influenced by very recent changes in demography or distribution, as is the case for several species of birds (e.g., Lynch 1996; Grant and Grant 1997; Weiner 2014). We currently recognize 11 different blue whale song types, across four (or five) subspecies, resulting in some subspecies being associated with multiple song types (Fig. 1). Genetic data support differentiation among Antarctic, pygmy, and the putative Chilean blue whale subspecies (Sremba et al. 2012; Torres-Florez et al. 2014; LeDuc et al. 2007), as well as among the Northeastern Pacific blue whale populations of Northern blue whales and the Chilean blue whale subspecies (Torres-Florez et al. 2014; Leduc et al. 2017). Likewise, the Southeastern Indian and Southwestern Pacific blue whale populations appear to be genetically distinct (Attard et al. 2010; Barlow et al. 2018), although not sufficiently so to be presently considered a different subspecies (Barlow et al. 2018). Each of these groups has a unique song type supporting a link between sub-speciation and song type. Apart from long physical distances and distinct seasonal migration patterns, there are few geographic barriers to blue whale mating, suggesting that other factors provide a mechanism for assortative mating. As has been well-documented in birds, behavioral plasticity may lead to population divergence, with song's role as a male reproductive display providing insight into possible population divergence before the separation is apparent in the genome (e.g., Baker and Cunningham 1985; MacDougall-Shackleton and MacDougall-Shackleton 2001; Danner et al. 2017).

But song type is not always a clear indicator of genetic distinction among blue whale populations. For example, it has been suggested there are two blue whale populations in the North Atlantic (Sears and Perrin 2018). Only a single song type has been reported there, although regional variation in phrasing (Berchok et al. 2006) may be an important indicator of divergence if it is consistently correlated with a specific region or group of blue whales. Careful examination of recordings from the eastern North Atlantic would be valuable in evaluating the likelihood of two song types in this basin. However, possibly two song types, or at least largely differing variants of the same song, have been recorded off Chile (Buchan et al. 2014), and genetic and morphological data indicate a single subspecies and population in this area (Torres-Florez et al. 2014; Pastene et al. 2020). In the case of Antarctic blue whales, recent genetic analyses have yielded different conclusions regarding population structure, indicating either a single population or three sympatric circumpolar populations (Attard et al. 2012, 2016, 2018; Sremba et al. 2012). Given song similarities within the population, the mechanism for maintaining population diversity, if it exists within this feeding ground, would have to be related to migratory destination and timing. Attard et al. (2012) suggested the year-round detection of Southern Ocean blue whale song off Antarctica could be the result of a year-round population that resides along the ice edge, in addition to two latitudinally migrating populations. Exact links between song and genetic differences are likely complex.

Current evidence generally supports the notion of at least one common song for a population, as this is the current pattern observed for at least eight populations in the Indian, North Pacific, and even South Pacific Ocean. Further detailed evaluation of the levels of song variability and genetic differentiation would be particularly valuable for songs and populations from the Southern Ocean and the North Atlantic, as the data from those locations to date are inconclusive. Additionally, a future study quantifying the variability in the Southeastern Pacific songs might help further our understanding of blue whale song plasticity.

9.3.3 Relationship Between Foraging and Calling Behaviors

Blue whales are obligate filter feeders that have evolved highly specialized foraging strategies to take advantage of patchy, ephemeral and seasonally abundant prey resources (Hazen et al. 2015). During summers in Southern California, blue whales have been observed breaking from foraging dives to produce both D calls and singular A or B calls, which have the same structure as song units but are produced individually, not as part of a song (Lewis et al. 2018; Oleson et al. 2007b). As calling also incurs an energetic cost, either from call production or from lost opportunity while foraging, it is interesting to consider the prey conditions under which whales produce these call types, or switch to singing behavior. Downswept D calls have been considered an indicator of foraging activity, even if not emitted during foraging dives (Oleson et al. 2007b; Lewis et al. 2018). Multi-sensor tags deployed on foraging blue whales in the Corcovado Gulf, Chile recorded both song units and D calls (Saddler et al. 2017), suggesting the association between foraging and D-calling behavior extends beyond Northeastern Pacific blue whales (Širović et al. 2020). Off the Antarctic Peninsula, however, D calls were detected in areas with no krill (Širović and Hildebrand 2011). Before we can conclude whether D calls can be considered a general indicator of foraging activity, information on spatial scales of the relationship between krill occurrence and calling whale presence across regions is necessary.

The relationship between singing, or the production of single song phrases, and foraging may be even more complex. Oleson et al. (2007b) suggested that blue whales may sing while traveling between known foraging locations in a feeding area, optimizing both foraging efficiency and opportunity to advertise to listening conspecifics. This is supported by a higher incidence of singing at night (Wiggins et al. 2005; Lewis and Širović 2018; Oleson et al. 2007a), potentially following the cessation of foraging on dense daytime prey patches at depth. Lewis et al. (2018) reported that 85% of all sounds (song, singular and D calls) attributed to tagged individuals were produced at shallow depth during non-lunging dives, regardless of overall dive depth. AB phrase production was highest during shallow dives, including both shallow dives with and without lunges (Lewis et al. 2018), suggesting that reduced lunging effort during these shallow foraging dives may allow opportunity to acoustically communicate at the beginning or the end of the dive to maintain contact with other whales. Observations of singing Southeastern Indian blue whales in Perth Canyon suggested that during austral autumn, whales were intermittently singing while engaged in milling behavior, likely indicative of foraging (Beck 2019). The apparent difference between movement and foraging patterns of singing blue whales could be indicative of different levels of productivity; it is possible that productivity in the Perth Canyon region may not require travel to other foraging locations. More dedicated studies of acoustic behaviors and prey availability across different populations will be needed to test whether productivity and prey availability impact daily singing behavior.

9.4 Future Directions

9.4.1 Song Variability and Population Structure

The character, occurrence, and variability in blue whale songs can be used to examine a number of basic, yet important unresolved questions regarding blue whale population biology and ecology. Broadly, the idea that songs can be used to delineate populations has gained ground, with new songs described since the McDonald et al. (2006) review, and has led to insights into understudied blue whale populations in the western Pacific and northwestern Indian Oceans (Fig. 1; McDonald et al. 2017; Cerchio et al. 2020). However, discoveries of potentially multiple songs in Southeastern Pacific population (Buchan et al. 2014), or possible sympatric populations off Antarctica with the same song (Attard et al. 2016) challenge a strict relationship between song type and population identity. Additionally, hybrids between pygmy and Antarctic blue whale subspecies are believed to exist (Attard et al. 2012), but hybrid call types have not been identified, suggesting that hybrid whales adopt the song of one of the parental populations. These recent findings do not necessarily negate the overall relationship between song and population structure but suggest that this initial conclusion of a simple relationship may not hold, and that further examination of the relationship between song characters and other biological, ecological, and behavioral factors (genetics, morphology, foraging, and movement patterns) are still needed. The temporal scale over which genetic and song differentiation occurs is very different, so it is reasonable to expect some divergence between the patterns of population structure as measured by genetics versus those represented by songs, which might also offer insights into the divergence process.

At a finer geographic scale, song phrasing can also be highly variable within a foraging ground (Lewis and Širović 2018; Jolliffe et al. 2019). Might fine-scale variations in song structure within a population be early indicators of potential future divergence? Alternatively, does such fine-scale variability represent a whale's capacity to adjust its song in response to local whale density, foraging conditions, or other ecological factors? Better understandings of the function of song would help to elucidate the biological significance of this variability.

Blue whale populations occur in highly productive areas with high krill abundances and may selectively seek out habitats with consistently high productivity (Abrahms et al. 2019). The relationship between blue whale occurrence and high prey density has been established off California, Chile, and in the Southern Ocean (Buchan and Quiñones 2016; Fiedler et al. 1998; Miller et al. 2019). Identifying other areas of high krill abundance could lead to additional new discoveries of blue whale occurrence. Are there remote locations of seasonal or ephemeral krill aggregation that could be examined further with targeted passive acoustic monitoring? Interestingly, an area of higher productivity in the western Arabian Sea (Qasim 1977) was recently suggested as a location of a newly documented, or possibly recovered, blue whale population (Cerchio et al. 2020). More focused effort in other areas of high productivity, but with relatively little past survey effort, could lead to discoveries of other new or under-explored populations. The southern Atlantic Ocean might be a perfect candidate for such a targeted study, especially as it features the once abundant whaling ground off South Georgia that recently shows renewed evidence of blue whale presence (Calderan et al. 2020).

The Indian Ocean features a surprising diversity of blue whale populations delineated through differences in song (Stafford et al. 2011). Among these, the Southeastern Indian blue whale produces a song that is more complex than has been noted until now in any other blue whale population (Jolliffe et al. 2019). The sympatry of blue whale populations in this region (Samaran et al. 2013) could be the result of diverse prey assemblages that enabled divergent prey preferences among the different populations. This sympatry may have driven the complexity of songs in this region, as predicted by the acoustic competition theory (Naugler and Ratcliffe 1994). The greater diversity and complexity of songs within the Indian Ocean may also suggest that song function could extend beyond a breeding context, for example, in social organization or relaying information on resource availability, as we also find complex song in populations that are not sympatric with other blue whale populations.

Many blue whale populations that occur primarily offshore appear to use simple songs consisting entirely of only tonal units, although detailed examination of the variability in song syntax has not been carried out. Central Northern Pacific, Southern Ocean, and Northern Atlantic blue whales all sing relatively simple songs with only tonal units, but there have been no detailed studies of fine-scale variability in their features. Northeastern Pacific blue whales tend to use more tonal units when singing in deep water than when closer to shore. Extensive passive acoustic recording effort has been conducted on the western side of the Atlantic Ocean, as well as along the mid-Atlantic ridge (e.g., Nieukirk et al. 2004; Risch et al. 2014; Clark and Gagnon 2002; Davis et al. 2020; Clark 1996), but the paucity of published information on recordings from the eastern Atlantic in the twenty-first century (Clark and Charif 1998; Romagosa et al. 2020) has made it impossible to examine the current diversity and variability in blue whale acoustic behavior across the North Atlantic to evaluate if two acoustic populations may be found there as suggested by Sears and Perrin (2018).

Similarly, extensive recording effort across the North Pacific provides an opportunity for detailed exploration of the song structure, particularly across such a vast region where song structure is broadly similar. Small variations were discovered (Širović et al. 2016), and may offer further insights into whether blue whales found there form a single population or if there is a finer-scale division among those animals (Carbaugh-Rutland et al. 2021). Although some areas have been well-monitored (the Gulf of Alaska, the Hawaiian Islands) much remains poorly explored (Stafford 2003; PIFSC 2019), including the far northwest Pacific off Japan and Russia and remote regions of the central Pacific. Whale surveys in the North Pacific have found consistent blue whale aggregations in the offshore waters along the convergence zone in the western North Pacific (Konishi et al. 2018) and point to an area where long-term passive acoustic effort would be valuable. Additionally, a broad recording effort along the tropical and subtropical Pacific may yield information on the possible breeding grounds for these animals.

9.4.2 Blue Whale Mating Strategy and Song

Two of the great unresolved mysteries of blue whale natural history, indeed of balaenopterid biology, are: Where do they mate and what is their mating system? Research so far indicates numerous similarities across balaenopterid species' behaviors, suggesting there may also be similarities in their mating strategies. For example, many balaenopterids produce regionally distinct songs and broadly shared social calls (Oleson et al. 2007b; Širović et al. 2013; Sayigh 2014). It is also clear that blue whale songs and D calls are produced throughout the year (Oleson et al. 2007c; Širović et al. 2015; Paniagua-Mendoza et al. 2017; Širović et al. 2004; Clark and Gagnon 2002). However, the rate of song production may change over the course of the year, peaking in the fall, possibly under hormonal influence (Vu et al. 2015) and indicative of mating season. Unlike balaenid and gray whales, balaenopterid males have small testes relative to their body size so are less likely to engage in sperm competition or multi-mate strategy (Brownell and Ralls 1986). Hence, seasonal song may be an alternate way to advertise their quality.

As some balaenopterids do not undergo complete seasonal migrations (Watkins et al. 2000; Širović et al. 2004, 2013), mating may be occurring across a broader latitudinal range than the traditional migration theory suggests (Simon et al. 2010). Unlike some other baleen whales, blue whales forage not just at higher-latitude feeding grounds, but also during their migrations and on their breeding grounds (Bailey et al. 2010), indicating plasticity in their migratory and foraging behavior, perhaps most notably exemplified by the resident blue whales within New Zealand waters (Barlow et al. 2018). For non-calving individuals, the added cost of travel to lower latitudes may not be adaptive if breeding can be accomplished at higher latitudes, especially if it is yielding higher foraging efficiency (Goldbogen et al. 2011; Guilpin et al. 2019) and can result in a better body condition.

9.4.3 D Calls, Songs, and Foraging Behavior

A better understanding of D call variability within and across populations could help elucidate the function of these calls. Are different types of D calls associated with particular behaviors and behavioral contexts (Jolliffe et al. 2019; Schall et al. 2019), and are these associations consistent across populations? Do they contain acoustic features that allow animals to identify populations or even individuals, as has been implied by Saddler et al. (2017)? Early studies into the behavioral context of these calls indicated they are produced by individuals in groups during foraging (Oleson et al. 2007b). Do these calls help with locating food or do they serve to advertise the presence of food? Are they contact calls? Further targeted studies, including playback experiments and using acoustic tags, would shed light on those questions. This line of questioning on the communicative function of D calls could be extended beyond blue whale populations. The similarity in duration and frequency of these calls with fin and sei whale (*B. borealis*) calls (Rankin and Barlow 2007; Baumgartner et al. 2008; Širović et al. 2013; Bass and Clark 2003) may point to it being a common congeneric behavior that is co-opted by eavesdroppers to facilitate locating productive areas in a patchy environment. In a high-productivity region like the Southern Ocean (Schofield et al. 2018; Laws 1985) where different baleen whale species partition resources (Friedlaender et al. 2009), taking advantage of similarity in signaling might be a mutually beneficial approach to a difficult task of searching for ephemeral prey patches. Social foraging theory provides a framework describing how this type of behavior can be evolutionarily adaptive in a patchy environment (Giraldeau and Caraco 2018) such as the Southern Ocean.

In a complex and variable marine acoustic environment, communicating information about resource availability may require more than a single call type. In the Northeastern Pacific population, it has been suggested that singular production of song units may differ from the function of song itself and may be intended for individuals close by to the male producing the unit (Oleson et al. 2007b). The production of individual song units that are not associated with singing behavior has not been studied in detail across other blue whale populations and general prevalence of this behavior is unknown. Is there a relationship between occurrence of singular units and complexity of song? Could it be related to resource availability? Or is there some additional driver which determines the variability in the way whales use song units? More detailed comparison of song unit production worldwide would start to shed light on some of these very basic questions about sound form, occurrence, context, and function.

9.4.4 Population Monitoring and Response to Change

Long-term trends of acoustic metrics (e.g., counts of sound types, aggregate acoustic energy) have been used as indicators of blue whale population trends (McCauley et al.

2018; Širović et al. 2015). Further study of how best to estimate population growth from passive acoustic data would be very valuable for allowing the development of trend estimates in areas where acoustic data collection is common, particularly when visual survey data are lacking, such as within the Indian Ocean or the vast expanses of the Southern Ocean (Van Opzeeland et al. 2014). A better understanding of trends among received (or source) levels and population numbers, behavioral context of calling, as well as impact of density on call rates are necessary. Understanding population trends will be one important component of unraveling the impact of climate change on blue whales.

Blue whale populations may respond differently to large-scale changes in ocean climate. Over the past decade, acoustic records have indicated the occurrence of blue whale songs atypical for areas where they were recorded: Antarctic blue whale songs near the equator, Southeastern Pacific song off Southern California, and Northeastern Pacific song in the western Pacific (Samaran et al. 2019; Širović unpublished data). Due to the lack of acoustic data across these broad areas in the past, we do not know if a small number of blue whales are commonly found beyond their typically assumed population range, or if this could be an indication of novel movements outside of typical areas in search of better habitat, higher prey abundance, or otherwise favorable conditions. As the world's oceans continue to warm, the ability to adapt by expanding their range or adjusting the timing of their migrations as they search for suitable habitat may become important for blue whales (Robinson et al. 2009). Additionally, movement to new areas may lead to increased risk from interaction with anthropogenic threats, further exacerbating a stressful situation. Understanding the past and current ranges, patterns, and occurrences of blue whale acoustic behaviors will be critical for understanding potential risks resulting from those movements.

Over the last several decades, our knowledge and understanding of blue whale acoustics, ecology, and genetics have increased tremendously. At the same time, there is an increasing awareness of the need to better understand the impacts that human activities have on individuals and populations. In the near future, the challenges blue whales will face include not just impacts from and influences of aggregate human activities, but also a warming and increasingly variable ocean environment resulting from climate change. Fine-scale variability in their acoustic behaviors and relationships to their habitat and prey variability may be the key that can help us explain the long-term drivers of blue whale population health. If we can understand how these factors interact and contribute to the success of individual populations, we will be able to use acoustic methods for more comprehensive monitoring of the population status. Ultimately, that understanding should bring us closer to the implementation of measures that can successfully mitigate deleterious impacts from human-caused changes to their environment.

References

Abrahms B, Hazen EL, Aikens EO, Savoca MS, Goldbogen JA, Bograd SJ, Jacox MG, Irvine LM, Palacios DM, Mate BR (2019) Memory and resource tracking drive blue whale migrations. Proc Natl Acad Sci USA 116:5582–5587

Aroyan JL, McDonald MA, Webb SC, Hildebrand JA, Clark D, Laitman JT, Reidenberg JS (2000) Acoustic models of sound production and propagation. In: Au WWL, Popper AN, Fay RR (eds) Hearing by whales and dolphins. Springer, New York, pp 409–469

Attard CR, Beheregaray LB, Jenner C, Gill P, Jenner M, Morrice M, Bannister J, LeDuc R, Möller L (2010) Genetic diversity and structure of blue whales (*Balaenoptera musculus*) in Australian feeding aggregations. Conserv Genet 11:2437–2441

Attard CR, Beheregaray LB, Jenner KCS, Gill PC, Jenner MN, Morrice MG, Robertson KM, Möller LM (2012) Hybridization of Southern Hemisphere blue whale subspecies and a sympatric area off Antarctica: impacts of whaling or climate change? Mol Ecol 21:5715–5727

Attard CR, Beheregaray LB, Möller LM (2016) Towards population-level conservation in the critically endangered Antarctic blue whale: the number and distribution of their populations. Sci Rep 6:22291

Attard CR, Beheregaray LB, Sandoval-Castillo J, Jenner KCS, Gill PC, Jenner M-NM, Morrice MG, Möller LM (2018) From conservation genetics to conservation genomics: a genome-wide assessment of blue whales (*Balaenoptera musculus*) in Australian feeding aggregations. R Soc Open Sci 5:170925

Bailey H, Mate BR, Palacios DM, Irvine L, Bograd SJ, Costa DP (2010) Behavioural estimation of blue whale movements in the Northeast Pacific from state-space model analysis of satellite tracks. Endanger Spec Res 10:93–106

Baker MC, Cunningham MA (1985) The biology of bird-song dialects. Behav Brain Sci 8:85–100

Balcazar NE, Tripovich JS, Klinck H, Nieukirk SL, Mellinger DK, Dziak RP, Rogers TL (2015) Calls reveal population structure of blue whales across the southeast Indian Ocean and the southwest Pacific Ocean. J Mammal 96:1184–1193

Barlow DR, Torres LG, Hodge KB, Steel D, Baker CS, Chandler TE, Bott N, Constantine R, Double MC, Gill P (2018) Documentation of a New Zealand blue whale population based on multiple lines of evidence. Endanger Spec Res 36:27–40

Bass AH, Clark CW (2003) The physical acoustics of underwater sound communication. In: Simmons AM, Popper AN, Fay RR (eds) Acoustic communication. Springer, New York, pp 15–64

Baumgartner MF, Van Parijs SM, Wenzel FW, Tremblay CJ, Carter Esch H, Warde AM (2008) Low frequency vocalizations attributed to sei whales (*Balaenoptera borealis*). Antarctic J 124:1339–1349

Beck CD (2019) Vocal behaviour of the Eastern Indian Ocean pygmy blue whale and its changes over time and between aggregation areas. PhD Dissertation, Curtin University, Perth, Australia

Berchok CL, Bradley DL, Gabrielson TB (2006) St. Lawrence blue whale vocalizations revisited: characterization of calls detected from 1998 to 2001. Antarctic J 120:2340–2354

Branch TA, Stafford K, Palacios D, Allison C, Bannister J, Burton C, Cabrera E, Carlson C, Galletti Vernazzani B, Gill P (2007) Past and present distribution, densities and movements of blue whales *Balaenoptera musculus* in the Southern Hemisphere and northern Indian Ocean. Mammal Rev 37:116–175

Brownell R, Ralls K (1986) Potential for sperm competition in baleen whales. Rep Int Whaling Comm 8(Spec Iss):97–112

Buchan SJ, Quiñones RA (2016) First insights into the oceanographic characteristics of a blue whale feeding ground in northern Patagonia, Chile. Mar Ecol Prog Ser 554:183–199

Buchan SJ, Hucke-Gaete R, Rendell L, Stafford KM (2014) A new song recorded from blue whales in the Corcovado Gulf, Southern Chile, and an acoustic link to the Eastern Tropical Pacific. Endanger Spec Res 23:241–252

Calambokidis J, Barlow J (2004) Abundance of blue and humpback whales in the eastern North Pacific estimated by capture-recapture and line-transect methods. Mar Mamm Sci 20:63–85
Calambokidis J, Schorr GS, Steiger GH, Francis J, Bakhtiari M, Marshall G, Oleson EM, Gendron D, Robertson K (2007) Insights into the underwater diving, feeding, and calling behavior of blue whales from a suction-cup-attached video-imaging tag (CRITTERCAM). Mar Tech Soc J 41(4):19–29
Calambokidis J, Barlow J, Ford JKB, Chandler TE, Douglas AB (2009) Insights into the population structure of blue whales in the Eastern North Pacific from recent sightings and photographic identification. Mar Mamm Sci 25:816–832
Calderan S, Black A, Branch TA, Collins M, Jackson J, Kelly N, Leaper R, Lurcock S, Miller BS, Moore M, Olson PA, Širović A, Wood A (2020) South Georgia blue whales five decades after the end of whaling. Endanger Spec Res 43:359–373
Campbell GS, Thomas L, Whitaker K, Douglas A, Calambokidis J, Hildebrand JA (2014) Inter-annual and seasonal trends in cetacean distribution, density and abundance in waters off southern California. Deep-Sea Res II 112:143–157
Carbaugh-Rutland A, Rasmussen JH, Sterba-Boatwright B, Širovi´c A (2021) Geographically distinct blue whale call variants in the Northeast Pacific. Endanger Spec Res 46:19–33
Caruso F, Hickmott L, Warren JD, Segre P, Chiang G, Bahamonde P, Español-Jiménez S, Li S, Bocconcelli A (2020) Diel differences in blue whale (*Balaenoptera musculus*) dive behavior increase nighttime risk of ship strikes in northern Chilean Patagonia. Integr Zool. https://doi.org/10.1111/1749-4877.12501
Cerchio S, Willson A, Leroy EC, Muirhead C, Al Harthi S, Baldwin R, Cholewiak DM, Collins T, Minton G, Rasoloarijao T, Rogers TL, Sarrouf Willson M (2020) A new blue whale song-type described for the Arabian Sea and Western Indian Ocean. Endanger Spec Res 43:495–515
Clark C (1996) Whales' 93: hot tracks and greatest hits; blue whales' 93; fin whales' 93; minke whales' 93, North Atlantic. A seven-CD set Cornell Laboratory of Ornithology, Bioacoustics Research Program, vol 1, p 59
Clark CW, Charif RA (1998) Acoustic monitoring of large whales to the west of Britain and Ireland using bottom-mounted hydrophone arrays, October 1996–September 1997. JNCC Report No. 281, Aberdeen, UK
Clark CW, Gagnon GC (2002) Low-frequency vocal behaviors of baleen whales in the North Atlantic: insights from IUSS detections, locations and tracking from 1992 to 1996. J Underwater Acous 52(3):609–640
Clark CW, Ellison WT (2004) Potential use of low-frequency sounds by baleen whales for probing the environment: evidence from models and empirical measurements. In: Moss C, Vater M (eds) Thomas J. Echolocation in bats and dolphins, University of Chicago Press, pp 564–582
Committee on Taxonomy (2020) List of marine mammal species and subspecies. Society for Marine Mammalogy www.marinemammalscience.org. Consulted on February 14, 2020
Croll DA, Marinovic B, Benson S, Chavez FP, Black N, Ternullo R, Tershy BR (2005) From wind to whales: trophic links in a coastal upwelling system. Mar Ecol Prog Ser 289:117–130
Croll DA, Tershy BR, Hewitt RP, Demer DA, Fiedler PC, Smith SE, Armstrong W, Popp JM, Kiekhefer T, Lopez VR, Urban J, Gendron D (1998) An integrated approach to the foraging ecology of marine birds and mammals. Deep-Sea Res II 45:1353–1371
Danner J, Fleischer R, Danner R, Moore I (2017) Genetic population structure in an equatorial sparrow: roles for culture and geography. J Evo Bio 30(6):1078–1093
Davis GE, Baumgartner MF, Corkeron PJ, Bell J, Berchok C, Bonnell JM, Thornton JB, Brault S, Buchanan GA, Cholewiak DM (2020) Exploring movement patterns and changing distributions of baleen whales in the western North Atlantic using a decade of passive acoustic data. Glob Chang Biol 26(9):4812
Edds-Walton PL (1997) Acoustic communication signals of mysticete whales. Bioacoustics 8:47–60
Fiedler PC, Reilly SB, Hewitt RP, Demer D, Philbrick VA, Smith S, Armstrong W, Croll DA, Tershy BR, Mate BR (1998) Blue whale habitat and prey in the California Channel Islands. Deep-Sea Res II 45:1781–1801

Friedlaender AS, Lawson GL, Halpin PN (2009) Evidence of resource partitioning between humpback and minke whales around the western Antarctic Peninsula. Mar Mamm Sci 25:402–415
Garcia-Rojas MI, Jenner KCS, Gill PC, Jenner MNM, Sutton AL, McCauley RD (2018) Environmental evidence for a pygmy blue whale aggregation area in the Subtropical Convergence Zone south of Australia. Mar Mamm Sci 34:901–923
Gavrilov AN, McCauley RD (2013) Acoustic detection and long-term monitoring of pygmy blue whales over the continental slope in southwest Australia. J Acoust Soc Am 134:2505–2513
Gavrilov AN, McCauley RD, Salgado-Kent C, Tripovich J, Burton C (2011) Vocal characteristics of pygmy blue whales and their change over time. J Acoust Soc Am 130:3651–3660
Gavrilov AN, McCauley RD, Gedamke J (2012) Steady inter and intra-annual decrease in the vocalization frequency of Antarctic blue whales. J Acoust Soc Am 131:4476–4480
Gill PC (2002) A blue whale (*Balaenoptera musculus*) feeding ground in a southern Australian coastal upwelling zone. J Cetacean Res Manag 4(2):179–184
Giraldeau L-A, Caraco T (2018) Social foraging theory, vol 73. Princeton University Press, Princeton
Goldbogen J, Madsen P (2018) The evolution of foraging capacity and gigantism in cetaceans. J Exp Biol 221(11):jeb166033
Goldbogen JA, Calambokidis J, McKenna M, Oleson E, Potvin J, Pyenson N, Schorr G, Shadwick R (2011) Mechanics, hydrodynamics and energetics of blue whale lunge feeding: predatory strategies and efficiency dependence on krill density. Integr Comp Biol 51:E49–E49
Grant PR, Grant BR (1997) Genetics and the origin of bird species. Proc Natl Acad Sci USA 94:7768–7775
Guilpin M, Lesage V, McQuinn I, Goldbogen JA, Potvin J, Jeanniard-du-Dot T, Doniol-Valcroze T, Michaud R, Moisan M, Winkler G (2019) Foraging energetics and prey density requirements of western North Atlantic blue whales in the Estuary and Gulf of St. Lawrence, Canada. Mar Ecol Prog Ser 625:205–223
Hazen EL, Friedlaender AS, Goldbogen JA (2015) Blue whales (*Balaenoptera musculus*) optimize foraging efficiency by balancing oxygen use and energy gain as a function of prey density. Sci Adv 1(9):e1500469
Hucke-Gaete R, Bedrinana-Romano L, Viddi FA, Ruiz JE, Torres-Florez JP, Zerbini AN (2018) From Chilean Patagonia to Galapagos, Ecuador: novel insights on blue whale migratory pathways along the Eastern South Pacific. Peer J 6:e4695
Jolliffe CD, McCauley RD, Gavrilov AN, Jenner KCS, Jenner M-NM, Duncan AJ (2019) Song variation of the South Eastern Indian Ocean pygmy blue whale population in the Perth Canyon, Western Australia. PLoS One 14(1):e0208619
Kawamura A (1980) A review of food of Balaenopterid whales. Sci Rep Whales Res Inst 32:155–197
Kawamura A (1994) A review of baleen whale feeding in the Southern Ocean. Rep Int Whale Comm 44:261–271
Kellogg R (1929) What is known of the migrations of some of the whalebone whales. Smithsonian Inst Ann Rep 1928:467–496
Konishi K, Isoda T, Nakai K, Oikawa H, Kanbayashi Y, Uchida M, Tsunekawa M, Ueda Y, Kominami T, Kawabe S, Eguchi H, Tamura T (2018) Results of the first cruise of the New Scientific Whale Research Program in the western North Pacific (NEWREP-NP) in the 2017 summer season—Off shore component. Paper SC/67B/SCSP06 presented to IWC Scientific Committee, April 2018, Bled, Slovenia
Laws RM (1985) The ecology of the Southern Ocean. Am Sci 73:26–40
Leduc R, Dizon A, Goto M, Pastene L, Kato H, Nishiwaki S, LeDuc C, Brownell R (2007) Patterns of genetic variation in Southern Hemisphere blue whales and the use of assignment test to detect mixing on the feeding grounds. J Cetacean Res Manag 9(1):73–80
Leduc R, Archer F, Lang A, Martien K, Hancock-Hanser B, Torres-Florez J, Hucke-Gaete R, Rosenbaum H, Van Waerebeek K, Brownell R Jr (2017) Genetic variation in blue whales in the eastern pacific: implication for taxonomy and use of common wintering grounds. Mol Ecol 26:740–751

Leroy EC, Royer JY, Bonnel J, Samaran F (2018a) Long-term and seasonal changes of large whale call frequency in the Southern Indian Ocean. J Geophys Res Oceans 123:8568–8580

Leroy EC, Samaran F, Stafford KM, Bonnel J, Royer J-Y (2018b) Broad-scale study of the seasonal and geographic occurrence of blue and fin whales in the Southern Indian Ocean. Endanger Spec Res 37:289–300

Leslie MS, Perkins-Taylor CM, Durban JW, Moore MJ, Miller CA, Chanarat P, Bahamonde P, Chiang G, Apprill A (2020) Body size data collected non-invasively from drone images indicate a morphologically distinct Chilean blue whale (*Balaenoptera musculus*) taxon. Endanger Spec Res 43:291–304

Lewis LA, Širović A (2018) Variability in blue whale acoustic behavior off southern California. Mar Mamm Sci 34(2):311–329

Lewis LA, Calambokidis J, Stimpert AK, Fahlbusch J, Friedlaender AS, McKenna MF, Mesnick SL, Oleson EM, Southall BL, Szesciorka AR (2018) Context-dependent variability in blue whale acoustic behaviour. R Soc Open Sci 5 (8):180241

Ljungblad D, Stafford KM, Shimada H, Matsuoka K (1997) Sounds attributed to blue whales recorded off the southwest coast of Australia in December 1995. Rep Int Whaling Comm 47:435–439

Lomac-MacNair K, Smultea MA (2016) Blue whale (*Balaenoptera musculus*) behavior and group dynamics as observed from an aircraft off Southern California. Anim Behav Cognition 3(1):1–21

Lynch A (1996) The population memetics of birdsong. In: Kroodsma DE, Miller EH (eds) Ecology and evolution of acoustic communication in birds. Comstock Publishing, Ithaca, NY, pp 181–197

MacDougall-Shackleton EA, MacDougall-Shackleton SA (2001) Cultural and genetic evolution in mountain white-crowned sparrows: song dialects are associated with population structure. Evolution 55:2568–2575

Malige F, Patris J, Buchan SJ, Stafford KM, Fannie S, Findlay K, Rodrigo H-G, Neira S, Clark CW, Hervé G (2020) Inter-annual decrease in pulse rate and peak frequency of Southeast Pacific blue whale song types. Sci Rep 10:8121

Mate BR, Lagerquist BA, Calambokidis J (1999) Movements of North Pacific blue whales during the feeding season off Southern California and their southern fall migration. Mar Mamm Sci 15:1246–1257

Matteson RS (2010) The Costa Rica Dome: a study of physics, zooplankton and blue whales. Master's Thesis, Oregon State University, Corvalis, OR, USA

McCauley RD, Gavrilov AN, Jolliffe CD, Ward R, Gill PC (2018) Pygmy blue and Antarctic blue whale presence, distribution and population parameters in southern Australia based on passive acoustics. Deep Sea Res II 157:154–168

McDonald MA, Calambokidis J, Teranishi AM, Hildebrand JA (2001) The acoustic calls of blue whales off California with gender data. J Acoust Soc Am 109:1728–1735

McDonald MA, Mesnick SL, Hildebrand JA (2006) Biogeographic characterisation of blue whale song worldwide: using song to identify populations. J Cetacean Res Manag 8:55–65

McDonald MA, Hildebrand JA, Mesnick S (2009) Worldwide decline in tonal frequencies of blue whale songs. Endanger Spec Res 9:13–21

McDonald MA, Širović A, Sugioka H, Kato H, Yoshida R, Kyo N (2017) Preliminary analysis of blue and fin whale acoustic presence off Hokkaido Japan. Paper presented at the Paper SC/67A/NH01 presented to the Scientific Committee of the International Whaling Commission, Bled, Slovenia

Mellinger DK, Clark CW (2003) Blue whale (*Balaenoptera musculus*) sounds from the North Atlantic. J Acoust Soc Am 114:1108–1119

Miksis-Olds JL, Nieukirk SL, Harris DV (2018) Two unit analysis of Sri Lankan pygmy blue whale song over a decade. J Acoust Soc Am 144:3618–3626

Miller BS, Collins K, Barlow J, Calderan S, Leaper R, McDonald M, Ensor P, Olson PA, Olavarria C, Double MC (2014a) Blue whale vocalizations recorded around New Zealand: 1964-2013. J Acoust Soc Am 135:1616–1623

Miller BS, Leaper R, Calderan S, Gedamke J (2014b) Red shift, blue shift: Investigating Doppler shifts, blubber thickness, and migration as explanations of seasonal variation in the tonality of Antarctic blue whale song. PLoS One 9 (9):e107740

Miller EJ, Potts J, Cox M, Miller B, Calderan S, Leaper R, Olson P, O'Driscoll R, Double M (2019) The characteristics of krill swarms in relation to aggregating Antarctic blue whales. Sci Rep 9:1–13

Miller BS, Calderan S, Leaper R, Miller EJ, Širović A, Stafford KM, Bell E, Double MC (2021) Source level of Antarctic blue and fin whale sounds recorded on sonobuoys deployed in the deep-ocean off Antarctica. Front Mar Sci 8:792651

Morano JL, Salisbury DP, Rice AN, Conklin KL, Falk KL, Clark CW (2012) Seasonal and geographical patterns of fin whale song in the western North Atlantic Ocean. J Acoust Soc Am 132:1207–1212

Naugler CT, Ratcliffe L (1994) Character release in bird song: a test of the acoustic competition hypothesis using American Tree Sparrows *Spizella arborea*. J Avian Biol 25:142–148

Nemoto T (1957) Foods of baleen whales in the Northern Pacific. Sci Rep Whales Res Inst 12:33–89

Nickels CF, Sala LM, Ohman MD (2018) The morphology of euphausiid mandibles used to assess selective predation by blue whales in the southern sector of the California Current System. J Crust Biol 38(5):563–573

Nieukirk SL, Stafford KM, Mellinger DK, Dziak RP, Fox CG (2004) Low-frequency whale and seismic airgun sounds recorded in the mid-Atlantic Ocean. J Acoust Soc Am 115:1832–1843

Oleson EM, Calambokidis J, Barlow J, Hildebrand JA (2007a) Blue whale visual and acoustic encounter rates in the southern California bight. Mar Mamm Sci 23:574–597

Oleson EM, Calambokidis J, Burgess WC, McDonald MA, LeDuc CA, Hildebrand JA (2007b) Behavioral context of Northeast Pacific blue whale call production. Mar Ecol Prog Ser 330:269–284

Oleson EM, Wiggins S, Hildebrand JA (2007c) Temporal separation of blue whale call types on a southern California feeding ground. Anim Behav 74:881–894

Oleson EM, Sirovic A, Bayless AR, Hildebrand JA (2014) Synchronous seasonal change in fin whale song in the North Pacific. PLoS One 9(12):e115678

Olson PA, Kinzey D, Double MC, Matsuoka K, Pastene LA, Findlay K (2018) Capture-recapture estimated of abundance of Antarctic blue whales. Paper presented at the Paper SC/67B/SH08_rev2 presented to the Scientific Committee of the International Whaling Commission, Bled, Slovenia

Paniagua-Mendoza A, Gendron D, Romero-Vivas E, Hildebrand JA (2017) Seasonal acoustic behavior of blue whales (*Balaenoptera musculus*) in the Gulf of California, Mexico. Mar Mamm Sci 33:206–218

Pastene LA, Acevedo J, Branch TA (2020) Morphometric analysis of Chilean blue whales and implications for their taxonomy. Mar Mamm Sci 36:116–135

Payne RS, McVay S (1971) Songs of humpback whales. Science 173:585–597

PIFSC (2019) Passive acoustic monitoring. https://origin-apps-pifsc.fisheries.noaa.gov/cetacean/passive_acoustic_monitoring.php. Accessed 4 March 2020

Qasim S (1977) Biological productivity of the Indian Ocean. Indian J Mar Sci 6:122–137

Rankin S, Barlow J (2007) Vocalizations of the Sei Whale *Balaenoptera borealis* off the Hawaiian Islands. Bioacoustics 16(2):137–145

Rankin S, Ljungblad D, Clark CW, Kato H (2005) Vocalizations of Antarctic blue whales, Balaenoptera musculus intermedia, recorded during the 2001–2002 and 2002–2003 IWC-SOWER circumpolar cruises, Area V, Antarctica. J Cetacean Res Manag 7:13–20

Rice A, Širović A, Trickey JT, Debich AJ, Gottlieb RS, Wiggins SM, Hildebrand JA, Baumann-Pickering S (2021) Cetacean occurrence in the Gulf of Alaska from long-term passive acoustic monitoring. Mar Biol 168:1–29

Risch D, Castellote M, Clark CW, Davis GE, Dugan PJ, Hodge LE, Kumar A, Lucke K, Mellinger DK, Nieukirk SL, Popescu CM, Ramp C, Read AJ, Rice AN, Silva MA, Siebert U, Stafford KM, Verdaat H, Van Parijs SM (2014) Seasonal migrations of North Atlantic minke whales: novel insights from large-scale passive acoustic monitoring networks. Mov Ecol 2:24

Robinson RA, Crick HQ, Learmonth JA, Maclean IM, Thomas CD, Bairlein F, Forchhammer MC, Francis CM, Gill JA, Godley BJ (2009) Travelling through a warming world: climate change and migratory species. Endanger Spec Res 7:87–99

Rolland RM, Schick RS, Pettis HM, Knowlton AR, Hamilton PK, Clark JS, Kraus SD (2016) Health of North Atlantic right whales *Eubalaena glacialis* over three decades: from individual health to demographic and population health trends. Mar Ecol Prog Ser 542:265–282

Romagosa M, Baumgartner M, Cascão I, Lammers MO, Marques TA, Santos RS, Silva MA (2020) Baleen whale acoustic presence and behaviour at a Mid-Atlantic migratory habitat, the Azores Archipelago. Sci Rep 10(1):1–11

Saddler MR, Bocconcelli A, Hickmott LS, Chiang G, Landea-Briones R, Bahamonde PA, Howes G, Segre PS, Sayigh LS (2017) Characterizing Chilean blue whale vocalizations with DTAGs: a test of using tag accelerometers for caller identification. J Exp Biol 220:4119–4129

Samaran F, Stafford KM, Branch TA, Gedamke J, Royer J-Y, Dziak RP, Guinet C (2013) Seasonal and geographic variation of Southern Blue Whale subspecies in the Indian Ocean. PLoS ONE 8(8):1–10

Samaran F, Berne A, Leroy EC, Moreira S, Stafford KM, Maia M, Royer JY (2019) Antarctic blue whales (*Balaenoptera musculus intermedia*) recorded at the Equator in the Atlantic Ocean. Mar Mamm Sci 35:641–648

Sayigh LS (2014) Cetacean acoustic communication. In: Witzany G (ed) Biocommunication of animals. Springer, Dordrecht, pp 275–297

Schall E, Di Iorio L, Berchok C, Filún D, Bedriñana-Romano L, Buchan SJ, Van Opzeeland I, Sears R, Hucke-Gaete R (2019) Visual and passive acoustic observations of blue whale trios from two distinct populations. Mar Mamm Sci 36:365–374

Schall E, Di Iorio L, Berchok C, Filún D, Bedriñana-Romano L, Buchan SJ, Van Opzeeland I, Sears R, Hucke-Gaete R (2020) Visual and passive acoustic observations of blue whale trios from two distinct populations. Mar Mamm Sci 36:365–374

Schofield O, Brown M, Kohut J, Nardelli S, Saba G, Waite N, Ducklow H (2018) Changes in the upper ocean mixed layer and phytoplankton productivity along the West Antarctic Peninsula. Phil Trans R Soc A 376:20170173

Sears R, Larsen F (2002) Long range movements of a blue whale (*Balaenoptera musculus*) between the Gulf of St. Lawrence and West Greenland. Mar Mamm Sci 18:281–285

Sears R, Perrin WF (2018) Blue whale: *Balaenoptera musculus*. In: Wursig B, Thewissen JGM, Kovacs KM (eds) Encyclopedia of marine mammals. Academic Press, San Diego, CA, pp 110–114

Shabangu FW, Findlay KP, Yemane D, Stafford KM, van den Berg M, Blows B, Andrew RK (2019) Seasonal occurrence and diel calling behaviour of Antarctic blue whales and fin whales in relation to environmental conditions off the west coast of South Africa. J Mar Syst 190:25–39

Simon M, Stafford KM, Beedholm K, Lee CM, Madsen PT (2010) Singing behavior of fin whales in the Davis Strait with implications for mating, migration and foraging. J Acoust Soc Am 128:3200–3210

Širović A (2016) Variability in the performance of the spectrogram correlation detector for North-east Pacific blue whale calls. Bioacoustics 25(2):145–160

Širović A, Hildebrand JA (2011) Using passive acoustics to model blue whale habitat off the Western Antarctic Peninsula. Deep-Sea Res II 58:1719–1728

Širović A, Hildebrand JA, Wiggins SM, McDonald MA, Moore SE, Thiele D (2004) Seasonality of blue and fin whale calls and the influence of sea ice in the Western Antarctic Peninsula. Deep-Sea Res II 51:2327–2344

Širović A, Hildebrand JA, Wiggins SM (2007) Blue and fin whale call source levels and propagation range in the Southern Ocean. J Acoust Soc Am 122:1208–1215

Širović A, Williams LN, Kerosky SM, Wiggins SM, Hildebrand JA (2013) Temporal separation of two fin whale call types across the eastern North Pacific. Mar Biol 160:47–57

Širović A, Rice A, Chou E, Hildebrand JA, Wiggins SM, Roch MA (2015) Seven years of blue and fin whale call abundance in the Southern California Bight. Endanger Spec Res 28:61–76

Širović A, Oleson EM, Favilla A, FIsher-Pool P (2016) Blue whale song variability in the North Pacific Ocean. Paper presented at the Paper SC/66B/IA12 presented to the Scientific Committee of the International Whaling Commission, Bled, Slovenia

Širović A, Oleson EM, Buccowich J, Rice A, Bayless AR (2017) Fin whale song variability in southern California and the Gulf of California. Sci Rep 7:11

Širović A, Wood M, Warren JD, Stafford K, Miller B (2020) Mesoscale dynamics of blue and fin whales and Antarctic krill off East Antarctica. In: Ocean Sciences Meeting 2020, San Diego, CA, USA

Sremba AL, Hancock-Hanser B, Branch TA, LeDuc RL, Baker CS (2012) Circumpolar diversity and geographic differentiation of mtDNA in the critically endangered Antarctic blue whale (*Balaenoptera musculus intermedia*). PLoS One 7(3):e32579

Stafford KM (2003) Two types of blue whale calls recorded in the Gulf of Alaska. Mar Mamm Sci 19:682–693

Stafford KM, Nieukirk SL, Fox CG (2001) Geographic and seasonal variation of blue whale calls in the North Pacific. J Cetacean Res Manag 3:65–76

Stafford KM, Bohnenstiehl DR, Tolstoy M, Chapp E, Mellinger DK, Moore SE (2004) Antarctic-type blue whale calls recorded at low latitudes in the Indian and eastern Pacific Oceans. Deep-Sea Res I 51:1337–1346

Stafford KM, Citta JJ, Moore SE, Daher MA, George JE (2009) Environmental correlates of blue and fin whale call detections in the North Pacific Ocean from 1997 to 2002. Mar Ecol Prog Ser 395:37–53

Stafford KM, Chapp E, Bohnenstiel DR, Tolstoy M (2011) Seasonal detection of three types of "pygmy" blue whale calls in the Indian Ocean. Mar Mamm Sci 27:828–840

Szesciorka AR, Ballance LT, Širović A, Rice A, Ohman MD, Hildebrand JA, Franks PJS (2020) Timing is everything: drivers of interannual variability in blue whale migration. Sci Rep 10:7710

Thode AM, D'Spain GL, Kuperman WA (2000) Matched-field processing, geoacoustic inversion, and source signature recovery of blue whale vocalizations. J Acoust Soc Am 107:1286–1300

Thomisch K, Boebel O, Clark CW, Hagen W, Spiesecke S, Zitterbart DP, Van Opzeeland I (2016) Spatio-temporal patterns in acoustic presence and distribution of Antarctic blue whales *Balaenoptera musculus intermedia* in the Weddell Sea. Endanger Spec Res 30:239–253

Torres L (2013) Evidence for an unrecognised blue whale foraging ground in New Zealand. New Zealand J Mar Freshw Res 47(2):235–248

Torres-Florez J, Hucke-Gaete R, LeDuc R, Lang A, Taylor B, Pimper L, Bedriñana-Romano L, Rosenbaum H, Figueroa C (2014) Blue whale population structure along the eastern South Pacific Ocean: evidence of more than one population. Mol Ecol 23:5998–6010

Van Opzeeland I, Samaran F, Stafford KM, Findlay K, Gedamke J, Harris D, Miller BS (2014) Towards collective circum-Antarctic passive acoustic monitoring: the Southern Ocean hydrophone network (SOHN). Polarforschung 83(2):47–61

Vu ET, Clark C, Catelani K, Kellar NM, Calambokidis J (2015) Seasonal blubber testosterone concentrations of male humpback whales (*Megaptera novaeangliae*). Mar Mamm Sci 31:1258–1264

Watkins WA, Daher MA, Reppucci GM, George JE, Martin DL, DiMarzio NA, Gannon DP (2000) Seasonality and distribution of whale calls in the North Pacific. Oceanography 13(1):62–67

Weiner J (2014) The beak of the finch: a story of evolution in our time. Vintage, New York

Wiggins SM, Oleson EM, McDonald MA, Hildebrand JA (2005) Blue whale (*Balaenoptera musculus*) diel calling patterns offshore of Southern California. Aquat Mamm 31:161–168

Chapter 10
Humpback Whales: A Seemingly Socially Simple Whale with Communicative Complexity

Rebecca A. Dunlop

Abstract Humpback whales (*Megaptera novaeangliae*) occur in all major oceans. Given this worldwide distribution, and since they tend to migrate along coastlines, they are one of the best known of the baleen whales. Humpbacks are relatively easy to find and easy to observe. This, along with their surface behaviors and attraction to vessels, makes them popular with whale-watching businesses. In the scientific world, their song and behaviors have been studied since the 1970s, producing hundreds of scientific papers. Despite this, there are still many unsolved mysteries. Why do humpbacks sing (Chaps. 8 and 11), how do they locate their prey (Chap. 5), and how do they navigate when migrating (Chap. 4)? In this chapter, we focus on the mysteries of their social communication. Communication and social complexities often go hand in hand. Animals with more complex social structures tend to have more complex vocal repertoires, perhaps peaking, with humans. Within the baleen whales, humpback whales are considered an exception. Present knowledge indicates that humpbacks have a relatively complex acoustic repertoire but work on their breeding social system has considered them to be socially simple. Animals regarded as having a simple social structure tend to have small group sizes and a lack of repeat associations between individuals over time. Humpback whales meet this criteria in that they form temporary associations between a small number of individuals, and these associations are not repeated over time, leading to the conclusion that their social structure is simple and individually based. Why then do humpbacks have what could be considered a complex acoustic repertoire? Is the conclusion that they possess a simple social structure supported by best available scientific evidence? This chapter illustrates that humpbacks may in fact have a complex social structure, with complexity defined differently than traditional definitions of complexity (number in a group, number of repeat associations). Rather than forming large permanent groups with repeated interactions between individuals, humpback whales during the breeding season form networks that encompass multiple groups. These groups are frequently changing membership, and animals are constantly moving into, and out of this network. Whales must therefore continuously assess, and respond to, a changing

R. A. Dunlop (✉)
School of Biological Sciences, University of Queensland, St. Lucia, Brisbane, Australia
e-mail: r.dunlop@uq.edu.au

C.W. Clark and E. C. Garland (eds.), *Ethology and Behavioral Ecology of Mysticetes*, Ethology and Behavioral Ecology of Marine Mammals,
https://doi.org/10.1007/978-3-030-98449-6_10

social environment. Given that humpbacks likely rely on acoustic communication to manage these interactions, this added layer of social complexity may go toward explaining their large and varied vocal repertoire. Perhaps communicative and social complexities do go hand in hand for the humpback whale.

Keywords Baleen whale · Breeding behavior · Communication repertoire · Social complexity · Social behavior · Vocal repertoire · Vocal complexity

10.1 Introduction

Humpback whales reach sexual maturity between five and ten years old, depending on their sex, population, and individual body condition (Chittleborough 1958; Clapham 1992); they can live up to about 95 years (Chittleborough 1965). As with other "great whale" species, humpback whales were heavily harvested with only a small percentage of the global population remaining at the end of the commercial whaling era (Chapman 1974; Clapham et al. 1999). Despite low rates of exchange between neighboring populations (Dawbin 1966; Garrigue et al. 2004; Olavarria et al. 2007; Jackson et al. 2014), only the Arabian Sea population appears to be suffering from the effects of a small founder population, i.e., reduced reproductive fitness due to inbreeding (Mikhalev 1997; Pomilla et al. 2014), with most other populations flourishing or at least stabilizing. Humpback whales were listed by the IUCN as of "least concern" in 2008 (Reilly et al. 2008).

The life history of many baleen whale species involves migration. Humpback whales undertake long annual migrations from high-latitude summer feeding grounds to low-latitude warm water winter breeding grounds (Chaps. 4 and 11). To mate and give birth (Chittleborough 1965; Tyack and Whitehead 1983; Payne and Payne 1985; Clapham 1996). The speed and timing of their migration to breeding grounds depend on their age and reproductive status (Chittleborough 1965; Brown et al. 1995; Craig and Herman 1997; Dawbin 1997). Females in late lactation, along with their soon-to-be-weaned calf, are the first to migrate to the breeding grounds during late autumn or early winter. They are followed by sexually immature males and females, then mature males, and finally, pregnant females that give birth in warm water breeding grounds. This arrival order is reversed for the migration back toward high-latitude feeding grounds, with newly pregnant females the first to depart tropical waters in late winter/early spring (Clapham 1996; Craig et al. 2003). Males appear to spend more time than females in the migratory corridor, as well as on the breeding grounds, so a breeding ground can contain more males than females (Clapham 1996; Craig et al. 2003; Cartwright and Sullivan 2009; Herman et al. 2011). This male-dominated social environment makes humpback whales an interesting species to study in terms of socially mediated breeding interactions. As discussed later in this chapter, having a large number of males competing for a smaller number of breeding females leads to some interesting social dynamics.

The social system of humpback whales is, if using traditional classifiers of social complexity, is relatively "simple." This is because they do not live in large, permanent, or stable groups (Chap. 7). Nor is there evidence of long-term repeat associations between individuals over time, apart from some feeding groups. Most studies carried out on humpback whale social groups on feeding and breeding grounds, as well as during the migration, conclude that such groups are small and unstable (Baker and Herman 1984; Mobley Jr. and Herman 1985; Clapham et al. 1992; Clapham 1993, 1996; Valsecchi et al. 2002). Early work on group social structure in Northern Hemisphere feeding and breeding grounds found that humpbacks form large, but temporary aggregations (Tyack and Whitehead 1983; Tyack 1983; Baker and Herman 1984; Clapham et al. 1992, 1993). There was no evidence to suggest that their group social structure within these aggregations was any more complex than genetically unrelated animals coming together and temporarily associating. In other words, animals in these groups did not repeatedly re-associate with each other again, unless by chance, suggesting they did not form long-term associations.

Other work on summer feeding groups in the Gulf of Maine did find some evidence of stable, long-term group associations (e.g., Weinrich et al. 2006; Ramp et al. 2010). It was believed that these long-term associations help facilitate the formation of cooperative feeding groups, which are energetically beneficial as long as each group member obtains more calories when feeding together than when feeding alone. These longer-term associations were believed to be influenced by maternal lineages (Weinrich et al. 2006), because animals related maternally were more likely to occur together. Female-only feeding associations were more likely to occur than male-only, or male–female, associations (Weinrich and Kuhlberg 1991; Ramp et al. 2010). This may be because females require more energy due to pregnancy and therefore benefit from cooperating within a group to corral food. Group feeding in humpback whales often involves synchronous behaviors such as bubble netting (Sharpe and Dill 1997; Wiley et al. 2011). Here, one whale circles the prey and blows bubbles, while another emits a feeding call (Sharpe 2001; D'Vincent et al. 1985). Once the net is complete, all whales (as many as 12) lunge synchronously to engulf the prey. During bubble netting events, there are obvious role specializations and divisions, implying that these feeding interactions are complex and require cooperation. However, not all feeding aggregations involve complex interactions. More recently, some large feeding aggregations (up to 200 animals), termed "super-groups," have been observed off South Africa (Findlay et al. 2017). These are unusual given most humpback whales feed in near-polar regions rather than in low-latitude waters. So far, there is no evidence to suggest that these large super-groups are temporary aggregations of genetically unrelated individuals taking advantage of large aggregations of prey such as fish and/or invertebrates.

Therefore, although there may be some evidence of long-term associations in feeding groups, evidence of any long-term associations between humpbacks in the breeding grounds is still lacking. The breeding behavior of humpback whales is explored in more detail in Chaps. 6 and 11. Of note is that multiple studies on the Hawaiian breeding ground concluded that humpback whales have a fluid social structure comprised of short-term groups with little evidence of long-term associations

(Herman and Antinoja 1977; Mobley and Herman 1985). Humpback whale society within breeding areas is therefore classified as individually based, with the only stable (i.e., lasting more than a few hours) association being a female with her calf. Males temporarily join and escort females, including mothers with suckling calves (Craig et al. 2002; Morete et al. 2007). More males may then join this group to form competitive pods around a single female (Baker and Herman 1984). However, these interactions typically last no more than a few hours (Mobley and Herman 1985; Clapham et al. 1992), with no evidence that individuals during breeding interactions form long-term associations (Brown and Corkeron 1995; Valsecchi et al. 2002), and no evidence that individuals regularly interact with the same individuals.

In many mammals and bird species that are organized into social groups, vocal signals play a particularly important role in the regulation of social interactions and the coordination of group activities. Animals that have more complex social structures are often associated with higher levels of vocal complexity (Freeberg 2006), most notably in primates (Lemasson 2011) and humans. Highly social species should have a rich and diverse communication system to convey a wider range of information, including individual or group identity, behavior, and motivation (Freeberg et al. 2012). This concept is known as the "social complexity hypothesis for communicative complexity" (Freeberg et al. 2012). Following this hypothesis, the individually based social structure in humpback whales, especially during breeding and migrating, would suggest that their vocal repertoire would also be relatively simple. This does not seem to be the case because, as highlighted later in the chapter, humpback whales have a rich vocal repertoire. The following section introduces their acoustic communication system, demonstrates the complexity of this acoustic repertoire, and then explores some reasons why humpbacks produce a complex acoustic repertoire.

10.2 Acoustic Communication in Humpback Whales

Humpback whales use vocal signals for feeding, such as the calls emitted during bubble netting and song on the mating grounds. They use a large and varied repertoire, ranging from pulse sounds during feeding, such as paired bursts (Parks et al. 2014) and "megapclicks" (Stimpert et al. 2007), to loud ornate song signals during apparent mating attempts (Payne and McVay 1971). Humpback whale song is arguably one of the most well-studied communication signal in a baleen whale species (see Chaps. 8 and 11). This chapter focusses on a different set of sounds made by humpback whales; those that are likely used for social interactions, known as social sounds, or sometimes non-song calls to differentiate them from song.

The humpback whale acoustic repertoire includes many different types of non-song vocal calls, as well as surface-generated sounds such as breaching and slapping of pectoral fins and flukes (Dunlop et al. 2007). Their vocal repertoire spans a large frequency range, from infrasonic grumbles that are below human hearing range, to high-frequency bird-like chirps. It includes frequency-modulated (e.g., "moans," "trumpets," and "cries"), amplitude-modulated (e.g., "purrs" and "growls"), and broadband (e.g., "roars" and "underwater blows") sounds (Dunlop et al. 2007). One way to define vocal complexity is the size of the animal's repertoire, where increased communicative complexity can be inferred through an increase in the number of sound types used. Here, if the function of the sounds is not known, sounds that share similar structural characteristics (frequency content, duration) are grouped together. Defining the size of a species' repertoire, in this case the humpback whale helps researchers to make comparisons of vocal complexity among animals in different contexts, among populations, and among species.

It is difficult to assess the number of different sound types within the humpback repertoire. One study, using a combination of subjective (what the analyst can hear) and quantitative (using measured features of the sounds to statistically separate them) found more than 30 different sound types (Dunlop et al. 2007). This study used recordings from groups of whales migrating southward from their breeding grounds in the Great Barrier Reef of Australia, toward their Antarctic feeding grounds. The analysis method was similar to that used by others. First, they subjectively separated out units into different sound types by ear according to frequency and duration as those features are meaningful to us humans. Then, each sound type was given an onomatopoeia name; "grumble," "wop," "grunt". For any person listening to the sounds of humpback whales, a high-frequency "cry" sounds, to us, quite different from a low-frequency "grunt." This created a frequency gradient, which involved moving from lower-frequency "grumbles," "groans," "grunts," and "growls," to higher-frequency "whines," "cries," and "chirps." The duration gradient ranged, in the lower frequencies, from short duration "grunts" and "snorts" to long duration "grumbles and "growls," and in the higher-frequency range, from short duration "chirps" to long duration "cries" (e.g., Dunlop et al. 2007; Stimpert et al. 2011; Rekdahl et al. 2013, 2017; Fournet et al. 2015).

One problem with human categorization of sounds is that it is not objective and therefore may not be repeatable between observers. What one person calls a "growl," another may call a "grumble." Perhaps "grumbles" and "growls" are quite similar and therefore should be one sound type (you are known as a "lumper"). Perhaps they sound different to each other in that a "grumble" sounds more tonal and less amplitude modulated than a "growl" (you are known as a "splitter"). Quantitative methods provide standardization and objectivity, which can help researchers make comparisons among different populations and species. These techniques tend to rely on measures of multiple features about each sound (e.g., measures of frequency, duration, and bandwidth) and then use these measures to categorize the sounds into different sound types—those with similar measures to each other cluster and are categorized together. The use of quantitative analysis methods takes some of the human subjectivity out of the process, making the studies more objective and therefore repeatable. Such statistical techniques, however, still rely on subjective human input.

The use of a combination of subjective and quantitative analysis techniques to quantify a social sound repertoire has been the *modus operandi* for the majority of studies on the humpback whale social sound repertoire. However, even when including quantitative techniques, results can differ between studies. Studies on the

Eastern Australian population of humpback whales during their migration found approximately 30 (Dunlop et al. 2007 study) and 46 (Rekdahl et al. 2013) different sound types. Both studies noted they likely did not capture all of the different sound types, and both stipulated that each sound type did not necessarily have a different function. A study on the Southeast Atlantic population that migrates along the east coast of Africa found 11 different sound types (Rekdahl et al. 2017) noting that higher-frequency sound types were likely to have been missed due to the constraints of the recording equipment. Eight calls were found within the repertoire of the Northwest Atlantic population whilst feeding (Stimpert et al. 2011) compared to 16 in the Southeast Alaskan population whilst feeding (Fournet et al. 2015). The repertoire of a migrating female with her newborn calf included approximately eight to ten different sound types (Indeck et al. 2020). Even calves are quite vocal, using a range of frequency-modulated, amplitude-modulated, and pulsed sound types (Zoidis et al. 2008). From these studies, it is clear that humpback whale vocalizations are highly variable in structure, and this variability makes it difficult to definitely quantify all the different sound types produced by members of the species.

One reason for this variability is that part of the humpback whale sound repertoire is "graded" as opposed to "distinct." A humpback whale "wop," for example, is a distinct sound type, that sounds like no other, and tends to be heard year after year (Rekdahl et al. 2013) and in different populations (Fournet et al. 2018b). The "wop" has been recorded from humpback whales off the coast of Eastern Australia (Dunlop et al. 2007; Rekdahl et al. 2013), in the Western North Atlantic off the coast of New England (Stimpert et al. 2011; Fournet et al. 2018a), in the Southeast Atlantic off the coast of Angola (Rekdahl et al. 2017), and in Southeast Alaska (Fournet et al. 2015). It is relatively easy to categorize the "wop" as a discrete sound type as it is different in structure compared to the other sound types; there is low variation in structure between "wops," and it is commonly heard. A graded repertoire means that sound types tend to structurally merge into each other, making it difficult to separate one sound type from another. Quantitative methods used to categorize sounds into sound types are not well suited to account for gradation in call types, as the underlying assumption of these analyses is that calls are discrete. Soft classification methods, such as the "fuzzy k-means" method (Ferraro and Giordani 2015; Wadewitz et al. 2015), may be a better way of classifying sounds within a graded communication system, as the underlying assumption is that calls can have qualities that are typical of more than one cluster (Wadewitz et al. 2015). Work, again using the same migrating population of whales used in the Dunlop et al. (2007) and Rekdahl et al. (2013) studies, trialed this method. When using what is known as a "hard clustering" technique (one that assumes discrete categories), the repertoire size was 35 sound types. When using a "fuzzy clustering" technique (one that deals with the graded nature of calls), the repertoire size was 14 (Cusano 2020). These highly graded humpback whale vocal signals are contributing to their seemingly endless variety of different sounds.

Acoustic communication signals in baleen whales are likely to have a variety of functions. Through repeated observations of humpback whale groups behaving in a certain way and emitting particular sound types while performing those behaviors (e.g., a female migrating with her calf, a migrating female with one or more escorts, migrating individuals changing group membership), functions have been proposed for migrating humpback whale social sounds (Dunlop et al. 2008). Within the "stable" sounds, the "wop," for example, was thought to function as a contact call (Dunlop et al. 2008) as it was predominantly heard in groups containing a female and calf (Fig. 10.1a). This particular sound type has been recorded in multiple years, within multiple populations, and in multiple contexts, suggesting a common function for this particular sound type. "Snorts" (Fig. 10.1b) were also a common sound heard in groups that were not splitting or joining and again may be some way of keeping the group together (Dunlop et al. 2008). "Snorts" were also recorded in the Southern African population (Rekdahl et al. 2017), and though the function of these sounds was not a focus of the study, we do not know if these sounds were made in the same context as that found in Dunlop et al. (2008). "Grunts" (Fig. 10.1c) were produced more often when animals were joining a group and therefore may be used to coordinate group social interactions (Dunlop et al. 2008). Again, these sounds persisted in the repertoire year after year (Rekdahl et al. 2013) and were recorded in the Southern African population (Rekdahl et al. 2017). It stands to reason that a sound that has a basic function, such as for contact calling, should be conserved over time (year after year) and space (between and perhaps among populations).

Breaching and slapping the water surface are also likely to be communication signals (Dunlop et al. 2010). These sounds probably indicate something about the

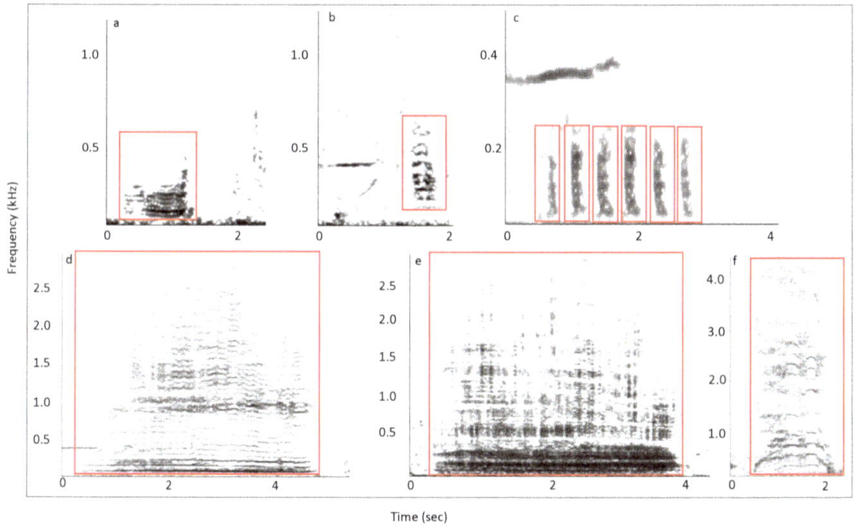

Fig. 10.1 Spectrograms (time in seconds) and frequency in kHz of six common humpback whale social sounds; "wop" (**a**), "snort" (**b**), "grunt train" (**c**), "growl" (**d**), "purr" (**e**), and "screech" (**f**). Sounds were recorded on the Eastern Australian coastline from the East Australia population of humpback whales during their return migration from breeding grounds in the Great Barrier Reef to feeding grounds in the Antarctic in 2002, 2003, and 2004

signaler; its size, location, and even motivational state, because seemingly "aggressive" male–male interactions involve "head slaps" and "peduncle throws" compared to the less energetic "pectoral slaps" and "tail slaps" that are commonly used by groups in which there is no obvious surface male–male competitive behavior. Therefore, there may be some information in the rate these behaviors are performed. Breaching and surface slapping behaviors are also commonly observed in all humpback whale populations.

Animal acoustic communication signals can encode the caller's motivation (Morton 1977). An animal's motivation is a combination of its arousal level and whether it is in a positive or negative state. For example, a high-arousal and negative state indicates the animal is likely in a "fearful" or "aggressive" state, while a high-arousal and positive state may indicate an "affiliative" or "playful" state. Can one tell which motivational state the animal is in based on the type of vocal signals it uses? Morton (1977) proposed a "motivation-structural rule" (M-S rule) for animal communication, based on his assumption that animals would produce signals with some similar acoustic features to express a motivation. This helps a receiver of the signal assess a signaler's motivation remotely without "getting involved." There are four parts to the M-S rule; frequency, or perceived pitch; bandwidth or "harshness"; rate of delivery, or how often a signal is produced; and amplitude or "loudness." Vocal signals that follow these rules are a good way for animals to separate "fearful" from "aggressive," where both are negative high-arousal states but are quite different from each other: flight versus fight. Aggressive signals tend to be lower in frequency, harsher (i.e., have a broader bandwidth) and are usually loud and produced at a high rate. These signals likely evolved to convey body size as larger animals are typically able to produce lower-frequency sounds more readily than smaller animals. To appear threatening, an individual would therefore attempt to produce a call at the lowest frequency possible, loudly, and as often as possible. Fearful signals are completely different, as they are higher in frequency and relatively tonal. This type of signal may serve to convey a smaller body size, or mimic infant-generated sounds, to prevent injury. These sounds would be produced much more quietly.

Many animals convey motivational information by varying one common sound type. Humans can usually tell the motivational state of a domestic dog (*Canis familiaris*) by listening to its bark (Yin and McCowan 2004); e.g., fearful bark sounds are quite different from aggressive or playful barks (Pongracz et al. 2006). We can do this because humans and dogs follow the same basic M-S code. Animals can also use different sound types depending on their motivational state. An aggressive dog uses aggressively sounding growls, a playful dog uses higher-frequency whines (Taylor et al. 2010). Humpback whales seem to take this latter option. Rather than varying the structure of specific sounds, they appear to use different sound types to express different motivational states. This has been shown in breeding interactions, which involve multiple males competing for access to a single female (Dunlop 2017; Cusano 2020). These groups exhibit high-arousal levels as indicated by their surface behavior—they move fast and erratically, take more surface breaths, and dive for short amounts of time. Males can be aggressive towards each other; the primary male escort attempts to chase other smaller secondary escort males, and males slam into each other and dive on top of each other. We would expect these aggressive males to be using aggressive acoustic signals, whereas the subdominant males, the ones being chased off, would use appeasement-type signals. In keeping with this expectation, sounds that were relatively low in frequency and "harsh" were commonly heard in the more erratic high-arousal groups ("growls," "purrs," "screeches," and "underwater blows"; Dunlop 2017; Cusano 2020; Fig. 10.1d–f) and rarely heard in groups that were not displaying such obvious competitive behaviors (Cusano 2020). It may be that producing a different variety of sounds to transmit motivational information, rather than varying the structure of a small number of discrete sound types, is one factor in explaining the size of the humpback whale social repertoire.

Behavioral response studies, where sounds are played back to the whales and their responses observed, are used to make inferences about the function of a sound. To date, there are only two published playback studies using humpback whale social sounds. Both used social sounds recorded from groups containing multiple adults and likely to contain males competing for access to a female. The first used a recording from a competitive group in which the males were observed to be fighting (Tyack 1983). The second used a recording from a group containing multiple males and one female, but no observed fighting (Dunlop et al. 2013c). Both studies found that some whale groups approached and even charged (Tyack 1983) the playback vessel when the sounds were played. These animals were likely to be other males noting that singing males also approached the playback vessel during both studies. In other words, males were likely attracted by the sounds of other males competing for a female. Both studies also found that female-calf pairs tended to avoid the playback vessel. It appears logical that a female with a newborn calf would want to avoid competitive groups. The problem with both experiments was that both recordings contained multiple sound types; therefore, it is impossible to know which of the sounds (or all of them together) produced these reactions. More work is needed.

An investigation of the relationship between a social system and the environment can provide valuable insight into the evolution of communication systems (Janik and Slater 1997). Terrestrial and marine mammal communication systems have evolved in different environments, with the marine environment providing opportunities for longer-range communication because sound propagates well underwater (Janik 2005). In a terrestrial system, we rely on what we can see. In the marine environment, we rely on what we hear, as sound is the most effective means of communicating (Richardson et al. 1995). Sounds produced by whales can travel for many tens of kilometers and are therefore likely available to many other whale receivers. A whale's ability to emit a signal that is available to a number of dispersed receivers that are out of sight may be the key to explaining the supposed disparity between vocal and social complexity in humpback whales. We must therefore change our primary observation modality from what we can see to what whales can hear.

10.3 The Effect of Physical Environment on Humpback Whale Acoustic Communication

The area over which an acoustic signal transmits and is audible to the intended receiver is known as the active or communication space (Brenowitz 1982; Janik 2000; Clark et al. 2009). The size of this area depends on the structure of the signal, the characteristics of the environment, and the receiver's ability to detect and discriminate the signal from background noise (Wiley and Richards 1978; Clark et al. 2009). Acoustic signal levels attenuate during signal transmission, meaning that as the distance from source to receiver increases, the received signal level decreases. A signal can propagate large distances, such as those used to broadcast information, or small distances, such as private signals. The environment can also change how well a signal propagates; a signal transmits much further if it reflects off the ocean floor, but disappears much more quickly if it is absorbed. Finally, how well the receiver can hear the signal in noise contributes toward the communication space of the signaler. A common term used for this concept is masking (Clark et al. 2009; Erbe et al. 2016). In the case of signalers and receivers in the underwater environment, the sound would be from the signaler, and the masking sound would be natural underwater noise, such as noise from breaking waves. A receiver that finds it difficult to hear the signal in noise (it is more likely to be masked) will need to be closer to the signaler than a receiver that can hear the signal well. If receivers are able to hear the signal well in noise, even from a signaler that is far away, it is less likely to be masked.

The signaler has some control over its communication space since it is up to the signaler to produce the signal at an appropriate amplitude for the intended receiver to detect and decode it. If the receiver is further away, the signaler can signal louder to reach the receiver. For example, if the person we are trying to communicate with is across the road, we vocalize louder compared to if the person is next to us. We might also use signals that travel better through the environment and are therefore easier to hear. We might whistle, for example, if the receiver is far away, and we might use the receiver's name if they are close-by. In other words, a signaler might use different signals according to the distance of the receiver. The amount of noise also changes how we signal. For us, the amount of traffic noise affects how well we hear each other, as does a noisy shore break, rustling leaves, a party full of people. In response to increased noise, humans change how they speak. Vocal characteristics, such as level, pitch, and/or rate of signal production, are modified in a noisy environment to improve signal detection, i.e., make them easier to hear (Lombard 1911). Humans do this reflexively, and presumably so do other animals. Studies designed to test for the Lombard effect in animals look for the same changes in vocal characteristics in response to increased broadband background noise levels.

In a shallow water (less than 50 m deep) environment, there are several natural abiotic and biotic sources of noise, such as wind, breaking waves, and rain (Cato 1976, 1991; Bass and Clark 2002), and these noise sources can limit the communication space available to animals (Clark et al. 2009). Humpback whales emit social sounds at relatively high source levels (Thompson et al. 1986; Dunlop et al. 2013a). Therefore, under certain conditions, social sounds can propagate quite far away (a few kms) from the signaler, even in relatively high noise environments when there are breaking waves on the surface of the ocean due to high winds. In a more detailed study on the migrating Eastern Australian population, the source levels of humpback sounds (how loud the sounds were) varied from 123 to 184 dB *re* 1 μPa @ 1 *m* (*rms*) (Dunlop et al. 2013a) in wind-dominated background noise levels ranging from 76 to 120 dB re 1 μPa. To put this into context, wind-dominated background noise at this study site ranged from 76 dB re 1 μPa (very quiet, flat, ocean, with very little wind and no noise from passing vessels) to 120 dB re 1 μPa (wind speeds of 15–20 knots, with surface breaking waves, with no noise from passing vessels). The noise was measured over a 40 Hz–2 kHz frequency band, which was the frequency range of the humpback sounds. This was assumed to be the noise band that was the most "meaningful" to the whales. A 60 dB signal excess above noise suggests these particular signals can be heard above the background noise out to a few kms from the signaler and are therefore audible to other whales in the area.

To determine how far away other humpbacks can hear, or detect, a signaling whale is difficult as we need to understand how the receiver (in this case another humpback whale) hears a conspecific signal in noise. Just because a signal can be detected does not necessarily mean that the signal can be recognized and understood. Some bird species, for example, require a higher amplitude signal for successful discrimination of conspecific calls compared to just detecting them and higher amplitude again for conspecific call recognition (Lohr et al. 2003; Dooling and Blumenrath 2016). For humpback whales, as we do not know how well they can detect their conspecific signals in noise, we make assumptions. One assumption is that the communication space of their social signals extends out to where the signal-to-noise ratio of their conspecific signals becomes zero. Using this assumption, one study carried out in a shallow water environment of less than 50 m depth found that the communication space of lower-frequency signals ("grumbles," "groans," "wops," "moans") was approximately 4 km and that of higher-frequency signals ("cries," "screams," "yaps") was approximately 2 km (Dunlop 2018). The estimated size of their communication space in the Dunlop (2018) study was confirmed behaviorally. Joining signaling groups began vocalizing when separated by approximately 4 km. This distance matched the measured communication space of the lower-frequency sounds, suggesting that groups could hear each other out to about 4 km, in this study site, during average wind noise conditions. The majority of social interactions, however, occurred among groups that were separated by 2 km. If higher-frequency signals are being used to mediate group joining interactions, then the estimated communication space of the higher-frequency sounds matches their proposed function.

When assessing a marine mammal's response to increased underwater noise, we assume that their signaling behavior has evolved to cope with elevated levels of natural noise. Humpback whales have two coping mechanisms for natural (wind-related) increases in noise. Both of these "anti-masking" strategies maintained their communication space out to approximately 4 km, even in high wind noise conditions (Dunlop 2018). First, they increased their vocal source levels (vocalized louder) with increasing levels of wind noise (Dunlop et al. 2014). In other words, this was a Lombard response (Dunlop et al. 2014). Second, they switched to using surface-generated percussive sounds such as breaching and slapping (Dunlop et al. 2010, 2013b), which tend to propagate further in noise (Dunlop 2018). Humans clap their hands together, or bang something, to get another's attention when it is noisy. Humpback whales may do the same, as it might be easier to hear a short loud bang in noise than emit a vocal signal. It is currently unknown what potential information these breaching and slapping signals may contain, but it is likely that they contain information on the presence and position of the signaler, the signaler's size, and, potentially, its motivation.

Up until now, we have only considered natural noise conditions. Underwater noise can also be anthropogenic in origin (e.g., from vessel engines, oil and gas exploration, naval sonar activity; Richardson et al. 1995). These sounds may result in further masking of the animals' acoustic signals compared to masking effects of natural noise. The presence of anthropogenic sound may therefore reduce the ability of a listening animal to detect and interpret biologically important sounds over and above its ability under natural noise conditions unless the signaler and/or receiver compensates in some manner. To measure this, many masking studies relate the signal level in noise (i.e., a measure of signal-to-noise, or *SNR*), to increasing distance from the receiver to determine the distance at which a signal is just detectable in noise. Then, they measure the reduction in this detectable distance with an increase in noise level; this equates to the loss of communication space. In vessel noise conditions, the communication space of humpback whales is reduced by about one-half (Cholewiak et al. 2018; Dunlop 2019). In addition, the likelihood of individuals joining together is also reduced (Dunlop 2019). We would initially interpret this change in joining behavior as a result of signal masking by the vessel noise, making the signals more difficult to detect and recognize. However, when measuring the signals in vessel noise, there was some evidence of masking, but not enough to explain the observed reduction in joining behavior (Dunlop 2019). In other words, the effect predicted based on sound transmission considerations did not fully explain the observed behavioral change. Perhaps the presence of the vessel had a behavioral effect on the whales such that the context of vessel presence has some influence on the likelihood of individuals joining.

Anthropogenic noise is a different source of noise compared to natural sounds. Even so, several studies have reported evidence of the Lombard effect in a variety of cetaceans relative to anthropogenic sound. The North Atlantic right whale (*Eubalaena glacialis*, Parks et al. 2011), beluga (*Delphinapterus leucas*, Scheifele et al. 2005), killer whale (*Orcinus orca*, Holt et al. 2009), and bowhead whale (*Balaena mysticetus*; Thode et al. 2020) all were found to have a Lombard response to elevated anthropogenic (shipping) noise. It makes sense that humpback whales would do the same. However, humpback whales neither increased the source levels of their vocalizations, nor switched to using surface-generated sounds, with increased noise due to passing fishing vessels (Dunlop 2016b). Contrary to these findings, a different study found that humpbacks showed a Lombard response to increased background (shipping) noise (Fournet et al. 2018c). Why the difference in results? One hypothesis was that whales have another mechanism for coping with noise called "spatial release from masking." This is where the receiver can compensate for increased noise coming from the direction of a source by audibly separating the sound of interest from the noise arriving from a different direction. Humans do this in crowded noisy situations (e.g., a cocktail party), where we focus on the individual we are talking to, and filter out the background noise. Little is known, however, about this strategy in cetaceans (Erbe et al. 2016).

The major difference between the humpback study (Dunlop 2016b) and many of the other studies on the response of baleen whales to noise (e.g., Parks et al. 2011; Fournet et al. 2018c) was that the noise environment in the latter studies was dominated by noise from ships that would have been relatively far from the group. In the Dunlop (2016b) study, the noise was from passing vessels that would have come relatively close to the vocalizing group (most came within 2–3 km). Proximity matters, since humpbacks consider how far away a source of noise is as well as how loud it is. A good example of this is the response of humpback whale groups to seismic air gun noise. Many studies on the effects of a noise source on whale behavior relate the "dose" (or received level of the noise source) to the likelihood they will respond in some negative way, or the magnitude (size) of their behavioral response (Antunes et al. 2014; Miller et al. 2014; Williams et al. 2014). The dose–response paradigm predicts that as the received level goes up, the likelihood of a whale responding, and the size of this response, increases (Harris et al. 2018). However, a second factor also plays a part in response—their proximity to the source. A vessel with a higher source level that produces a higher received level at the whale, but far away, might not elicit any response, while a vessel with a lower source level that produces a lower received level at the whale, but close-by, might do the opposite (see DeRuiter et al. 2013; Dunlop et al. 2017 for examples). Humpback whale groups, in response to seismic air gun noise, increased the magnitude of their avoidance response at higher received levels, but only if the source was within 3.5 km (Dunlop et al. 2017). The proximity of the source combined with the received level determined their likelihood of response, as well as the size of their response. Perhaps the proximity of these sources plays a part in dictating their response and explains the different responses observed.

If proximity was to predict how humpback whales respond to ships and close-by vessels, then they would be more likely to respond (e.g., increase the source level of their vocalizations, switch to surface-generated sounds) to the close-by passing vessels and less likely to respond to background shipping. If they consider close-by passing vessels something to avoid, vocal crypsis could occur. Vocal crypsis is when an animal vocalizes more quietly, or less often, to avoid detection by something it wishes to avoid. Perhaps humpback whales, rather than attempting to maintain the size of their communication space in response to vessel noise, are reducing it to avoid detection by these close-by vessels. In other words, they perceive the vessel as a threat. To recap, humpbacks also appear to reduce their breeding interactions in the presence of a vessel (Dunlop et al. 2020), and this change in behavior is not fully explained by signal masking. Again, this suggests that humpbacks are avoiding the vessel by "going quiet." Later in this chapter, I refer to results indicating that sometimes females are quiet to potentially avoid detection by nearby conspecifics, such as singing whales.

10.4 Communication Networks and the Social Environment

In water, sound travels much more efficiently compared to in air, and sound especially low-frequency sounds (<1000 Hz) can propagate over much greater distances. We have discussed how this results in potentially large communication spaces in cetaceans. Within such a space, for any given signaler, there are likely to be a number of other conspecifics; the larger the space, the greater the likely number of conspecifics is. In other words, signaling behavior is not just a dyadic exchange between a signaler and an intended receiver, but involves other eavesdropping animals such as other conspecifics, predators, or prey. Together these signaling interactions form what is known as a communication network (McGregor 1993; McGregor and Horn 2015). Though these signals may reach intended receivers (benefit), eavesdroppers can also obtain information from these signaling interactions, and this can come at a cost. If it was a predator, the cost could be injury or death. If it was another conspecific, the cost is that it provides direct benefits to the eavesdropping animal (McGregor 1993; Otter et al. 1999; Peake et al. 2001). Here, the eavesdropper gains information about the signaler/receiver relationship without having to directly take part in the interaction. Much of this work has been undertaken using Siamese fighting fish (*Betta splendens*), where male (Oliveira et al. 1998; Peake and McGregor 2004) and female (Doutrelant and McGregor 2000) eavesdropping fish observe aggressive interactions between fighting fish to gain information that could be used in future interactions. Therefore, the social knowledge acquired by eavesdropping, as well as providing individual benefits, can also influence the future actions of the eavesdropper. A losing male in a male–male contest, for example, would spend less time displaying to a female who witnessed the loss compared to a naïve female. It seems these males are taking into account the females' observations and how this might affect their future mating potential (Herb et al. 2003).

The fighting fish studies outlined above focused on the visual information eavesdroppers gained from observing interactions between signalers and receivers. Eavesdropping, however, can occur acoustically. Here, the presence of an audience, i.e., eavesdroppers, can affect communicative interactions in that signalers can change their signaling behavior according to who is "listening in." This is known as the "audience effect," first documented in a study on domestic chickens (*Gallus*; Evans and Marler 1991, 1994). Free-ranging chimpanzees (*Pan troglodytes*) that are victims of aggression will scream. They exaggerate the aggression experienced (their screams) if the audience consists of individuals that are capable of helping the victim; for example, they are a higher rank than the aggressor (Slocombe and Zuberbuhler 2007). Understanding your audience, and changing your signaling behavior accordingly, can therefore have benefits to the signaler.

Communication networks provide a means to facilitate the maintenance of group cohesion and group decision-making processes. Mutual eavesdropping between conspecifics can structure cooperation and alliance formation and, hence, contribute to long-term group stability. Therefore, social complexity seems to go hand in hand with communication network complexity. Since humpback whale sounds can propagate up to a few kilometers away from the signaler, it is likely these sounds are available to a network of listeners. Humpback whales within groups are likely signaling to each other, but by doing so, these signals are available to other eavesdropping groups in the area. Perhaps this is the source of some of the complexity within humpback whale signaling. Recent work on humpback communication networks has focused on groups exhibiting breeding behaviors. Therefore, the breeding behavior of humpback whales will first be outlined, before exploring signaling behavior within their breeding network.

Female humpbacks are seasonally polyestrous (Chittleborough 1954, 1965), gestation is generally 11 months (Chittleborough 1958; Nishiwaki 1959), and the modal birth interval is 2 years (Craig et al. 2002). Postpartum pregnancies were assumed to be rare due to the high energetic cost of producing calves in successive years (Chittleborough 1958, 1965; Baker et al. 1987). However, evidence from an Antarctic population indicates that this occurs often (Pallin et al. 2018) and likely explains the rapid population growth in the Australian populations (Bannister and Hedley 2001; Noad et al. 2019).

Humpback whale mating behavior has been described as female defense polygyny or promiscuity (Helweg and Herman 1994) or male dominance polygyny and lekking (Clapham 1996; Cerchio et al. 2005; Darling et al. 2006). Clapham (1996) suggested that humpback whale mating behavior could be described as a "floating lek," in that male humpbacks, when competing for reproductive access to females, do not defend territories, but likely display (sing) to attract females. Male humpback whales also form "competitive pods" around a single female (Baker and Herman 1984). The group is usually comprised of any number of "secondary escorts," jostling aggressively for the position of "primary escort" (Tyack and Whitehead 1983; Clapham et al. 1992; Brown and Corkeron 1995). The males in primary escort position have been observed to be marginally larger in body size to those in secondary positions (Darling et al. 2006), suggesting that females may select mates based, at least partially, on size. Males frequently escort mothers with suckling calves, and the rate at which this occurs seems to increase as the breeding season progresses (Dawbin 1966; Craig et al. 2002; Morete et al. 2007). It appears that males preferentially associate with females with high reproductive potential (no calf), but as the mating season progresses, they become less "choosy" and are seen more often in association with females with calves (Craig et al. 2002). It is also possible that males escorting females with calves may be an alternate mating strategy. Perhaps the male's ability to compete physically for access to highly valued females is such that females with a calf are an easier option (with less competition). These, along with song displays (Chap. 11), have been proposed as secondary mating strategies for males (Baker and Herman 1984; Cerchio et al. 2005). Given that copulation has never been observed in humpback whales, reproductive success is difficult to determine. It is therefore not known if each strategy is mutually exclusive, or if any strategy is more successful than the others.

Songs have a high source level, are continuous, and are likely to be a broadcast signal. In other words, the song has a large communication space probably made up of breeding females and other competitive males. Singing whales occur alone, or as escorts within a group containing a female (Smith et al. 2008). There is much debate as to whether song functions to provide fitness information used by females (female choice and mate attraction), to compete with other males (dominance sorting), or both (see review by Herman 2017). It is also unknown whether singers sing differently when they are alone, and likely broadcasting to other conspecifics, compared to when they are within a group, and perhaps signaling to other members within the group. Since the signal is of a high amplitude, even if they are signaling to other animals within their group, the signal is likely available to other eavesdroppers in the area. These eavesdroppers can use information contained within the song and respond to it however they choose. For example, song features likely contain information related to the fitness of male singers. Other males within audible range may use this information to determine whether or not to challenge the singing male. Females may use this information to determine whether or not they will breed with that male. For a male, there is a trade-off between singing and emitting information with the risk of attracting competitors. Males would therefore have to decide whether or not it is worth signaling. The risk is that he might attract competitors, and the benefit is that he may increase his chances of breeding. The decision to sing or not sing would therefore be dependent at least in part on the social environment. If there were many competitive (eavesdropping) males in the area, it may not be worth the risk.

To test if the presence of eavesdropping males changed male humpback signaling behavior, the behavior of males joining groups containing a female was studied (Dunlop and Noad 2016). These joining males could either join while singing, or join while producing social sounds (which sometimes contained a few song units which were not sung in a constant pattern; Dunlop et al. 2007). The groups either contained one female and her calf or a female-calf pair and a small number of escorts. Males significantly changed their signaling behavior in the presence of singing (potentially eavesdropping) males. If there were other singers in the area, signaling males joined a group containing a female using quieter social sounds. These were less likely to be "overheard" by other eavesdropping males. If there was no other singer in the area, males joined a female using song. In other words, if the risk of attracting other competitive males outweighed the benefits of singing, males did not sing. Further to this, if there were already males within the group (other escorts), these males did not join using song. The strategy here seemed to be that males would join to form a competitive group and rely on physical male–male competition, rather than on singing, to compete for breeding access to the female. These results suggest that male humpback whales are considering the presence of other males, both within the group they are joining, as well as in the surrounding area, when deciding on which breeding strategy to use. Since a male's social environment is constantly changing, with males and females moving in and out of an area in which reproductive behaviors occur, the growing body of recent evidence based on extensive observations under natural and experimental conditions supports the conclusion that social interactions are not simple.

It is not just males that consider their social environment when signaling. Another study considered the behavior of vocalizing groups in response to other non-singing and singing groups in the area (Dunlop 2016a). Vocalizing groups, in this study, consisted mostly of either female-calf pairs or female-calf-escort triads. With a close-by singer, these groups tended to decrease both the level of their social sounds (lower intensity) and the frequency (Hz) of some of their social sounds. This is another example of the "audience effect" in humpback whales, indicating that individuals modify their acoustic behaviors depending on the acoustic context. Studies have shown that males joining and harassing females with a newborn calf can cause energetic consequences to the calf as well as increase the risk of separation (Cartwright and Sullivan 2009; Craig et al. 2014). To mitigate this, females with a calf avoid these males by staying away from them (Indeck 2020). Those females that come into acoustic range of the singing male seem to produce sounds at lower intensity, perhaps to avoid acoustic detection by these males (Dunlop 2016a), an example of vocal crypsis and a similar result to what was found in the presence of nearby vessels. Note, however, that this strategy is not always successful, as males are seen joining with females with a calf quite often (Dunlop et al. 2008; Smith et al. 2008). Perhaps these females are in estrous and therefore tolerate the presence of at least one male.

10.5 Tridactic Awareness and Social Complexity

The Machiavellian/Social Intelligence Hypothesis proposes that a complex social structure has selected for advanced cognitive abilities in vertebrates (Vanschaik and Vanhooff 1989). A complex social structure often involves sophisticated social strategies, one of these being the knowledge of one's own relationships with other group members, as well as the relationships between the other group members (known as tridactic awareness). In other words, individuals can assess third-party social bonds (i.e., who is friends with whom?) and the competitive abilities of other group members (i.e., who is likely to win against whom?; see Cheney 2011). Evidence suggests that tridactic awareness is widespread in primate species (Silk 1999; Schino et al. 2006) and potentially other species that live in stable groups (Engh et al. 2005). How do group members obtain this information? One way is through "eavesdropping" on social interactions (McGregor 1993). Earlier, we explored potential eavesdropping behavior in humpback whales by assessing the "audience effect," where individuals changed their signaling behavior depending on the presence of potential (audible) eavesdroppers. Does this mean humpback whales also possess tridactic awareness? The short answer is: we do not know.

There is much to do within this area of humpback whale research. Firstly, for the most part, humpback groups and communication networks are not stable. Therefore, "keeping up with the neighbors" becomes more difficult, since the neighbors are constantly changing. Future studies should therefore consider the entire communication network, as well as the effects of repeated interactions within an individual, to fully understand an individual's signaling interactions. Secondly, we do not yet know the functional significance of sounds; i.e., what information they contain. Playback studies, and studies looking at the outcome of various signaling interactions (a mating, pregnancy, alliance formation between males, winner and loser of a competitive interaction, etc.) will help to decipher their meaning. Thirdly, one of the most important pieces of the puzzle, the internal state (physiology) of the signaling individual, is often not measured. Physiology drives behavior. Studies should also include more physiological information on the signaling individual, as well as (if possible) its audience. If concentrating on breeding interactions, the age, size, body condition, and testosterone levels of the males involved could help to quantify their fitness relative to each other. A larger male, in excellent body condition and with high levels of testosterone, could be considered fitter than a small male, in poorer condition and with reduced testosterone levels. For females, knowledge of their body condition, pregnancy status, stage in estrous cycle, age, and fecundity will help to quantify their breeding potential and likelihood of choosing to interact with, or avoid, males.

We know much more now about humpback whale signaling behavior than we did 50 years ago, but there is still much to explore. We are only beginning to unravel the complexity of their vocal communications, and by doing this, we may redefine how we view social complexity in animals. Social complexity gives rise to social intelligence and with that we enter the world of animal cognition. Given what we already know, of all the baleen whale species studied to date, humpback whales are most likely the species to offer the best entrée into revelations on baleen whale cognitive abilities.

Acknowledgements The author would like to thank everyone involved in the Humpback Acoustic Research Collaboration (HARC), in particular the numerous volunteers who donated their time and energy to the many studies mentioned in this chapter. The author would particularly like to acknowledge Professor Michael Noad for his input into these studies. Finally, the author would like to thank Christopher Clark, Ellen Garland, and Bernd Würsig for their valuable input into earlier drafts.

References

Antunes R, Kvadsheim PH, Lam FPA, Tyack PL, Thomas L, Wensveen PJ, Miller PJO (2014) High thresholds for avoidance of sonar by free-ranging long-finned pilot whales (*Globicephala melas*). Mar Pollut Bull 83(1):165–180

Baker CS, Herman LM (1984) Aggressive behavior between humpback whales (*Megaptera novaeangliae*) wintering in Hawaiian waters. Can J Zool 62(10):1922–1937

Baker CS, Perry A, Herman LM (1987) Reproductive histories of female humpback whales *Megaptera novaeangliae* in the North Pacific. Mar Ecol Prog Ser 41(2):103–114

Bannister JL, Hedley SL (2001) Southern Hemisphere Group IV humpback whales: their status from recent aerial survey. Mem Queensl Mus 47:587–598

Bass AH, Clark CW (2002) The physical acoustics of underwater sound communication. In: Simmons AM, Fay RR, Popper AN (eds) Acoustic communication. Springer, pp 15–64

Brenowitz EA (1982) Long-range communication of species identity by song in the red-winged blackbird. Behav Ecol Sociobiol 10(1):29–38

Brown M, Corkeron P (1995) Pod characteristics of migrating humpback whales (*Megaptera novaeangliae*) off the east Australian coast. Behaviour 132(3–4):163–179

Brown M, Corkeron P, Hale P, Schultz K, Bryden MM (1995) Evidence for a sex-segregated migration in the humpback whale (*Megaptera novaeangliae*). Proc R Soc Lond B 259:229–234

Cartwright R, Sullivan M (2009) Associations with multiple male groups increase the energy expenditure of humpback whale (*Megaptera novaeangliae*) female and calf pairs on the breeding grounds. Behaviour 146:1573–1600

Cato DH (1976) Ambient sea noise in waters near Australia. J Acoust Soc Am 60(2):320–328

Cato DH (1991) Theoretical and measured underwater noise from surface-wave orbital motion. J Acoust Soc Am 89(3):1096–1112

Cerchio S, Jacobsen JK, Cholewiak DM, Falcone EA, Merriwether DA (2005) Paternity in humpback whales, *Megaptera novaeangliae*: assessing polygyny and skew in male reproductive success. Anim Behav 70:267–277

Chapman DG (1974) Status of Antarctic rorqual stocks. In: Schevill WE (ed) The whale problem. Harvard University Press, Cambridge, MA, pp 218–238

Cheney DL (2011) Extent and limits of cooperation in animals. Proc Natl Acad Sci USA 108:10902–10909

Chittleborough R (1954) Studies on the ovaries of the humpback whale, *Megaptera nodosa* (Bonnaterre), on the Western Australian Coast. Mar Freshwater Res 5:35–63

Chittleborough R (1958) The breeding cycle of the female humpback whale, *Megaptera nodosa* (Bonnaterre). Aust J Mar Freshw Res 9(1):1–18

Chittleborough R (1965) Dynamics of two populations of the humpback whale, *Megaptera novaeangllae* (Borowski). Aust J Mar Freshw Res 16(1):33–128

Cholewiak D, Clark CW, Ponirakis D, Frankel A, Hatch LT, Risch D, Stanistreet JE, Thompson M, Vu E, Van Parijs SM (2018) Communicating amidst the noise: modeling the aggregate influence of ambient and vessel noise on baleen whale communication space in a national marine sanctuary. Endanger Species Res 36:59–75

Clapham PJ (1992) Age at attainment of sexual maturity in humpback whales, *Megaptera novaeangliae*. Can J Zool 70(7):1470–1472

Clapham PJ (1993) Social organization of humpback whales on a North Atlantic feeding ground. Symp Zool Soc Lond 66:131–145

Clapham PJ (1996) The social and reproductive biology of humpback whales: an ecological perspective. Mamm Rev 26(1):27–49

Clapham PJ, Palsboll PJ, Mattila DK, Vasquez O (1992) Composition and dynamics of humpback whale competitive groups in the West-Indies. Behaviour 122:182–194

Clapham PJ, Mattila DK, Palsboll PJ (1993) High-latitude-area composition of humpback whale competitive groups in Samana Bay—further evidence for panmixis in the North-Atlantic population. Can J Zool 71(5):1065–1066

Clapham PJ, Young SB, Brownell RL (1999) Baleen whales: conservation issues and the status of the most endangered populations. Mamm Rev 29(1):35–60

Clark CW, Ellison WT, Southall BL, Hatch L, Van Parijs SM, Frankel A, Ponirakis D (2009) Acoustic masking in marine ecosystems: intuitions, analysis, and implication. Mar Ecol Prog Ser 395:201–222

Craig AS, Herman LM (1997) Sex differences in site fidelity and migration of humpback whales (*Megaptera novaeangliae*) to the Hawaiian Islands. Can J Zool 75(11):1923–1933

Craig AS, Herman LM, Pack AA (2002) Male mate choice and male-male competition coexist in the humpback whale (*Megaptera novaeangliae*). Can J Zool 80(4):745–755

Craig AS, Herman LM, Gabriele CM, Pack AA (2003) Migratory timing of humpback whales (*Megaptera novaeangliae*) in the central North Pacific varies with age, sex and reproductive status. Behaviour 140(8–9):981–1001

Craig AS, Herman LM, Pack AA, Waterman JO (2014) Habitat segregation by female humpback whales in Hawaiian waters: avoidance of males? Behaviour 151(5):613–631

Cusano D (2020) Flexible information within social sounds of humpback whales (*Megaptera novaeangliae*). PhD thesis, School of Veterinary Science, The University of Queensland

Darling JD, Jones ME, Nicklin CP (2006) Humpback whale songs: do they organize males during the breeding season? Behaviour 143:1051–1101

Dawbin WH (1966) The seasonal migratory cycle of humpback whales. In: Whales, dolphins and porpoises. University of California Press, Berkeley, CA, pp 145–169

Dawbin WH (1997) Temporal segregation of humpback whales during migration in Southern Hemisphere waters. Mem Queensl Mus 42:105–138

DeRuiter SL, Southall BL, Calambokidis J, Zimmer WMX, Sadykova D, Falcone EA, Friedlaender AS, Joseph JE, Moretti D, Schorr GS, Thomas L, Tyack PL (2013) First direct measurements of behavioural responses by Cuvier's beaked whales to mid-frequency active sonar. Biol Lett 9(4):20130223

Dooling RJ, Blumenrath SH (2016) Masking experiments in humans and birds using anthropogenic noises. In: Popper A, Hawkins A (eds) The effects of noise on aquatic life II. Advances in experimental medicine and biology, vol 875. Springer, New York, NY, pp 239–243

Doutrelant C, McGregor PK (2000) Eavesdropping and mate choice in female fighting fish. Behaviour 137:1655–1669

Dunlop RA (2016a) Changes in vocal parameters with social context in humpback whales: considering the effect of bystanders. Behav Ecol Sociobiol 70(6):857–870

Dunlop RA (2016b) The effect of vessel noise on humpback whale, *Megaptera novaeangliae*, communication behaviour. Anim Behav 111:13–21

Dunlop RA (2017) Potential motivational information encoded within humpback whale non-song vocal sounds. J Acoust Soc Am 141(3):2204–2213

Dunlop RA (2018) The communication space of humpback whale social sounds in wind-dominated noise. J Acoust Soc Am 144(2):540–551

Dunlop RA (2019) The effects of vessel noise on the communication network of humpback whales. R Soc Open Sci 6(11):190967

Dunlop RA, Noad MJ (2016) The "risky" business of singing: tactical use of song during joining by male humpback whales. Behav Ecol Sociobiol 70(12):2149–2160

Dunlop RA, Noad MJ, Cato DH, Stokes D (2007) The social vocalization repertoire of east Australian migrating humpback whales (*Megaptera novaeangliae*). J Acoust Soc Am 122(5):2893–2905

Dunlop RA, Cato DH, Noad MJ (2008) Non-song acoustic communication in migrating humpback whales (*Megaptera novaeangliae*). Mar Mamm Sci 24(3):613–629

Dunlop RA, Cato DH, Noad MJ (2010) Your attention please: increasing ambient noise levels elicits a change in communication behaviour in humpback whales (*Megaptera novaeangliae*). Proc R Soc B 277(1693):2521–2529

Dunlop RA, Cato DH, Noad MJ, Stokes DM (2013a) Source levels of social sounds in migrating humpback whales (*Megaptera novaeangliae*). J Acoust Soc Am 134(1):706–714

Dunlop RA, Noad M, Cato D (2013b) Modification of humpback whale social sound repertoire and vocal source levels with increased noise. J Acoust Soc Am 133(5):3535

Dunlop RA, Noad MJ, Cato DH, Kniest E, Miller PJO, Smith JN, Stokes MD (2013c) Multivariate analysis of behavioural response experiments in humpback whales (*Megaptera novaeangliae*). J Exp Biol 216(5):759–770

Dunlop RA, Cato DH, Noad MJ (2014) Evidence of a Lombard response in migrating humpback whales (*Megaptera novaeangliae*). J Acoust Soc Am 135(1):430–437

Dunlop RA, Noad MJ, McCauley RD, Scott-Hayward L, Kniest E, Slade R, Paton D, Cato DH (2017) Determining the behavioural dose-response relationship of marine mammals to air gun noise and source proximity. J Exp Biol 220(16):2878–2886

Dunlop RA, McCauley RD, Noad MJ (2020) Ships and air guns reduce social interactions in humpback whales at greater ranges than other behavioral impacts. Mar Pollut Bull 154:111072

D'Vincent C, Nilson R, Hanna R (1985) Vocalization and coordinated feeding behavior of the humpback whale in southeastern Alaska. Sci Rep Whales Res Inst Tokyo 36:41–47

Engh AL, Siebert ER, Greenberg DA, Holekamp KE (2005) Patterns of alliance formation and post-conflict aggression indicate spotted hyenas recognize third-party relationships. Anim Behav 69:209–217
Erbe C, Reichmuth C, Cunningham K, Lucke K, Dooling R (2016) Communication masking in marine mammals: a review and research strategy. Mar Pollut Bull 103(1–2):15–38
Evans CS, Marler P (1991) On the use of video images as social stimuli in birds—audience effects on alarm calling. Anim Behav 41:17–26
Evans CS, Marler P (1994) Food calling and audience effects in male chickens, *Gallus gallus*—their relationships to food availability, courtship and social facilitation. Anim Behav 47(5):1159–1170
Ferraro MB, Giordani P (2015) A toolbox for fuzzy clustering using the R programming language. Fuzzy Sets Syst 279:1–16
Findlay KP, Seakamela SM, Meyer MA, Kirkman SP, Barendse J, Cade DE, Hurwitz D, Kennedy AS, Kotze PGH, McCue SA, Thornton M, Vargas-Fonseca OA, Wilke CG (2017) Humpback whale "super-groups"—a novel low-latitude feeding behaviour of Southern Hemisphere humpback whales (*Megaptera novaeangliae*) in the Benguela Upwelling System. PLoS ONE 12(3)
Fournet ME, Szabo A, Mellinger DK (2015) Repertoire and classification of non-song calls in Southeast Alaskan humpback whales (*Megaptera novaeangliae*). J Acoust Soc Am 137(1):1–10
Fournet MEH, Gabriele CM, Culp DC, Sharpe F, Mellinger DK, Klinck H (2018a) Some things never change: multi-decadal stability in humpback whale calling repertoire on Southeast Alaskan foraging grounds. Sci Rep 8:13186
Fournet MEH, Jacobsen L, Gabriele CM, Mellinger DK, Klinck H (2018b) More of the same: allopatric humpback whale populations share acoustic repertoire. PeerJ 6:e5365
Fournet MEH, Matthews LP, Gabriele CM, Haver S, Mellinger DK, Klinck H (2018c) Humpback whales *Megaptera novaeangliae* alter calling behavior in response to natural sounds and vessel noise. Mar Ecol Prog Ser 607:251–268
Freeberg TM (2006) Social complexity can drive vocal complexity: group size influences vocal information in Carolina chickadees. Psychol Sci 17(1):557–561
Freeberg TM, Dunbar RIM, Ord TJ (2012) Social complexity as a proximate and ultimate factor in communicative complexity. Phil Trans R Soc B 367:1785–1801
Garrigue C, Dodemont W, Steel D, Baker CS (2004) Organismal and 'gametic' capture-recapture using microsatellite genotyping confirm low abundance and reproductive autonomy of humpback whales on the wintering grounds of New Caledonia. Mar Ecol Prog Ser 274:251–262
Harris CM, Thomas L, Falcone EA, Hildebrand J, Houser D, Kvadsheim PH, Lam FPA, Miller PJO, Moretti DJ, Read AJ, Slabbekoorn H, Southall BL, Tyack PL, Wartzok D, Janik VM (2018) Marine mammals and sonar: dose-response studies, the risk-disturbance hypothesis and the role of exposure context. J App Ecol 55(1):396–404
Helweg DA, Herman LM (1994) Diurnal patterns of behaviour and group membership of humpback whales (*Megaptera novaeangliae*) wintering in Hawaiian waters. Ethology 98(3–4):298–311
Herb BM, Biron SA, Kidd MR (2003) Courtship by subordinate male Siamese fighting fish, *Betta splendens*: their response to eavesdropping and naive females. Behaviour 140:71–78
Herman LM (2017) The multiple functions of male song within the humpback whale (*Megaptera novaeangliae*) mating system: review, evaluation, and synthesis. Biol Rev 92(3):1795–1818
Herman LM, Antinoja RC (1977) Humpback whales in the Hawaiian breeding waters: population and pod characteristics. Sci Rep Whales Res Inst 29:59–85
Herman LM, Pack AA, Rose K, Craig A, Herman EYK (2011) Resightings of humpback whales in Hawaiian waters over spans of 10–32 years: site fidelity, sex ratios, calving rates, female demographics, and the dynamics of social and behavioral roles of individuals. Mar Mamm Sci 27(4):736–768
Holt MM, Noren DP, Veirs V, Emmons CK, Veirs S (2009) Speaking up: killer whales (*Orcinus orca*) increase their call amplitude in response to vessel noise. J Acoust Soc Am 125(1):27–32
Indeck KL (2020) Acoustic communication of female-calf humpback whales during migration. PhD thesis, School of Veterinary Science, The University of Queensland

Indeck KL, Girola E, Torterotot M, Noad MJ, Dunlop RA (2020) Adult female-calf acoustic communication signals in migrating east Australian humpback whales. Bioacoustics. https://doi.org/10.1080/09524622.2020.1742204
Jackson JA, Steel DJ, Beerli P, Congdon BC, Olavarria C, Leslie MS, Pomilla C, Rosenbaum H, Baker CS (2014) Global diversity and oceanic divergence of humpback whales (*Megaptera novaeangliae*). Proc R Soc B 281:1786
Janik VM (2000) Food-related bray calls in wild bottlenose dolphins (*Tursiops truncatus*). Proc R Soc B 267(1446):923–927
Janik VM (2005) Underwater acoustic communication networks in marine mammals. In: McGregor PK (ed) Animal communication networks. Cambridge University Press, pp 390–415
Janik V, Slater PJB (1997) Vocal learning in mammals. In: Jank V, Slater PJB, Bramwell J (eds) Advances in the study of behaviour, vol 26. Academic Press, pp 59–99
Lemasson A (2011) What can forest guenons "tell" us about the origin of language? In: Vilain A, Schwartz J-L, Abry C, Vauclair J (eds) Primate communication and human language: vocalisation, gestures, imitation and deixis in humans and non-humans, vol 1. John Benjamins Publishing Company, France, pp 39–70
Lohr B, Wright TF, Dooling RJ (2003) Detection and discrimination of natural calls in masking noise by birds: estimating the active space of a signal. Anim Behav 65:763–777
Lombard E (1911) Le signe de l'elevation de la voix. Ann Mal Oreil Larynx 37:101–199
McGregor PK (1993) Signaling in territorial systems—a context for individual identification, ranging and eavesdropping. Philos Trans R Soc 340(1292):237–244
McGregor PK, Horn AG (2015) Communication networks. In: Krause J, Croft D, James D (eds) Animal social networks: perspectives and challenges. Oxford University Press
Mikhalev YA (1997) Humpback whales *Megaptera novaeangliae* in the Arabian Sea. Mar Ecol Prog Ser 149(1–3):13–21
Miller PJO, Antunes RN, Wensveen PJ, Samarra FIP, Alves AC, Tyack PL, Kvadsheim PH, Kleivane L, Lam FPA, Ainslie MA, Thomas L (2014) Dose-response relationships for the onset of avoidance of sonar by free-ranging killer whales. J Acoust Soc Am 135(2):975–993
Mobley JR, Herman LM (1985) Transience of social affiliations among humpback whales (*Megaptera novaeangliae*) on the Hawaiian wintering grounds. Can J Zool 63(4):762–772
Morete ME, Bisi TL, Rosso S (2007) Temporal pattern of humpback whale (*Megaptera novaeangliae*) group structure around Abrolhos Archipelago breeding region, Bahia, Brazil. J Mar Biol 87(1):87–92
Morton ES (1977) Occurrence and significance of motivation structural rules in some bird and mammal sounds. Am Nat 111(981):855–869
Nishiwaki M (1959) Humpback whales in Ryukyuan waters. Sci Rep Whales Res Inst 14:49–87
Noad MJ, Kniest E, Dunlop RA (2019) Boom to bust? Implications for the continued rapid growth of the eastern Australian humpback whale population despite recovery. Popul Ecol 61(2):198–209
Olavarria C, Baker CS, Garrigue C, Poole M, Hauser N, Caballero S, Florez-Gonzalez L, Brasseur M, Bannister J, Capella J, Clapham P, Dodemont R, Donoghue M, Jenner C, Jenner MN, Moro D, Oremus M, Paton D, Rosenbaum H, Russell K (2007) Population structure of South Pacific humpback whales and the origin of the eastern Polynesian breeding grounds. Mar Ecol Prog Ser 330:257–268
Oliveira RF, McGregor PK, Latruffe C (1998) Know thine enemy: fighting fish gather information from observing conspecific interactions. Proc R Soc B 265(1401):1045–1049
Otter K, McGregor PK, Terry AMR, Burford FRL, Peake TM, Dabelsteen T (1999) Do female great tits (*Parus major*) assess males by eavesdropping? A field study using interactive song playback. Proc R Soc B 266(1426):1305–2130
Pallin LJ, Baker CS, Steel D, Kellar NM, Robbins J, Johnston DW, Nowacek DP, Read AJ, Friedlaender AS (2018) High pregnancy rates in humpback whales (*Megaptera novaeangliae*) around the Western Antarctic Peninsula, evidence of a rapidly growing population. R Soc Open Sci 5:180017

Parks SE, Johnson M, Nowacek D, Tyack PL (2011) Individual right whales call louder in increased environmental noise. Biol Lett 7(1):33–35

Parks SE, Cusano DA, Stimpert AK, Weinrich MT, Friedlaender AS, Wiley DN (2014) Evidence for acoustic communication among bottom foraging humpback whales. Sci Rep 4:7508

Payne RS, McVay S (1971) Songs of humpback whales. Science 173(3997):585–597

Payne K, Payne R (1985) Large-scale changes over 19 years in songs of humpback whales in Bermuda. J Comp Ethol 68(2):89–114

Peake TM, McGregor PK (2004) Information and aggression in fishes. Learn Behav 32(1):114–121

Peake TM, Terry AMR, McGregor PK, Dabelsteen T (2001) Male great tits eavesdrop on simulated male-to-male vocal interactions. Proc R Soc B 268(1472):1183–1187

Pomilla C, Amaral AR, Collins T, Minton G, Findlay K, Leslie MS, Ponnampalam L, Baldwin R, Rosenbaum H (2014) The world's most isolated and distinct whale population? Humpback whales of the Arabian Sea. PLoS ONE 9(12)

Pongracz P, Molnar C, Miklosi A (2006) Acoustic parameters of dog barks carry emotional information for humans. Appl Anim Behav Sci 100(3–4):228–240

Ramp C, Hagen W, Palsboll P, Berube M, Sears R (2010) Age-related multi-year associations in female humpback whales (*Megaptera novaeangliae*). Behav Ecol Sociobiol 64(10):1563–1576

Reilly SB, Bannister JL, Best PB, Brown M, Brownell Jr RL, Butterworth DS, Clapham PJ, Cooke J, Donovan GP, Urban J, Zerbini AN (2008) *Megaptera novaeangliae*. In: TUCN 2011. IUCN red list of threatened species. Version 2011.2. www.iucnredlist.org

Rekdahl ML, Dunlop RA, Noad MJ, Goldizen AW (2013) Temporal stability and change in the social call repertoire of migrating humpback whales. J Acoust Soc Am 133(3):1785–1795

Rekdahl M, Tisch C, Cerchio S, Rosenbaum H (2017) Common nonsong social calls of humpback whales (*Megaptera novaeangliae*) recorded off northern Angola, southern Africa. Mar Mamm Sci 33(1):365–375

Richardson W, Greene CR Jr, Malme CI, Thomson DH (1995) Marine mammals and noise. Academic Press, San Diego

Scheifele PM, Andrew S, Cooper RA, Darre M, Musiek FE, Max L (2005) Indication of a Lombard vocal response in the St. Lawrence River beluga. J Acoust Soc Am 117(3):1486–1492

Schino G, Tiddi B, Di Sorrentino EP (2006) Simultaneous classification by rank and kinship in Japanese macaques. Anim Behav 71:1069–1074

Sharpe FA (2001) Social foraging of the southeast Alaskan humpback whale, *Megaptera novaeangliae*. Ph.D. dissertation, Simon Fraser University, Vancouver, Canada

Sharpe FA, Dill LM (1997) The behavior of Pacific herring schools in response to artificial humpback whale bubbles. Can J Zool 75(5):725–730

Silk JB (1999) Male bonnet macaques use information about third-party rank relationships to recruit allies. Anim Behav 58:45–51

Slocombe KE, Zuberbuhler K (2007) Chimpanzees modify recruitment screams as a function of audience composition. Proc Natl Sci USA 104(43):17228–17233

Smith JN, Goldizen AW, Dunlop RA, Noad MJ (2008) Songs of male humpback whales, *Megaptera novaeangliae*, are involved in intersexual interactions. Anim Behav 76:467–477

Stimpert AK, Wiley DN, Au WWL, Johnson MP, Arsenault R (2007) 'Megapclicks': acoustic click trains and buzzes produced during night-time foraging of humpback whales (*Megaptera novaeangliae*). Biol Lett 3(5):467–470

Stimpert AK, Au WWL, Parks SE, Hurst T, Wiley DN (2011) Common humpback whale (*Megaptera novaeangliae*) sound types for passive acoustic monitoring. J Acoust Soc Am 129(1):476–482

Taylor AM, Reby D, McComb K (2010) Size communication in domestic dog, *Canis familiaris*, growls. Anim Behav 79(1):205–210

Thode AM, Blackwell SB, Conrad AS, Kim KH, Marques T, Thomas L, Oedekoven CS, Harris D, Broker K (2020) Roaring and repetition: how bowhead whales adjust their call density and source level (Lombard effect) in the presence of natural and seismic airgun survey noise. J Acoust Soc Am 147(3):2061–2080

Thompson PO, Cummings WC, Ha SJ (1986) Sounds, source levels, and associated behavior of humpback whales, southeast Alaska. J Acoust Soc Am 80(3):735–740

Tyack P (1983) Differential response of humpback whales, *Megaptera novaeangliae*, to playback of song or social sounds. Behav Ecol Sociobiol 13(1):49–55

Tyack P, Whitehead H (1983) Male competition in large groups of wintering humpbacks whales. Behaviour 83:132–154

Valsecchi E, Hale P, Corkeron P, Amos W (2002) Social structure in migrating humpback whales (*Megaptera novaeangliae*). Mol Ecol 11(3):507–518

Vanschaik CP, Vanhooff J (1989) Machiavellian intelligence—social expertise and the evolution of intellect in monkeys, apes, and humans. Nature 338(6209):700

Wadewitz P, Hammerschmidt K, Battaglia D, Witt A, Wolf F, Fischer J (2015) Characterizing vocal repertoires—hard vs. soft classification approaches. PLoS ONE 10(4)

Weinrich MT, Kuhlberg AE (1991) Short-term association patterns of humpback whale (*Megaptera novaeangliae*) groups on their feeding grounds in the southern Gulf of Maine. Can J Zool 69(12):3005–3011

Weinrich MT, Rosenbaum H, Baker C, Blackmer AL, Whitehead H (2006) The influence of maternal lineages on social affiliations among humpback whales (*Megaptera novaeangliae*) on their feeding grounds in the southern Gulf of Maine. J Hered 97(3):226–234

Wiley RH, Richards DG (1978) Physical constraints on acoustic communication in atmosphere—implications for evolution of animal vocalizations. Behav Ecol Sociobiol 3(1):69–94

Wiley D, Ware C, Bocconcelli A, Cholewiak D, Friedlaender A, Thompson M, Weinrich M (2011) Underwater components of humpback whale bubble-net feeding behaviour. Behaviour 148(5–6):575–602

Williams R, Erbe C, Ashe E, Beerman A, Smith J (2014) Severity of killer whale behavioral responses to ship noise: a dose-response study. Mar Pollut Bull 79(1–2):254–260

Yin S, McCowan B (2004) Barking in domestic dogs: context specificity and individual identification. Anim Behav 68:343–355

Zoidis AM, Smultea MA, Frankel AS, Hopkins JL, Day A, McFarland AS, Whitt AD, Fertl D (2008) Vocalizations produced by humpback whale (*Megaptera novaeangliae*) calves recorded in Hawaii. J Acoust Soc Am 123(3):1737–1746

Chapter 11
Humpback Whales: Exploring Global Diversity and Behavioral Plasticity in an Undersea Virtuoso

Danielle Cholewiak and Salvatore Cerchio

... it does give one pause to realize that the ocean has been resounding with whale song for millions of years, and that the ocean's inhabitants have long been swimming in the midst of the echoing choruses. It is not altogether surprising to find that humans nearly extinguished the whole source and only in the last few years have we begun to hear the music. (Payne and Payne 1985)

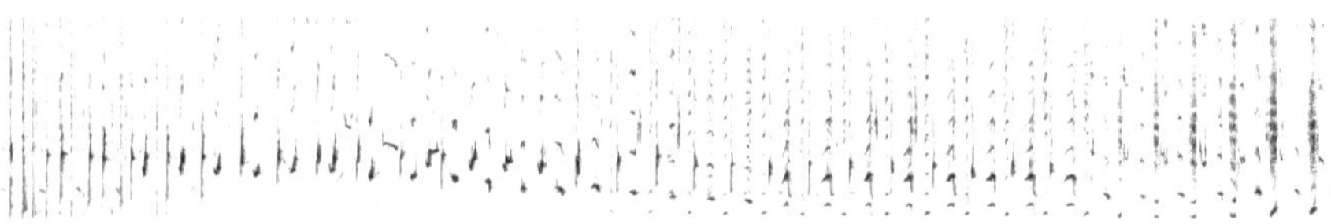

Abstract It has been fifty years since Payne and McVay's seminal publication on the strange and beautiful sounds of the humpback whale (*Megaptera novaeangliae*), the study which inspired decades of research into their complex, underwater acoustic world. In the subsequent five decades, there probably have been more research projects and publications on humpback whale song than on the vocal behavior of any other baleen whale. What makes humpback song so unique? What have we learned? What questions remain? With this chapter, we explore these questions with an eye toward the overarching theme of studies that address proximate mechanisms (*the "how do humpback whales do this?" questions*) versus those that inquire about ultimate causes (*the "why do humpback whales do this?" questions*). We draw a distinction between studies that focus on singing behavior, versus those that use song as a proxy to investigate other biological processes. We take the reader through a

D. Cholewiak (✉)
Northeast Fisheries Science Center, NOAA Fisheries, Woods Hole, MA 02543, USA
e-mail: danielle.cholewiak@noaa.gov

S. Cerchio
African Aquatic Conservation Fund, Chilmark, MA 02535, USA
e-mail: scerchio@africanaquaticconservation.org

C.W. Clark and E. C. Garland (eds.), *Ethology and Behavioral Ecology of Mysticetes*, Ethology and Behavioral Ecology of Marine Mammals,
https://doi.org/10.1007/978-3-030-98449-6_11

historical review of the literature, ending with a series of observations about the unanswered questions intended to provoke new ideas and new lines of inquiry. Through this exploration, we hope to synthesize a global body of research to identify common themes and probe the lingering gaps in our understandings of humpback whale song.

Keywords Humpback whale · Acoustics · Song · Sexual selection · Cultural transmission · Cultural evolution · Breeding behavior

11.1 Early Song Studies: Laying the Foundations from 1960s Through 1980s

Humpback whales (*Megaptera novaeangliae*) are almost cosmopolitan, occurring in all major ocean basins. Individuals in all but one or two populations are known or believed to undertake long-range seasonal migrations between high-latitude feeding and low-latitude breeding areas. By the mid-twentieth century, humpback whales had been hunted nearly to extinction throughout most of their range (reviewed in Baker et al. 1993). In 1955, commercial whaling for humpbacks was banned in the North Atlantic, followed by the Southern Hemisphere in 1963–64, and the North Pacific in 1966. How many animals were taken, and what the pre-whaling population sizes were, is not entirely clear. However, the populations that remained were a fraction of their original sizes, and in some cases occupied perhaps only a fraction of their original ranges (e.g., Reeves et al. 2001). The impacts of this population decimation on humpback whale movements, interchange, and communication structures are unknown. When the earliest acoustic recordings were made of humpback whales in the 1950s, US naval researchers and scientists were just beginning to learn about whale acoustic behavior by listening in on remnant populations left after post-industrial whaling. We are still unraveling the implications of impacts from the whaling era and considering how our interpretations of acoustic communication are affected by the study of populations that are recovering and expanding back into old and possibly new habitats. As we explore the processes that govern humpback whale singing behavior, it is important to keep in mind that our understanding of humpback communication dynamics may be influenced by these population-level changes.

By 1964, it was known that humpback whales make sounds during the breeding season, and also that these sounds were different than those heard on feeding grounds. As Schevill (1964) eloquently wrote, "The sonorous moans and screams associated with the migrations of Megaptera past Bermuda and Hawai'i may be an audible manifestation of more fundamental urges, for in New England waters and at other seasons we do not hear anything nearly so spectacular from this species." It would be a few more years before the breeding sounds of humpbacks would be recognized as "song", and by the time Roger Payne and Scott McVay published their seminal paper (Payne and McVay 1971), multiple researchers were already recording and exploring this newly documented acoustic phenomenon.

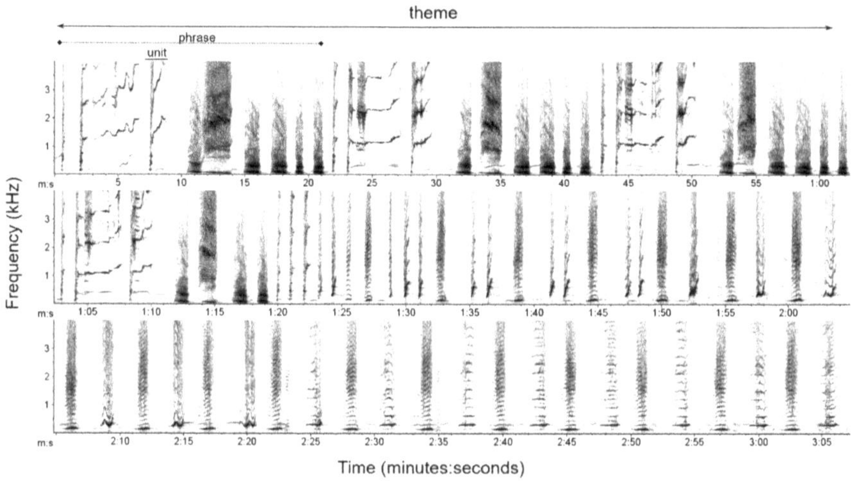

Fig. 11.1 Spectrogram showing a partial song sequence from one humpback whale, recorded off Hawai'i in 1990 (1024 pt FFT, Hann window, 80% overlap). This image reveals a diversity of song units as well as the hierarchically patterned structure of humpback song

The pioneering work in the 1970s developed a framework by which to analyze and understand humpback song structure. Payne and McVay (1971) described the well-structured hierarchy of song sequences, composed of individual units, organized into repeated patterns ("phrases"), repeated a variable number of times ("themes") before a singer switches to a new set of phrases and themes (Fig. 11.1). This type of singing had already been defined in the avian song literature as "eventual variety" (Kroodsma 1977; Fig. 11.2), which was also later applied to humpback whales (Cholewiak et al. 2013). Payne and McVay laid the groundwork for describing the hierarchical structure of humpback song, and explored variability in the singing behavior of individual singers, documenting, for example, that singers vary the number of times they repeat a phrase within a theme, resulting in themes and song sequences of varying lengths. These observations of plasticity in song production underlie an important behavioral mechanism that we continue to explore. Payne and McVay also speculated on the possible role of song in mate attraction or group cohesion and speculated that singing may be common in baleen whales. While their insights were seminal for humpback whale research, they also had clear influence on baleen whale song research worldwide.

The mid-1970s through the 1980s saw an explosion of research on humpback whales. By 1979, preliminary descriptions of acoustic signals had been produced for many baleen whale species (Thompson et al. 1979), but work on humpback whales was more advanced than on other species. Perhaps largely due to their accessibility in warm water coastal habitats and their charismatic vocal displays, humpback whales attracted dedicated and consistent study. Our understanding of humpback whale song and singing behavior during this era was shaped almost entirely by studies conducted in the Northern Hemisphere; most research was being conducted off Bermuda, around

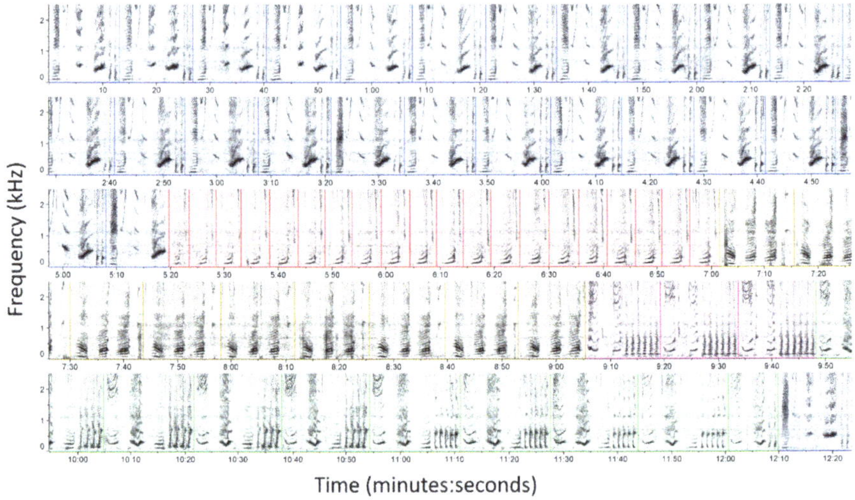

Fig. 11.2 Spectrogram showing partial song sequence from one humpback whale recorded off the Sultanate of Oman in 2012 (4096 pt FFT, Hann window, 50% overlap), with phrases delineated and colored by theme, showing how humpback singing behavior fits into the framework of "eventual variety" singing. Recording courtesy of the Environment Society of Oman

the Caribbean, and around the Hawaiian Islands. Scientists were exploring concepts related to individual identification based on song characteristics (Hafner et al. 1979), diel periodicity in song production (Thompson et al. 1979), and the use of song to assess population distribution and size (Winn et al. 1975; Levenson and Leapley 1978), recognizing that it was possible to monitor portions of a population in breeding areas using passive acoustics as a scientifically robust methodology. Work was also underway to assess geographic variation in song. In the first published comparison of song structure between populations, Winn et al. (1981) found strong similarities among songs in widely separated wintering grounds, recognizing that within the North Atlantic, songs recorded in the West Indies and Cape Verdes shared the same themes, as did songs recorded between Hawai'i and México in the North Pacific. In contrast, between ocean basins, North Atlantic songs shared no themes with those from the North Pacific, and both were different from one sample obtained from Tonga in the South Pacific.

In 1983, the book "Communication and behavior of whales" (Payne 1983) was published, which contained an important compilation of detailed studies on humpback whale song and singing behavior. These laid out the basic concepts that are still considered foundational to our understanding of humpback whale singing, that: singers are males (Glockner 1983); all individuals within a local breeding aggregation (e.g., the Hawaiian Islands) share the same phrases at any given time (Payne et al. 1983); there exists distinct individual variation such that some singers diverge from the typical ordering of themes and may be considered "aberrant" (Frumhoff 1983); songs change progressively over time (Payne et al. 1983); songs of individual singers change through time in the same ways that songs from their regional "population" of singers change through time (Guinee et al. 1983); and singers in a completely different breeding aggregation located across an ocean basin, e.g., in México, sing songs composed from some of the same phrases and change those phrases in a similar manner across seasons as the singers in the Hawaiian Islands (Payne and Guinee 1983). An important concept running through many of these studies was that humpback whales change their song in different ways over time, but the subtleties, dynamics, and scales of such changes were just beginning to be appreciated.

Two fundamental papers on song structure and changes in song were published by Katharine Payne and her colleagues (Payne et al. 1983; Payne and Payne 1985). These papers, built on multiyear studies of humpback song in two different ocean basins, provided profound insights that shaped the research community's understanding of humpback singing behavior. At their core, these studies examined the patterns and dimensions of variability of temporal change at the organizational level of song structure. Payne et al. (1983) observed that most of the structural changes in song do not occur between seasons, but rather during the breeding season, dismissing an early notion of "forgetfulness" as the root cause of song change. They observed that singers change their song structure at different rates over time, that "chaotic" themes could become organized, and that song structure goes through periods of instability and stability. Entire themes die out and new ones appear, and the birth of new themes most often occurred through a process by which stable themes become variable and later begin to differentiate. Payne and her colleagues described the humpback song system as "at once highly organized and also labile". Payne's observations and insights were influenced by contemporary research on oscine songbirds, in particular, the work in Peter Marler's lab at Rockefeller University, with which the Paynes were associated. This was an era of intensive study on song learning in avian species, utilizing captive-bird laboratory experiments and field manipulations to develop a better understanding of the innate and environmental influences on acoustic repertoire development (e.g., Marler and Peters 1981; Marler 1984, 1990).

The Paynes' observations were remarkable, as at the time the idea of song learning in baleen whales had not yet been considered. Importantly, Payne et al. (1983) recognized that progressive changes in humpback song, which imply that whales continuously learn song changes from conspecifics, represent a process of *cultural transmission* and *cultural evolution*. Cultural transmission and song evolution had already been well-documented in a number of oscine avian species (i.e., "song-birds") by the 1970s (e.g., Lemon 1975; Jenkins 1978; Slater and Ince 1979), and researchers were developing frameworks to describe the process conceptually and quantitatively (Mundinger 1980; Cavelli-Sforza and Feldman 1981). Cultural evolution of a vocal behavior is the process by which a song tradition is changed over time, and changes are transmitted between individuals. This is the consequence of an individual having the capability to learn, coupled with the transmission of acoustic elements between individuals. These changes in a song pattern, sometimes referred to as "cultural mutations" (Jenkins 1978; Mundinger 1980), are believed to result from an "error" in copying, improvisation, innovation, or some combination of each of these. In most songbirds, changes accumulate slowly across generations, such as the typical "vertical" transmission described in New Zealand's South Island saddlebacks (*Philesturnus carunculatus*; Jenkins 1978) and indigo buntings (*Passerina cyanea*; Payne et al. 1981). However, in exceptional cases, "horizontal" transmission between individuals within the same generation, coupled with continual copying of individual innovations within singing seasons and throughout an individual's lifetime, results in what has been termed "*rapid cultural evolution*", as displayed by humpback whales (Chap. 8). The remarkable phenomenon of population continuity for a form of song undergoing rapid cultural evolution, in which emergent changes are adopted among individuals within a population to maintain the same song patterns, is not unique to humpback whales. Several avian species also exhibit similar phenomena in their acoustic breeding displays, such as the village indigobird (*Vidua chalybeata*, Payne 1979), and the yellow-rumped cacique (*Cacicus cela*, Trainer 1989).

There are likely multiple mechanisms by which such changes can be maintained by individual humpback whales in breeding areas thousands of kilometers apart. At least three mechanisms were proposed for the exchange of song material between individuals from different breeding populations: Singers from different breeding areas may overlap on the feeding grounds; they may interact and hear one another during migration; and/or they may move between breeding areas within a season or inter-annually (Payne and Guinee 1983). However, song similarity within an ocean basin is more complicated than originally believed (explored in more detail in Sect. 11.2).

Payne and Payne (1985) described humpback whale song structure patterns from 19 years off Bermuda, defining the concept of "transitional phrases", as well as three types of themes: "static", "shifting", and "unpatterned" (Fig. 11.3). These categories are still used today in characterizing variation in humpback song. However, in many more recent papers, these distinctions are often not observed or described,

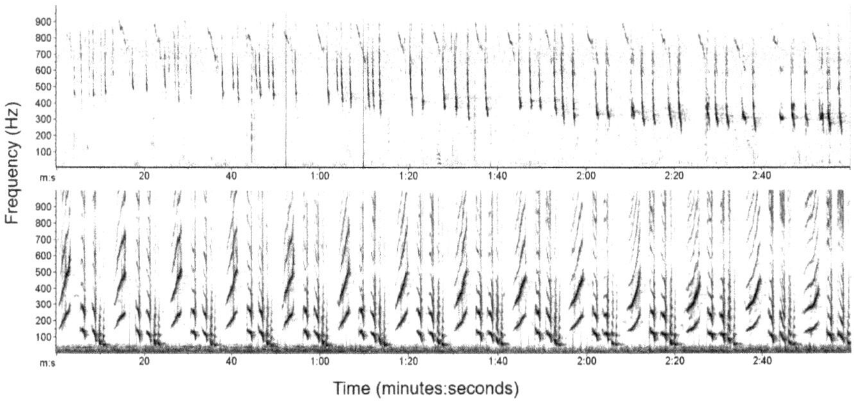

Fig. 11.3 Two excerpts from humpback song showing a shifting theme (top) and static theme (bottom). In the top panel, note both the progressively decreasing frequency and the increasing temporal spacing of the song units with each phrase repetition. In contrast, in the bottom panel, the spectral-temporal features of the song units are consistent among phrases. Recording sites: top panel: Madagascar, 2018; bottom panel: México, 2006

leaving us to wonder if some of the variation in song is either not accounted for or is misinterpreted. For example, without recognizing that transitional phrases are composed of components of different phrases from consecutive themes, and that in shifting themes the phrase structure changes progressively from the first to last repetition of the phrase, there exists the potential to "over-categorize" or split themes, leading to potentially erroneous interpretations.

The long time series of the samples used by the Paynes allowed them to recognize commonalities in humpback songs across years. Some of these commonalities have to do with the finer details in song units—such as the inclusion of both "simple" and "complex" sounds (defined as ranging from tonal to noisy bursts), and other types of contrasting sounds, such as those that are frequency-modulated and relatively constant in frequency, very long and very short in duration, and abrupt and gradual in onset. In examining the processes of theme evolution, the Paynes found that themes had dissimilar rates of extinction; while about half the themes were present in more than one year in their study, some were present for 7 or 8 years. Between adjacent years, the ratio of new to previous material varied; theme content in some adjacent years was very similar, while songs in other adjacent years had few themes in common. In fact, every 5–6 years, a period of stability seemed to be reached in which songs remained very similar between years. The Paynes also observed variation among themes and years in the "rhythm" of the song, the song tempo; a feature that students of humpback whale song studies recognize, but is difficult to capture in words. Finally, the Paynes recognized that despite inter- and intra-individual variation, overall, all songs from one breeding season are more similar to other songs from that same season than to songs from other seasons. Many of these insights were possible due to the expansive sample analyzed by the Paynes, and their insights have had lasting impacts on our understanding of the strengths and limitations of using song analyses to evaluate population structure.

In his book "Sociobiology", E.O. Wilson (1975) stated "the most elaborate single display known in any animal species may be the song of the humpback whale *Megaptera novaeangliae*." Among the balaenopterid whales, this observation holds true. None of the other rorquals produces a song that comes close to rivaling the complexity of humpback whale song. It has become clear that most (if not all) mysticete species sing; therefore, song can be considered an ancestral synapomorphic trait for this branch of cetacean evolution. When considering the acoustic frequency range and variety of unit types, and the complexity of song structure, humpback whale song is the most highly derived among the balaenopterid clade. What are the evolutionary drivers for this elaboration and complexity?

Early studies of humpback breeding behavior considered the question of evolutionary drivers. The existence of a trait that is highly derived and elaborate within a comparative phylogenetic framework is a typical indication of sexually selected exaggeration. In recognizing that the choruses of singers that they observed on the breeding grounds could function as a local "communal" display, Herman and Tavolga (1980) made the first connection between the humpback mating system and the lek mating systems described in terrestrial taxa (Emlen and Oring 1977). In lek mating systems, males aggregate and display to attract females, and females visit display arenas for the purposes of evaluating males and choosing a mate based on their displays. Leks evolve in situations where males cannot directly control females or access to resources, where female movements to an area are predictable, and where females are relatively asynchronous in their periods of sexual receptivity, thereby skewing the operational sex ratio toward males. Many of these features are good descriptors of the typical humpback whale breeding system. Females arrive on the breeding grounds over a protracted period of several months (Dawbin 1966), and have an average two-year reproductive cycle (Mackintosh 1972), thus strongly skewing the operational sex ratio among sexually active adults toward males. Furthermore, during the breeding season, individuals are primarily or exclusively fasting (Chittleborough 1965); therefore, selective pressures with respect to defending food resources are unlikely to play a role in breeding behavior. Tyack (1981) put sexual selection at the forefront when considering the evolution of humpback singing behavior, noting that female choice may have been a driving force for song complexity (Chap. 7). Payne et al. (1983) took this concept one step further, drawing the analogy to the process by which female choice in some songbird species has led to the evolution of complexity in the form of large vocal repertoires.

Perhaps the most important insights to come out of this early period of work have to do with not only characterization of the intricate nature of humpback song structure and patterns of song sharing and rapid cultural evolution, but the exploration of hypotheses related to the underlying evolutionary drivers of song complexity. Therein lies the distinction between asking questions about the proximate processes that explain how an observed pattern develops, versus asking questions about the ultimate reasons for why these processes evolved. The foundational work of this early period of research, with its descriptions of song structure and the dynamics of humpback whale singing behavior, falls primarily into the proximate category. Based on these explorations, these authors cited above began to speculate on the ultimate questions of why humpback whales have evolved such a complex and dynamic singing behavior, which set the stage for further considerations of the selective forces and evolutionary drivers of the complexity in humpback singing behavior.

11.2 The Next Generation: Further Description of Patterns During the 1990s–2010s

The first phase of research on humpback whale song and singing behavior was conducted primarily by just a few research groups. Starting in the 1990s, humpback whale research in the ensuing three decades has focused on a broad range of topics by an ever-expanding number of researchers. Much of this more "modern" work has focused on exploring proximate mechanisms of how song patterns change and are transmitted between populations. Often song similarity has been used as a proxy metric to describe potential interactions among populations, supplementing early knowledge derived from whaling data and photo-ID mark recapture studies.

While earlier work concentrated on the Northern Hemisphere, Cato (1991), Dawbin and Eyre (1991), and Mednis (1991) confirmed that Southern Hemisphere humpback whales exhibit similar song structure and variable rates of song change to those of the Northern Hemisphere, undergoing periods of stability and of substantial change. Initial studies indicated that songs from east and west Australia showed little similarity, indicating that the geographic barrier of Australia may be more important than overall distances apart, as shown by the greater distances between breeding assemblages within the South Pacific with similar songs.

Passive acoustic studies have shown that patterns of physical movement (i.e., interchange) of individuals within and between populations can be more complex than previously thought. Within the North Pacific Ocean, initial studies showed that songs recorded from the breeding assemblages in Hawai'i and México shared all phrase types (Winn et al. 1981; Payne and Guinee 1983). Further ocean-basin-scale studies in the North Pacific revealed fluctuations in similarity among songs recorded from different breeding areas, sometimes resulting in the occurrence of temporary, regional differences in the number of shared versus "private" themes (Helweg 1990; Cerchio et al. 2001; Darling et al. 2014, 2019a). For example, Darling et al. (2019a) found that some themes were present across the entire North Pacific from the Philippines to México during a 3-year study, but others were only present in certain regions (e.g., "private themes"), and ocean-basin-scale similarity in song content varied across years. The variation in rates of song change and development of temporary, regional differences in song composition likely reflect variable levels of interchange and mixing of singers from different breeding areas over time. Within an ocean basin, therefore, humpback population structure could be described as a "metapopulation" (Levins 1969; Hanski and Gilpin 1991), within which there exists a complex pattern of site fidelity and overlap of individuals between feeding and breeding areas that varies across years. Similarly detailed song studies have not been conducted between the main breeding areas of the North Atlantic West Indies and Cape Verde Islands, though comparable fluctuations in similarity of song content between breeding regions may be present.

North Pacific and North Atlantic populations are separated by a continental land mass, with no movement of individuals between oceans. But the Southern Hemisphere is more open, with the high-latitude (>60° S) circumpolar "Southern Ocean" connecting separate ocean basins, and large expanses of lower latitudes dotted by islands. This allows for enhanced opportunities for whales mixing both between ocean basins across seasons, and within ocean basins during a season. The first example of consequent song sharing came from Noad et al. (2000), where humpback whales migrating along the east coast of Australia (western South Pacific Ocean), adopted the songs of animals from the west coast of Australia (southeast Indian Ocean) across two singing seasons. The process by which higher level song features (i.e., themes) are transferred between populations or aggregations is termed "*cultural diffusion*" (Mundinger 1980; Whiten et al. 2016). Mundinger (1980) applied this term to describe the process by which novel behaviors are introduced to one population by immigrants from a different population, as a cultural analog to gene flow between different populations. In the case of the Australian humpback whale populations, immigrant individuals from the Indian Ocean (the origin population) introduced their song (the novel behavior) into the South Pacific population, which then adopted it. This is distinct from rapid cultural evolution that occurs incrementally and progressively as individuals from *within* a population gradually modify their existing song as they copy innovations made by one other. Noad et al. (2000) coined the term "cultural revolution" (perhaps due to the appeal of rhyming with cultural evolution, rather than drawing a parallel with other biological phenomena) for the transmission of song traditions between different populations. We maintain the original term "cultural diffusion" in this chapter.

During cultural diffusion from western to eastern Australia, there was complete replacement of all phrase types previously sung by the eastern Australian South Pacific population. There are also examples of cultural diffusion of partial songs, with only some novel phrase types being introduced and adopted, as described above in the North Pacific (e.g., Darling et al. 2019a). The phenomenon of cultural diffusion and complete replacement of songs has also been occasionally observed in other species, such as between social groups with distinct dialects in village indigobirds (Payne 1985). As in humpback whales, village indigobirds otherwise exhibit progressive rapid cultural evolution of songs during the singing season via individuals making innovations within songs that are copied and transmitted within their social group. However, Payne (1985) documented at least one case in which males from a neighboring social group permanently emigrated and introduced their origin social group's song dialect into their new social group, whose previous dialect entirely disappeared.

Since 2000, the cultural diffusion of changes in song content between humpback breeding populations in the western and central South Pacific has been demonstrated repeatedly through a number of studies (e.g., Eriksen et al. 2005; Garland et al. 2011, 2015; Owen et al. 2019; Chap. 8). Other studies have shown limited song sharing across the Indian Ocean (Murray et al. 2012), and variable rates of song sharing across years between the southeast Atlantic and southwest Indian Oceans (Rekdahl et al. 2018). Song comparisons between breeding regions in the western and eastern South Atlantic (Darling and Sousa-Lima 2005) have shown that there is some song sharing between populations on opposite sides of the Atlantic Ocean, similar to patterns in the Northern Hemisphere (e.g., Darling et al. 2019a). Taken together, these studies reveal complex patterns of interchange between individuals both within and between breeding populations and subpopulations. These studies extend our early understanding of population movements derived from the whaling period (e.g., Dawbin 1966) as well as what has been shown through satellite telemetry and photo-ID studies (e.g., Calambokidis et al. 2001; Hauser et al. 2010; Garrigue et al. 2011, 2015).

While it is not clear which mechanisms for song transmission (as first proposed by Payne and Guinee 1983) are at work, it is likely that multiple mechanisms exist. Numerous studies have documented the occurrence of singing activity on feeding grounds in Northern Hemisphere and Southern Hemisphere (e.g., Gabriele and Frankel 2002; Clark and Clapham 2004; Vu et al. 2012; Stimpert et al. 2012; Garland et al. 2013; Van Opzeeland et al. 2013; Magnúsdóttir et al. 2014; Kowarski et al. 2018). Several studies have also documented singing in the open ocean, away from breeding aggregations, presumably from migrating animals (Norris et al. 1999; Clark and Gagnon 2002; Darling et al. 2019b), as well as potential mixing between populations during migration (Owen et al. 2019). Finally, interannual movements of individuals between breeding areas have also been documented by individual recapture studies in both hemispheres (Dawbin 1966; Darling and Jurasz 1983; Darling and Cerchio 1993; Pomilla and Rosenbaum 2005; Stevick et al. 2016). All of these are likely mechanisms by which song sharing and cultural diffusion of song elements occur within and between populations. Levels of interchange between individuals appear to vary over time; thus, relatively short-term studies of only one or a few years cannot capture the full picture. Long-term, broad-scale studies (e.g., Garland et al. 2011) are needed to appropriately interpret what the patterns in song structure reveal about population interactions.

The 1990–2010 period also showed that changes in singing behavior can be used as an indicator of behavioral disturbance from human activities. Whereas numerous studies have assessed the impact of acoustic disturbance on movement patterns of individuals or distribution of populations (e.g., Frankel and Clark 2000, 2002; Dunlop et al. 2018), comparatively few have addressed behavioral impacts on singing activity or an acoustic response in the form of alteration of a vocalization. Humpbacks respond acoustically to a wide variety of anthropogenic sounds, such as vessel noise (Norris 1995; Sousa-Lima and Clark 2008; Tsujii et al. 2018), naval sonar (Fristrup et al. 2003; Risch et al. 2012), and seismic surveys (Cerchio et al. 2014), by altering song structure or decreasing singing activity. Subsequent studies modeling the potential communication space for humpbacks engaged in vocal interactions (both song and non-song) have demonstrated how noise from human activities can reduce the area over which individuals can communicate with one another (Cholewiak et al. 2018a; Dunlop 2019), a disturbing consequence of the expansion of human use of the oceans. Collectively, these studies have important implications, particularly considering cases when intense noise is generated in breeding habitat during the peak of reproductive activities; for example, seismic surveys occurring in tropical African waters during the austral winter. Since singing is a breeding display, disturbance in breeding habitats has the potential to impact the reproductive success of males and the ability of females to make important mate choice decisions.

11.3 The Next Phase of Research: Progress Toward Ultimate Questions

Many studies since the 1990s have documented cultural transmission of songs between humpback whale populations. Song is used in these cases as an indicator variable to study individual movements and population dynamics, rather than for the study of singing behavior itself. These studies reveal some of the proximate mechanisms of song transmission, but rarely address ultimate reasons for the evolution of song complexity, or the evolution of a system that favors a changing song along with the rapid assimilation of changes in song units and patterns by individual males within a population. Compared to the numerous studies examining patterns of geographic variation in song, relatively few have been conducted in the past five decades that explore ultimate evolutionary hypotheses.

Why did the complexity that we observe in humpback whale song and singing behavior evolve in the first place? Evolutionary theory can give us some clues. While natural selection shapes signals that are important for survival, sexual selection shapes signals that are important within the context of reproduction. In some species, females show preferences for song characteristics even though these do not seem related to direct benefits to a female. The preference for traits that indicate indirect benefit is traditionally thought to arise through one of three types of selection: (1) traits that indicate "good genes", which might result in higher offspring survivorship (Hamilton and Zuk 1982), (2) traits for which males incur a cost, thereby indicating male viability (the "handicap principle", Zahavi 1975), or (3) indirect selection for an arbitrary trait that has no particular benefit other than increasing reproductive success for males who possess the trait (Fisher 1930, 1958). The latter involves the linkage between the male trait and female preference for the trait and is often referred to as "self-reinforcing" selection (Andersson 1994); this may lead to the evolution of highly exaggerated traits, such as the extravagant plumage of male peacocks (e.g., *Pavo cristatus*) or birds-of-paradise (family *Paradisaeidae*, e.g., *Parotia lawesii*), or the elaborate vocal performance of the common nightingale (*Luscinia megarhynchos*).

To address the adaptive function of rapid cultural evolution of songs in passerine birds, Trainer (1989) contrasted the reproductive behaviors of caciques and village indigobirds (species that are flexible learners with songs that are culturally transmitted horizontally and undergo rapid evolution), with the reproductive behaviors of chaffinches (*Fringilla coelebs*) and indigo buntings (*Passerina cyanea*) (species with stable songs that are transmitted vertically and change slowly across generations). Caciques and village indigobirds are polygynous, males are not territorial and frequently engage in competitive contact, males disperse widely, and songs can be learned as adults. In contrast, chaffinches and indigo buntings tend to be monogamous, males are territorial and engage in less social contact, males usually return to the same territory throughout their lifetimes, and learning ability is restricted to a period in early life. Trainer (1989), therefore, suggested that rapid cultural evolution of bird song occurs in species with a high degree of male-male competition. Payne (1985) suggested that the male village indigobirds with the most matings were those that sometimes initiated changes in song composition and were imitated most by other males. Trainer (1989) did not directly observe this in caciques, but proposed a process based on the preferential adoption of songs of successful males by less successful males, in which dominant males innovate on an existing song model to produce a song distinguishable from other males. This song variant from a high-ranking male acts as a strong stimulus to either threaten males or attract females. Other males try to imitate this song to confer its selective advantage upon themselves. Once all males share the song, it is no longer effective, and the dominant male makes a new improvisation. Therefore, there is continual incentive to improvise, and the repeating cycle results in non-random, directional change in the details of the song composition. In humpback whales, this specific mechanism of copying dominant males cannot be the primary driving force behind song evolution; thousands of males spread across an ocean basin sing similar songs, and it is unlikely that a male's reproductive success in one breeding aggregation is affected by males in another breeding aggregation. However, humpback whales exhibit similar processes to some passerine songbirds in the rapid cultural evolution in song characteristics, suggesting that there may be similarities in the general underlying selective pressures.

In a comparative phylogenetic context and relative to other balaenopterids, the song system of the humpback whale clearly fits into the category of exaggerated traits, shaped by strong sexual selection. Singing is one of a suite of alternative mating tactics employed by males on the breeding grounds (Clapham 1996; Cerchio 2003; Cerchio et al. 2005). One of the main hypotheses is that female choice is the selective force driving both rapid cultural evolution and the complexity observed in humpback song (Tyack 1981; Payne et al. 1983; Clapham 1996; Cerchio et al. 2001; Cholewiak et al. 2018b); however, few studies have yet to document the role of intersexual selection. Medrano et al. (1994) documented two cases of females joining male singers, and Darling and Bérubé (2001) documented one case of a male–female pair in which the male was singing, and several instances of males singing in the presence of a female with calf. Smith et al. (2008) found that singers off eastern Australia were more likely to join mother–calf pairs than other types of groups and sang longer in the presence of a mother–calf pair than other group types. Anecdotal evidence from Kaua'i suggests that mothers accompanied by singers altered their respiratory behavior in response to the singers' presence by surfacing with the male (whose surfacings were timed to occur during a specific theme in the song), thus suggesting the female was listening to the singer and synchronizing her behavior with him (Cerchio 2003). Successful males participate in a variety of breeding tactics, and singing to mothers with calves may be one of them (Cerchio 2003; Cerchio et al. 2005). Molecular determination of paternity of calves off Socorro, México, matched the paternity of a female's second calf in consecutive years to a male observed escorting her in the previous year (Cerchio 2003). Estimated rates of post-partum estrus range from 3 to 20% of females giving birth in consecutive years (Clapham and Mayo 1987; Glockner-Ferrari and Ferrari 1990; Cerchio 2003) to 40%, based on whaling data (Chittleborough 1965); therefore, courtship of a mother with calf may be an important male breeding tactic.

These studies provide a glimpse of insight into potential intersexual selection. Studying the behavior and interactions of non-parous females with males on the breeding grounds could be quite revealing, but is unfortunately very challenging. Females are often the center of attention in competitive groups, which can involve many males vying for the position closest to her (Tyack and Whitehead 1982; Clapham et al. 1992), precluding the opportunity to study their behavior in the absence of interference by males. When unaccompanied, single individuals can be difficult to visually locate and track, as they can be relatively cryptic compared to other types of groups (singers being a notable exception). Therefore, although lone females may well be within acoustic detection range and listening to (or even interacting acoustically with) male singers, their behavior is difficult to document. These challenges have thus far limited our ability to study potential female choice relative to singing males.

Studying singer behavior to examine the potential role of intrasexual selection has been a more fruitful line of inquiry. From the intrasexual perspective, using song to mediate interactions may allow males to assess potential rivals and avoid physical conflict. In taxa from ungulates to birds, escalation or de-escalation of agonistic interactions can be based on an individual's motivation and perception of one's own quality or competitive ability compared to that of a rival. Several studies have examined the spatial and acoustic responses of singers to interactions with other males. Frankel et al. (1995) found that singers generally maintained greater separation distances between one another than non-singing individuals in a breeding area off Hawai'i, leading to the idea that males may use song to maintain spacing between other singers (also observed in minke whales, Gedamke 2004). Darling et al. (2006) found that 89% of singers in their study sang until they were "joined" by non-singing males (143/161 observed interactions), often leading to brief interactions (<10 min) that typically appeared to be non-agonistic. This is consistent with other observations of lone males interacting with singers, who then often ceased to sing (Smith et al. 2008). Darling et al. (2012), using an experimental playback design, also found that a singing male approached a playback of song that was similar to his own song 7 out of 9 times (78%), and stopped singing in 8 out of 9 trials (89%). In considering these types of interactions, Darling et al. (2006) hypothesized that males may assess one another using details of song content as an index of association, and that this information may allow individuals to determine whether to interact antagonistically or otherwise with another singer. Only one study has examined the response of singers to one another by evaluating not only their spatial movements and singing persistence, but also by examining whether males are responding acoustically—specifically, changing the details of their song structure in response to other singers (Cholewiak et al. 2018b). If male singers are using song interactively to mediate interactions with one another, then it should be possible to measure the responses of one individual singer relative to the actions of another singer. With song, this could take the form of acoustically "pointing" at specific receivers, via timing or pattern changes (Todt and Naguib 2000). Cholewiak et al. (2018b) found that 11 out of 12 singers modified their song presentation in the presence of a singing rival, changing the rate at which they switched between phrase types, and the overall "evenness" of their song presentation. A similar result was found via a controlled playback study, in which singing males were presented simplified versions of their own songs (Cholewiak 2008). Similarly to previous studies, Cholewiak et al. (2018b) also found that some singers (5 of 12) approached other singers. Rapid approaches were associated with one or both males ceasing to sing (in 3 out of 3 cases), suggesting that the combination of singing while approaching could be a form of threat. This hypothesis suggests that males are interrupting the display of their rivals as a form of competition, as is commonplace on bird leks (Wiley 1991; Höglund and Alatalo 1995).

If humpback whale males are using their singing behavior and song characteristics to mediate potential competitive interactions, then there should be strong selection for song characteristics that allow males to assess one another, and these characteristics should be both costly and an honest reflection of male quality (i.e., difficult to cheat). Costs may be incurred either through production (e.g., energetic costs) or through enforcement of honesty by conspecifics (Bradbury and Vehrencamp 1998). Song elements or behavior that are difficult to produce, or those that indicate body size might be useful for male–male assessment. For example, differences in the spectral or amplitude characteristics of song units (Frankel 1994), or speed of swimming while singing as observed in fin whales (*Balaenoptera physalus*; Clark et al. 2019), could be features that convey male quality. The quality of a male's song itself may also reflect experience and age of the singer, if some features of signal production or adherence to the "optimal" pattern require a skill level only acquired through repetitive practice and experience. Additionally, graded song characters (such as switching rates) that allow males to predict the likelihood of escalating aggression are also important in some species (e.g., Searcy et al. 2000).

It is likely that humpback whale song may have evolved in response to both inter- and intrasexual selective pressures (Herman 2017; Cholewiak et al. 2018b). In many avian species, song has been shown to serve both functions within the breeding system, where males may use different song types or change presentation style depending on the primary audience (e.g., Kroodsma et al. 1989; Searcy and Yasukawa 1990). Furthermore, both males and females may eavesdrop on acoustic interactions, which may also play a strong role in sexual selection (Naguib et al. 1999; Mennill et al. 2002). Individuals can gain information about another individual's quality or motivation through eavesdropping and use this information to make subsequent decisions (Bartsch et al. 2014). Females may use the same types of song traits to assess male quality. The combination of stereotypy and complexity in humpback song may indicate that it is a multimessage display (Murray et al. 2018), and both males and females may be listening in on male–male interactions (Cholewiak 2008; Dunlop and Noad 2016; Cholewiak et al. 2018b).

A potential approach to exploring selective forces that have shaped humpback whale singing behavior is to use comparative studies from non-interacting populations and different periods of time (e.g., different decades), to distinguish between aspects of the singing behavior and song that are heritable versus entirely learned (Chap. 2). Several researchers have suggested that some aspects of singing behavior may follow innate templates, as opposed to those features of songs (e.g., the patterns themselves) that appear to be learned and transmitted between individuals and thus subject to the processes of cultural evolution. The continuity of humpback whale song structure and singing behavior on a global scale was first recognized by Payne and Guinee (1983). Populations in all ocean basins follow the same hierarchical song structure, change their songs progressively through time, and have similar levels of complexity, so Payne and Payne (1985) argued that there are innate constraints guiding the form of song. This is not entirely surprising, as all species of whales have their own recognizable species-specific song structure that is clearly a heritable trait.

Humpbacks are set apart from at least the other balaenopterids by the predisposition for flexible learning, apparent innovation, and aptitude for copying conspecifics; these traits are also clearly heritable, found in all populations currently studied, and apparently unique within the Balaenopteridae clade (Chap. 12). Therefore, it is clear that some evolutionary force has selected for the process of dynamic temporal change in humpback song. Another level of questioning is to ask whether these same evolutionary forces may have selected for specific ways in which singers change their songs over time, and thus constraining the observed variation in song patterns. Cerchio et al. (2001), during a study of geographic variation between Hawai'i and México, observed that some quantitative variables of unit and phrase structure changed in parallel in the two distant breeding areas over a single singing season. Since it was concluded that singers in these distant breeding areas could not interact during the study time period, the authors suggested that this may be evidence for an innate template that could govern some aspects of how singers change song over time, in the same manner as innate processes determine song structure. It was argued that progressive change therefore could be explained as a combination of cultural transmission overlaid on a set of innate rules for change. The prediction of such a system would be recurring patterns of how songs are changed over time, when comparing songs from distant populations or different periods of time (e.g., different decades). Other more recent studies have examined song samples from such non-interacting populations and similarly suggested "constraints" on what song elements are copied (Mercado III et al. 2003) and persistent recurring patterns of both phrase structure and how songs are changed over time (Green et al. 2011). Together, these studies expand upon the concept of genetic predispositions for song structure and flexible learning that were first proposed in the early studies (Payne and Guinee 1983; Payne and Payne 1985) and extend it to propose innate components to how these patterns change over time. Studies exploring a heritable component to song structure and temporal change have been relatively few in comparison to studies of cultural transmission and the resultant patterns of geographic variation. If in fact there are constraints on song variation and behavior, and definable patterns of temporal change, the description of these patterns could inform what has been selected during the evolution of the behavior, and in turn reveal those aspects of song that are important to reproductive success and inclusive fitness, driving the development of the observed complex system.

11.4 Where Do We Go Now? Open Questions for Future Work

In the fifty years since humpback whale song was first described, there have been well over 100 papers published in the scientific literature describing the content and patterns of song across space and time, the occurrence of singing activity, the behavior of singers, as well as speculations about the form and function of song within the humpback breeding system. The beauty and complexity of humpback whale song have captivated many hearts and minds.

Yet, despite decades of research, fundamental questions remain. What are the evolutionary drivers that have resulted not only in a highly exaggerated acoustic display, but also the predisposition for life-long learning and incorporation of new features into an individual's song? Which features of humpback song and singing behavior are culturally transmitted, and which are genetically constrained? How do individual whales develop and change their own songs, and what are the social factors and thresholds for hearing new material that prompts an individual male to adopt new song patterns into his own song (e.g., Beecher 2017)?

For the most part, these questions have not been addressed because it is extremely difficult to study sexual selection. To do so would require documenting mate choice by females, quantifying the reproductive success of individual males, and identifying specific quantifiable male traits that contribute to reproductive success. Unlike songbirds, where individuals can be raised in captivity and the influences on a male's song development (and his subsequent reproductive success) can be tested in a controlled environment, no such research opportunity exists with free-ranging, long-lived large whales (or other large mammals, for that matter). Mating has been nearly impossible to observe in humpback whales, and there have been few molecular genetic studies of paternity and/or male reproductive success. Only three such studies have been published (Clapham and Palsboll 1997; Nielsen et al. 2001; Cerchio et al. 2005). The insights from these studies have revealed only mild skew in male reproductive success beyond what would be expected from a random mating model (with many males siring single calves and few males siring multiple calves), and indicate that successful males use a variety of mating tactics, singing being among them (Cerchio 2003; Cerchio et al. 2005). Studies of male reproductive success are particularly challenging in species with large populations that cover broad ranges, where the chances of sampling successful males are low.

There are multiple avenues for further study that have not been fully explored and could reveal new insights into both the proximate mechanisms and ultimate drivers of humpback whale singing behavior. From a comparative perspective, only one other species of cetacean has evolved a similarly complex song—the bowhead whale (*Balaena mysticetus*; Chap. 12). Bowhead whales are part of the family Balaenidae, which represents a separate evolutionary lineage within mysticetes, having diverged from the lineage that gave rise to the balaenopterids over 30 million years ago (Marx and Fordyce 2015). Within the four species of extant balaenids, it was previously thought that right whales (*Eubalaena glacialis*, *E. japonica*, *E. australis*) did not exhibit male acoustic reproductive displays, although recent studies have revealed that right whales do produce patterned sequences of gunshot sounds that have been characterized as song (Crance et al. 2019). Bowhead whales, on the other hand, are well known for their varied song repertoire, with extreme diversity in song types reported in some populations (Stafford et al. 2018; Chap. 12). Male bowhead song is generally comprised of one to three themes, but many different song types have been recorded within and between breeding seasons, suggesting potentially strong selective pressure for novelty (Stafford et al. 2018). In addition, bowheads incorporate biphonation into some of their song units (Tervo et al. 2011), essentially singing with two voices. Given the phylogeny of mysticete whales, the occurrence of an exaggerated, complex acoustic display in both bowhead and humpback whales likely represents convergent evolution rather than shared ancestry. Of further consideration, the breeding systems of these two species are quite different; the bowhead whale mating system likely relies on sperm competition, given their larger-than-predicted testes size (Brownell and Ralls 1986) and observed mating behavior (Everitt and Krogman 1979), whereas humpback whales are believed to have a form of a polygynous mating system with alternative male mating tactics (Clapham 1996; Cerchio 2003). Further exploration of bowhead behavior and song within a comparative framework may reveal commonalities between the species, indicating similarities in sexual selection.

Within studies of humpback song and humpback singing behavior, several topics remain understudied. One of these is the ontogeny of song, both developmental, as a juvenile male matures into adulthood, and seasonal, i.e., with the increase and decrease in song production at the beginning/end of the breeding season. Studying the developmental ontogeny of song within individual males would require directed effort to record vocal production by juvenile males at various stages during their growth and maturation, an undertaking which has not been attempted, to our knowledge. Studies of calves with their mothers on the breeding grounds have revealed the production of non-song sounds by calves (Zoidis et al. 2008; Saloma 2018), but documentation of song production and song structure of sexually immature males is lacking. Such studies would reveal insights into how individual males develop and learn their songs. Areas where juvenile humpbacks occur during winter months, such as the mid-Atlantic coast of the US eastern seaboard (Swingle et al. 1993), could provide study areas where such targeted research could take place.

Recent work describing the seasonal ontogeny of song structure on feeding grounds in the western North Atlantic (Kowarski et al. 2019; Kowarski 2020) has revealed patterns in the onset and cessation of song production prior to and post breeding season. In the autumn, there appeared to be a gradual onset of singing behavior, with the production of song fragments prior to full song sequences. Song fragments were composed of individual phrases or subphrases, repeated a variable number of times, and were detected for several weeks prior to the detection of full song sequences. In contrast, during spring months, presumably when males would have been returning from the breeding grounds, the cessation of singing behavior was abrupt, with fewer instances of song fragments being detected. The authors discuss these patterns within the context of singing and hormonal development in oscine birds, suggesting that the production of song fragments is similar to the production of "subsong" (or perhaps more accurately, "plastic song") in songbirds (Kowarski et al. 2019). More detailed analysis of complete songs throughout the autumn on the feeding ground, into the winter on Caribbean breeding grounds, indicated a steady increase in repetition number of phrases and consistency in order of themes, suggesting an analogous progression from "plastic song" to "crystalized song" (Kowarski 2020). While these studies were not able to discern the seasonal ontogeny of song development for individual males, they provide information on population-level patterns. Interestingly, the most common forms of song fragments detected by Kowarski et al. (2019) were comprised of single or multiple song themes, with singular phrases or subphrases rarely detected.

Another line of inquiry that has not been well explored is that of the commonalities in the compositional organization and spectral content of humpback song found in songs worldwide. The Paynes first noted that all songs include "a remarkable range of contrasting frequencies, dynamics, and rhythms" (Payne and Payne 1985). Beyond simply the inclusion of contrasting frequencies and rhythms, our own observations are that completely unrelated songs (i.e., songs from different ocean basins and different years) may share similar compositional characteristics when examined from a broad view; for example, the production of themes with relatively higher frequency units (e.g., 300–500 Hz) followed by themes with consistently lower frequency units (e.g., <200 Hz), as though the themes are somewhat "ordered" by spectral content. These organizational features can be seen in example spectrograms of songs recorded from two different ocean basins (North Pacific vs. Indian Ocean) across a nearly 40-year time period (1990 vs. 2019) (Fig. 11.4). A similar pattern in overall spectral composition has been observed in other ocean basins, such as in the songs recorded in the North Atlantic (pers. obs.) as well as in the South Pacific (Mercado III and Perazio 2021) in different time periods. While the finer details of the spectral features and song composition are clearly different between these regions and periods, there appear to be similarities in the broad pattern that is greater than the individual components. These attributes describe what could be considered an emergent property of humpback whale song, which is often (though not always) observed. This observation may be one of the indications distinguishing what may be innate structural rules governing humpback song organization, versus the details of song composition that may be culturally transmitted (Cerchio et al. 2001).

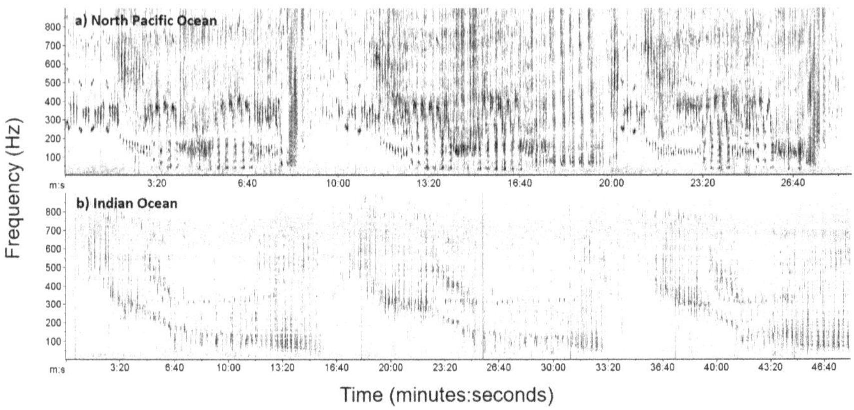

Fig. 11.4 Spectrograms showing examples of humpback song from two different ocean basins (**a** North Pacific Ocean, **b** Indian Ocean), and two different years (1990, 2019, respectively). Note the general similarity in the spectral-temporal organization of frequency content in both examples, demonstrating what appears to be a periodically recurring phenomenon in thematic organization in humpback song structure

The cultural diffusion and song convergence episodes documented in the South Pacific (e.g., Noad et al. 2000; Garland et al. 2011) provide another opportunity to study song learning, cultural transmission, and patterns of change. As individual males in one breeding aggregation appear to have learned and sung the songs from males in a different breeding aggregation, how is the new material incorporated into an individual's own song, and consequently the song of the population as a whole? One study has provided a glimpse into these mysteries, examining the song sequences of four males during periods of cultural diffusion (Garland et al. 2017). In each example, males incorporated complete phrases from new themes into their existing song pattern, and in some cases sang transition phrases that incorporated material from both the old and new themes. This suggests that the salient component for males incorporating new material into their song may be song phrases, rather than individual units. The observations of song hybridization and song convergence raise a new set of questions related to the cultural evolution of song patterns. After a song convergence episode between two breeding aggregations that had previously sung different songs (i.e., different sets of themes), there appears to be a period of isolation of the two aggregations during which song divergence occurs once again (Darling et al. 2019a). While some studies have examined measures of song "complexity" within a population before and after such convergence episodes (Allen et al. 2018), no studies have yet examined the differential patterns of song change between breeding aggregations after such convergence. Since cultural diffusion and song convergence are reported to occur repeatedly over decadal scales in both the Southern Hemisphere and Northern Hemisphere (Darling et al. 2019a; Garland and McGregor 2020), it is apparent that when males in one breeding aggregation are isolated from males in a different aggregation, some song changes must occur in different directions between the two separate aggregations. Therefore, studies on the temporal progression of song changes after convergence could provide valuable insights into cultural evolution, allowing one to potentially tease apart changes that are purely the result of cultural transmission of individual innovations (which one would predict to be divergent between aggregations), from those types of changes that may occur according to an innate heritable template, that would appear as similar parallel trends in the two diverging populations (Cerchio et al. 2001; Green et al. 2011).

As humpback whale populations increase after their decimation during the industrial whaling era, there may be increased opportunities for mixing of individuals from different breeding aggregations during migration or on the feeding grounds. Climate change and the resulting effects on the distribution and abundance of prey will continue to impact the movements and potential interchange between individuals from what had been different breeding aggregations on the foraging grounds. Observed variations in primary productivity have been correlated with fluctuations in humpback singing activity (Ryan et al. 2019), and the seasonal distributions of multiple species of large whales have already shifted in recent years in some regions (e.g., Davis et al. 2020). In addition, some breeding areas near to but north of the equator are utilized by individuals from both the Northern and Southern Hemisphere populations, in opposing breeding seasons. In the Cape Verde Islands and western African coast in the North Atlantic, and along the coast of Costa Rica in the North Pacific, both Northern and Southern Hemisphere populations have been documented at different times of year (Acevedo and Smultea 1995; Rasmussen et al. 2007; Hazevoet et al. 2011; Van Waerebeek et al. 2013). While simultaneous utilization of those breeding areas by individuals from opposite hemispheres has not been documented, overlap of breeding males at the extreme temporal limits of their respective breeding seasons could be a possibility. Each of these factors (population recovery, shifts in distribution and potential overlap in breeding habitats) could facilitate increased cultural transmission of song features between males from different breeding aggregations. What is the threshold of exposure before an individual singer decides to incorporate a new song type into his own, and how does that relate to what is happening at the level of the breeding aggregation? Some breeding aggregations have appeared more "acoustically dynamic" than others (e.g., Darling et al. 2019a), and in the South Pacific, cultural transmission of new song types appears to spread from larger to smaller breeding groups (Garland et al. 2011, 2015). However, in some cases new song types were detected but failed to be adopted by the breeding aggregation as a whole (Garland et al. 2011). We are reminded that the changing dynamics of populations that are growing in a post-whaling era provide opportunities, but also complications, to our understanding of the factors driving the observed patterns in humpback song structure.

Finally, we reflect on a unique population of humpback whales that inhabits the Arabian Sea. Unlike all other humpback populations, the animals in the Arabian Sea comprise an extremely small population of year-round residents. They follow a Northern Hemisphere breeding cycle (Mikhalev 1997), but have no physical access to high-latitude feeding areas: Biannual periods of monsoon-driven productivity allow them to remain in the Arabian Sea and feed throughout the year (Minton et al. 2011). Therefore, they do not undertake the latitudinal migrations typical of other populations. This population was subjected to intensive illegal Soviet whaling in the 1960s that reduced a population that was thought to be already small in size, on the order of hundreds of animals (Mikhalev 1997); current estimates suggest the remaining population could be less than 100 animals (Minton et al. 2011) and has significantly lower genetic diversity than Southern Hemisphere populations (Pomilla et al. 2014). What can studies of an isolated, non-migratory population of humpback whales tell us about the mechanisms and underlying drivers for song complexity and change? Passive acoustic monitoring over a two-year period off the Arabian Sea coast of Oman indicated that singing activity followed the same boreal seasonal pattern typical of other Northern Hemisphere populations (Cerchio et al. 2016), despite the lack of latitudinal migration and the population having been derived from a Southern Hemisphere population approximately 70,000 years ago (Pomilla et al. 2014). Initial analyses of acoustic data from the region found that while patterns in song and phrase structure at any given time fit within the range of variation seen in other populations globally, temporally the population exhibited an unusually low rate of song change over a 3-year period, possibly representing a form of temporal "stasis" (Cerchio et al. 2018). The authors speculate that one potential explanation could be that the small population size has resulted in a low level of innovation in song patterns. Furthermore, during the study period, songs from the southwest

Indian Ocean humpback population were recorded on several days during the Boreal summer (Austral winter) off the coast of Oman, indicating a low level of occurrence of animals from the Southern Hemisphere population. However, the Arabian Sea animals did not incorporate or adopt any elements from the southwest Indian Ocean song into their own song, similar to the observation by Garland et al. (2011) in French Polynesia (where new songs were recorded but failed to be adopted by the breeding aggregation as a whole). Cerchio et al. (2018) propose three hypotheses for their observation: that no Arabian Sea animals were in the area during the time that the Southern Hemisphere animals appeared, that the number of the Southern Hemisphere singers was too few to result in cultural diffusion (i.e., for the local males to adopt the new elements into their song), or that the Arabian Sea population was either not receptive or possibly has partially lost the trait for adoption of novel material. The combined observation of comparatively slow change of songs over three years and the lack of cultural diffusion despite the presence of the immigrants' songs suggest that there may be some unique selective drivers for this population's singing behavior.

11.5 Afterword

Looking back on the past half century of research, it is clear that we have uncovered fascinating mysteries about the captivating songs of humpbacks whales. The growing availability of acoustic recording and analytical technologies in the past few decades has facilitated our ability to listen in on the ocean in many remote parts of the world, revealing new and interesting insights into the acoustic behavior of not only humpback whales, but many marine species. However, as our ability to remotely collect passive acoustic data has increased, there seems to have been a corresponding decrease in field efforts dedicated to the focal study of individual animal behavior; a line of inquiry that is essential if we are to further unravel some of the lingering questions that we have described in this chapter. Without continued study of humpback breeding behavior, and in particular the behavior of singers and how individual males change their songs over time, our understanding of the processes and forces that drive the beautiful complexity in humpback song will always be constrained. It is only through combining our collective efforts and observations, and retaining a deep connection to the animals that we study, that we can reflect on the most interesting and in-depth questions that inspire us, and pay homage to the wonders of the natural world.

Acknowledgements We acknowledge and appreciate over half a century of work by dozens of researchers around the globe that have contributed to the body of knowledge that has allowed us to explore the intricacies of humpback whale song in so many different ways. In particular, we thank Dr. Christopher Clark for many years of support, thought-provoking discussions, and insightful reflections that have profoundly influenced our thought processes and understanding of not just humpback whale song, but baleen whale acoustic ecology as a whole. Finally, we dedicate this chapter to Katharine Payne, for inspiring generations of curious minds, including our own, to listen a little more closely to the world around us.

References

Acevedo A, Smultea M (1995) First records of humpback whales including calves at Golfo Dulce and Isla del Coco, Costa Rica, suggesting geographical overlap of northern and southern hemisphere populations. Mar Mamm Sci 11:554–560

Allen JA, Garland EC, Dunlop RA, Noad MJ (2018) Cultural revolutions reduce complexity in the songs of humpback whales. Proc R Soc B Publ 285(1891). https://doi.org/10.1098/rspb.2018.2088

Andersson M (1994) Sexual selection. Princeton University Press, Princeton, NJ

Baker CS, Perry A, Bannister JL, Weinrich MT, Abernethy RB, Calambokidis J, Lien J, Lambertsen RH, Urban Ramirez J, Vasquez O, Clapham PJ, Alling A, O'Brien SJ, Palumbi SR (1993) Abundant mitochondrial DNA variation and world-wide population structure in humpback whales. Proc Natl Acad Sci 90:8239–8243

Bartsch C, Wenchel R, Kaiser A, Kipper S (2014) Singing onstage: female and male common nightingales eavesdrop on song type matching. Behav Ecol Sociobiol 68(7):1163–1171

Beecher M (2017) Birdsong learning as a social process. Anim Behav 124:233–246

Bradbury JW, Vehrencamp SL (1998) Principles of animal communication. Sinauer Associates Inc.

Brownell RL, Ralls K (1986) Potential for sperm competition in baleen whales. Rep Int Whaling Comm 8:97–112

Calambokidis J, Steiger GH, Straley JM, Herman LM, Cerchio S, Salden DR, Jorge UR, Jacobsen JK, Ziegesar OV, Balcomb KC, Gabriele CM (2001) Movements and population structure of humpback whales in the North Pacific. Mar Mamm Sci 17:769–794

Cato DH (1991) Songs of humpback whales: the Australian perspective. Mem Queensland Museum 30:277–290

Cavelli-Sforza LL, Feldman M (1981) Cultural transmission and evolution: a quantitative approach. Princeton University Press, Princeton, NJ

Cerchio S (2003) Paternity, polygyny and alternative mating tactics in humpback whales (*Megaptera novaeangliae*)

Cerchio S, Jacobsen JK, Norris TF (2001) Temporal and geographical variation in songs of humpback whales, *Megaptera novaeangliae*: synchronous change in Hawaiian and Mexican breeding assemblages. Anim Behav 62(2):313–329

Cerchio S, Jacobsen JK, Cholewiak DM, Falcone EA, Merriwether DA (2005) Paternity in humpback whales, *Megaptera novaeangliae*: assessing polygyny and skew in male reproductive success. Anim Behav 70(2):267–277. https://doi.org/10.1016/j.anbehav.2004.10.028

Cerchio S, Strindberg S, Collins T, Bennett C, Rosenbaum H (2014) Seismic surveys negatively affect humpback whale singing activity off northern Angola. PLoS ONE 9(3):e86464–e86464. https://doi.org/10.1371/journal.pone.0086464

Cerchio S, Willson A, Muirhead C, Minton G, Collins T, Baldwin R, Sarrouf Willson M, Al Harthi S (2016) Preliminary report on long-term detection of Arabian Sea humpback whale vocalizations off Oman

Cerchio S, Willson A, Muirhead C, Al Harthi S, Baldwin R, Bonato M, Collins T, Di Clemente J, Dulau V, Estrade V, Latha G (2018) Geographic variation in song indicates both isolation of Arabian Sea humpback whales and presence of Southern Hemisphere whales off Oman

Chittleborough RG (1965) Dynamics of two populations of the humpback whale, *Megaptera novaeangliae* (Borowski). Mar Freshw Res 16(1):33–128. https://doi.org/10.1071/MF9650033

Cholewiak D (2008) Evaluating the role of song in the humpback whale (*Megaptera novaeangliae*) breeding system with respect to intra-sexual interactions

Cholewiak DM, Sousa-Lima RS, Cerchio S (2013) Humpback whale song hierarchical structure: historical context and discussion of current classification issues. Mar Mamm Sci 29(3):E312–E332. https://doi.org/10.1111/mms.12005
Cholewiak D, Clark CW, Ponirakis D, Frankel A, Hatch LT, Risch D, Stanistreet JE, Thompson M, Vu E, Van Parijs SM (2018a) Communicating amidst the noise: modeling the aggregate influence of ambient and vessel noise on baleen whale communication space in a national marine sanctuary. Endang Species Res 36:59–75
Cholewiak DM, Cerchio S, Jacobsen JK, Urban RJ, Clark CW (2018b) Songbird dynamics under the sea: acoustic interactions between humpback whales suggest song mediates male interactions. R Soc Open Sci 5(2):171298. https://doi.org/10.1098/rsos.171298
Clapham PJ (1996) The social and reproductive biology of humpback whales: an ecological perspective. Mamm Rev 26(1):27–49
Clapham PJ, Mayo CA (1987) The attainment of sexual maturity in two female humpback whales. Mar Mamm Sci 3(3):279–283
Clapham P, Palsboll PJ (1997) Molecular analysis of paternity shows promiscuous mating in female humpback whales (*Megaptera novaeangliae*, Borowski). Proc R Soc Lond B 264(264):95–98
Clapham P, Palsboll PJ, Matilla D, Vasquez O (1992) Composition and dynamics of humpback whale competitive groups in the West Indies. Behaviour 122(3–4):182–194
Clark CW, Clapham PJ (2004) Acoustic monitoring on a humpback whale (*Megaptera novaeangliae*) feeding ground shows continual singing into late spring. Proc R Soc Lond B 271(1543):1051–1057. https://doi.org/10.1098/rspb.2004.2699
Clark CW, Gagnon C (2002) Insights from IUSS detections, locations and tracking from 1992 to 1996. J Underw Acoust (USN) 52:609–640
Clark CW, Gagnon GJ, Frankel AS (2019) Fin whale singing decreases with increased swimming speed. R Soc Open Sci 6(6):180525. https://doi.org/10.1098/rsos.180525
Crance JL, Berchok CL, Wright DL, Brewer AM, Woodrich DF (2019) Song production by the North Pacific right whale, *Eubalaena japonica*. J Acoust Soc Am 145(6):3467–3479. https://doi.org/10.1121/1.5111338
Darling JD, Bérubé M (2001) Interactions of singing humpback whales with other males. Mar Mamm Sci 17(3):570–584
Darling JD, Cerchio S (1993) Movement of a humpback whale (*Megaptera novaeangliae*) between Japan and Hawaii. Mar Mamm Sci 9(1):84–88
Darling JD, Jurasz C (1983) Migratory destinations of North Pacific humpback whales. In: Payne R (ed) Communication and behavior of whales. Westview Press, pp 359–368
Darling JD, Sousa-Lima RS (2005) Songs indicate interaction between humpback whale (*Megaptera novaeangliae*) populations in the western and eastern South Atlantic Ocean. Mar Mamm Sci 21(3):557–566
Darling JD, Jones ME, Nicklin CP (2006) Humpback whale songs: do they organize males during the breeding season? Behaviour 143:1051–1101
Darling JD, Jones ME, Nicklin CP (2012) Humpback whale (*Megaptera novaeangliae*) singers in Hawaii are attracted to playback of similar song. J Acoust Soc Am 132:2955–2958
Darling JD, Acebes JMV, Yamaguchi M (2014) Similarity yet a range of differences between humpback whale songs recorded in the Philippines, Japan and Hawaii in 2006. Aquat Biol 21(2):93–107
Darling JD, Acebes JMV, Frey O, Jorge Urban R, Yamaguchi M (2019a) Convergence and divergence of songs suggests ongoing, but annually variable, mixing of humpback whale populations throughout the North Pacific. Sci Rep 9(1):7002. https://doi.org/10.1038/s41598-019-42233-7
Darling JD, Goodwin B, Goodoni MK, Taufmann AJ, Taylor MG (2019b) Humpback whale calls detected in tropical ocean basin between known Mexico and Hawaii breeding assemblies. J Acoust Soc Am 145(6):EL534–EL550
Davis GE, Baumgartner MF, Corkeron PJ, Bell J, Berchok C, Bonnell JM, Bort Thornton J, Brault S, Buchanan GA, Cholewiak DM, Clark CW, Delarue J, Hatch LT, Klinck H, Kraus SD, Martin B, Mellinger DK, Moors-Murphy H, Nieukirk S, Nowacek DP, Parks SE, Parry D, Pegg N, Read

AJ, Rice AN, Risch D, Scott A, Soldevilla MS, Stafford KM, Stanistreet JE, Summers E, Todd S, Van Parijs SM (2020) Exploring movement patterns and changing distributions of baleen whales in the western North Atlantic using a decade of passive acoustic data. Glob Change Biol 26:29. https://doi.org/10.1111/gcb.15191

Dawbin WH (1966) The seasonal migratory cycle of humpback whales. In: Norris KS (ed) Whales, dolphins, and porpoises. University of California Press, pp 147–170

Dawbin WH, Eyre EJ (1991) Humpback whale songs along the coast of Western Australia and some comparison with east coast songs. Mem Queensland Museum 30:249–254

Dunlop RA (2019) The effects of vessel noise on the communication network of humpback whales. R Soc Open Sci 6(11):190967. https://doi.org/10.1098/rsos.190967

Dunlop RA, Noad MJ (2016) The "risky" business of singing: tactical use of song during joining by male humpback whales. Behav Ecol Sociobiol 70(12):2149–2160

Dunlop RA, Noad MJ, McCauley RD, Kniest E, Slade R, Paton D, Cato DH (2018) A behavioural dose-response model for migrating humpback whales and seismic air gun noise. Mar Pollut Bull 133:506–516. https://doi.org/10.1016/j.marpolbul.2018.06.009

Emlen ST, Oring LW (1977) Ecology, sexual selection, and the evolution of mating systems. Science 197(4300):215–223

Eriksen N, Miller LA, Tougaard J, Helweg DA (2005) Cultural change in the songs of humpback whales (*Megaptera novaeangliae*) from Tonga. Behaviour 142:305–328

Everitt RD, Krogman BD (1979) Sexual behavior of bowhead whales observed off the north coast of Alaska. Arctic 32(3):277–280

Fisher RA (1930) The genetical theory of natural selection, 1st edn. Clarendon Press, Oxford

Fisher RA (1958) The genetical theory of natural selection, 2nd edn. Dover, New York

Frankel A (1994) Acoustic and visual tracking reveals distribution, song variability and social roles of humpback whales in Hawaiian waters. University of Hawaii

Frankel AS, Clark CW (2000) Behavioral responses of humpback whales (*Megaptera novaeangliae*) to full-scale ATOC signals. J Acoust Soc Am 108(4):1930–1937

Frankel AS, Clark CW (2002) ATOC and other factors affecting the distribution and abundance of humpback whales, *Megaptera novaeangliae*, off the north shore of Kauai. Mar Mamm Sci 18(3):644–662

Frankel AS, Clark CW, Herman L, Gabriele CM (1995) Spatial distribution, habitat utilization, and social interactions of humpback whales, *Megaptera novaeangliae*, off Hawai'i, determined using acoustic and visual techniques. Can J Zool 73:1134–1146

Fristrup KM, Hatch LT, Clark CW (2003) Variation in humpback whale (*Megaptera novaeangliae*) song length in relation to low-frequency sound broadcasts. J Acoust Soc Am 113(6):3411–3424. https://doi.org/10.1121/1.1573637

Frumhoff P (1983) Aberrant songs of humpback whales (*Megaptera novaeangliae*): clues to the structure of humpback songs. In: Communication and behavior of whales, pp 81–127

Gabriele C, Frankel A (2002) The occurrence and significance of humpback whale songs in glacier bay, southeastern Alaska. In: Arctic research of the United States, pp 42–47

Garland EC, McGregor PK (2020) Cultural transmission, evolution, and revolution in vocal displays: insights from bird and whale song. Front Psychol 11:2387. https://doi.org/10.3389/fpsyg.2020.544929

Garland EC, Goldizen AW, Rekdahl ML, Constantine R, Garrigue C, Hauser ND, Poole MM, Robbins J, Noad MJ (2011) Dynamic horizontal cultural transmission of humpback whale song at the ocean basin scale. Curr Biol 21(8):687–691. https://doi.org/10.1016/j.cub.2011.03.019

Garland EC, Gedamke J, Rekdahl M, Noad MJ, Garrigue C, Gales N (2013) Humpback whale song on the southern ocean feeding grounds: implications for cultural transmission. PLoS ONE 8(11):e79422

Garland EC, Goldizen AW, Lilley MS, Rekdahl M, Garrigue C, Constantine R, Hauser ND, Poole MM, Robbins J, Noad MJ (2015) Population structure of humpback whales in the western and central South Pacific Ocean as determined by vocal exchange among populations. Conserv Biol 29(4):1198–1207

Garland EC, Rendell L, Lamoni L, Poole MM, Noad MJ (2017) Song hybridization events during revolutionary song change provide insights into cultural transmission in humpback whales. PNAS 114(30)

Garrigue C, Constantine R, Poole M, Hauser N, Clapham P, Donoghue M, Russell K, Paton D, Mattila DK, Robbins J, Baker CS (2011) Movement of individual humpback whales between wintering grounds of Oceania (South Pacific), 1999 to 2004. J Cetac Res Manage Spec Issue 3:275–291

Garrigue C, Clapham PJ, Geyer Y, Kennedy AS, Zerbini AN (2015) Satellite tracking reveals novel migratory patterns and the importance of seamounts for endangered South Pacific humpback whales. R Soc Open Sci 2

Gedamke J (2004) Minke whale song, spacing, and acoustic communication on the Great Barrier Reef, Australia. Ph.D., University of California, Santa Cruz

Glockner DA (1983) Determining the sex of humpback whales (*Megaptera novaeangliae*) in their natural environment. In: Payne R (ed) Communication and behavior of whales, vol 76. Westview Press, Boulder, CO, pp 447–464

Glockner-Ferrari DA, Ferrari M (1990) Reproduction in the humpback whale (*Megaptera novaeangliae*) in Hawaiian waters, 1975–1988: the life history, reproductive rates and behavior of known individuals identified through surface and underwater photography. Rep Int Whaling Comm (spec Issue) 12:161–169

Green SR, Mercado E, Pack AA, Herman LM (2011) Recurring patterns in the songs of humpback whales (*Megaptera novaeangliae*). Behav Proc 86(2):284–294. https://doi.org/10.1016/j.beproc.2010.12.014

Guinee LN, Chu K, Dorsey EM (1983) Changes over time in the songs of known individual humpback whales (*Megaptera novaeangliae*). In: Payne R (ed). Westview Press, Boulder, CO, pp 59–80

Hafner GW, Hamilton CL, Steiner WW, Thompson TJ, Winn HE (1979) Signature information in the song of the humpback whale. J Acoust Soc Am 66(1):1–1. https://doi.org/10.1121/1.383072

Hamilton WD, Zuk M (1982) Heritable true fitness and bright birds: a role for parasites? Science 218:384–387

Hanski I, Gilpin M (1991) Metapopulation dynamics: brief history and conceptual domain. Biol J Lin Soc 42:3–16

Hauser N, Zerbini A, Geyer Y, Heide-Jorgensen MP, Clapham P (2010) Movements of satellite-monitored humpback whales, *Megaptera novaeangliae*, from the Cook Islands. Mar Mamm Sci 26(3):679–685

Hazevoet CJ, Gravanita B, Suárez PL, Wenzel FW (2011) Seasonality of humpback whale *Megaptera novaeangliae* (Borowski, 1781) records in Cape Verde seas: evidence for the occurrence of stocks from both hemispheres? Zool Caboverdiana 2(1):25–29

Helweg DA (1990) Comparison of songs of humpback whales (*Megaptera novaeangliae*) recorded in Japan, Hawaii and Mexico during in the winter of 1989. Sci Rep Cetac Res 1:1–20

Herman LM (2017) The multiple functions of male song within the humpback whale (*Megaptera novaeangliae*) mating system: review, evaluation, and synthesis. Biol Rev 92(3):1795–1818

Herman LM, Tavolga W (1980) The communication systems of cetaceans. In: Herman LM (ed). Wiley Interscience, New York, pp 149–209

Höglund J, Alatalo RV (1995) Leks. Princeton University Press, Princeton

Jenkins PF (1978) Cultural transmission of song patterns and dialect development in a free-living bird population. Anim Behav 26:50–78

Kowarski K (2020) Humpback whale singing behavior in the western North Atlantic: from methods for analysing passive acoustic monitoring data to understanding humpback whale song ontogeny. Dalhousie University, Halifax, Nova Scotia

Kowarski K, Evers C, Moors-Murphy H, Martin B, Denes SL (2018) Singing through winter nights: seasonal and diel occurrence of humpback whale (*Megaptera novaeangliae*) calls in and around the Gully MPA, offshore eastern Canada. Mar Mamm Sci 34(1):169–189. https://doi.org/10.1111/mms.12447

Kowarski K, Moors-Murphy H, Maxner E, Cerchio S (2019) Western North Atlantic humpback whale fall and spring acoustic repertoire: insight into onset and cessation of singing behavior. J Acoust Soc Am 145(4):2305. https://doi.org/10.1121/1.5095404

Kroodsma DE (1977) Correlates of song organization among North American wrens. Am Nat 111(981):995–1008. https://doi.org/10.1086/283228

Kroodsma DE, Bereson RC, Byers BE, Minear E (1989) Use of song types by the chestnut-sided warbler: evidence for both intra- and inter-sexual functions. Can J Zool 67:447–456

Lemon RE (1975) How birds develop song dialects. The Condor 77(4):385–406

Levenson C, Leapley WT (1978) Distribution of humpback whales (*Megaptera novaeangliae*) in the Caribbean determined by a rapid acoustic method. J Fish Res Board Can 35(8):1150–1152

Levins R (1969) Some demographic and genetic consequences of environmental heterogeneity for biological control. Bull Entomol Soc Am 15(3):237–240

Mackintosh NA (1972) Biology of the populations of large whales. Sci Prog 60(240):449–464

Magnúsdóttir EE, Rasmussen MH, Lammers MO, Svavarsson J (2014) Humpback whale songs during winter in subarctic waters. Polar Biol 37(3):427–433. https://doi.org/10.1007/s00300-014-1448-3

Marler P (1984) Song learning: innate species differences in the learning process. In: Marler P, Terrance HS (eds) The biology of learning. Springer, pp 289–309

Marler P (1990) Song learning: the interface between behaviour and neuroethology. Philos Trans R Soc Lond B Biol Sci 329(1253):109–114

Marler P, Peters S (1981) Sparrows learn adult song and more from memory. Science 213(4509):780–782

Marx FG, Fordyce RE (2015) Baleen boom and bust: a synthesis of mysticete phylogeny, diversity and disparity. R Soc Open Sci 2(4):140434. https://doi.org/10.1098/rsos.140434

Mednis A (1991) An acoustic analysis of the 1988 song of the humpback whale, *Megaptera novaeangliae*, off Eastern Australia. Mem Queensland Museum 30:323-332

Medrano L, Salinas M, Salas I, Deguevara PL, Aguayo A, Jacobsen JK, Baker CS (1994) Sex identification of humpback whales, *Megaptera novaeangliae*, on the wintering grounds of the Mexican Pacific Ocean. Can J Zool 72:1771–1774

Mennill DJ, Ratcliffe LM, Boag PT (2002) Female eavesdropping on male song contests in songbirds. Science 296:873

Mercado E III, Perazio CE (2021) Similarities in composition and transformations of songs by humpback whales (*Megaptera novaeangliae*) over time and space. J Comp Psychol 135(1):28–50

Mercado E III, Herman LM, Pack AA (2003) Stereotypical sound patterns in humpback whale songs: usage and function. Aquat Mamm 29(1):37–52

Mikhalev YA (1997) Humpback whales *Megaptera novaeangliae* in the Arabian Sea. Mar Ecol Prog Ser 149:13–21

Minton G, Collins T, Findlay K, Ersts P, Rosenbaum H, Berggren P, Baldwin R (2011) Seasonal distribution, abundance, habitat use and population identity of humpback whales in Oman. J Cetac Res Manag 3:185–198

Mundinger PC (1980) Animal cultures and a general theory of cultural evolution. Ethol Sociobiol 1(3):183–223

Murray A, Cerchio S, McCauley R, Jenner CS, Razafindrakoto Y, Coughran D, McKay S, Rosenbaum H (2012) Minimal similarity in songs suggests limited exchange between humpback whales (*Megaptera novaeangliae*) in the southern Indian Ocean. Mar Mamm Sci 28(1):E41–E57. https://doi.org/10.1111/j.1748-7692.2011.00484.x

Murray A, Dunlop RA, Noad MJ, Goldizen AW (2018) Stereotypic and complex phrase types provide structural evidence for a multi-message display in humpback whales (*Megaptera novaeangliae*). J Acoust Soc Am 143(2):980–994

Naguib M, Fichtel C, Todt D (1999) Nightingales respond more strongly to vocal leaders of simulated dyadic interactions. Proc B R Soc Publ 266(1419):537–542

Nielsen R, Mattila DK, Clapham PJ, Palsbøll PJ (2001) Statistical approaches to paternity analysis in natural populations and applications to the North Atlantic humpback whale. Genetics 157(4):1673–1682

Noad MJ, Cato DH, Bryden MM, Jenner M-N, Jenner KCS (2000) Cultural revolution in whale songs: humpbacks have picked up a catchy tune sung by immigrants from a distant ocean. Nature 408:537–537. https://doi.org/10.1038/35046199

Norris TF (1995) Effects of boat noise on the singing behavior of humpback whales (*Megaptera novaeangliae*)

Norris TF, McDonald M, Barlow J (1999) Acoustic detections of singing humpback whales (*Megaptera novaeangliae*) in the eastern North Pacific during their northbound migration. J Acoust Soc Am 106(1):506–514

Owen C, Rendell L, Constantine R, Noad MJ, Allen J, Andrews O, Garrigue C, Michael Poole M, Donnelly D, Hauser N, Garland EC (2019) Migratory convergence facilitates cultural transmission of humpback whale song. R Soc Open Sci 6(9):190337. https://doi.org/10.1098/rsos.190337

Payne RB (1979) Song structure, behaviour, and sequence of song types in a population of village indigobirds, *Vidua chalybeata*. Anim Behav 27:997–1013

Payne R (ed) (1983) Communication and behavior of whales, vol 76. 1 edn. Westview Press for the American Association for the Advancement of Science, Boulder, Colorado, 643 pp

Payne RB (1985) Behavioral continuity and change in local song populations of village indigobirds, *Vidua chalybeata*. Z Tierpsychol 70:1–44

Payne R, Guinee LN (1983) Humpback whale (*Megaptera novaeangliae*) songs as an indicator of "stocks". In: Payne R (ed) Communication and behavior of whales, vol 76, 1st edn. Westview Press for the American Association for the Advancement of Science, Boulder, CO, pp 333–358

Payne RS, McVay S (1971) Songs of humpback whales. Science 173(3997):585–597

Payne K, Payne R (1985) Large scale changes over 19 years in songs of humpback whales in Bermuda. Z Tierpsychol 68:89–114

Payne RB, Thompson WL, Fiala KL, Sweany LL (1981) Local song traditions in indigo buntings: cultural transmission of behavior patterns across generations. Behaviour 77(4):199–221

Payne K, Tyack P, Payne R (1983) Progressive changes in the songs of humpback whales (*Megaptera novaeangliae*): a detailed analysis of two seasons in Hawaii. In: Payne R (ed), 1 edn. Westview Press, Boulder, CO, pp 9–57

Pomilla C, Rosenbaum HC (2005) Against the current: an inter-oceanic whale migration event. Biol Lett 1(4):476–479. https://doi.org/10.1098/rsbl.2005.0351

Pomilla C, Amaral AR, Collins T, Minton G, Findlay KP, Leslie MS, Ponnampalam L, Baldwin R, Rosenbaum H (2014) The world's most isolated and distinct whale population? Humpback whales of the Arabian Sea. PLoS ONE 9(12):e114162

Rasmussen K, Palacios DM, Calambokidis J, Saborío MT, Dalla Rosa L, Secchi ER, Steiger GH, Allen JM, Stone GS (2007) Southern Hemisphere humpback whales wintering off Central America: insights from water temperature into the longest mammalian migration. Biol Lett 3:302–305. https://doi.org/10.1098/rsbl.2007.0067

Reeves RR, Swartz SL, Wetmore SE, Clapham PJ (2001) Historical occurrence and distribution of humpback whales in the eastern and southern Caribbean Sea, based on data from American whaling logbooks. J Cetac Res Manage 3(2):117–129

Rekdahl M, Garland EC, Carvajal GA, King CD, Collins T, Razafindrakoto Y, Rosenbaum H (2018) Culturally transmitted song exchange between humpback whales (*Megaptera novaeangliae*) in the southeast Atlantic and southwest Indian Ocean basins. R Soc Open Sci 5(11):172305

Risch D, Corkeron PJ, Ellison WT, Van Parijs SM (2012) Changes in humpback whale song occurrence in response to an acoustic source 200 km away. PLoS ONE 7(1):e29741–e29741. https://doi.org/10.1371/journal.pone.0029741

Ryan JP, Cline DE, Joseph JE, Margolina T, Santora JA, Kudela RM, Chavez FP, Pennington JT, Wahl C, Michisaki E, Benoit-Bird K, Forney KA, Stimpert AK, DeVogelaere A, Black N, Fischer M (2019) Humpback whale song occurrence reflects ecosystem variability in feeding and migratory habitat of the northeast Pacific. PLoS ONE e0222456

Saloma A (2018) Humpback whales (*Megaptera novaeangliae*) mother-calf interactions. Doctoral dissertation, Université Paris Saclay (COmUE), Université d'Antananarivo
Schevill WE (1964) Underwater sounds of cetaceans. In: Tavolga WN (ed). Pergamon Press Inc., New York, NY, pp 307–316
Searcy WA, Yasukawa K (1990) Use of the song repertoire in intersexual and intrasexual contexts by male red-winged blackbirds. Behav Ecol Sociobiol 27(2):123–128
Searcy WA, Nowicki S, Hogan C (2000) Song type variants and aggressive context. Behav Ecol Sociobiol 48(5):358–363
Slater PJB, Ince SA (1979) Cultural evolution in chaffinch song. Behaviour 71(1–2):146–166
Smith JN, Goldizen AW, Dunlop RA, Noad MJ (2008) Songs of male humpback whales, *Megaptera novaeangliae*, are involved in intersexual interactions. Anim Behav 76(2):467–477. https://doi.org/10.1016/j.anbehav.2008.02.013
Sousa-Lima RS, Clark CW (2008) Modeling the effect of boat traffic on the fluctuation of humpback whale singing activity in the Abrolhos National Marine Park, Brazil. Can Acoust 36(1):175–181
Stafford KM, Lydersen C, Wiig O, Kovacs KM (2018) Extreme diversity in the songs of Spitsbergen's bowhead whales. Biol Lett 14(4). https://doi.org/10.1098/rsbl.2018.0056
Stevick PT, Berrow SD, Bérubé M, Bouveret L, Broms F, Jann B, Kennedy A, López Suárez P, Meunier M, Ryan C, Wenzel F (2016) There and back again: multiple and return exchange of humpback whales between breeding habitats separated by an ocean basin. J Mar Biol Assoc UK 96(4):885–890. https://doi.org/10.1017/S0025315416000321
Stimpert AK, Peavey LE, Friedlaender AS, Nowacek DP (2012) Humpback whale song and foraging behavior on an Antarctic feeding ground. PLoS ONE 7(12):e51214–e51214. https://doi.org/10.1371/journal.pone.0051214
Swingle WM, Barco SG, Pitchford TD, Mclellan WA, Pabst DA (1993) Appearance of juvenile humpback whales feeding in the nearshore waters of Virginia. Mar Mamm Sci 9(3):309–315
Tervo OM, Christoffersen MF, Parks SE, Kristensen RM, Madsen PT (2011) Evidence for simultaneous sound production in the bowhead whale (*Balaena mysticetus*). J Acoust Soc Am 130(4):2257–2262. https://doi.org/10.1121/1.3628327
Thompson TJ, Winn HE, Perkins PJ (1979) Mysticete sounds. In: Winn HE, Olla BL (eds). Plenum Press, New York, London, pp 403–431
Todt D, Naguib M (2000) Vocal interactions in birds: the use of song as a model in communication. In: Advances in the study of behavior, vol 29, pp 247–296
Trainer JM (1989) Cultural evolution in song dialects of yellow-rumped caciques in Panama. Ethology 80:190–204
Tsujii K, Akamatsu T, Okamoto R, Mori K, Mitani Y, Umeda N (2018) Change in singing behavior of humpback whales caused by shipping noise. PLoS ONE 13(10):e0204112
Tyack P (1981) Interactions between singing Hawaiian humpback whales and conspecifics nearby. Behav Ecol Sociobiol 8(2):105–116. https://doi.org/10.1007/BF00300822
Tyack PL, Whitehead H (1982) Male competition in large groups of wintering humpback whales. Behaviour 83(1/2):132–154
Van Opzeeland I, Van Parijs S, Kindermann L, Burkhardt E, Boebel O (2013) Calling in the cold: pervasive acoustic presence of humpback whales (*Megaptera novaeangliae*) in Antarctic coastal waters. PLoS ONE 8(9):1–7
Van Waerebeek K, Djiba A, Krakstad J, Samba Ould Bilal A, Bamy I, Almeida AM, Mbye EM (2013) New evidence for a South Atlantic stock of humpback whales wintering on the Northwest African continental shelf. Afr Zool 48:177–186
Vu ET, Risch D, Clark CW, Gaylord S, Hatch LT, Thompson MA, Wiley DN, Van Parijs SM (2012) Humpback whale song occurs extensively on feeding grounds in the western North Atlantic Ocean. Aquat Biol 14(2):175–183. https://doi.org/10.3354/ab00390
Whiten A, Caldwell CA, Mesoudi A (2016) Cultural diffusion in humans and other animals. Curr Opin Psychol 8:15–21. https://doi.org/10.1016/j.copsyc.2015.09.002
Wiley RH (1991) Lekking in birds and mammals: behavioral and evolutionary issues. In: Advances in the study of behavior, vol 20, pp 201–291

Wilson EO (1975) Sociobiology: the new synthesis. Harvard University Press
Winn HE, Edel RK, Taruski AG (1975) Population estimate of the humpback whale (*Megaptera novaeangliae*) in the West Indies by visual and acoustic techniques. J Fish Res Board Can 32:499–506
Winn HE, Thompson TJ, Cummings WC, Hain JHW, Hudnall J, Hays H, Steiner WW (1981) Song of the humpback whale—population comparisons. Behav Ecol Sociobiol 8:41–46
Zahavi A (1975) Mate selection—a selection for a handicap. J Theor Biol 53(1):205–214
Zoidis AM, Smultea MA, Frankel AS, Hopkins JL, Day A, McFarland AS, Whitt AD, Fertl D (2008) Vocalizations produced by humpback whale (*Megaptera novaeangliae*) calves recorded in Hawaii. J Acoust Soc Am 123(3):1737–1746. https://doi.org/10.1121/1.2836750

Chapter 12
Singing Behavior in the Bowhead Whale

Kathleen M. Stafford

> *Captain Kelley was the first to discover this singing ... In 1882, several ships lay at anchor under Indian Point (Cape Chaplino,* Chukotka, *Russia.). As usual the masters got together, and in the midst of their conversation, Captain Kelley broke in "There's a bowhead!" Everybody laughed about "Kelley's band," but he insisted that there were whales nearby, and he was going to give chase. One master suggested that it was the copper on the ship, another that it was seals, another that it was the ice, and so on. But when Captain Kelley took up anchor and set sail, every ship followed him. One whale was caught. Soon more singing was heard. The result was the capture of several whales. After having attention thus forcibly called to the singing, it was not long before the masters were on their guard for it. As singing is almost never heard in the Arctic, it is inferred that it is a sort of a call, or signal for whales when making a passage through Behring Sea, to notify each other that they are bound north, and perhaps that the Straits are clear of ice.*—Aldrich (1889, pp. 32–33)

Abstract Bowhead whales are the only baleen whale endemic to the Arctic. There are four recognized populations globally, all of which were driven to commercial extinction before the twentieth century. As an ice-adapted species which spends much of its life under the ice in polar night, acoustic communication is critical to bowhead whale life history, but only recently have we been able to listen to them year-round. The acoustic behavior of bowhead whales consists of calls which are produced year-round and are low-frequency, short duration signals, and song, which is a seasonal, likely reproductive, behavior that occurs primarily in winter. Bowhead whales sing complex, varying songs that change completely from week to week and year to year. Individual songs are short (45 s–2 min) and are repeated over many hours in long bouts. Some individual whales share the same song within a winter, but others do not. Why song varies so much within and between years, and within and between individuals remains a puzzle to be solved.

K. M. Stafford (✉)
Marine Mammal Institute, Oregon State University, Newport, OR 97365, USA
e-mail: kate.stafford@oregonstate.edu

C.W. Clark and E. C. Garland (eds.), *Ethology and Behavioral Ecology of Mysticetes*,
Ethology and Behavioral Ecology of Marine Mammals,
https://doi.org/10.1007/978-3-030-98449-6_12

Keywords Bowhead whale · Acoustic behavior · Arctic · Song · Reproductive tactic

A pair of bowhead whales (*Balaena mysticetus*) surfacing in a spring lead, surrounded by *sikuliaq* (the Inupiat word for young ice) northwest of Utqiaġvik Alaska, April 29, 2015. The whales are navigating through a lead system with some heavy sea ice. Once the whales pass Point Barrow, about 16 kilometers to the northeast, they will enter the Beaufort Sea, which is mostly 100% ice-covered. They will continue to migrate eastward through the ice and spend the summer foraging in the Canadian Beaufort Sea. *Photo Credit* Kathleen Stafford

12.1 Introduction

The bowhead whale (*Balaena mysticetus*) is a whale of superlatives. It is the only baleen whale to live year-round in the Arctic, and as such, is capable of breaking up through ice half a meter thick to breathe (George et al. 1989). In the high Arctic, three populations of bowheads are recognized, and each is named largely by its known distribution: the Bering-Chukchi-Beaufort (BCB) or western Arctic population, the East Canada-West Greenland (ECWG) population, and the East Greenland-Svalbard-Barents (also referred to as the Spitsbergen, SP) population. The Okhotsk Sea population (SO) occurs south of the Arctic circle in the geographically isolated Sea of Okhotsk (Baird and Bickham 2020). Each population was heavily depleted by commercial whaling beginning as early as the 1500s, such that by the advent of the twentieth century, all four populations were commercially extinct (George and Thewissen 2020). Whalers targeted bowheads for two of their exceptional attributes: bowheads have the thickest blubber and the longest baleen of any baleen whale species (Bockstoce 1986). Perhaps most remarkable, bowheads can live to be over 200 years old (George et al. 1999; George and Bockstoce 2008). Lastly, and most importantly for this chapter, bowhead whales are prolific singers with populations frequently changing songs within a season and between years (Stafford et al. 2018). "Kelley's band" may be the first direct attribution of song to any species of whale (Aldrich 1889), and the confusion of the sounds heard through the wooden hull of a whaleship perfectly encapsulate the variability of the songs of the bowhead which can screech and cry like ice, warble like a seal, and moan like the timbers of a wooden sailing ship.

Of all species of baleen whale, only the humpback (*Megaptera novaeangliae*) and the bowhead sing complex, varying songs that change rapidly over time: from song to song in the bowhead (Stafford et al. 2008, 2012, 2018) and from population to population across ocean basins in the humpback (Noad et al. 2000; Garland et al. 2011; Chap. 11). In both species, songs differ dramatically from the sounds they typically produce during summer and fall (Ljungblad et al. 1982; Blackwell et al. 2007; Dunlop; Chap. 9). This is not the case for other species of baleen whale to whom singing behaviors have been attributed. Any seasonally produced repeated patterned signals are often termed "song." Blue (*Balaenoptera musculus*), fin (*B. physalus*), and minke (*B. acutorostrata* and *B. bonaerensis*) whales, the rorquals that sing, all share aspects of their singing behavior, including conservation of most song structures over time. Although only a few populations have been sampled, it seems likely that, like humpback whales, rorqual singers are male (Croll et al. 2002; Oleson et al. 2007). Blue and fin whales produce very low-frequency songs that overlap each other in frequency range (Chaps. 2 and 9). Fin whales produce what appear at the outset to be relatively simple songs composed of one to three short duration sounds (often referred to as pulses, units, or notes) repeated in a series, with the "20-Hz" sound type being the most common. Globally, this 20-Hz sound is characteristic of the species, while song features such as unit center frequencies, presence of other units, and intervals between units (i.e., inter-note intervals) vary geographically (Watkins et al. 1987; Thompson et al. 1992; Hatch 2004; Archer et al. 2020). Blue whales produce songs composed of repeated phrases of 1–6 much longer units that may be amplitude- and/or frequency-modulated (Chap. 9). Blue whale songs are considered geographically distinct, stereotyped and remain relatively stable over long-time (decades). Minke whale song appears to vary geographically in frequency band and song structure, but within a region frequency features and song structure are conserved (Gedamke et al. 2001; Rankin and Barlow 2005; Risch et al. 2014; Chap. 14). The uniting characteristic of *Balaenoptera* songs as presently described is that they exhibit little variability among individuals or over time. A recent report of North Pacific right whale (*Eubalaena japonica*) "song" found multiple different patterns in the production of gunshot sounds, with evidence that the same patterns were repeated within and between years (Crance et al. 2019; Chap. 13). For all these species, some of the units that may be part of songs are also produced throughout the year, either as single calls or as part of calling bouts. For instance, northeast Pacific blue whales produce singular "A" and/or "B" sounds during different behavioral contexts than when these same sounds are produced in

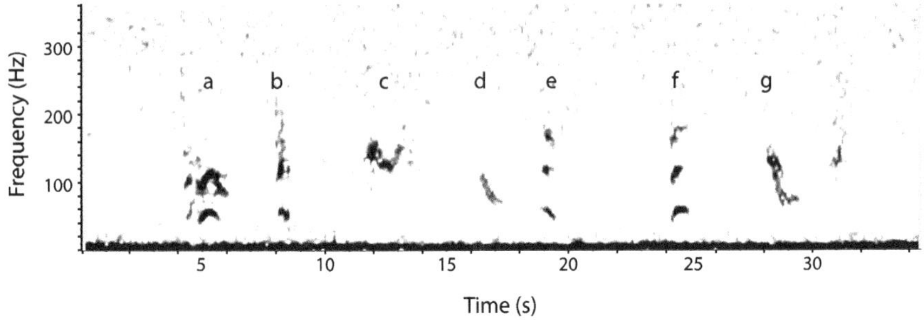

Fig. 12.1 Examples of eight different bowhead whale calls recorded in the Beaufort Sea on October 1, 2009. From left to right: **a** complex growl, **b** upsweep, **c** inflected, **d** downsweep, **e** complex, **f** complex, **g** downsweep (2048 sample FFT, 50% overlap, Hanning window)

songs (Oleson et al. 2007; Chap. 9). In contrast, bowhead and humpback whales change their songs within and between years, and song units are often distinct from calls, particularly in their bandwidth and frequency composition (but see Rekdahl et al. 2013 for a discussion of humpback "song-unit calls").

Bowhead whale calls are produced year-round, although in mid-winter they are often drowned out by low-frequency song units (reviewed in Stafford and Clark 2020). Calls are generally low-frequency (50–500 Hz), short duration signals that may be frequency- or amplitude-modulated (Clark and Johnson 1984). Most calls are simple frequency-modulated signals, and there appears to be a relatively limited repertoire of call types: frequency-modulated downsweeps and upsweeps, tonals, inflected calls as well as calls that are more complex and are less readily characterized (Fig. 12.1). Calls are produced relatively infrequently. Indeed, a recent study of the BCB population during the fall migration found that the population-wide mean call rate was only 1.3 calls/whale/hour (range 0.5–5.4, Blackwell et al. 2021).

In contrast to calls, songs are sung seasonally (generally November through April and into May depending on location), and the continuum of units that compose the different song types is extensive and has not been fully defined. In the two populations for which inter-annual comparisons have been studied (ECWG and SP), the units that make up each season's songs appear to change annually with no between-year repetitions (Tervo et al. 2011b; Stafford et al. 2018). This is one distinction between bowhead song and humpback song; humpbacks within a population produce songs with similar units and phrases between years (Darling et al. 2019; Allen et al. 2019; Chap. 11). Further, adjacent humpback populations in the Southern Hemisphere acquire songs from each other through cultural transmission on shared migratory routes (Garland et al. 2011; Owen et al. 2019; Chap. 8). Until very recently, all four bowhead whale populations were geographically isolated from each other by areas of heavy ice, or by distance in the case of SO bowheads, precluding the possibility that one population has learned songs from another. Other distinctions include humpback whale songs being of much longer duration than bowhead whale songs, composed of numerous phrases and themes, and all males in a population are assumed to sing the same song within a singing season (Payne and McVay 1971; Payne and Payne 1985).

Bowhead songs, in contrast to those of humpbacks, are composed of a single theme of 1–3 phrases repeated over and over (Fig. 12.2; Clark and Johnson 1984; Würsig and Clark 1993). Further, bowhead whales within a population sing dozens of

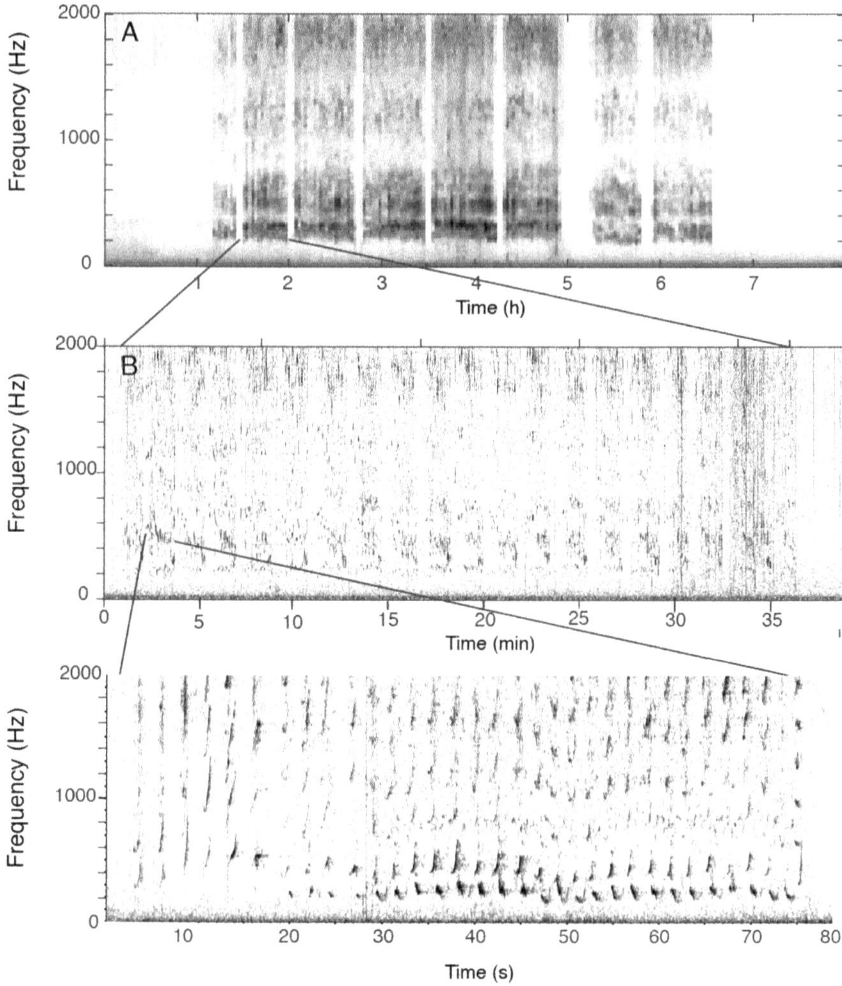

Fig. 12.2 Singing behavior of a bowhead whale recorded north of Point Barrow, Alaska, May 15, 2016. **a** An 8 h spectrogram showing a song session from a single bowhead whale. Each dark band represents many (8–20) songs sung repeatedly with long gaps ("rests") from 4 to 18 min between song bouts (120 s average, 256 sample FFT); **b** one song bout from A showing 20 complex songs separated by ~15 s (256 sample FFT, 0% overlap, Hanning window); **c** single song from song bout shown in **b**. Two-voiced singing can be seen in the lowest frequency downsweeps that occur simultaneously with the harmonic upsweeps from 25 to 75 s (512 sample FFT, 0% overlap, Hanning window)

song types composed of different phrases which consist of individual units that vary by song type during a season, and these song types are not repeated from one season to the next (Stafford et al. 2012, 2018; Johnson et al. 2014). This within-year and between-year variability is a relatively new discovery, with only recent observations of this feature due to the difficulty of studying this Arctic species year-round.

Most baseline information on bowhead whale singing behavior was the result of joint visual–acoustic censuses of BCB bowhead whales as they migrated north past Point Barrow, Alaska, in the Chukchi Sea in spring, in addition to earlier studies (Cummings and Holliday 1987; Zeh et al. 1993). These early efforts documented that, like humpback whales, the song of the bowhead whale changed annually (Würsig and Clark 1993). These data, which were recorded from early–mid-April into mid-late May suggested that all singing whales sang the same overall song type during the migration, although there was variation between and within individuals (Würsig and Clark 1993). Further, a unique characteristic of bowhead whale song was described whereby bowheads are capable of producing two very different sounds simultaneously (Würsig and Clark 1993). This suggests that bowhead whales use two separate sound sources while singing (Würsig and Clark 1993; Tervo et al. 2011a), an observation that implies that there must be some unique features in the bowhead whale laryngeal sound production mechanism (Schoenfuss et al. 2014; Chap. 3).

Before 2008, despite extensive recording effort during the spring migration off Barrow (now Utqiaġvik), Alaska, detections of bowhead whale singing were infrequent and when heard, only one singer was recorded at a time, and the perception was that all whales in the same year sang the same general song (Clark 1990). From two studies elsewhere (northwest of Spitsbergen, Norway and Disko Bay, Greenland), there was some indication that bowhead whales might sing more than a single song in a season (Clark et al. 1990; Stafford et al. 2008). The initiation of year-round monitoring of bowhead whales in Fram Strait beginning in 2008 showed that Spitsbergen bowhead whales sang nearly continuously from late October through early April and that dozens of distinct songs were produced by animals in this population (Stafford et al. 2012, 2018). In Fram Strait, distinct song types appeared and then disappeared somewhat sequentially over a season, where "season" refers to the presumed annual breeding season (Tarpley et al. 2020), in this case from about November to the following April or May (Stafford et al. 2012, 2018). As an example, a song type first heard in November might also be heard the next month in December, but it is unlikely to be recorded after this time. A song type heard for the first time in March will not have been noted in prior months. Furthermore, based on four years of duty-cycled recordings (12-min every hour), no song type was ever recorded in more than one season (Stafford et al. 2018). While similar long-term, year-round data from other populations of bowhead whales either do not exist or have not been analyzed, the data that do exist from two of these (BCB and ECWG) indicate that each population sings more than one distinct song each season. Data recorded in the northeastern Chukchi Sea (BCB population) from 2007 to 2008 detected song nearly continuously from October 30 to January 1, during which time six distinct song types were recorded (Delarue et al. 2009). In contrast, from April 10 to May 15, 2011, one dozen different songs were recorded in just six weeks during the spring migration off

Utqiaġvik, Alaska (Johnson et al. 2014). Recordings from late winter and early spring months from Disko Bay, Greenland (ECWG population) included several distinct songs from February to April in the same year, and entirely different repertoires of units that compose songs during those months in different years (Stafford et al. 2008; Tervo et al. 2011b).

For the three populations of bowhead whales for which song has been described (only the SO is lacking), all songs shared similar characteristics: They were composed of 1–2 repeated phrases, some of which were "two voiced," and included highly variable combinations of frequency-modulated and amplitude-modulated units (Clark and Johnson 1984; Würsig and Clark 1993; Stafford et al. 2008, 2012; Delarue et al. 2009; Tervo et al. 2011a; Johnson et al. 2014). What is unknown, or has not yet been explored, is whether there are population-level differences in singing behavior. For instance, are there differences in seasonal patterns in singing behavior, does each population sing many different song types within and between seasons, and are song structures similar among the different populations?

Our understanding of singing behavior in the bowhead whale is decades behind our understanding of humpback whale singing behavior (although even for that species new discoveries are being made, e.g., Garland et al. 2011; Cholewiak et al. 2018; Allen et al. 2019; Darling et al. 2019; Owen et al. 2019; Chap. 11). Unlike humpback whales, which mostly spend winter periods in tropical climes with clear water and warm temperatures, and are therefore easier to observe, bowhead whales winter in heavy pack ice and in polynyas in the Arctic and sing during frigid atmospheric temperatures of the polar night. For the BCB and ECWG populations, satellite tracking data have provided some indication of wintering and presumed breeding areas (Heide-Jørgensen et al. 2006; Citta et al. 2011), while passive acoustic data have done so for SP bowheads (Stafford et al. 2012, 2018). For the SP population, northern Fram Strait appears to be a wintering ground, and several years of passive acoustic data exist for this region. For the BCB population, Anadyr Strait and Saint Lawrence Island, Alaska, are wintering grounds (Noongwook et al. 2007; Citta et al. 2011); some long-term passive acoustic data exist from the northern Bering and southern Chukchi Seas but have been largely unanalyzed to date. The ECWG population appears to winter in Hudson Strait, Canada (Heide-Jørgensen et al. 2006), where passive acoustic monitoring data show the presence of acoustically active bowheads from December through April (Simard, unpublished data). All of the information on ECWG bowhead whale song has come from Disko Bay, Greenland, where the sex ratio is highly skewed toward females (~85%, Heide-Jørgensen et al. 2007). There are still no publications on acoustic data from the SO population.

12.2 Song Variability

For species such as blue and fin whales, song structure can be diagnostic of the distribution of males, and by association, females, such that song structure characteristics can be used to identify where the animals reside (e.g., Stafford et al. 2001; Castellote et al. 2011). In the Southern Hemisphere, humpback whale song can be traced from one population to another over a span of a few years (Garland et al. 2011). For bowheads, at least based on data from Fram Strait and Disko Bay, Greenland, the within-population monthly and inter-season variability in song unit types and song composition is so great that distinguishing between populations by using some form of acoustic identification could be a remarkably difficult undertaking. We know that songs recorded from all bowhead whale populations share similar characteristics. That is, a song lasts from 45 to 180 s and is composed of a song type with repetitions of 1–3 phrases, each containing relatively few (less than 10 on average) sounds. Individual song types are generally broadband (50 Hz–5 kHz) and repeated over and over again with short gaps between them, and singing can last for many hours at a time (Fig. 12.2).

Songs have been classified as "simple" or "complex" (Stafford et al. 2008; Delarue et al. 2009). "Simple songs" (Fig. 12.3) are composed of fewer than four relatively low-frequency units (<2 kHz) that are frequency-modulated, do not have harmonics, and no simultaneous production of two different sounds (Stafford et al. 2008; Delarue

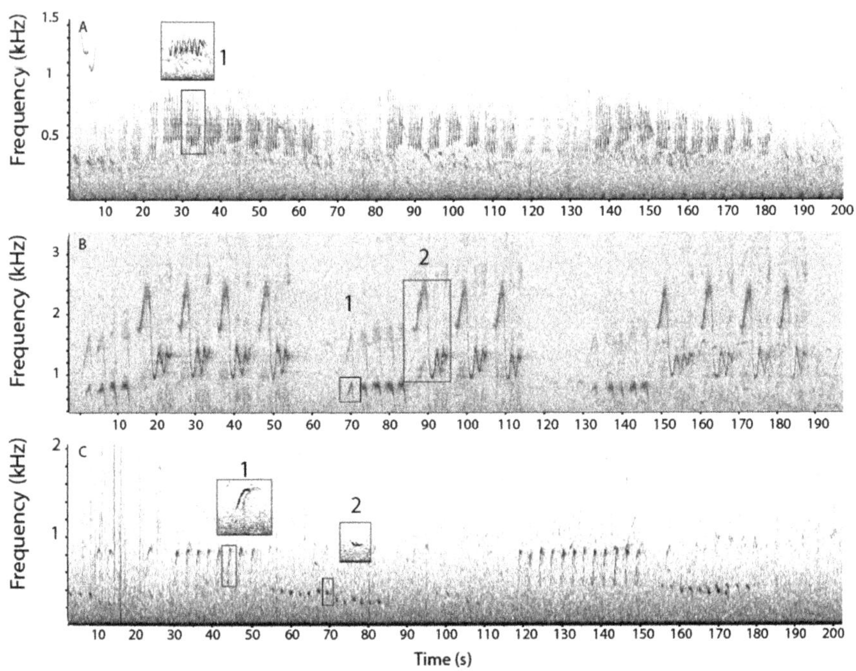

Fig. 12.3 Three examples of simple songs recorded in Fram Strait. **a** Three iterations of a single phrase song composed of repetitions of one sound unit, recorded December 20, 2010; **b** three iterations of a phrase incorporating 2 units, recorded January 3, 2013; **c** two iterations of a single phrase song composed of two units, recorded December 18, 2013 (2048 pt. FFT, 50% overlap, Hann window). Expanded units for song types in **a** and **c** are shown as insets in those spectrograms

et al. 2009). In some instances, a simple song may consist of a single unit type repeated over and over again (Fig. 12.3a).

The distinction between these simple songs and repeated calls (Ljungblad et al. 1982; Würsig and Clark 1993) is that simple songs repeat for long periods of time (e.g., hours), whereas a repeated call event is not repeated over and over. "Complex song" (Fig. 12.4) has broader bandwidth (with some units having fundamental frequencies over 2 kHz), most often includes two-voiced singing, and is composed of a greater number of units, many of which are considered complex, and may be frequency- and/or amplitude-modulated (Würsig and Clark 1993; Stafford et al. 2008, 2018; Delarue et al. 2009). Presently, the distinction between simple and complex songs is subjective as it is based only on human perception, and it is unknown if these different song forms serve different roles in bowhead whale behavioral ethology.

Based on three years of year-round data from Fram Strait, the numbers of simple versus complex songs recorded in each month from October to April were similar among years and showed similar seasonal patterns, with the greatest number of all song types in December, January, and February (Fig. 12.5). Further, there were no clear differences among the number of units per song in complex songs over a season, with complex songs averaging 5 units each (range 3–9). What is unknown is whether there were differences in song length or in the acoustic features of units that might indicate seasonal changes in songs over time. The only study to examine this in bowhead whales (Tervo et al. 2009) documented changes in the vocal behavior of whales in Disko Bay, Greenland. In that study, the bowhead whale songs recorded in February and March included more, and more complex, song units in each song type than were heard in song types recorded in April and May, suggesting that overall song complexity decreased from February to May.

A detailed analysis of mid-winter singing for the BCB population has not yet been undertaken to examine seasonal differences in song complexity, but a relatively short-term study during the spring migration off Point Barrow in April and May 2011 documented 12 different song types, many of which were complex (Johnson et al. 2014). Unlike later season song recorded in Disko Bay, Greenland, there was no decrease in the number of units or duration of songs as spring progresses (Tervo et al. 2009; Johnson et al. 2014). What was clear in the Alaska data, similar to Fram Strait and Disko Bay, was an overall decrease in singing behavior from winter through spring, whether measured in units/hour (Tervo et al. 2009) or in hourly and daily occurrences of song (Tervo et al. 2009; Johnson et al. 2014) from winter through spring (Stafford et al. 2018). This seasonality, where singing increases in fall, peaks in winter, and decreases in spring, is presumably reflective of breeding season. This is similar to what is known for other singing whales and aligns with what is known of the testosterone cycle, and related changes in testes size, in bowheads and other whale species (Chittleborough 1955; Vu et al. 2014; Hunt et al. 2018).

With the establishment of long-term, year-round recordings in wintering regions of the BCB, ECWG, and SP populations, there are now ample opportunities to study in detail the singing behavior of bowhead whales throughout the Arctic. Current information gaps include studying within-song variation within song bouts of individual singers, intra-individual and inter-individual differences in songs for whales that

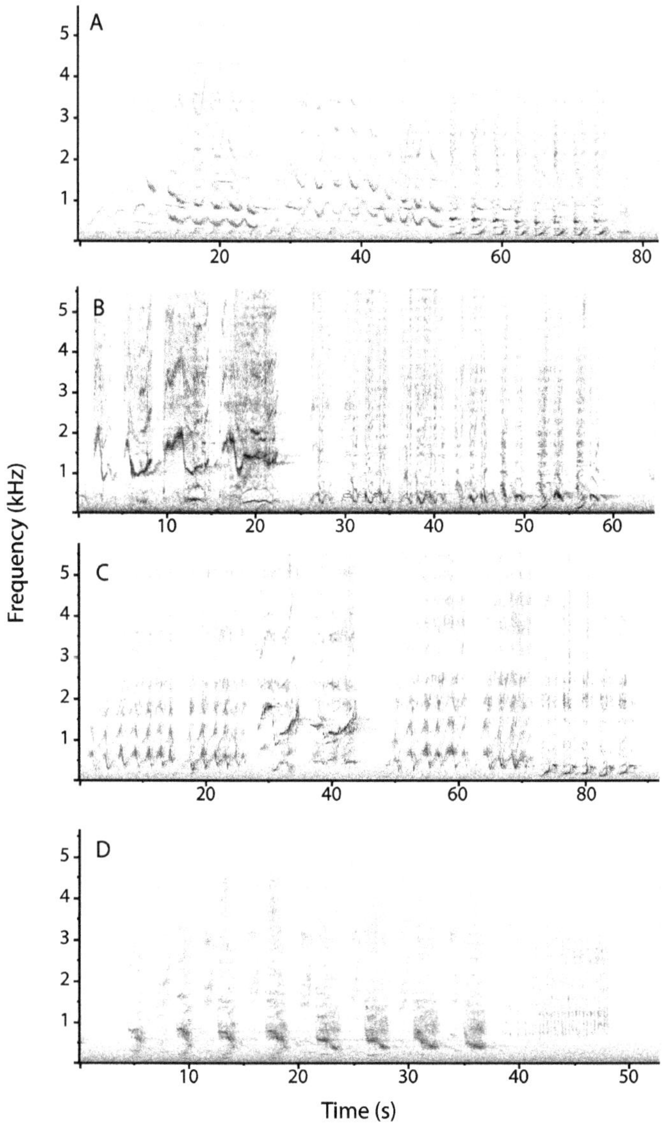

Fig. 12.4 Examples of single iterations of four different complex songs recorded in Fram Strait. **a** 5-unit song recorded January 19, 2013; **b** 8^+-unit song recorded October 23, 2012; **c** 4-unit song December 15, 2012; **d** 3-unit song recorded November 19, 2012 (2048 pt. FFT, 50% overlap, Hann window)

Fig. 12.5 Total number of simple (gray) and complex (black) song types for 2010–11, 2012–13, and 2013–14 from Fram Strait. Whiskers show the minimum and maximum number of song types recorded by month over the three years

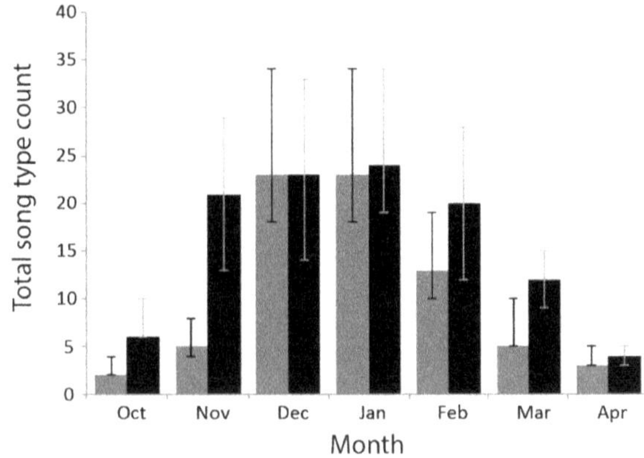

share song types, and the extent of song sharing within populations; for example, how often do different singers share units or phrases (Fig. 12.6); how do songs "evolve" throughout the singing season from late October through the following April or May; do simple versus complex songs serve different biological functions (or is this division simply a human construct?); how do song characteristics (e.g., song duration, number of units, the use of two-voiced singing) differ between different populations? As sea ice continues to decline in the Arctic, essentially opening borders between

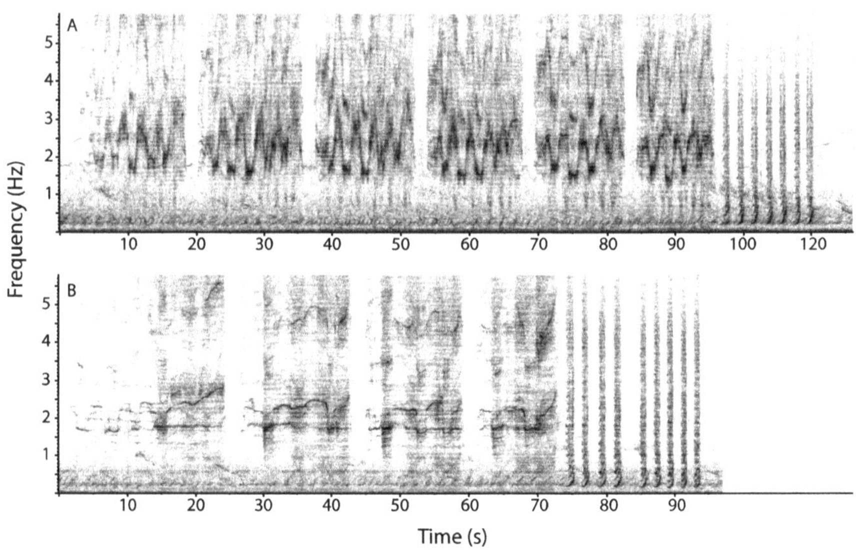

Fig. 12.6 Example of two songs recorded 10 h apart in the northern Bering Sea. Each song begins with distinctly different high frequency-modulated and amplitude-modulated long units but ends with multiple repetitions of the same unit. **a** Song recorded on March 6, 2018; **b** song recorded on March 7, 2018 (4096 pt. FFT, 50% overlap, Hann window)

populations, will we hear (and could we discern) cultural transmission of song types between western and eastern Arctic populations?

12.3 Reproductive Tactics in the Right Whales: Why Do Bowheads Sing?

The only other genus of Balaenidae is the right whale (*Eubalaena* spp.). Bowheads and right whales have similar feeding ecologies, morphologies, acoustic call types, and, in the males, extremely large testes (Clark and Johnson 1984; Brownell and Ralls 1986; Chap. 13). Unlike bowhead whales, right whales are not known to sing, although a recent study of a remnant population composed of >70% males in the Bering Sea has attributed song characteristics to right whale gunshot sequences (Crance et al. 2019). Female right whales in surface active groups (SAGs), produce sequences of sounds called "screams" or "high calls" which attract one or more males to the female (Clark 1982; Parks 2003; Parks and Tyack 2005). These SAGs are hypothesized to play a role in reproduction, as copulation is frequently observed, sometimes including more than one male copulating with a single female at a time (Cummings et al. 1974; Mate et al. 2005; Chap. 13). As right whale SAGs occur in many months of the year and can involve combinations of whales of different ages and sex, including juvenile animals, this behavior likely serves multiple purposes (Payne 1986; Best et al. 2003; Parks et al. 2007). It is worth noting that North Atlantic (*E. glacialis*) and South Atlantic (*E. australis*) right whales have been acoustically monitored for decades with no reports of singing behavior. While it is possible, it seems unlikely that such behavior would have been missed by researchers studying the sounds of right whales. Similar to right whales, bowheads aggregate in groups that often are social and/or sexual in nature (Würsig and Clark 1993). Like right whales, in bowheads sperm competition has been proposed as a reproductive tactic (Brownell and Ralls 1986). Unlike right whales, however, bowhead whales sing profusely from late fall into spring.

Song throughout the animal kingdom is often considered to be a male reproductive tactic, produced to mediate male–male interactions and provide information to females about the physical quality or social status (fitness) of the singer (e.g., Clutton-Brock and Albon 1979; Catchpole and Leisler 1996; Davidson and Wilkinson 2004; Byers and Kroodsma 2009; Chabout et al. 2015; Cholewiak et al. 2018; Clark et al. 2019). Although available data are limited, bowheads (and other baleen whales) show clear annual seasonal peaks in levels of testosterone in their baleen (Hunt et al. 2018). It seems reasonable to hypothesize that the seasonality of these peaks should match the seasonality of singing behavior and that changes in testosterone drive the seasonal changes in song as observed in birds (Galeotti et al. 1997; Smith et al. 1997; Brenowitz 2004). Since bowhead whales both sing and have large testes, this suggests that bowheads employ inter- and intra-sexual reproductive tactics and have multiple mating strategies. But why?

One clue may lie in the Arctic habitat of the bowhead whale. This species spends most of its life associated with sea ice, particularly during the time of year that singing occurs (October–April) and when bowheads occur in >90% sea ice cover (Stafford et al. 2012). Sea ice is dynamic, and open water cracks and leads are constantly shifting. This habitat is clearly not conducive to the formation or maintenance of SAGs, nor to escalation of male-male competition as has been documented in humpback whales (Tyack and Whitehead 1982). Therefore, song may be the most efficient, or only, reproductive tactic under heavy ice conditions. Not only might song convey the location of potential mates and their overall fitness, perhaps this dynamic environment has also led to song serving to advertise a different resource to females (e.g., Croll et al. 2002), which in this case would be open water, i.e., oxygen to breathe. It has long been assumed that some of the longer time gaps in cetacean songs (sometimes called "rests," Fig. 12.2a) occur when animals surface to breathe (e.g., fin whales: Watkins et al. 1987; Clark et al. 2019; Clark pers. comm.). Perhaps a singer's pattern of singing, and not singing, relays information about the ice environment overhead. As leads open in bowhead whale habitat and whales begin to migrate in the late winter and early spring when sea ice cover becomes looser, sperm competition in sexually active groups may become a more straightforward reproductive strategy than singing. This speculation about the drivers of multiple mating strategies in bowhead whales is untested, and perhaps untestable, but highlights one of the many areas where advances in our understanding of bowhead whale ethology can be made.

12.4 Why Such Intra-annual Diversity in Song?

One of the most interesting puzzles with regard to bowhead whale singing behavior as we currently understand it, is why does song vary so much within and between years and within and between individual bowheads? Song diversity could be linked to population size, if the number of different songs in a season is related to the number of male whales in a population. During the bowhead whale census efforts off Utqiaġvik, Alaska in the 1980s–2000s only single song types were identified during the spring migration (Würsig and Clark 1993). More recent data from this region clearly show multiple songs recorded during spring migration and in the fall (Delarue et al. 2009; Johnson et al. 2014). This increase in song types may in some way be related to a consistent increase in the size of this population of bowhead whales (George et al. 2004; Givens et al. 2016), although the extremely high song diversity from the Spitsbergen population, which may number in the low hundreds, seems to contradict this explanation (Vacquié-Garcia et al. 2017; Stafford et al. 2018). The increase in song diversity could also be a result of immigration of whales from one population to another, if songs are population specific. The three high-Arctic populations, BCB, ECWG, SP were geographically isolated from each other for millennia due to extensive, impenetrable sea ice cover in the High Arctic, but recent extreme reduction in sea ice extent and thickness has permitted at least a few animals from two of these populations to migrate between regions with the possibility of introducing novel acoustic behaviors (Heide-Jørgensen et al. 2012). However, this does not adequately explain the high diversity in song types sung each season nor the lack of recurrence of songs from one year to the next.

While some bowhead whales in a population appear to share the same song within a year (Clark 1990; Johnson et al. 2014), others do not, as evidenced by the overlapping singing of different songs from Greenland, Spitsbergen, and Alaska (Stafford et al. 2008, 2018; Tervo et al. 2009; Johnson et al. 2014). Whether individuals switch songs within a season, between years, or perhaps choose which song they sing from a repertoire of songs, is currently unknown (Würsig and Clark 1993; Stafford et al. 2008, 2012, 2018; Johnson et al. 2014). Perhaps long-term cultural innovation in songs, or song novelty, might somehow confer a reproductive advantage to the singer as an indicator of male quality (e.g., Noad et al. 2000; Laiolo et al. 2008; Byers and Kroodsma 2009; Chabout et al. 2015). It has been hypothesized that song characteristics such as song duration, types of phrases used, or amplitude of the song might serve as indicators of a humpback singer's physical condition (e.g., Chu and Harcourt 1986; Frankel et al. 1995). Likewise, in fin whales the combination of swimming speed and amount of singing has been proposed as an indicator of a singer's physical stamina (Clark et al. 2019). For bowheads, perhaps a singer's quality can be assessed by other bowheads via song complexity under the ice.

Innovation in song could also be driven by the acoustic habitat in which bowheads sing. Singing peaks in the middle of winter under heavy ice cover and the polar night and overlaps with times when other Arctic marine mammal species are vocalizing (Stafford et al. 2012, 2018). Perhaps novelty in bowhead song is in response to the dynamic abiotic sounds of ice and less variable sounds of other marine mammals, including walrus, belugas, and bearded seals (Clark et al. 2015). Although some bowhead whale sounds are quite distinct from these other Arctic endemic species, at times, bowhead whales seem to be imitating said species (beluga whales in particular). Likewise, units of some songs sound so much like some sounds produced by ice that only the clear pattern and repetition of those units convinces the listener that they are, in fact, produced by a singing bowhead. What purpose this might serve is a mystery, but the bowhead whale's vocal plasticity strongly suggests that this species has a remarkable ability to learn and reproduce novel sounds in their acoustic environment. Bowhead whales are constantly innovating new songs and new song units, they imitate conspecific calls and even appear to mimic other Arctic species and ice noise (Clark 1990; Rendell and Whitehead 2001).

The singing behavior of bowhead whales, a species that lives in tight association with sea ice, certainly suggests that there has been strong evolutionary pressure, possibly via sexual selection, driving extreme complexity in song. The very environment in which they live, and the polar night during which they sing, has in the past limited so much of our understanding and interpretation of bowhead whale acoustic behavior to those few locations and times of year when they can be studied. The technology for studying polar environments is rapidly evolving, and, as bowhead populations recover from whaling, there is tremendous potential for new discoveries about the acoustic ecology, especially the singing behavior, of the bowhead. What is presently known about singing behavior is only "the tip of the iceberg" for this polar whale.

References

Aldrich HL (1889) Arctic, Alaska, and Siberia, or eight months with the Arctic whalemen. Rand McNally Press, Chicago, IL
Allen JA, Garland EC, Dunlop RA, Noad MJ (2019) Network analysis reveals underlying syntactic features in a vocally learnt mammalian display, humpback whale song. Proc R Soc B Biol Sci 286:20192014
Archer FI, Rankin S, Stafford KM, Castellote M, Delarue J (2020) Quantifying spatial and temporal variation of North Pacific fin whale (*Balaenoptera physalus*) acoustic behavior. Mar Mamm Sci 36:224–245
Baird AB, Bickham JW (2020) The stocks of bowheads. In: George JC, Thewissen JGM (eds) The bowhead whale: *Balaena mysticetus*: biology and human interactions. Elsevier, pp 19–29
Best PB, Schaeff CM, Reeb D, Pallsbøll PJ (2003) Composition and possible function of social groupings of southern right whales in South African waters. Behaviour 140:1469–1494
Blackwell SB, Richardson WJ, Greene CR, Streever JAB (2007) Bowhead whale (*Balaena mysticetus*) migration and calling behaviour in the Alaskan Beaufort Sea, autumn 2001–04: an acoustic localization study. Arctic 60:255–270
Blackwell SB, Thode AM, Conrad AS, Ferguson MC, Berchok CL, Stafford KM, Marques TA, Kim KH (2021) Estimating acoustic cue rates in bowhead whales, *Balaena mysticetus*, during their fall migration through the Alaskan Beaufort Sea. J Acoust Soc Am 149:3611–3625
Bockstoce JR (1986) Whales, ice and men. University of Washington Press, Seattle
Brenowitz EA (2004) Plasticity of the adult avian song control system. Ann N Y Acad Sci 1016:560–585
Brownell R Jr, Ralls K (1986) Potential for sperm competition in baleen whales. Rep Int Whaling Comm Spec Issue 8:97–112
Byers BE, Kroodsma DE (2009) Female mate choice and songbird song repertoires. Anim Behav 77:13–22
Castellote M, Clark CW, Lammers MO (2011) Fin whale (*Balaenoptera physalus*) population identity in the western Mediterranean Sea. Mar Mamm Sci 28:325–344
Catchpole CK, Leisler B (1996) Female aquatic warblers (*Acrocephalus paludicola*) are attracted by playback of longer and more complicated songs. Behaviour 133:1153–1164
Chabout J, Sarkar A, Dunson DB, Jarvis ED (2015) Male mice song syntax depends on social contexts and influences female preferences. Front Behav Neurosci 9:1–16
Chittleborough RG (1955) Aspects of reproduction in the male humpback whale, *Megaptera nodosa* (Bonnaterre). Mar Freshw Res 6:1–29
Cholewiak DM, Cerchio S, Jacobsen JK, Urbán RJ, Clark CW (2018) Songbird dynamics under the sea: acoustic interactions between humpback whales suggest song mediates male interactions. R Soc Open Sci 5:171298
Chu K, Harcourt P (1986) Behavioral correlations with aberrant patterns in humpback whale songs. Behav Ecol Sociobiol 19:309–312
Citta JJ, Quakenbush L, George J, Small R, Heide-Jørgensen M, Brower H, Adams B (2011) Winter movements of bowhead whales (*Balaena mysticetus*) in the Bering Sea. Arctic 1–64

Clark CW (1982) The acoustic repertoire of the southern right whale, a quantitative analysis. Anim Behav 30:1060–1071

Clark CW (1990) Acoustic behaviour of mysticete whales. In: Thomas J, Kastelein R (eds) Sensory abilities of cetaceans. Plenum Press

Clark CW, Johnson JH (1984) The sounds of the bowhead whale, *Balaena mysticetus*, during the spring migrations of 1979 and 1980. Can J Zool 62:1436–1441

Clark CW, Brown LM, von der Heydt K, Baggeroer A, Dyer I (1990) Songs of the bowhead whale from the western Arctic and Spitsbergen stocks. In: Fifth conference on the biology of the bowhead whale *Balaena mysticetus*, Anchorage, AK, 1–3 Apr 1990, pp 209–212

Clark CW, Berchok CL, Blackwell SB, Hannay DE, Jones J, Ponirakis D, Stafford KM (2015) A year in the acoustic world of bowhead whales in the Bering, Chukchi and Beaufort Seas. Prog Oceanogr 136:223–240

Clark CW, Gagnon GJ, Frankel AS (2019) Fin whale singing decreases with increased swimming speed. R Soc Open Sci 6:180525

Clutton-Brock TH, Albon SD (1979) The roaring of red deer and the evolution of honest advertisement. Behaviour 69:145–170

Crance JL, Berchok CL, Wright DL, Brewer AM, Woodrich DF (2019) Song production by the North Pacific right whale, *Eubalaena japonica*. J Acoust Soc Am 145:3467–3479

Croll DA, Clark CW, Acevedo A, Tershy B, Flores S, Gedamke J, Urban J (2002) Only male fin whales sing loud songs. Nature 417:809–809

Cummings W, Holliday D (1987) Sounds and source levels from bowhead whales off Pt. Barrow, Alaska. J Acoust Soc Am 82:814–821

Cummings W, Thompson PO, Fish JF (1974) Behavior of southern right whales: R/V Hero cruise 72-1. Antarct J US 9:33–38

Darling JD, Acebes JMV, Frey O, Urbán RJ, Yamaguchi M (2019) Convergence and divergence of songs suggests ongoing, but annually variable, mixing of humpback whale populations. Sci Rep 9(7002):1–14

Davidson SM, Wilkinson GS (2004) Function of male song in the greater white-lined bat, *Saccopteryx bilineata*. Anim Behav 67:883–891

Delarue J, Laurinolli M, Martin B (2009) Bowhead whale (*Balaena mysticetus*) songs in the Chukchi Sea between October 2007 and May 2008. J Acoust Soc Am 126:3319–3328

Frankel AS, Clark CW, Herman L, Gabriele CM (1995) Spatial distribution, habitat utilization, and social interactions of humpback whales, *Megaptera novaeangliae*, off Hawai'i, determined using acoustic and visual techniques. Can J Zool 73:1134–1146

Galeotti P, Saino N, Sacchi R, Moller AP (1997) Song correlates with social context, testosterone and body condition in male barn swallows. Anim Behav 53:687–700

Garland EC, Goldizen AW, Rekdahl ML, Constantine R, Garrigue C, Hauser ND, Poole MM, Robbins J, Noad MJ (2011) Dynamic horizontal cultural transmission of humpback whale song at the ocean basin scale. Curr Biol 21:687–691

Gedamke J, Costa DP, Dunstan A (2001) Localization and visual verification of a complex minke whale vocalization. J Acoust Soc Am 109:3038–3047

George JC, Bockstoce JR (2008) Two historical weapon fragments as an aid to estimating the longevity and movements of bowhead whales. Polar Biol 31:751–754

George JC, Thewissen JGM (2020) Commercial whaling. In: George JC, Thewissen JGM (eds) The bowhead whale: *Balaena mysticetus*: biology and human interactions. Elsevier, pp 537–547

George JC, Clark C, Carroll GM, Ellison WT (1989) Observations on the ice-breaking and ice navigation behavior of migrating bowhead whales (*Balaena mysticetus*) near Point Barrow, Alaska, Spring 1985. Arctic 42:24–30

George JC, Bada J, Zeh J, Scott L, Brown SE, O'Hara T, Suydam R (1999) Age and growth estimates of bowhead whales (*Balaena mysticetus*) via aspartic acid racemization. Can J Zool 77:571–580

George J, Zeh J, Suydam RS, Clark C (2004) Abundance and population trend (1978–2001) of western Arctic bowhead whales surveyed near Barrow, Alaska. Mar Mamm Sci 20:755–773

Givens GH, Edmondson SL, George JC, Suydam R, Charif RA, Rahaman A, Hawthorne D, Tudor B, DeLong RA, Clark CW (2016) Horvitz-Thompson whale abundance estimation adjusting for uncertain recapture, temporal availability variation, and intermittent effort. Environmetrics 27:134–146

Hatch LT (2004) Male genes and male songs: integrating genetic and acoustic data in defining fin whale, *Balaenoptera physalus*, management units. Ph.D. dissertation, Cornell University, Ithaca, NY

Heide-Jørgensen M, Laidre K, Jensen M, Dueck L, Postma L (2006) Dissolving stock discreteness with satellite tracking: bowhead whales in Baffin Bay. Mar Mamm Sci 22:34–45

Heide-Jørgensen MP, Laidre K, Borchers D, Samarra F, Stern H (2007) Increasing abundance of bowhead whales in West Greenland. Biol Lett 3:577–580

Heide-Jørgensen MP, Laidre KL, Quakenbush LT, Citta JJ (2012) The northwest passage opens for bowhead whales. Biol Lett 8:270–273

Hunt KE, Lysiak NSJ, Matthews CJD, Lowe C, Fernández Ajó A, Dillon D, Willing C, Heide-Jørgensen MP, Ferguson SH, Moore MJ, Buck CL (2018) Multi-year patterns in testosterone, cortisol and corticosterone in baleen from adult males of three whale species. Conserv Physiol 6:coy049

Johnson HD, Stafford KM, George JC, Ambrose WG Jr, Clark CW (2014) Song sharing and diversity in the Bering-Chukchi-Beaufort population of bowhead whales (*Balaena mysticetus*), spring 2011. Mar Mamm Sci 31:902–922

Laiolo P, Vögeli M, Serrano D, Tella JL (2008) Song diversity predicts the viability of fragmented bird populations. PLoS ONE 3:e1822–e1822

Ljungblad DK, Thompson PO, Moore SE (1982) Underwater sounds recorded from migrating whales, *Balaena mysticetus*, in 1979. J Acoust Soc Am 71:477–482

Mate B, Duley P, Lagerquist B, Wenzel F, Stimpert A, Clapham P (2005) Observations of a female North Atlantic right whale (*Eubalaena glacialis*) in simultaneous copulation with two males: supporting evidence for sperm competition. Aquat Mamm 31:157–160

Noad MJ, Cato DH, Bryden MM, Jenner MN, Jenner KCS (2000) Cultural revolution in whale songs. Nature 408:537

Noongwook G, The Native Village of Savoonga, The Native Village of Gambell, Huntington HP, George JC (2007) Traditional knowledge of the bowhead whale (*Balaena mysticetus*) around St. Lawrence Island, Alaska. Arctic 60:47–54

Oleson EM, Calambokidis J, Burgess WC, McDonald MA, LeDuc CA, Hildebrand JA (2007) Behavioral context of call production by eastern North Pacific blue whales. Mar Ecol Prog Ser 330:269–284

Owen C, Rendell L, Constantine R, Noad MJ, Allen J, Andrews O, Garrigue C, Michael Poole M, Donnelly D, Hauser N, Garland EC (2019) Migratory convergence facilitates cultural transmission of humpback whale song. R Soc Open Sci 6:190337

Parks SE (2003) Response of North Atlantic right whales (*Eubalaena glacialis*) to playback of calls recorded from surface active groups in both the North and South Atlantic. Mar Mamm Sci 19:563–580

Parks SE, Tyack PL (2005) Sound production by North Atlantic right whales (*Eubalaena glacialis*) in surface active groups. J Acoust Soc Am 117:3297–3306

Parks SE, Clark CW, Tyack PL (2007) Short- and long-term changes in right whale calling behavior: the potential effects of noise on acoustic communication. J Acoust Soc Am 122:3725–3731

Payne R (1986) Long term behavioral studies of the southern right whale (*Eubalaena australis*). Rep Int Whaling Comm Spec Issue 10:161–167

Payne RS, McVay S (1971) Songs of humpback whales. Sci New Ser 173:585–597

Payne K, Payne R (1985) Large-scale changes over 19 years in songs of humpback whales in Bermuda. Z Tierpsychol 68:89–114

Rankin S, Barlow J (2005) Source of the North Pacific "boing" sound attributed to minke whales. J Acoust Soc Am 118:3346–3351

Rekdahl ML, Dunlop RA, Noad MJ, Goldizen AW (2013) Temporal stability and change in the social call repertoire of migrating humpback whales. J Acoust Soc Am 133:1785–1795

Rendell L, Whitehead H (2001) Culture in whales and dolphins. Behav Brain Sci 24:309–382

Risch D, Van Parijs SM, Siebert U (2014) Individual calling behaviour and movements of North Atlantic minke whales (*Balaenoptera acutorostrata*). Behaviour 151:1335–1360

Schoenfuss HL, Bragulla HH, Schumacher J, Henk WG, George JC, Hillmann DJ (2014) The anatomy of the larynx of the bowhead whale, *Balaena mysticetus*, and its sound-producing functions. Anat Rec 297:1316–1330

Smith G, Brenowitz E, Beecher M, Wingfield J (1997) Seasonal changes in testosterone, neural attributes of song control nuclei, and song structure in wild songbirds. J Neurosci 17:6001–6010

Stafford KM, Clark CW (2020) Acoustic behavior. In: George JC, Thewissen JGM (eds) The bowhead whale: *Balaena mysticetus*: biology and human interactions. Elsevier, pp 323–338

Stafford KM, Nieukirk SL, Fox CG (2001) Geographic and seasonal variation of blue whale calls in the North Pacific. J Cetac Res Manage 3:65–76

Stafford KM, Moore SE, Laidre KL, Heide-Jørgensen MP (2008) Bowhead whale springtime song off West Greenland. J Acoust Soc Am 124:3315–3323

Stafford KM, Moore SE, Berchok CL, Wiig Ø, Lydersen C, Hansen E, Kalmbach D, Kovacs KM (2012) Spitsbergen's endangered bowhead whales sing through the polar night. Endang Species Res 18:95–103

Stafford KM, Lydersen C, Wiig Ø, Kovacs KM (2018) Extreme diversity in the songs of Spitsbergen's bowhead whales. Biol Lett 14:20180056

Tarpley RJ, Hillmann DJ, George JC, Thewissen JGM (2020) Female and male reproduction. In: George JC, Thewissen JGM (eds) The bowhead whale: *Balaena mysticetus*: biology and human interactions. Elsevier, pp 185–211

Tervo OM, Parks SE, Miller LA (2009) Seasonal changes in the vocal behavior of bowhead whales (*Balaena mysticetus*) in Disko Bay, Western-Greenland. J Acoust Soc Am 126:1570–1580

Tervo OM, Christoffersen MF, Parks SE, Kristensen RM, Madsen PT (2011a) Evidence for simultaneous sound production in the bowhead whale (*Balaena mysticetus*). J Acoust Soc Am 130:2257–2262

Tervo OM, Parks SE, Christoffersen MF, Miller LA, Kristensen RM (2011b) Annual changes in the winter song of bowhead whales (*Balaena mysticetus*) in Disko Bay, Western Greenland. Mar Mam Sci 27:E241–E252

Thompson P, Findley L, Vidal O (1992) 20-Hz pulses and other vocalizations of fin whales, *Balaenoptera physalus*, in the Gulf of California, Mexico. J Acoust Soc Am 92:3051–3057

Tyack P, Whitehead H (1982) Male competition in large groups of wintering humpback whales. Behaviour 83:132–154

Vacquié-Garcia J, Lydersen C, Marques TA, Aars J, Ahonen H, Skern-Mauritzen M, Oien N, Kovacs KM (2017) Late summer distribution and abundance of ice-associated whales in the Norwegian High Arctic. Endang Species Res 32:59–70

Vu ET, Clark C, Catelani K, Kellar NM, Calambokidis J (2014) Seasonal blubber testosterone concentrations of male humpback whales (*Megaptera novaeangliae*). Mar Mamm Sci 31:1258–1264

Watkins W, Tyack P, Moore K, Bird J (1987) The 20-Hz signals of finback whales (*Balaenoptera physalus*). J Acoust Soc Am 82:1901–1912

Würsig B, Clark CW (1993) Behavior. In: Burns JJ, Montague JJ, Cowles CJ (eds) The bowhead whale. Special publication number 2. The society for marine mammalogy. Allen Press, Lawrence, KS, pp 157–199
Zeh JE, Clark CW, George JC, Withrow DE, Carroll GM, Koski WR (1993) Current population size and dynamics. In: Burns JJ, Montague JJ, Cowles CJ (eds) The bowhead whale. Special publication number 2. The society for marine mammalogy. Allen Press, Lawrence KS, pp 409–489

Chapter 13
Right Whales from North to South: Similarities and Differences in Acoustic Communication

Susan E. Parks

Abstract Right whales have a long history of interactions with humans, being arguably the first commercially hunted species of whale. There are currently three recognized species of right whales including the North Pacific right whale (*Eubalaena japonica)*, the North Atlantic right whale (*Eubalaena glacialis)*, and the southern right whale (*Eubalaena australis*). The conservation status of these three species varies. Both of the northern hemisphere species are endangered and have only a few hundred individuals remaining, while the southern hemisphere species have multiple growing breeding populations that include thousands of individuals. Studies of right whale behavioral ecology and acoustic communication started in the 1960s and 1970s with studies off Massachusetts for North Atlantic right whales, and off Argentina for southern right whales. Behavioral studies of the less accessible North Pacific right whales have lagged behind, with most information on their behavior being published in the past 20 years. All right whale species share similarities in their acoustic repertoires, with a stereotyped contact call, the upcall and a loud broadband impulsive sound, the gunshot, being described for all species. Beyond these stereotyped sound types, right whales all share a graded repertoire of signals that range from pulsive to tonal in structure. Sound type usage varies by the behavioral context of the communication, and sound types appear to serve similar behavioral roles in all three species. Studies across species suggest that there are differences in sound production that reflect the sex and age class of the signaling individual. Additional studies have demonstrated behavioral responses of right whales to exposure from noise in their environments. Emerging areas for future research include the expansion of the use of passive acoustic monitoring to aid in the conservation of these whales.

Keywords Right whale · *Eubalaena* · Baleen whale · Behavior · Endangered species · Acoustics · Passive acoustic monitoring

---

S. E. Parks (✉)
Department of Biology, Syracuse University, Syracuse, NY, USA
e-mail: sparks@syr.edu

C.W. Clark and E. C. Garland (eds.), *Ethology and Behavioral Ecology of Mysticetes*, Ethology and Behavioral Ecology of Marine Mammals,
https://doi.org/10.1007/978-3-030-98449-6_13

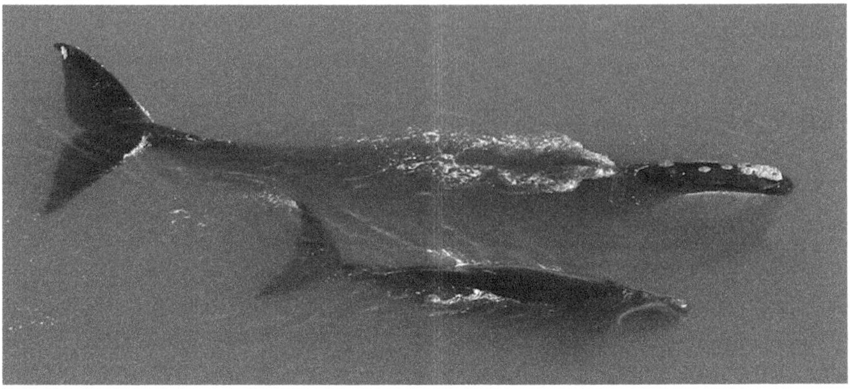

Aerial view North Atlantic right whale (*Eubalaena glacialis*, Eg #3860, Bocce) skim feeding alongside her 2021 calf in Cape Cod Bay on March 21, 2021. (*Photo Credit* Center for Coastal Studies, taken under NOAA permit #19,315-01)

13.1 Right Whale Species

Right whales comprise three recognized baleen whale species of the genus *Eubalaena*, the North Atlantic right whale (*E. glacialis*), North Pacific right whale (*E. japonica*), and southern right whale (*E. australis*) (Kenney 2018). Right whales have a global distribution and are present in most of the oceans of the world, including the Atlantic, Pacific, Indian, and southern oceans, and occasional sightings in the Arctic and tropical waters (Harcourt et al. 2019). These three species are very similar in external appearance. Adult right whales are large, 14 m in length, weighing between 36,000 and 72,000 kg with long plates of baleen that are an adaptation for filter feeding (Omura 1958; Kraus and Rolland 2007; Chap. 5). Right whales are characterized by stout bodies that lack a dorsal fin, distinct wide paddle shaped flippers, and keratinized growths referred to as callosities on the rostrum that provide individually distinct patterns and allow for distinguishing among individuals (Hamilton et al. 2007 and Fig. 13.1). Generally, individuals have black skin, with varying amounts of white patches on the body, often on the underside and/or chin. Species vary in the occurrence and size of dorsal and ventral white patches, with southern right whales having more gray or leucistic individuals with white on their dorsal side than the two northern hemisphere species (Omura et al. 1969; Schaeff 1999; Kraus and Rolland 2007).

Right whales were one of the first baleen whales to be described in some detail, as they were among the first species that hunted for meat, oil and baleen, given their nearshore distributions in coastal waters, positive buoyancy, high commercial value, and "docile" behavior (Scoresby 1820; Ellis 2018; Norris 1967; Reeves et al. 2007). Due to whaling pressure, starting as early as the eleventh century from Basque whalers (Aguilar 1986), all right whale species around the globe were reduced to

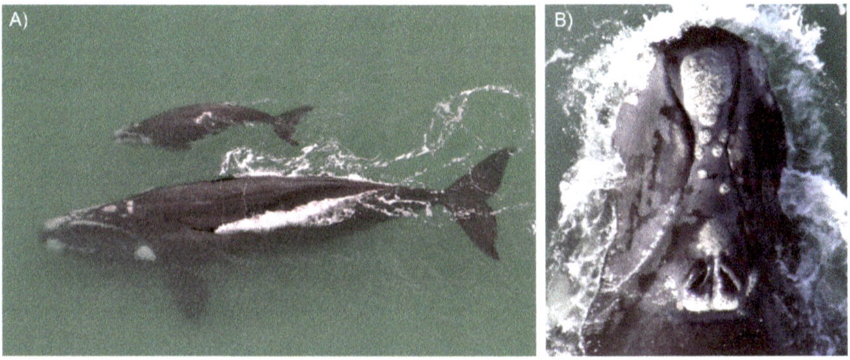

Fig. 13.1 Aerial images of southern right whales photographed off Brazil. **a** Mother-calf pair, Photographer: Karina Groch.; **b** Close up of callosities on the rostrum. Photographer: Paulo A.C. Flores

low numbers by the early twentieth century (IWC 2013; Crespo et al. 2019). Right whales were among the first cetacean species to be protected globally from whaling, through the Convention for Regulation of Whaling in 1931 by the League of Nations (Harcourt et al. 2019). Since that period of protection, for over 90 years, the three separate right whale species have shown variable levels of recovery, most likely due to varying levels of human-caused mortality in the form of illegal hunting, collisions with vessels, and entanglements in fishing gear. All three right whale species are listed as endangered under the US Endangered Species Act. Based on the International Union for Conservation of Nature (IUCN) redlist, *E. glacialis* are listed as critically endangered, *E. japonica* as endangered (Cooke and Clapham 2018a; Cooke 2020), and *E. australis* listed as a species of least concern due to their increasing population size over the past 40 years (Cooke and Zerbini 2018b). The North Atlantic right whale is the most endangered of the three species with a current estimate of fewer than 415 individuals (Pettis et al. 2020). Despite some increase in numbers for this species in the early 2000s, North Atlantic right whale reproduction has seen an abrupt decline since 2014, with no calves born in 2018 (Pettis et al. 2020). This collapse in reproduction, coupled with sustained and increasing levels of incidental human-caused mortality, has raised grave concerns about the future sustainability of this species (Corkeron et al. 2018; Davies and Brillant 2019). The North Pacific right whales are also endangered, with an estimated few hundred remaining, including approximately only 30 individuals in the separate breeding eastern stock in the Bering Sea (Wade et al. 2011) with only three calves sighted in the past century (Wade et al. 2006). In contrast, southern right whales, which have at least five distinct breeding populations, have shown a much stronger recovery in population size after whaling than northern hemisphere species. Some of these southern right whale populations have shown periods of rapid growth rate over the past twenty years, with estimated annual population growth rates between 7 and 13%/year in the South Atlantic, Southwest Pacific and Australia (Best et al. 2001; Groch et al. 2005; Carroll et al. 2013; Stamation et al. 2020). Other southern right whale breeding populations remain small, with limited data on population size (e.g., eastern South Pacific) (Vernazzani et al. 2014). The overall southern hemisphere population size was estimated to be at approximately 13,900 whales in 2009, with some evidence of slowing population growth in recent years (IWC 2013; Crespo et al. 2019).

Right whales were also among the first species of baleen whales to receive extensive scientific study of their behaviors and communication. Despite small populations, right whales are predominantly a coastal species, with individuals easily sighted from shore in the North Atlantic and southern hemisphere, particularly during their calving season. Some of the earliest detailed descriptions of right whale acoustics and behavior were published in the 1960s and 1970s for North Atlantic right whales (Schevill and Watkins 1962), and southern right whales (Payne and Payne 1971; Cummings et al. 1972). Behavioral descriptions for the North Pacific right whales have lagged behind the other right whale species, due to low numbers of individuals and challenging offshore conditions making their habitat more difficult to access. Descriptions for North Pacific right whale acoustic behavior did not begin until the early 2000s, almost 50 years after the first acoustic behavioral descriptions from other right whale species.

Due to this combination of conservation concerns and accessibility issues, the three right whale species have received variable levels of research attention. The North Atlantic right whale, while an endangered species, has been extensively researched on both the feeding and calving grounds (e.g., Kraus et al. 1986; Winn et al. 1986; Kraus and Rolland 2007). This has been driven by major conservation concerns for whales off the coasts of the USA and Canada combined with their nearshore habitat, which makes them accessible for study throughout most of their range. Southern right whale populations have primarily been studied on the calving grounds, due to the more remote offshore nature of their feeding grounds in the South Atlantic and southern oceans. Foundational studies of right whale behavior and acoustics were conducted on the Southwest Atlantic population in Argentina (e.g., Payne and Payne 1971; Cummings et al. 1972; Clark 1982; Taber and Thomas 1982). Extensive studies of the Southeast Atlantic population off South Africa started soon after (Donnelly 1967; Best 1990), though more limited data on their acoustic behavior are available (Hofmeyr-Juritz and Best 2011). Studies of southern right whales from other habitats have expanded with their growing populations in the past decade, with several new studies of behavior and acoustic communication coming out of other populations in Australia, New Zealand and Chile (Webster et al. 2016; Carroll et al. 2016; Jacobs et al. 2019; Nielsen et al. 2019). North Pacific right whales, due to their inaccessibility and low numbers, are relatively less known from detailed behavioral and acoustic studies compared to the other two species. Very few individuals in this population have been observed in recent years, particularly in the Eastern Bering Sea population, though several studies of their acoustic behavior are available (McDonald and Moore 2002; Munger et al. 2008; Rone et al. 2012). The following section summarizes the current state of the science related to the acoustic behavioral ecology of each of these species.

13.2 Right Whale Acoustic Repertoire

The earliest comprehensive acoustic repertoire description associated with behavior was published for southern right whales on the calving grounds in Argentina (Clark 1982, 1983). These studies were based on extensive behavioral observations and acoustic recordings taken over seven months in 1976–1977. A random subset of 1005 calls was used to describe a graded vocal repertoire along a continuum of quantifiable call features from tonal (i.e., "simple") to pulsive (i.e., "complex") in structure. Six different call types received categorical labels within the repertoire: up calls, down calls, constant calls, high calls, hybrid calls, and pulsive calls. Two additional sound categories included blows (exhalations) and slaps (Clark 1982). Within these different categories, there were some sound types that showed high stereotypy in their structure (up calls, down calls, constant calls and slaps), while others were highly variable within and between individuals (high calls, hybrid calls and pulsive calls). Call type usage varied by behavioral state, with the stereotyped up calls being produced more frequently by resting or traveling individuals, often within the context of use as a contact call, which Clark proposed. The highly variable call types were most often associated with higher energy social interactions (Clark 1983).

These first quantitative studies highlighted the challenge of classifying sounds into discrete call types. The intra and inter-individual variation in signal production, combined with a graded repertoire, reveals that the right whale sound production repertoire is composed of a rich assembly of sound types that cannot always be neatly separated into discrete call types. The variation in the usage of many of the sound types, as well, suggests that other variables, including timing, sequencing, emphasis, and intonation, also likely convey meaning to receivers. The full function of these signals in communication can only be understood through studies of the behavioral response whales to playbacks of different sound types.

Research studies of right whale acoustic behavior from other right whale populations and species have typically focused on the similarities found among the repertoires, with various attempts to characterize call types. Most studies have attempted to match call types to these original labels defined by Clark (1982), with variable success (Table 13.1). For the North Atlantic right whales, calls consistent with all of the basic categories described by Clark (1982) have been well described (Matthews et al. 2001; Vanderlaan et al. 2003; Parks and Tyack 2005; Parks et al. 2011b; Trygonis et al. 2013). For North Pacific right whales, only a subset of the full repertoire known for other right whale species has been described, including the upcall, down call, and constant calls (McDonald and Moore 2002; Munger et al. 2008; Širović et al. 2014; Crance et al. 2017). Additionally, a variant of the upcall, known as the "down-up" call (McDonald and Moore 2002) as well as gunshot sounds which correspond to underwater slaps (Rone et al. 2012; Crance et al. 2017) have been described. This "down-up" call variant has also been described from southern right whales (Dombroski et al. 2016) and highlights variation in the labeling of sounds produced along a gradient of call parameters. These findings suggest that North Pacific right whales likely

Table 13.1 Comparison of call types from Clark (1982), with call types described from other right whale species

Clark (1982) defined E. australis call types[a]	Clark (1982) described characteristics[a]	E. glacialis call types[b, c, d]	E. japonica call types[e, f, g, h, i]	Example spectrograms
Up calls	Tonal FM upsweeps 50–200 Hz, 0.5–1.5 s, FM rate 20–100 Hz/s	Upcall	Upcall	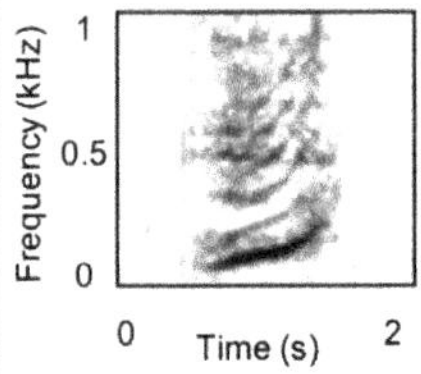
Down calls	Tonal FM downsweeps, 100–200 Hz, 0.5–1.5 s, FM rate −70–20 Hz/s	Down call	Down	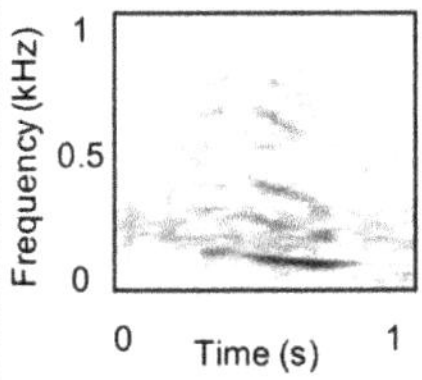
Constant calls	Tonal with little FM, 50–500 Hz, 0.5–6 s	Constant, LF Call, Foghorn	Constant	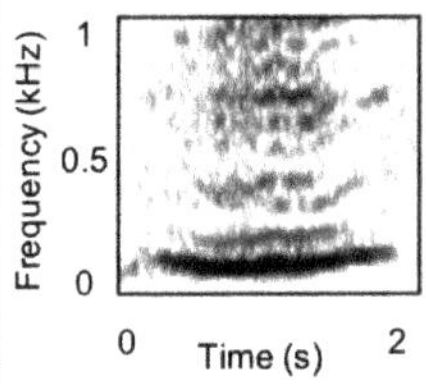
High calls	Tonal FM sweeps, 200–500 Hz, 0.5–2.5 s, FM rate-700–100 Hz/s	Moan, Scream, Warble		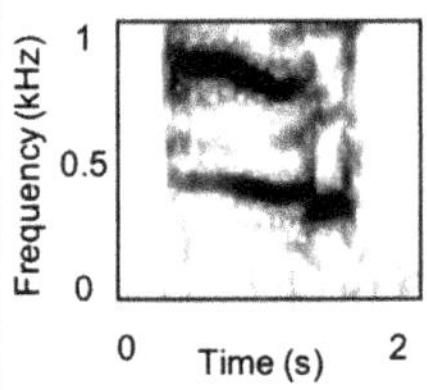
Hybrid calls	Complex mix of FM sweep and amplitude modulation, 50–500 Hz, 0.5–2.5 s, often pulsive at end	Moan, Scream, Hybrid		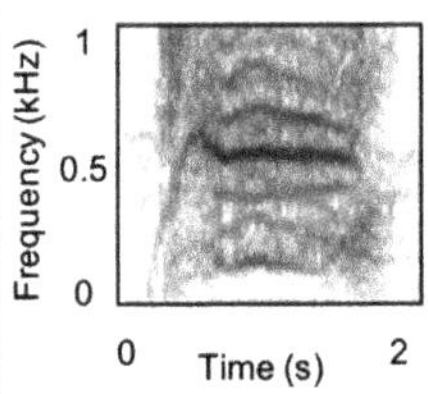

(continued)

Table 13.1 (continued)

Clark (1982) defined E. australis call types[a]	Clark (1982) described characteristics[a]	E. glacialis call types[b, c, d]	E. japonica call types[e, f, g, h, i]	Example spectrograms
Pulsive calls	Complex mixture with amplitude modulation of noise and/or an FM signal, 50–200 Hz, 0.5–3.5 s	Moan, Scream, Pulsive		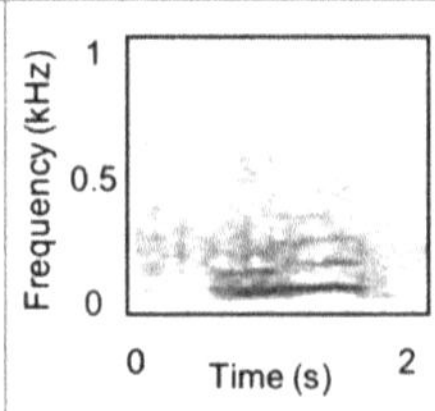
Blows	Noisy, broadband 100–400 Hz, 0.5–26 s, sometimes tonal like a long moan, sometimes noisy/pulsive	Blow		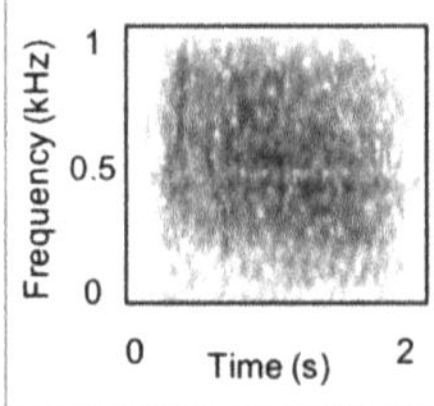
Slaps	Noisy, broadband sharp onset, 50–1000 Hz, 0.2 s, when produced underwater very intense	Gunshot	Gunshot	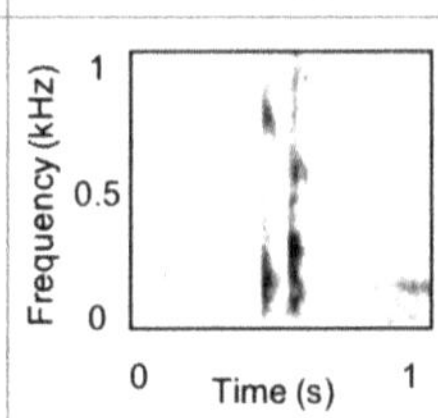

[a] Clark(1982), [b] Matthews et al. (2001), [c] Parks and Tyack (2005), [d] Trygonis et al. (2013), [e] McDonald and Moore (2002), [f] Mellinger et al. (2004), [g] Munger et al. (2008), [h] Širović et al. (2014), [I] Crance et al. (2017)

share much of their repertoire of sounds as the other two right whale species, but the behavioral contexts of some of the more complex sounds have not yet been described. Given the very small population size of North Pacific right whales, and all right whale species, it is likely that the currently observed behaviors may differ from ancestral behaviors due to decreases in population size resulting from commercial whaling. Therefore, it is worth noting that differences in acoustic behavior across right whale species could be a result of very limited genetic (and presumably behavioral) diversity in some species (Rosenbaum et al. 2000).

Whether there are true differences among the acoustic repertoires of these species is difficult to assess. For both North Atlantic right whales and North Pacific right whales, studies are typically published based on either a small subset of behavioral observations coincident with acoustic recordings (Matthews et al. 2001; McDonald and Moore 2002; Parks et al. 2005; Parks and Tyack 2005; Trygonis et al. 2013; Crance et al. 2017, 2019) or a large dataset of passive acoustic recordings for which no visual behavioral observations exist (Mellinger et al. 2004; Munger et al. 2008;

Van Parijs et al. 2009; Matthews et al. 2014; Soldevilla et al. 2014; Bort et al. 2015; Davis et al. 2017). For the long-term passive acoustic recordings, the potential for misidentification of some of the more variable call types as belonging to the repertoire of other species, most notably humpback whales (McDonald and Moore 2002; Waite et al. 2003; Clark et al. 2007; Mellinger et al. 2007a), limits the ability to confidently assign sounds to right whales. This has generally led to limiting long-term acoustic analyses to distinct sound types such as the more stereotyped upcall or gunshot signals (Mellinger et al. 2007a; Munger et al. 2008; Bort et al. 2015). This emphasis on only one or two distinct call types improves confidence for accurately detecting right whale presence from acoustic recordings. However, this approach does decrease the potential information about right whale ecology that can be extracted from these recordings. Previous studies have demonstrated seasonal and spatial variation in repertoire usage in right whales (Van Parijs et al. 2009; Matthews et al. 2014; Crance et al. 2019), and inclusion of additional call types in these analyses could provide valuable inference about variation in the behavioral activities of right whales in particular locations and different times of year. More studies that explore the full acoustic repertoire in association with behavioral, environmental, and oceanographic context for each species would be beneficial to make more direct comparisons of the sound produced by these three species and would allow for more insight into seasonal changes in behaviors in different habitats.

13.3 Seasonal and Spatial Variation in Sound Production

Behaviors and the related usage of acoustic signals vary seasonally and spatially in right whales. Right whale species undergo seasonal migrations where a portion of the population migrates between foraging and calving areas, providing both spatial and temporal separation between calving grounds and foraging grounds (Chap. 4 and Chap. 8). The following summarizes what is currently known in regards to sound production for each of the three species on the calving grounds and on the foraging grounds, emphasizing similarities, differences, and data gaps that warrant further exploration.

13.3.1 Southern Right Whale Acoustics on Calving Grounds

The early descriptions of the southern right whale acoustic repertoire were exclusively made from recordings in nearshore, shallow water habitats centered around the calving season from austral winter through austral summer in the Argentine population. The seasonal and spatial focus of these early studies could not capture the full behavioral repertoire of right whales across all seasons and over their entire home range; however, they provided fundamental data on right whale behavior on the calving grounds. More recently, studies have expanded to study additional southern right whale populations on other calving grounds (e.g., South Africa, Brazil, New Zealand and Australia) (Hofmeyr-Juritz and Best 2011; Webster et al. 2016; Dombroski et al. 2016; Nielsen et al. 2019). In addition to confirming similarities in acoustic repertoire to the early repertoire descriptions by Clark (1982) (Hofmeyr-Juritz and Best 2011; Webster et al. 2016; Dombroski et al. 2016), recent studies have provided additional insights into diel and seasonal trends (Dombroski et al. 2017; Rayment et al. 2017; Webster et al. 2019) in sound production on the calving grounds. Behavioral observations coupled with acoustic recordings (Hofmeyr-Juritz and Best 2011; Dombroski et al. 2016) and acoustic recording tags attached with suction cups (Nielsen et al. 2019; Dombroski et al. 2020) have been used to explore the acoustic communication of mother-calf pairs on the calving grounds in greater detail. Acoustic tag data from southern right whale mothers with young calves have provided evidence for a form of acoustic crypsis by nursing females (Nielsen et al. 2019; Dombroski et al. 2020). These studies reported low overall call rates of mothers and calves and call levels at relatively low amplitudes when compared to the ambient background noise levels associated with their nearshore distribution in the surf zone (Nielsen et al. 2019). Dombroski et al. (2020) also described a shift in repertoire by mothers with young calves, with very quiet paired pulsed sounds only detected from tags deployed on mothers with calves. The combination of lower call rates of long-distance communication sounds, combined with reduced call amplitude and a shift in repertoire to quieter call types, would make mothers with young calves more difficult for conspecifics or predators to detect.

Based on the protracted calving season and data from seasonal variation in fetal sizes from stranded animals, southern right whale gestation has been estimated to be between 12 and 13 months (Best 1994). Female right whales are conspicuously absent on the calving grounds the year prior to parturition, suggesting that conception takes place away from the calving grounds (Best 1994; Cole et al. 2013). Therefore, descriptions of sound production associated with conceptive mating behavior are unlikely to have been described by the acoustic studies on the calving grounds from multiple southern right whale populations (Clark 1982; Hofmeyr-Juritz and Best 2011; Webster et al. 2016; Dombroski et al. 2016). There is evidence from North Atlantic right whales that suggests that conception in this species currently occurs at higher latitudes, near the foraging grounds, rather than on the calving grounds for this species (Cole et al. 2013). While the small species population size may be a factor driving this observed behavior, similar observations regarding the absence of most females from the calving grounds the year before giving birth suggests that other right whale species may also have higher latitude or offshore breeding grounds that are separate from the calving grounds (Best 1994; Burnell and Bryden 1997; Cole et al. 2013). Therefore, comparative data of the repertoire from the calving grounds and other habitat areas may shed light on signals important for mating, as mating related signals are most likely to occur outside the nearshore calving grounds.

13.3.2 North Atlantic Right Whale Acoustics on the Calving Grounds

There is currently a single known calving ground for the North Atlantic right whales off the Southeastern USA in shallow coastal waters off Georgia and Florida (Kraus and Rolland 2007). Descriptions of acoustic behavior of North Atlantic right whales on the calving ground have lagged behind the southern right whale studies. In large part, this is due to the very low calving rates of North Atlantic right whales until the early 2000s (Corkeron et al. 2018). The first descriptions of acoustic behavior of North Atlantic right whales on the calving grounds came from a study describing acoustic behavior with concurrent behavioral observations of right whales in surface active groups (Trygonis et al. 2013) and long-term passive acoustic monitoring studies on the calving ground (Parks et al. 2009; Soldevilla et al. 2014). The term surface active group refers to interactions among two or more right whales at the surface, with frequent physical contact (Kraus and Hatch 2001). These groups were described for southern right whales on the calving grounds as courtship groups (Donnelly 1967; Payne 1995), due to the frequent observation of copulation of whales in the group at the surface. However, detailed analysis of group composition and related reproduction in southern right whales and North Atlantic right whales on the calving grounds revealed that females in these groups were unlikely to give birth the following year, but returned with calves two to three years later (Payne 1986; Best et al. 2003; Parks et al. 2007a). In southern right whales, females observed copulating in surface active groups off South Africa were never sighted in the following year with a calf, suggesting that the sexual activity of focal females in the groups did not generally result in conception (Best et al. 2003). For North Atlantic right whales, group compositions vary from groups of potentially reproductive animals (i.e., an adult female with at least one adult male) to non-conceptive groups of animals of all one sex (all male or all female) or groups of all juvenile animals (<4 years of age) (Parks et al. 2007a), indicating the behavioral functions of these groups likely vary and do not always involve fertilization.

The sounds described in the North Atlantic right whale surface active groups on the calving ground included all of the major call type categories described by Clark (1982). An additional call type, referred to as a "foghorn", was documented four times. This foghorn sound was a low frequency (35–45 Hz), long duration (6.5–9.2 s) signal that was produced by an individual at the surface, audible both in air and underwater (Trygonis et al. 2013). This sound likely corresponds to sounds included in the constant call category from Clark (1982), which included call durations of up to six seconds. These sounds have also been referred to as "moos" from observations in the Bay of Fundy, where nursing North Atlantic right whale mothers would produce these signals audible both in air and underwater immediately preceding reunion events with their calves (Parks and Clark 2007).

Additional studies of North Atlantic right whale acoustics on the calving ground have explored seasonal and diel trends in sound production from passive acoustic recordings (Soldevilla et al. 2014), descriptions of the acoustic behavior of mother-calf pairs through visual observations with a towed hydrophone array (Cusano et al. 2018), and through suction cup tags on the mother in mother-calf pairs and on lone pregnant or juvenile individuals (Parks et al. 2019a, b). Passive acoustic monitoring on the calving grounds showed a seasonal trend of right whale call detections that coincided with the peak in calving between November and March (Soldevilla et al. 2014; Davis et al. 2017). Behavioral focal follows and data from tags on individual whales on the calving ground indicated that mother-calf pairs produce calls at much lower rates than other juvenile or pregnant individuals (Cusano et al. 2018; Parks et al. 2019b), similar to behavioral descriptions from southern right whales (Nielsen et al. 2019; Dombroski et al. 2020). Analysis of acoustic tag data collected from North Atlantic right whales on the calving grounds indicated that mother-calf pairs produced previously undescribed very short, low amplitude pulsed sounds that were not recorded from juvenile or pregnant tagged whales. These data suggest that mother-calf pairs may use a different repertoire of very low amplitude sounds to maintain contact on the calving grounds, further reducing the risk of their detection through eavesdropping by other whales or predators (Parks et al. 2019a, b). Much of the behavioral focal follow data and tag data were collected during daylight hours, so night time call rates may increase for mother-calf pairs in this habitat; however, night time tag data from two mothers with calves also yielded very low call rates (Parks et al. 2019a).

13.3.3 North Pacific Right Whale Calving Grounds

Only two sightings of mother-calf pairs have been reported for North Pacific right whales in the past two decades (Wade et al. 2006), and the location of their calving grounds is unknown. Therefore, there are very limited data available regarding the acoustic behavior of North Pacific right whales on the calving grounds. It may be possible to predict suitable calving habitats for North Pacific right whales based on the characteristics of the calving grounds for North Atlantic and southern right whales, which would promote seasonal exploration of those areas in the winter and early spring (Good 2008). Despite low individual call rates by mother-calf pairs in southern and North Atlantic right whales (Parks et al. 2019a; Nielsen et al. 2019; Dombroski et al. 2020), passive acoustic monitoring of calving grounds suggest that long-term acoustic recordings can detect seasonal right whale presence in these habitats (Soldevilla et al. 2014; Webster et al. 2019).

13.3.4 Southern Right Whales on the Feeding Grounds

While the acoustic behavior of southern right whales has been extensively studied in multiple, separate breeding populations, much less is known about their acoustic behavior during the summer months on the feeding grounds or during migration between calving and feeding areas. The migratory corridors for southern right whales take many populations far off shore, where visual and behavioral observations are difficult to collect (William 1988; Bannister et al. 1999; IWC 2001; Valenzuela et al. 2009; Mate et al. 2011; Mackay et al. 2020; Chap. 4). As a result, data on the acoustic behavior while migrating to or from the feeding grounds or during social interactions on the foraging grounds are limited for most southern right whale populations. A study presented at an International Whaling Commission Scientific Committee meeting described upcall and gunshot sound detections based on concurrent visual and acoustic observations of 15 southern right whales on the feeding grounds near South Georgia (Jackson et al. 2018). Other published acoustic data for southern right whales on their migratory routes or feeding grounds are available from passive acoustic recordings with no concurrent visual data to relate sound types to the demographic or behavior variation of right whales (Širović et al. 2006). There is evidence that some southern right whales feed nearshore. One such area is off the western coast of South Africa with year-round right whale presence and evidence of feeding behavior during the summer months (Barendse and Best 2014). Opportunistic sightings of right whales feeding on plankton have also been reported in the austral spring on the calving grounds in Argentina (Hoffmeyer et al. 2010). These data suggest the potential to investigate and describe the acoustic behavior of southern right whales during nearshore foraging in more detail. Filling this data gap will better enable comparisons of acoustic behavior among right whale populations and species.

13.3.5 North Atlantic Right Whales on the Feeding Grounds

For North Atlantic right whales, most acoustic behavioral studies have focused on the feeding grounds from early spring into late summer (Matthews et al. 2001; Parks et al. 2005, 2011b; Parks and Tyack 2005). These studies indicate variation in repertoire usage and sound types when compared to sound types recorded on the calving grounds. The highest diversity of signal types described for North Atlantic right whales occurs during social interactions, or surface active groups (SAGs), which are most commonly sighted during summer (Parks et al. 2007a). The acoustic signals described for surface active groups of North Atlantic right whales on the feeding grounds include the full range of known right whale signal types, which in combination with behavioral evidence indicated that North Atlantic right whale females produce most of the tonal and pulsive sounds in SAGs (Kraus and Hatch 2001; Parks and Tyack 2005). Individuals of both sexes and all ages have been documented producing the upcall on the foraging grounds (Parks et al. 2011b). Additionally, stereotyped sequences of gunshot sounds, corresponding to underwater slaps described by Clark (1982), coupled with variable high and hybrid tonal calls were documented from behavioral observations of older (>12 y), sexually mature males in the late summer on the foraging grounds (Parks et al. 2005). Based on behavioral observations of the stereotyped, repetitive patterns of sound production during this acoustic behavior, it has been proposed that these gunshot signals are used as a form of male reproductive advertisement, either to attract females or compete with other males (Parks et al. 2005; Parks and Clark 2007).

Based on estimates of a 12–13 month gestation in right whales (Best 1994), North Atlantic right whale mating is thought to peak in the late autumn and early winter (November–February) (Cole et al. 2013). Limited behavioral observations are available based on concurrent visual and acoustic data during the autumn season for all right whale species, but data from long-term passive acoustic recordings of North Atlantic right whales have been used to determine migration timing during the full year along the US east coast (Davis et al. 2017). Aerial visual surveys for right whales have identified aggregations of right whales in the Gulf of Maine during the late autumn and early winter, suggesting this region may serve as the mating grounds for this species (Cole et al. 2013). Long-term passive acoustic recordings indicated a strong seasonal increase in gunshot sound production during the autumn and early winter months (Van Parijs et al. 2009; Matthews et al. 2014). The seasonal production of gunshot sounds peaks in late autumn (Matthews et al. 2014) coinciding with the presumed start of mating activity (Cole et al. 2013). These seasonal trends in gunshot sound production, coupled with production by adult males, support the hypothesis that the gunshot sound functions as a reproductive display in North Atlantic right whales (Parks et al. 2005). Mobile real-time passive acoustic survey technology deployed on gliders (Baumgartner et al. 2013, 2020) could be used to detect regions with high rates of gunshot sound occurrence during the boreal winter to aid visual surveys in locating the offshore mating grounds for North Atlantic right whales.

13.3.6 North Pacific Feeding Grounds

All current acoustic data for North Pacific right whales comes from the known feeding grounds in the Bering Sea and the Gulf of Alaska (McDonald and Moore 2002; Waite et al. 2003; Mellinger et al. 2004). The most commonly reported sound types were the upcall, the variant down-upcall (McDonald and Moore 2002) and the gunshot sound (Rone et al. 2012). Similar to the North Atlantic right whale, a seasonal increase in gunshot sound occurrence has been observed in the late summer and early autumn (Wright et al. 2019; Crance et al. 2019). Analysis of recordings of North Pacific right whales has revealed structured, patterned sequences of gunshot signal production combined with some tonal sounds. These stable patterned sequences of acoustic signals have been suggested to be a form of song in North Pacific right whales (Crance et al. 2019). Notably, this study identified similar timing patterns of gunshot sequences recorded on the same days at locations up to 700 km apart, indicating multiple individuals produced nearly identical patterned gunshot sequences (Crance et al. 2019). The stereotyped, rhythmic production of gunshot sounds by multiple individuals has only been described for North Pacific right whales but appears similar to the gunshot sound sequences described for North Atlantic right whales (Parks et al. 2005).

13.4 Differences in Sound Production Between the Sexes

Based on published acoustic and behavioral observations, both males and females appear to be capable of producing the entire currently documented right whale acoustic repertoire. The upcall, which functions as a contact call, appears to be used broadly by both sexes and across age classes (Parks et al. 2011b). The primary differences in sound production between the sexes appear to be in the seasonal production of signals thought to be produced in a reproductive context. In both the North Atlantic (Parks et al. 2005) and North Pacific (Crance et al. 2019), right whales documented producing long patterned sequences of gunshots that have been sexed have all been adult males. In southern right whales on the calving ground, males produced higher rates of gunshot sounds (Clark 1983). The seasonal timing of gunshot sound peak production aligns with the presumed breeding season, suggesting that these sequences of gunshots are in some way related to a reproductive context (Matthews et al. 2014; Crance et al. 2017). Both southern right whale (Clark 1983) and North Atlantic right whale (Parks et al. 2019a) females have been documented producing isolated gunshot signals on the calving grounds. In the North Pacific, an adult female has been observed producing a bout of multiple gunshot sounds in an unpatterned sequence (Crance et al. 2017). It is currently unknown whether the observation of a female producing bouts of gunshots is unique to North Pacific right whales, or simply a rare occurrence that has yet to be recorded in the other species.

Only adult North Atlantic females in surface active groups have been documented producing long bouts of complex acoustic tonal, pulsive, and hybrid signals, often collectively referred to as "screams"(Parks and Tyack 2005). The mating system for right whales has been hypothesized to involve sperm competition in fertilization (Brownell and Ralls 1986), based on the disproportionately large testes size of sexually mature males (Omura et al. 1969), and the observation of multiple males achieving intromission with a single female in surface active groups (Kraus and Hatch 2001; Mate et al. 2005). Broadcasts of recordings of the scream calls from North Atlantic right whale surface active groups on the feeding grounds were attractive to adult male whales, who approached the playback speaker and searched the vicinity for a prolonged period after the playback ended (Parks 2003). In contrast, playback of surface active group sounds recorded on the southern right whale calving grounds and played back on the North Atlantic right whale feeding grounds attracted both sexes and whales of all age classes (Parks 2003). Whether this difference in attractiveness was a result of seasonal differences in the behavioral function of surface active groups, or an incidental finding due to playing southern right whale sounds to a different species of right whale remains to be resolved.

Female right whales appear to modify their signal production behavior when nursing young calves. Studies of mother-calf acoustics on the calving grounds indicate low call rates by mother-calf pairs (Cusano et al. 2018; Nielsen et al. 2019) that are significantly lower than the call rates of juveniles or pregnant females in the same habitat (Parks et al. 2019a, b). Low call rates are thus the result of modified behavior by mothers with dependent young calves when compared to other individuals.

Females with dependent calves also produce very low amplitude single and double pulse sounds, much like a soft guttural grunt (Parks et al. 2019a). These sounds have only been detected using suction cup attached acoustic recording tags in both North Atlantic (Parks et al. 2019a) and southern right whales (Dombroski et al. 2020). Due to their low amplitude, these signals have not been detected from boat-based hydrophone recordings even at close ranges to mother-calf pairs (Parks et al. 2019a). This combination of reduced call rates of well-known signals in the right whale acoustic repertoire, along with the production of very quiet "whisper-like" signals that have not been recorded in other contexts, has been taken as evidence that right whale mothers may rely upon acoustic crypsis. Through reduced detectability of their sounds, mothers may be able to avoid drawing the attention of conspecifics or predators that may pose a risk to a young calf (Nielsen et al. 2019; Parks et al. 2019b).

Additional research is needed to explore the variation in acoustic repertoire usage by sex in all three species. General emergent trends include higher seasonal rates of gunshot sound production by adult males for North Atlantic and North Pacific right whales and a decrease in calling activity by females with young calves on the calving grounds for North Atlantic and southern right whales. Data gaps on gunshot sound production by southern right whales when not on the calving grounds may be the easiest to fill, given current ongoing studies off South Georgia (Jackson et al. 2018). A lack of calving in North Pacific right whales, with few calves seen in the past 50 years (Wade et al. 2006), make this data gap more challenging to fill. Other variations in calling behavior by sex likely exist but will require additional visual observations and concurrent acoustic data collection to reveal any trends.

13.5 Differences in Sound Production During Development

Vocal ontogeny of mammals typically involves the gradual development and refinement of repertoire with age. For example, many primates, including humans have a distinct "babbling" stage in infants that gradually resolve into the adult acoustic repertoire upon maturation (Elowson et al. 1998), with few species showing evidence of vocal learning (Janik and Slater 1997). A study examining variation in signal production in right whales from young calves to mature adults (>30 years of age) identified a similar pattern in the vocal development of North Atlantic right whales (Root-Gutteridge et al. 2018). North Atlantic right whale calves, less than five months of age, produce pulsive or hybrid signals that were significantly shorter in duration than adult signals and only gradually show production of more stereotyped tonal signals (Root-Gutteridge et al. 2018). Female North Atlantic right whale calves between seven and nine months of age recorded in surface active groups on the feeding grounds showed refined structure of calls that mimic the adult tonal signals but were generally higher in frequency and so referred to as "warbles"(Parks and Tyack 2005). One intriguing finding of the Root-Gutteridge et al. (2018) study was that refinement in the stability and duration of right whale calls appeared to continue past the time of maximum adult length and/or sexual maturity, with individuals >25 years of age showing greater vocal control than all younger animals (Root-Gutteridge et al. 2018). This may indicate that baleen whales, like humans (Sataloff et al. 1997), carry acoustic markers of age across their life spans.

Gunshot sound production in bouts also seems to be tied to age, with only mature adults visually observed while recordings were made. For North Atlantic right whales, only adult males >12 years of age have been recorded producing these signals in bouts (Parks et al. 2005). Genetic analyses of paternity in North Atlantic right whales between 1981 and 2001 indicate that no males <15 years old successfully fathered calves (Frasier et al. 2007). Physical reproductive maturity of right whales is generally thought to occur around eight years of age. Female right whales as young as five years have conceived and successfully produced calves, with an average age of first calving for females of ~eight years (Knowlton et al. 1994). Hormone analysis of North Atlantic right whales indicates that male reproductive hormone levels are significantly higher in males > nine years of age than juveniles between 2 and 9 years, suggesting that males are physiologically mature by age nine (Rolland et al. 2005). The age gap between females (8 years) and males (15 years) at first successful reproduction suggests there may be additional social or physical maturation required for males to successfully father a calf. Genetic analysis shows a relatively high reproductive skew in North Atlantic right whales, with a small number of males achieving more paternities than expected by chance, and a large number of males having fathered only zero or one calf (Frasier et al. 2007). These observations suggest that either male–male competition, female choice, or both affect male reproductive success. The production of gunshot bouts only by older males may indicate that this signal is related to the reproductive status of the individual males. These observations are consistent with the hypothesis that some aspect of additional physical development and/or social status plays a role in successful reproduction for males. This level of acoustic data linked to individual identity including known age and sex of the individual is currently limited to North Atlantic right whales, but future studies should explore this trend for known aged individuals in other right whales species.

13.6 Variation in Call Parameters with Background Noise

Right whale individuals match the frequency content of their signals to variations in background noise. Clark (1982) was the first to emphasize the correspondence between the average peak frequency of long-range communication signals and the local minimum in ambient background noise. This observation led to the hypothesis that selective pressures had resulted in the production of low-frequency signals that were matched to maximize propagation distance in natural background noise conditions (Clark and Ellison 2004; Clark et al. 2007). The frequency of the most broadly used stereotyped right whale call, the upcall, revealed significant differences in frequency range of the call between species. Notably, southern right whale upcalls were almost a full octave lower in pitch than those of North Atlantic right whales (Parks et al. 2007b). To explore whether this difference in frequency was due to species level differences arising gradually through time, or the result of differences in background noise conditions, recordings for both species were examined through time. For North Atlantic right whales, recordings obtained by William Schevill and William Watkins (from 1956) were compared with recordings obtained in 2000. Southern right whale recordings from Argentina were compared between 1977 and 2000. These comparisons revealed that in 1956, the starting frequency of North Atlantic right whale upcalls was indistinguishable from those of southern right whales in 1977 (Parks et al. 2007b). However, by 2000, North Atlantic right whale calls were shifted to higher frequencies. Southern right whales had also increased the average call frequency slightly, but not as much as the North Atlantic right whales (Parks and Clark 2007). Despite the long periods of time between these recordings in both species (30–40 years), these shifts represent population level behavioral changes through time. Given the long lifespan of this species, known to exceed 70 years (Hamilton et al. 1998), many of the same individuals would have been alive across the recording periods in both species. This indicates that individuals were changing their behavior, resulting in population level changes in the characteristics of their acoustic repertoire.

Antarctic populations of killer or orca whales (*Orcinus orca*) shift their call frequencies in the presence of biological background noise from chorusing leopard seals (Mossbridge and Thomas 1999). Studies of both southern and North Atlantic right whales indicate that matching the frequency of call parameters to the background noise is also behaviorally plastic in right whales, with individuals changing their calling behavior in response to the background noise they experience on a day-to-day basis. Recordings of southern right whales on the calving grounds in Brazil showed that the minimum start frequency of right whale upcalls shifted based on major changes in background noise. When low-frequency vessel noise was present, the average call start frequency shifted to a higher frequency band; during higher frequency fish choruses, the upcall frequency shifted to a lower frequency band (Parks et al. 2016) (Fig. 13.2). Similarly, over longer temporal and spatial scales, North Atlantic right whale upcall frequency differed between foraging habitats and the calving ground in predictable ways, with higher frequency calls produced in areas with higher levels of low-frequency noise from shipping (Parks et al. 2009). In both species, right whales appear to modify their upcall signal production to track the frequency band with a local minimum background noise level. This adaptive response is consistent with the hypothesis that right whales modify the characteristics of their upcalls to maximize communication space in the highly dynamic background noise found in the ocean environment.

While the studies of North Atlantic and southern right whales have mapped population level changes in call frequency content with background noise, one additional study has explored how an individual whale modifies its calling behavior in response to short-term changes in background noise. This study of short-term behavioral plasticity in individuals in response to changes in background noise was conducted using suction cup acoustic recording tags (Johnson and Tyack 2003). Individual

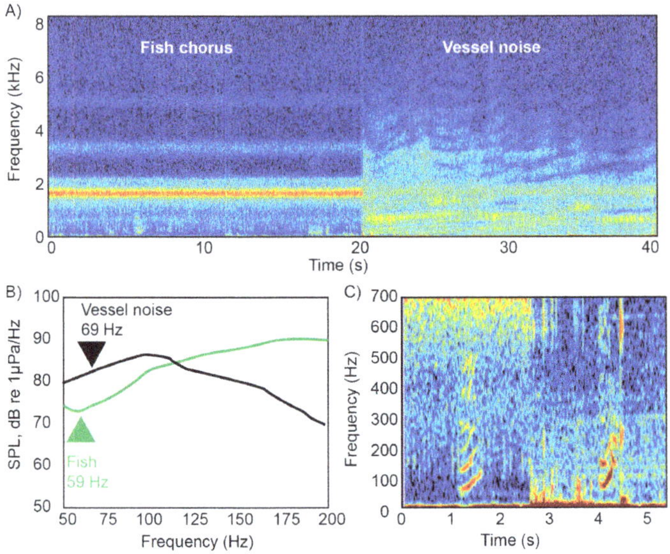

Fig. 13.2 **a** Spectrogram showing clips of fish chorus and vessel noise for southern right whales on the Brazilian calving grounds in 2011 (8 **kHz** sampling rate, Hann window, 512 pt. FFT, 50% overlap). **b** Mean spectrum noise level for southern right whales during a period of fish chorusing (black line) and vessel noise (green line) (Parks et al. 2016). Arrows matching the color of the lines mark the mean starting frequency for right whale long-distance upcalls in each habitat, showing a shift in the mean frequency in response to changing background acoustic conditions. **c** Spectrogram showing an upcall during a fish chorus (left) and ship noise (right) conditions (8 kHz sampling, Hann 1024, 50% overlap)

whales produced higher intensity calls in higher background noise level conditions, providing the first evidence of a Lombard effect (Lombard 1911; Hotchkin and Parks 2013) in a baleen whale (Parks et al. 2011a). Individual whales did modify the frequency content of their calls as a function of noise. Some individuals shifted call frequency higher and some shifted call frequency lower under the higher noise level condition (Parks et al. 2011a). This analysis did not consider the spectral distribution of the noise; for example, whether the frequency of the background noise was predominantly in a lower or higher frequency band than the upcalls. However, behavioral data from southern right whales in Brazil suggest that the direction of the behavioral shift reflected changes in the ambient background noise spectra (Parks et al. 2016), with whales shifting the frequency content of their calls to avoid the peak frequency of the background noise. In higher frequency noise conditions, the whales shifted the frequency content of their calls into a lower frequency band.

Observations of plasticity in vocal behavior by individual whales make it challenging to compare frequency, duration and amplitude of calls among different right whale species, as ambient background noise conditions are rarely included in descriptions of call types. Within a single species of right whale, the variation in upcall start frequency can be substantial. In a description of the mother-calf acoustic repertoire from southern right whales in Brazil, Dombroski et al. (2016) provided a comparison of the published upcall frequency and duration parameters for North Atlantic and southern right whales across different habitats. This comparison revealed an overlapping range of duration and frequency parameters that varied as much within a single species (southern right whales) as between species. It is likely that the observed variation in acoustic parameters in different right whale species and populations, to a large degree, relates to the ambient noise and propagation characteristics of the habitats where the recordings took place. It has been clearly demonstrated that individual right whales can vary the acoustic characteristic of their calls in response to noise (Parks et al. 2011a). Without information on the acoustic characteristics of background noise conditions, comparisons among the three species are not as informative about true underlying behavioral differences. Many of the reported differences in call characteristics may simply be a result of behavioral plasticity on the part of individuals matching their calls to their current background noise environment. An interesting approach would be to combine datasets from different species and populations and do a wholistic analysis, attempting to identify recordings with similar ambient background noise characteristics (e.g., frequency spectra, amplitudes) and comparing call features from those time periods. This would be a way to make robust comparisons of call characteristics (source level, frequency, and duration) of calls to identify which changes in parameters reflect short-term behavioral plasticity in response to changing environmental conditions vs. true variation in call parameters among species. This is a critical point to consider when analyzing acoustic repertoires of all species that show behavioral plasticity in response to ambient noise, particularly when acoustic parameters are proposed as a means by which to distinguish between subpopulations within a species.

13.7 Emerging Topics and Future Directions

With over fifty years of history in the study of right whale sound production, a surprisingly large number of questions remain regarding right whale acoustic communication behaviors. Many of my own experiences in the field hint at remarkable acoustic abilities in these species, including evidence for individual recognition of signals from other individuals and long-term memory of those calls. Based on observed responses to playbacks of right whale sounds, they have demonstrated an ability to locate a sound source many kilometers away and travel to get to within 100 m of the location a half-hour after the sound broadcast stops. All of the observations of their use of sound, both in terms of sound production and responses to sounds in the environment, hint at an acoustic awareness of the world around them that is very different than our own. While there are many lines of ongoing and fascinating research, one of the most significant and rapidly developing areas of research for right whale acoustics is the use of passive acoustic recordings to aid in right whale conservation by helping us understand where they go, how many whales are present, what they are doing, and how long they stay in an area.

The sounds produced by right whales can be utilized to aid in monitoring and protecting these endangered and threatened populations through passive acoustic monitoring (Clark et al. 2007; Van Parijs et al. 2009). In its simplest form, passive acoustic monitoring consists of detecting sounds to determine whale presence in a particular location at a particular time. Passive acoustic monitoring has been used to study all three right whale species and has provided valuable insights into their seasonal habitat usage and diel trends in acoustic signal production (McDonald and Moore 2002; Mellinger et al. 2007a; Munger et al. 2008; Širović et al. 2014; Davis et al. 2017; Rayment et al. 2017; Webster et al. 2019).

The technological complexity of passive acoustic monitoring of right whales has grown substantially in scope over the past sixty years. The earliest passive acoustic studies relied on direct recording from hydrophones deployed from vessels and from shore (Schevill and Watkins 1962; Payne and Payne 1971; Cummings et al. 1972; Watkins and Schevill 1972; Clark 1980, 1982) with direct observations of right whales present. The methods of hydrophone deployment from shore or vessel have remained a staple of acoustic behavioral data collection (Matthews et al. 2001; Parks et al. 2005; Dombroski et al. 2017). Technological developments in the 1990s allowed for development of archival acoustic recorders capable of recording for multi-day to multi-month deployments to detect right whale presence through detection of right whale calls (Gillespie and Leaper 2001; Waite et al. 2003; Mellinger et al. 2004, 2007a; Clark et al. 2007). These archival acoustic data recorders required that the recorder be recovered to access the data, allowing for a retrospective analysis of the acoustic behavior of right whales during the time the recorder was deployed. Long-term passive acoustic recordings within (Simard et al. 2019; Charif et al. 2020) and across (Davis et al. 2017) North Atlantic right whale habitats have revealed changes in the seasonal migratory timing of this species and demonstrated the substantial value of these long-term recording tools for understanding right whale behavior by "eavesdropping" on their acoustic communication.

More recently, technology has developed to allow for near-real-time detection of North Atlantic right whale upcalls. Near-real-time detectors record and detect right whale calls and then transmit the detection data back to analysts on shore either via cellphone or satellite technology (Spaulding et al. 2010; Baumgartner et al. 2013). The first successful demonstration of stationary detection units took place off the coast of Massachusetts to detect right whales in the shipping lanes off Boston (Clark et al. 2007; Van Parijs et al. 2009; Spaulding et al. 2010). These detections allowed for more rapid management responses to right whale seasonal presence and to alert captains of commercial vessels of potential right whales in the area. Additional near-real-time listening stations have been set up in known migratory and feeding habitats to expand the regional coverage of these types of detectors (Baumgartner et al. 2019). The latest technological developments now include mobile platforms of autonomous vehicles that allow for moving passive acoustic monitoring offshore. These devices can transmit detections of right whale calls back to shore in close to real time and be redirected while at sea to conduct passive acoustic surveys of a region remotely (Baumgartner et al. 2013, 2019; Meyer-Gutbrod et al. 2018). This suite of technologically advanced systems is making acoustics a viable real-time management tool for the study of right whale populations, to answer both basic biological questions and determine presence in high risk areas such as shipping lanes or active fishing areas for timely implementation of protective measures.

There is growing interest in gaining additional insight into right whale habitat usage from passive acoustic recordings beyond simple presence information. A primary focus has been to use passive acoustics to estimate the number of whales, or the density of whales present in a particular area. This research topic is referred to as acoustic density estimation (Marques et al. 2011). Visual surveys for marine mammals are the current standard for density estimation of marine mammal populations (Mellinger et al. 2007b) but are limited in temporal and spatial coverage. These visual survey limitations stem from a combination of high cost and dependence on daylight and suitable weather conditions to detect marine mammals at the ocean's surface. There have been several approaches to use passive acoustic recordings for density estimation developed for marine mammal systems (Marques et al. 2013). Broadly, these approaches can be categorized according to the type of object counted (individual animal, animal group, or "cue," i.e., a species specific sound) and the type of statistical method used to estimate how many objects are missed, also known as a false negative rate (e.g., plot sampling, distance sampling, spatial capture recapture, trial-based method or acoustic modeling).

While passive acoustic monitoring has been advantageous for identifying the occurrence of acoustically active right whales, passive acoustic density estimation of right whales is still in development. One major challenge in the acoustic density estimation of right whales has been the lack of data on vocal cue rates of individuals in all three species (Clark et al. 2010; Hofmeyr-Juritz and Best 2011; Marques et al. 2011). In Clark et al. (2010), acoustic detections of North Atlantic right whale upcalls within Cape Cod Bay were compared to the number of whales detected in coincident visual surveys using a total count method approach. Despite using a range of acoustic "windows" for acoustic sampling, little correlation was detected between the number of whales present and call rates (Clark et al. 2010). Marques et al. (2011), analyzed a small set of acoustic data from North Pacific right whales using right whale upcalls for a cue-counting distance sampling approach. A cue rate of ~1.7 upcalls/h/individual combined with an estimated range of detection resulted in an estimate of a population size of 25 whales (13–47), which is consistent with visual survey estimates for the species in this habitat (Wade et al. 2006, 2011). For southern right whales, one study examined call rates and the number of whales present in an area on the calving grounds in South Africa (Hoffmeyer-Juritz and Best 2011). Based on short-term boat-based recordings across the season, a complex relationship between total number of calls and total number of whales was detected, with an increase in call rate with increasing number of whales at relatively low densities, followed by a decrease in call rate at high densities of whales. When examining individual call types in southern right whales, upcalls and whale densities showed the strongest correlation, which is consistent with the findings of Marques et al. (2011) for North Pacific right whales. The stability of right whale cue rates across habitat, location, and time of day needs further refinement to allow for robust acoustic density estimation in this species and is an active area of research across all three species.

13.8 Closing Thoughts

The three recognized right whale species all have had different rates of "recovery" from over-exploitation from whaling, with North Atlantic and North Pacific right whale populations remaining extremely low almost a century after an end of legal commercial whaling. Coupled with this, all right whales face an uncertain long-term future due to steadily increasing anthropogenic threats from fishing activity, commercial shipping, and climate change (Harcourt et al. 2019). The North Pacific right whale species is on the brink of extinction, and it is questionable whether this species will be able to recover, given the risks of accidental mortality from entanglement in fishing gear and vessel strikes for this already tiny population (Wade et al. 2011). The eastern population of North Pacific right whales is so small, and individual right whales so long-lived, that it is arguable that this population is functionally extinct with only three calves documented in the past century (Wade et al. 2006). The North Atlantic right whale is facing a recent rapid decline in numbers due to a rapid collapse in reproduction coupled with a sharp rise in incidental death from shipping and fishing gear entanglements as whales have shifted their foraging distribution and migratory timing in response to large scale climate changes (Davis et al. 2017; Corkeron et al. 2018; Meyer-Gutbrod et al. 2018). This precipitous drop in population size is combined with evidence of declining overall health of individual whales based on visual health assessments (Pettis et al. 2004; Kraus et al. 2016). Southern right whales are arguably in the best shape of the three right whale species, but even among the more robust populations, such as the Western South Atlantic calving grounds in Argentina, there have been mass die offs in recent years, potentially linked to increases in harmful algal blooms (Wilson et al. 2016) or variations in prey availability (Rowntree et al. 2013).

Given variations in the overall health and populations sizes of the three species coupled with significant differences in their accessibility for study in different habitats, it is unclear how many reported "differences" in acoustic behaviors among the species are real and how many are a function of the lack of available data. For example, structured production of gunshot sounds with the hypothesis that this is a form of "song" has only been proposed recently in North Pacific right whales (Crance et al. 2019). Do only North Pacific right whales produce these structured patterned gunshot signals, or does the inaccessibility of southern right whale breeding grounds preclude data collection to explore this? North Atlantic right whale acoustic recordings of gunshots show a marked increase in occurrence coincident with the presumed breeding season (Matthews et al. 2014), and sequences of sounds have only been attributed to older adult males (Parks et al. 2005). However, these sounds have not been explored in detail to look for repetitive structure or patterns in gunshot production that might be shared across individuals. This is just one example of differences in the literature that may, in fact, be similarities that have yet to be adequately documented and described.

Beyond lack of data that may prevent us from identifying some similarities among the species, there are clear differences in the population sizes and gene pools that likely influence observed behaviors in each species. What is the influence of population size and gene pool on the variability of acoustic behavior? Could differences in frequency or signal production patterns be explained, in part, through genetic drift or cultural changes that may have resulted from rapid removal of older individuals during the whaling era? Based on genetic evidence, North Atlantic right whales have extremely low genetic diversity, with higher diversity among most of the southern right whale populations (Malik et al. 2000; Rosenbaum et al. 2000; Patenaude et al. 2007) and all three species have histories of rapid population decreases as a result of whaling activities (Reeves et al. 1999; Patenaude et al. 2007; Ivashchenko and Clapham 2012; Jackson et al. 2016). If learning plays a role in right whale acoustic communication, it seems likely that North Atlantic right whales, the eastern population of North Pacific right whales, and the Indo-Pacific southern right whales may show significant differences in their behavior due to a drastic reduction in population size and the associated reduction of potential learning opportunities to the few calves born into each population. One potential test for this would be to explore variation in behavior among southern right whale populations, which show a wide variation in genetic diversity and population size across the separate breeding populations.

This chapter summarizes our current state of knowledge regarding acoustic behavior of right whales and in the process identifies several obvious gaps in existing data that limit comparisons among species. There remain data gaps for right whale acoustic behavior related to seasonal and spatial variation in call rates and calling behavior as they move between habitats over the course of a year. For each species, one season or geographic region is typically better studied, increasing the challenge of comparing acoustic behavior among the species. For the North Pacific right whales and North Atlantic right whales, most data are available from the higher latitude feeding areas. For southern right whales, far more data are available from calving grounds and nearshore lower latitude habitats than the more remote, offshore feeding areas. These data gaps are, in large part, due to the lack of accessibility during different portions of the right whale annual migration pathway or due to extremely small population sizes. An improved understanding of acoustic behavior and how it varies across habitats and seasons will provide greater insight into right whale habitat usage through passive acoustic data collection. Passive acoustic monitoring is starting to shed light on the seasonality and spatial variation in call usage for North Atlantic (Davis et al. 2017) and North Pacific right whales (Munger et al. 2008). Similar data for southern right whales are available for relatively nearshore areas (Webster et al. 2019), but more passive acoustic data from more offshore regions would provide a better understanding of the movements of southern right whales throughout their migratory range. Individual, age and habitat level variations in sound production concurrent with behavioral observations have only been extensively described for North Atlantic right whales, with some studies looking at specific age/sex classes or acoustic behaviors in both southern and North Pacific right whales. More data from concurrent behavioral and acoustic focal follows or acoustic tag data from all right whale species will improve our detailed understanding of behavioral functions and encoded information contained in right whale sounds. Finally, recording and reporting the background noise characteristics when reporting measurements of right whale sound characteristics will greatly strengthen the ability to make inferences about the similarities and differences in acoustic behavior among the three species. Ongoing studies are tackling each of these data gaps, and results from these studies will provide us with a richer understanding of the acoustic communication and behavioral interactions of right whales.

Acknowledgements This chapter was only possible due to the mentoring, collaboration, and inspiration from many outstanding scientists, both past and present, who have contributed to the large body of knowledge about right whales. I acknowledge and thank Scott Kraus and the North Atlantic right whale consortium, my first introduction to right whales and a scientific home for me for the past 20 years studying these fascinating whales. The dedication of right whale biologists, researchers, and conservationists around the world working to understand and conserve these whales is truly inspirational, and I feel honored to have the opportunity to be part of this global community.

References

Aguilar A (1986) A review of old Basque whaling and its effect on the right whales (*Eubalaena glacialis*) of the North Atlantic. Rep Int Whaling Comm 10:191–199

Bannister JL, Pastene LA, Burnell SR (1999) First record of movement of a southern right whale (*Eubalaena australis*) between warm water breeding grounds and the antarctic ocean, South of 60°S. Mar Mammal Sci 15:1337–1342. https://doi.org/10.1111/j.1748-7692.1999.tb00895.x

Barendse J, Best PB (2014) Shore-based observations of seasonality, movements, and group behavior of southern right whales in a non-nursery area on the South African west coast. Mar Mammal Sci 30:1358–1382. https://doi.org/10.1111/mms.12116

Baumgartner MF, Bonnell J, Corkeron PJ, Van Parijs SM, Hotchkin C, Hodges BA, Bort Thornton J, Mensi BL, Bruner SM (2020) Slocum gliders provide accurate near real-time estimates of baleen whale presence from human-reviewed passive acoustic detection information. Front Mar Sci 7. https://doi.org/10.3389/fmars.2020.00100

Baumgartner MF, Bonnell J, Parijs SMV, Corkeron PJ, Hotchkin C, Ball K, Pelletier L-P, Partan J, Peters D, Kemp J, Pietro J, Newhall K, Stokes A, Cole TVN, Quintana E, Kraus SD (2019) Persistent near real-time passive acoustic monitoring for baleen whales from a moored buoy: system description and evaluation. Methods Ecol Evol 10:1476–1489. https://doi.org/10.1111/2041-210X.13244

Baumgartner MF, Fratantoni DM, Hurst TP, Brown MW, Cole TVN, Van Parijs SM, Johnson M (2013) Real-time reporting of baleen whale passive acoustic detections from ocean gliders. J Acoust Soc Am 134:1814–1823

Best PB (1990) Trends in the inshore right whale population off South Africa, 1969–1987. Mar Mammal Sci 6:93–108. https://doi.org/10.1111/j.1748-7692.1990.tb00232.x

Best PB (1994) Seasonality of reproduction and the length of gestation in southern right whales *Eubalaena australis*. J Zool 232:175–189

Best PB, Brandão A, Butterworth DS (2001) Demographic parameters of southern right whales off South Africa. J Cetacean Res Manag Spec Issue 161–169

Best P, Schaeff CM, Reeb D, Palsbøll P (2003) Composition and possible function of social groupings of southern right whales in South African waters. Behaviour 140:1469–1494. https://doi.org/10.1163/156853903771980675

Bort J, Van Parijs SM, Stevick PT, Summers E, Todd S (2015) North Atlantic right whale *Eubalaena glacialis* vocalization patterns in the central gulf of maine from October 2009 through October 2010. Endanger Species Res 26:271–280

Brownell RL, Ralls K (1986) Potential for sperm competition in baleen whales. Rep Int Whaling Comm 8(Spec Iss):97–112

Burnell SR, Bryden MM (1997) Coastal residence periods and reproductive timing in southern right whales, *Eubalaena australis*. J Zool 241:613–621. https://doi.org/10.1111/j.1469-7998.1997.tb05736.x

Carroll EL, Childerhouse SJ, Fewster RM, Patenaude NJ, Steel D, Dunshea G, Boren L, Baker CS (2013) Accounting for female reproductive cycles in a superpopulation capture–recapture framework. Ecol Appl 23:1677–1690. https://doi.org/10.1890/12-1657.1

Carroll EL, Fewster RM, Childerhouse SJ, Patenaude NJ, Boren L, Baker CS (2016) First direct evidence for natal wintering ground fidelity and estimate of juvenile survival in the New Zealand Southern right whale *Eubalaena australis*. PLoS ONE 11:e0146590-e146617

Charif RA, Shiu Y, Muirhead CA, Clark CW, Parks SE, Rice AN (2020) Phenological changes in North Atlantic right whale habitat use in Massachusetts Bay. Glob Change Biol 26:734–745. https://doi.org/10.1111/gcb.14867

Clark CW (1980) A real-time direction finding device for determining the bearing to the underwater sounds of Southern Right Whales, *Eubalaena australis*. J Acoust Soc Am 68:508–511. https://doi.org/10.1121/1.384762

Clark CW (1982) The acoustic repertoire of the southern right whale, a quantitative analysis. Anim Behav 30:1060–1071

Clark CW (1983) Acoustic communication of southern right whales. In: Payne RS (ed) Communication and behavior of whales. Westview Press, Boulder, CO, pp 163–198

Clark CW, Brown MW, Corkeron P (2010) Visual and acoustic surveys for North Atlantic right whales, *Eubalaena glacialis*, in Cape Cod Bay, Massachusetts, 2001–2005: management implications. Mar Mammal Sci 26:837–854

Clark CW, Ellison WT (2004) Potential use of low-frequency sounds by baleen whales for probing the environment, evidence from models and empirical measurements. In: Thomas J, Moss C, Vater M (eds) Echolocation in bats and dolphins. The University of Chicago Press, pp 564–582

Clark CW, Gillespie D, Nowacek DP, Parks SE (2007) Listening to their world: acoustics for monitoring and protecting right whales in an urbanized ocean. In: The urban whale: North Atlantic right whales at the crossroads. Harvard University Press, pp 333–357

Cole T, Hamilton P, Henry AG, Duley P, Pace RM III, White BN, Frasier T (2013) Evidence of a North Atlantic right whale *Eubalaena glacialis* mating ground. Endanger Species Res 21:55–64

Cooke JG (2020) *Eubalaena glacialis*. The IUCN red list of threatened species 2020:e.T41712A162001243

Cooke JG, Clapham PJ (2018a) *Eubalaena japonica*. The IUCN red list of threatened species 2018: e/T8153A50354147

Cooke JG, Zerbini AN (2018b) *Eubalaena australis*. The IUCN red list of threatened species 2018: e.T8153A50354147

Corkeron P, Hamilton P, Bannister J, Best P, Charlton C, Groch KR, Findlay K, Rowntree V, Vermeulen E, Pace RM (2018) The recovery of North Atlantic right whales, *Eubalaena glacialis*, has been constrained by human-caused mortality. R Soc Open Sci 5:180892. https://doi.org/10.1098/rsos.180892

Crance JL, Berchok CL, Keating JL (2017) Gunshot call production by the North Pacific right whale *Eubalaena japonica* in the southeastern Bering Sea. Endanger Species Res 34:251–267. https://doi.org/10.3354/esr00848

Crance JL, Berchok CL, Wright DL, Brewer AM, Woodrich DF (2019) Song production by the North Pacific right whale, *Eubalaena japonica*. J Acoust Soc Am 145:3467–3479. https://doi.org/10.1121/1.5111338

Crespo EA, Pedraza SN, Dans SL, Svendsen GM, Degrati M, Coscarella MA (2019) The southwestern Atlantic southern right whale, *Eubalaena australis*, population is growing but at a decelerated rate. Mar Mammal Sci 35:93–107. https://doi.org/10.1111/mms.12526

Cummings WC, Cummings WC, Fish JF, Thompson PO (1972) Sound production and other behavior of southern right whales *Eubalaena glacialis*. Trans San Diego Soc Nat Hist 17:1–14. https://doi.org/10.5962/bhl.part.19957

Cusano DA, Conger LA, Van Parijs SM, Parks SE (2018) Implementing conservation measures for the North Atlantic right whale: considering the behavioral ontogeny of mother-calf pairs. Anim Conserv 237:301–310

Davies KTA, Brillant SW (2019) Mass human-caused mortality spurs federal action to protect endangered North Atlantic right whales in Canada. Mar Policy 104:157–162. https://doi.org/10.1016/j.marpol.2019.02.019

Davis GE, Baumgartner MF, Bonnell JM, Bell J, Berchok C, Bort Thornton J, Brault S, Buchanan G, Charif RA, Cholewiak D, Clark CW, Corkeron P, Delarue J, Dudzinski K, Hatch L, Hildebrand J, Hodge L, Klinck H, Kraus S, Martin B, Mellinger DK, Moors-Murphy H, Nieukirk S, Nowacek DP, Parks S, Read AJ, Rice AN, Risch D, Širović A, Soldevilla M, Stafford K, Stanistreet JE, Summers E, Todd S, Warde A, Van Parijs SM (2017) Long-term passive acoustic recordings track the changing distribution of North Atlantic right whales (*Eubalaena glacialis*) from 2004 to 2014. Nat Publ Group 7:13460

Dombroski JRG, Parks SE, Flores PAC, Martín López LM, Shorter KA, Groch KR (2020) Animal-borne tags provide insights into the acoustic communication of southern right whales (*Eubalaena australis*) on the calving grounds. J Acoust Soc Am 147:EL498–EL503. https://doi.org/10.1121/10.0001391

Dombroski JRG, Parks SE, Groch KR, Flores PAC, Sousa-Lima RS (2016) Vocalizations produced by southern right whale (*Eubalaena australis*) mother-calf pairs in a calving ground off Brazil. J Acoust Soc Am 140:1850–1857

Dombroski JRG, Parks SE, Groch KR, Flores PAC, Sousa-lima RS (2017) Upcall production by southern right whale (*Eubalaena australis*) mother-calf pairs may be independent of diel period in a nursery area. Mar Mammal Sci 33:669–677. https://doi.org/10.1111/mms.12382

Donnelly BG (1967) Observations on the mating behaviour of the southern right whale *Eubalaena australis*. South Afr J Sci 63:176

Ellis R (2018) Whaling, aboriginal and western traditional. In: Würsig B, Thewissen JGM, Kovacs KM (eds) Encyclopedia of marine mammals 3rd edn. Academic Press, pp 1054–1063

Elowson AM, Snowdon CT, Lazaro-Perea C (1998) 'Babbling' and social context in infant monkeys: parallels to human infants. Trends Cogn Sci 2:31–37. https://doi.org/10.1016/S1364-6613(97)01115-7

Frasier TR, Hamilton PK, Brown MW, Conger LA, Knowlton AR, Marx MK, Slay CK, Kraus SD, White BN (2007) Patterns of male reproductive success in a highly promiscuous whale species: the endangered North Atlantic right whale. Mol Ecol 16:5277–5293

Gillespie D, Leaper R (2001) Report of the workshop on right whale acoustics: Practical applications in conservation. International Fund for Animal Welfare, Yarmouth Port, MA

Good CP (2008) Spatial ecology of the North Atlantic right whale (*Eubalaena glacialis*). Duke University. Ph.D. Dissertation

Groch KR, Palazzo JJT, Flores PAC, Adler FR, Fabian ME (2005) Recent rapid increases in the right whale (*Eubalaena australis*) population off southern Brazil. Lat Am J Aquat Mamm 4:41–47. https://doi.org/10.5597/lajam00068

Hamilton PK, Knowlton AR, Marx MK (2007) Right whales tell their own stories: the photo-identification catalog. The urban whale: North Atlantic right whales at the crossroads. Harvard University Press, Cambridge, MA, pp 75–104

Hamilton PK, Knowlton AR, Marx MK, Kraus SD (1998) Age structure and longevity in North Atlantic right whales *Eubalaena glacialis* and their relation to reproduction. Mar Ecol Prog Ser 171:285–292. https://doi.org/10.3354/meps171285

Harcourt R, van der Hoop J, Kraus S, Carroll EL (2019) Future directions in *Eubalaena* spp.: Comparative research to inform conservation. Front Mar Sci 5. https://doi.org/10.3389/fmars.2018.00530

Hoffmeyer MS, Lindner MS, Carribero A, Fulco VK, Menéndez MC, Fernández Severini MD, Diodato SL, Berasategui AA, Biancalana F, Berrier E (2010) Planktonic food and foraging of *Eubalaena australis*, on Peninsula Valdés (Argentina) nursery ground. Rev Biol Mar Oceanogr 45. https://doi.org/10.4067/S0718-19572010000100013

Hofmeyr-Juritz LH, Best PB (2011) Acoustic behaviour of southern right whales in relation to numbers of whales present in Walker Bay, South Africa. Afr J Mar Sci 33:415–427

Hotchkin C, Parks SE (2013) The Lombard effect and other noise-induced vocal modifications: insight from mammalian communication systems. Biol Rev 88(4):809–824

Ivashchenko Y, Clapham P (2012) Soviet catches of right whales *Eubalaena japonica* and bowhead whales *Balaena mysticetus* in the North Pacific Ocean and the Okhotsk Sea. Endanger Species Res 18:201–217. https://doi.org/10.3354/esr00443

International Whaling Commission (IWC) (2001) Report of the workshop on the comprehensive assessment of right whales: A worldwide comparison. J Cetacean Res Manage 2:1–35

International Whaling Commission (IWC) (2013) Report of the IWC workshop on the assessment of southern right whales. J Cetacean Res Manag 149:439–462

Jackson JA, Carroll EL, Smith TD, Zerbini AN, Patenaude NJ, Baker CS (2016) An integrated approach to historical population assessment of the great whales: case of the New Zealand southern right whale. R Soc Open Sci 3:150669. https://doi.org/10.1098/rsos.150669

Jackson JA, Carroll EL, Kennedy A, Leaper R, Calderan S, Leslie MS, Andriolo A, Stepien E, Zerbini AN, Miller BS, Kelly N, Stowasser G, Cheeseman T, Moore M (2018) Sightings and acoustic records of right whales collected in South Georgia (Islas Georgias Del Sur) waters January-February 2018. Rep SC67BSH20 Submitt Sci Comm Int Whal Comm Camb UK Sci Comm IWC Available Httpsiwcinthome. https://doi.org/10.13140/RG.2.2.33700.17285

Jacobs E, Duffy M, Magolan J, Vernazzani BG, Cabrera E, Landea R, Buchan S, Sayigh L (2019) First acoustic recordings of critically endangered eastern South Pacific southern right whales (*Eubalaena australis*). Mar Mammal Sci 35:284–289. https://doi.org/10.1111/mms.12519

Janik VM, Slater PJ (1997) Vocal learning in mammals. Adv Study Behav 26:59–100

Johnson MP, Tyack PL (2003) A digital acoustic recording tag for measuring the response of wild marine mammals to sound. IEEE J Ocean Eng 28:3–12

Kenney RD (2018) Right whales: *Eubalaena glacialis, E. japonica, and E. australis.* In: Würsig B, Thewissen JGM, Kovacs KM (eds) Encyclopedia of marine mammals (3rd edn). Academic Press, pp 817–822

Knowlton AR, Kraus SD, Kenney RD (1994) Reproduction in North Atlantic right whales (*Eubalaena glacialis*). Can J Zool 72:1297–1305. https://doi.org/10.1139/z94-173

Kraus SD, Hatch JJ (2001) Mating strategies in the North Atlantic right whale (*Eubalaena glacialis*). J Cetacean Res Manag Spec Issue 2:237–244

Kraus SD, Moore KE, Price CA, Crone MJ, Watkins WA, Winn HE, Prescott JH (1986) The use of photographs to identify individual North Atlantic right whales (*Eubalaena glacialis*). Rep Int Whal Comm Spec Issue 10:145–151

Kraus SD, Kenney RD, Mayo CA, McLellan WA, Moore MJ, Nowacek DP (2016) Recent scientific publications cast doubt on North Atlantic right whale future. Front Mar Sci 3:1401–1403

Kraus SD, Rolland RM (eds) (2007) The urban whale: North Atlantic right whales at the crossroads. Harvard University Press

Lombard É (1911) Le signe de l'élévation de la voix. Ann Des Mal De L'oreille Et Du Larynx 37:101–119

Mackay AI, Bailleul F, Carroll EL, Andrews-Goff V, Baker CS, Bannister J, Boren L, Carlyon K, Donnelly DM, Double M, Goldsworthy SD, Harcourt R, Holman D, Lowther A, Parra GJ, Childerhouse SJ (2020) Satellite derived offshore migratory movements of southern right whales (*Eubalaena australis*) from Australian and New Zealand wintering grounds. PLOS ONE 15:e0231577. https://doi.org/10.1371/journal.pone.0231577

Malik S, Brown MW, Kraus SD, White BN (2000) Analysis of mitochondrial DNA diversity within and between North and South Atlantic right whales. Mar Mammal Sci 16:545–558. https://doi.org/10.1111/j.1748-7692.2000.tb00950.x

Marques TA, Munger L, Thomas L, Wiggins S, Hildebrand JA (2011) Estimating North Pacific right whale *Eubalaena japonica* density using passive acoustic cue counting. Endanger Species Res 13:163–172

Marques TA, Thomas L, Martin SW, Mellinger DK, Ward JA, Moretti DJ, Harris D, Tyack PL (2013) Estimating animal population density using passive acoustics. Biol Rev 88:287–309. https://doi.org/10.1111/brv.12001

Mate B, Duley P, Lagerquist B, Wenzel F, Stimpert A, Clapham P (2005) Observations of a Female North Atlantic right whale (*Eubalaena glacialis*) in simultaneous copulation with two males: supporting evidence for sperm competition. Aquat Mamm 31:157–160. https://doi.org/10.1578/AM.31.2.2005.157

Mate BR, Best PB, Lagerquist BA, Winsor MH (2011) Coastal, offshore, and migratory movements of South African right whales revealed by satellite telemetry. Mar Mammal Sci 27:455–476. https://doi.org/10.1111/j.1748-7692.2010.00412.x

Matthews JN, Brown S, Gillespie D, Johnson M, McLanaghan R, Moscrop A, Nowacek D, Leaper R, Lewis T, Tyack P (2001) Vocalisation rates of the North Atlantic right whale (*Eubalaena glacialis*). J Cetacean Res Manag 3:271–282

Matthews LP, McCordic JA, Parks SE (2014) Remote acoustic monitoring of North Atlantic right whales (*Eubalaena glacialis*) reveals seasonal and diel variations in acoustic behavior. PLoS ONE 9:e91367–e91377

McDonald MA, Moore SE (2002) Calls recorded from North Pacific right whales (*Eubalaena japonica*) in the eastern Bering Sea. 261–266

Mellinger DK, Nieukirk SL, Matsumoto H, Heimlich SL, Dziak RP, Haxel J, Fowler M, Meinig C, Miller HV (2007a) Seasonal occurrence of North Atlantic right whale (*Eubalaena glacialis*) vocalizations at two sites on the Scotian Shelf. Mar Mammal Sci 23:856–867

Mellinger DK, Stafford KM, Moore SE, Dziak RP (2007b) An overview of fixed passive acoustic observation methods for cetaceans. Oceanography 20(4):36–45

Mellinger DK, Stafford KM, Moore SE, Munger L, Fox CG (2004) Detection of North Pacific Right Whale (*Eubalaena japonica*) Calls in the Gulf of Alaska. Mar Mammal Sci 20:872–879. https://doi.org/10.1111/j.1748-7692.2004.tb01198.x

Meyer-Gutbrod EL, Greene CH, Davies KTA (2018) Marine species range shifts necessitate advanced policy planning: The case of the North Atlantic right whale. Oceanography 31:19–23

Mossbridge JA, Thomas JA (1999) An Acoustic niche for antarctic killer whale and leopard seal sounds. Mar Mammal Sci 15:1351–1357. https://doi.org/10.1111/j.1748-7692.1999.tb00897.x

Munger LM, Wiggins SM, Moore SE, Hildebrand JA (2008) North Pacific right whale (*Eubalaena japonica*) seasonal and diel calling patterns from long-term acoustic recordings in the southeastern Bering Sea, 2000–2006. Mar Mammal Sci 16:530–620

Nielsen MLK, Bejder L, Videsen SKA, Christiansen F, Madsen PT (2019) Acoustic crypsis in southern right whale mother-calf pairs: infrequent, low-output calls to avoid predation? J Exp Biol 222:jeb190728–6

Norris KS (1967) Aggressive behavior in Cetacea. In: Clemente CD, Lindsley DB (eds) Aggression and defense: neural mechanisms and social patterns. University of California Press, pp 225–242

Omura H (1958) North Pacific right whale. Sci Rep Whales Res Inst 13:1–52

Omura H, Ohsumi S, Nemoto T, Nasu K, Kasuya T (1969) Black right whales in the North Pacific. Sci Rep Whales Res Inst 21:96

Parks SE (2003) Response of North Atlantic right whales (*Eubalaena glacialis*) to playback of calls recorded from surface active groups in both the North and South Atlantic. Mar Mammal Sci 19:563–580. https://doi.org/10.1111/j.1748-7692.2003.tb01321.x

Parks SE, Tyack PL (2005) Sound production by North Atlantic right whales (*Eubalaena glacialis*) in surface active groups. J Acoust Soc Am 117:3297–3306

Parks SE, Hamilton PK, Kraus SD, Tyack PL (2005) The gunshot sound produced by male North Atlantic right whales (*Eubalaena glacialis*) and its potential function in reproductive advertisement. Mar Mammal Sci 21:458–475. https://doi.org/10.1111/j.1748-7692.2005.tb01244.x

Parks SE, Clark CW (2007) Acoustic communication: Social sounds and the potential impacts of noise. In: The urban whale: North Atlantic right whales at the crossroads. Harvard University Press, pp 310–332

Parks SE, Brown MW, Conger LA, Hamilton PK, Knowlton AR, Kraus SD, Slay CK, Tyack PL (2007a) Occurrence, composition, and potential functions of North Atlantic right whale (*Eubalaena glacialis*) surface active groups. Mar Mammal Sci 23:868–887

Parks SE, Clark CW, Tyack PL (2007b) Short- and long-term changes in right whale calling behavior: The potential effects of noise on acoustic communication. J Acoust Soc Am 122:3725–3731

Parks SE, Urazghildiiev I, Clark CW (2009) Variability in ambient noise levels and call parameters of North Atlantic right whales in three habitat areas. J Acoust Soc Am 125:1230–1239

Parks SE, Johnson M, Nowacek D, Tyack PL (2011a) Individual right whales call louder in increased environmental noise. Biol Lett 7:33–35

Parks SE, Searby A, Célérier A, Johnson MP, Nowacek DP, Tyack PL (2011b) Sound production behavior of individual North Atlantic right whales: implications for passive acoustic monitoring. Endanger Species Res 15:63–76

Parks SE, Groch K, Flores P, Sousa-Lima R, Urazghildiiev IR (2016) Humans, fish, and whales: how right whales modify calling behavior in response to shifting background noise conditions. The effects of noise on aquatic life II. Springer, New York, pp 809–813

Parks SE, Cusano DA, Van Parijs SM, Nowacek DP (2019a) North Atlantic right whale (*Eubalaena glacialis*) acoustic behavior on the calving grounds. JASA-EL

Parks SE, Cusano DA, Van Parijs SM, Nowacek DP (2019b) Acoustic crypsis in communication by North Atlantic right whale mother–calf pairs on the calving grounds. Biol Lett 15:20190485. https://doi.org/10.1098/rsbl.2019.0485

Patenaude NJ, Portway VA, Schaeff CM, Bannister JL, Best PB, Payne RS, Rowntree VJ, Rivarola M, Baker CS (2007) Mitochondrial DNA diversity and population structure among Southern right whales (*Eubalaena australis*). J Hered 98:147–157. https://doi.org/10.1093/jhered/esm005

Payne R (1986) Long term behavioral studies of the Southern right whale (*Eubalaena australis*). Rep Int Whal Comm Spec Issue 10:161–167

Payne R, Payne K (1971) Underwater sounds of southern right whales. Zoologica 56:159–165

Payne R (1995) Among whales. Scribner. 432p

Pettis HM, Rolland RM, Hamilton PK, Brault S, Knowlton AR, Kraus SD (2004) Visual health assessment of North Atlantic right whales (*Eubalaena glacialis*) using photographs. Can J Zool 82:8–19

Pettis HM, Pace III RM, Hamilton PK (2020) North Atlantic right whale consortium 2019 annual report card. Report to the North Atlantic Right Whale Consortium

Rayment W, Webster T, Brough T, Jowett T, Dawson S (2017) Seen or heard? A comparison of visual and acoustic autonomous monitoring methods for investigating temporal variation in occurrence of southern right whales. Mar Biol 165:12. https://doi.org/10.1007/s00227-017-3264-0

Reeves RR, Breiwick JM, Mitchell ED (1999) History of whaling and Estimated Kill of Right Whales, *Balaena glacialis*, in the Northeastern United States, 1620–1924. Mar Fish Rev 61:1–36

Reeves RR, Smith TD, Josephson EA (2007) Near-annihilation of a species: right whaling in the North Atlantic. In: The Urban Whale: North Atlantic right whales at the crossroads. pp 39–74

Rolland RM, Hunt KE, Kraus SD, Wasser SK (2005) Assessing reproductive status of right whales (*Eubalaena glacialis*) using fecal hormone metabolites. Gen Comp Endocrinol 142:308–317

Rone BK, Berchok CL, Crance JL, Clapham PJ (2012) Using air-deployed passive sonobuoys to detect and locate critically endangered North Pacific right whales. Mar Mammal Sci 28:E528–E538. https://doi.org/10.1111/j.1748-7692.2012.00573.x

Root-Gutteridge H, Cusano DA, Shiu Y, Nowacek DP, Van Parijs SM, Parks SE (2018) A lifetime of changing calls: North Atlantic right whales, *Eubalaena glacialis*, refine call production as they age. Anim Behav 137:21–34

Rosenbaum HC, Brownell RL, Brown MW, Schaeff C, Portway V, White BN, Malik S, Pastene LA, Patenaude NJ, Baker CS, Goto M, Best PB, Clapham PJ, Hamilton P, Moore M, Payne R, Rowntree V, Tynan CT, Bannister JL, Desalle R (2000) World-wide genetic differentiation of

Eubalaena: questioning the number of right whale species. Mol Ecol 9:1793–1802. https://doi.org/10.1046/j.1365-294x.2000.01066.x

Rowntree VJ, Uhart MM, Sironi M, Chirife A, Di Martino M, La Sala L, Musmeci L, Mohamed N, Andrejuk J, McAloose D, Sala JE, Carribero A, Rally H, Franco M, Adler FR, Brownell RL Jr, Seger J, Rowles T (2013) Unexplained recurring high mortality of southern right whale *Eubalaena australis* calves at Península Valdés, Argentina. Mar Ecol Prog Ser 493:275–289

Sataloff RT, Rosen DC, Hawkshaw M, Spiegel JR (1997) The aging adult voice. J Voice 11:156–160. https://doi.org/10.1016/S0892-1997(97)80072-0

Schaeff CM (1999) Dorsal skin color patterns among southern right whales (*Eubalaena australis*): genetic basis and evolutionary significance. J Hered 90:464–471. https://doi.org/10.1093/jhered/90.4.464

Schevill WE, Watkins WA (1962) Whale and porpoise voices, a phonograph record. Woods Hole Oceanographic Institution

Scorseby W (1820) An account of the Arctic regions: with a history and description of the northern whale-fisheries. *Edinburgh: Archibald Constable and co. (Facsimile edition with an introduction by Alister Hardy 1969, Newton Abbot: David & Charles Reprints.)*

Simard Y, Roy N, Giard S, Aulanier F (2019) North Atlantic right whale shift to the Gulf of St. Lawrence in 2015, revealed by long-term passive acoustics. Endanger Species Res 40:271–284. https://doi.org/10.3354/esr01005

Širović A, Hildebrand JA, Thiele D (2006) Baleen whale spatial patterns in the Scotia Sea during January and February 2003. J Cetacean Res Manag 8:161–171

Širović A, Johnson SC, Roche LK, Varga LM, Wiggins SM, Hildebrand JA (2014) North Pacific right whales (*Eubalaena japonica*) recorded in the northeastern Pacific Ocean in 2013. Mar Mammal Sci 31:800–807

Soldevilla MS, Rice AN, Clark CW, Garrison LP (2014) Passive acoustic monitoring on the North Atlantic right whale calving grounds. Endanger Species Res 25:115–140

Spaulding E, Robbins M, Calupca T, Clark CW, Tremblay C, Waack A, Warde A, Kemp J, Newhall K (2010) An autonomous, near-real-time buoy system for automatic detection of North Atlantic right whale calls. J Acoust Soc Am 125(4):2615

Stamation K, Watson M, Moloney P, Charlton C, Bannister J (2020) Population estimate and rate of increase of southern right whales *Eubalaena australis* in southeastern Australia. Endanger Species Res 41:373–383. https://doi.org/10.3354/esr01031

Taber S, Thomas P (1982) Calf development and mother-calf spatial relationships in Southern right whales. Anim Behav 30:1072–1083

Trygonis V, Gerstein E, Moir J, McCulloch S (2013) Vocalization characteristics of North Atlantic right whale surface active groups in the calving habitat, southeastern United States. J Acoust Soc Am 134:4518–4531

Valenzuela LO, Sironi M, Rowntree VJ, Seger J (2009) Isotopic and genetic evidence for culturally inherited site fidelity to feeding grounds in southern right whales (*Eubalaena australis*). Mol Ecol 18:782–791

Van Parijs SM, Clark CW, Sousa-Lima RS, Parks SE, Rankin S, Risch D, Van Opzeeland IC (2009) Management and research applications of real-time and archival passive acoustic sensors over varying temporal and spatial scales. Mar Ecol Prog Ser 395:21–36

Vanderlaan ASM, Hay AE, Taggart CT (2003) Characterization of North Atlantic right whale (*Eubalaena glacialis*) sounds in the bay of fundy. IEEE J Ocean Eng 28:164–173

Vernazzani BG, Cabrera E, Brownell RL (2014) Eastern South Pacific southern right whale photo-identification catalog reveals behavior and habitat use patterns. Mar Mammal Sci 30:389–398. https://doi.org/10.1111/mms.12030

Wade P, Heide-Jorgensen MP, Shelden K, Barlow J, Carretta J, Durban J, LeDuc R, Munger L, Rankin S, Sauter A, Stinchcomb C (2006) Acoustic detection and satellite-tracking leads to discovery of rare concentration of endangered North Pacific right whales. Biol Lett 2:417–419

Wade PR, Kennedy A, LeDuc R, Barlow J, Carretta J, Shelden K, Perryman W, Pitman R, Robertson K, Rone B, Salinas JC, Zerbini A, Brownell RL, Clapham PJ (2011) The world's smallest whale population? Biol Lett 7:83–85. https://doi.org/10.1098/rsbl.2010.0477

Waite JM, Wynne K, Mellinger DK (2003) Documented sighting of a North Pacific right whale in the gulf of Alaska and post-sighting acoustic monitoring. Northwest Nat 84:38–43. https://doi.org/10.2307/3536721

Watkins WA, Schevill WE (1972) Sound source location by arrival-times on a non-rigid three-dimensional hydrophone array. Deep Sea Res Oceanogr Abstr 19:691–706

Webster TA, Dawson SM, Rayment WJ (2016) Quantitative analysis of the acoustic repertoire of southern right whales in New Zealand. J Acoust Soc Am 140:322–333

Webster TA, Van Parijs SM, Rayment WJ, Dawson SM (2019) Temporal variation in the vocal behaviour of southern right whales in the Auckland Islands, New Zealand. R Soc Open Sci 6:181487. https://doi.org/10.1098/rsos.181487

William M (1988) Behavior of southern right whales, *Eubalaena australis*, feeding on the Antarctic krill, *Euphausia superba*. Fish Bull 86:143–150

Wilson C, Sastre AV, Hoffmeyer M, Rowntree VJ, Fire SE, Santinelli NH, Ovejero SD, D'Agostino V, Marón CF, Doucette GJ, Broadwater MH, Wang Z, Montoya N, Seger J, Adler FR, Sironi M, Uhart MM (2016) Southern right whale (*Eubalaena australis*) calf mortality at Península Valdés, Argentina: Are harmful algal blooms to blame? Mar Mammal Sci 32:423–451. https://doi.org/10.1111/mms.12263

Winn HE, Price CA, Sorensen PW (1986) The distributional biology of the right whale (*Eubalaena glacialis*) in the western North Atlantic. Rep Int Whal Comm Spec Issue 10:129–138

Wright DL, Berchok CL, Crance JL, Clapham PJ (2019) Acoustic detection of the critically endangered North Pacific right whale in the northern Bering Sea. Mar Mammal Sci 35:311–326. https://doi.org/10.1111/mms.12521

Chapter 14
Mysterious Minke Whales: Acoustic Diversity and Variability

Denise Risch

Abstract The acoustic behavior of minke whale populations worldwide has been a mystery for the better part of the twentieth century. Several likely biological sound sources such as the 'boing' recorded in the North Pacific, or the 'bio-duck' with its ubiquitous distribution in the Southern Ocean, had been described by seafarers since the middle of the twentieth century. However, the origin of these sounds could only be revealed once technological advances allowed scientists to simultaneously acoustically and visually track the elusive species producing them. The current data show that, like other baleen whales, most minke whale populations produce long song sequences presumably in a reproductive context. Over the past two decades, by extending our listening efforts into remote habitats, we have learned much about minke whales and can assume that many more mysteries are waiting to be unlocked.

Keywords Minke whale · Antarctic minke whale · Vocal repertoire · Boing sound · Bio-duck sound · Star-wars sound · Pulse train

---

Dedicated to Thomas F Norris

D. Risch (✉)
Scottish Association for Marine Science (SAMS), Oban P37 1PU, Argyll, UK
e-mail: denise.risch@sams.ac.uk

C.W. Clark and E. C. Garland (eds.), *Ethology and Behavioral Ecology of Mysticetes*, Ethology and Behavioral Ecology of Marine Mammals,
https://doi.org/10.1007/978-3-030-98449-6_14

Antarctic minke whale (*Balaenoptera bonaerensis*) with multi-sensor acoustic recording tag. Photo courtesy of Ari S. Friedlaender

14.1 Introduction

Although they are one of the most widely distributed baleen whale species, often found in coastal waters during summer, most minke whale populations have been little studied. Most of our knowledge about the species' life history originates from scattered observations in coastal habitats and whaling records (Horwood 1990), and much of their behavioral ecology remains unknown.

Common minke whales (*Balaenoptera acutorostrata*) inhabit all oceans from the tropics to the poles and are closely related to their sister species, the Antarctic minke whale (*Balaenoptera bonaerensis*), also referred to as the 'southern minke whale' (Rychel et al. 2004). Based on genetic evidence and geographic distribution, the common minke whale is split into three subspecies: *Balaenoptera acutorostrata acutorostrata* in the North Atlantic, *Balaenoptera acutorostrata scammoni* in the North Pacific, and an unnamed subspecies of dwarf minke whale in the Southern Hemisphere (Rice 1998). The exact placement of the dwarf minke whale in baleen whale taxonomy is still unclear. At least two different populations of dwarf minke whales appear to exist in the Southern Hemisphere, one in the South Atlantic where they were first described in the 1980s (Best 1985) and one in the South Pacific. Genetically, the South Atlantic dwarf minke whale is more closely related to the North Atlantic minke whale than to the South Pacific dwarf minke whale (Pastene et al. 2010).

The lack of data on minke whales is partly due to the difficulties in observing them. At an average adult body length of 7–9 m, minke whales are one of the smallest of the baleen whales, only larger than the pygmy right whale (*Caperea marginata*). Due to their small body size, inconspicuous blows, and brief surfacings, minke whales have been described as elusive, especially in their more pelagic winter habitats. In addition, except in higher latitudes, where aggregations of up to a few hundred animals have been visually observed, larger aggregations are less prevalent than in other species (Edds and Macfarlane 1987). Indeed, across much of their range minke whales are often visually encountered as solitary individuals and exhibit a wide range of individually distinctive foraging behaviors (Hoelzel et al. 1989).

In contrast to this general reclusive demeanor, in some areas, common minke whales are known for their inquisitive behavior and tendency to actively associate with vessels. This more curious behavior is, for example, commonly observed in Australian's Great Barrier Reef region, where dwarf minke whales approach and spend extended time with swimmers and divers (Valentine et al. 2004). These interactions are particularly remarkable given that the winter distributions and behaviors of nearly all other minke whale populations are less understood than their distributions and behaviors in summer foraging habitats.

14.2 Mystery in All Oceans: A Brief Overview of Global Minke Whale Vocalizations

Comprehensive descriptions and behavioral functions of minke whale sounds remain incomplete. Vocalizations produced by the North Pacific, dwarf and Antarctic minke whales were only unequivocally assigned to these species in the early 2000s (Gedamke et al. 2001; Gedamke 2004; Rankin and Barlow 2005; Risch et al. 2014b), despite being recorded and described in parts of the world's oceans (Fig. 14.1) for decades (Wenz 1964; Thompson and Friedl 1982; Matthews et al. 2004; Dolman et al. 2005). It is now clear that minke whale populations produce a variety of sounds throughout their geographic ranges. However, given that the full acoustic repertoire and functional significance of their many different types of sounds are still largely unknown, many discoveries about their vocal behaviors are yet to be made.

New technologies are helping to fill some of the existing data gaps. For example, small, multi-sensor tags can now be equipped with acoustic recorders, allowing measurements of fine-scale movement behavior while simultaneously recording sounds from tagged and nearby animals (Risch et al. 2014b). Networks of bottom-mounted acoustic recorders, deployed for several months to years, are being used to investigate migratory routes and changes in minke whale distribution and to identify previously unknown habitats (Clark and Gagnon 2002; Risch et al. 2014a). This chapter aims to summarize what is currently known about common and Antarctic minke whale vocalizations worldwide.

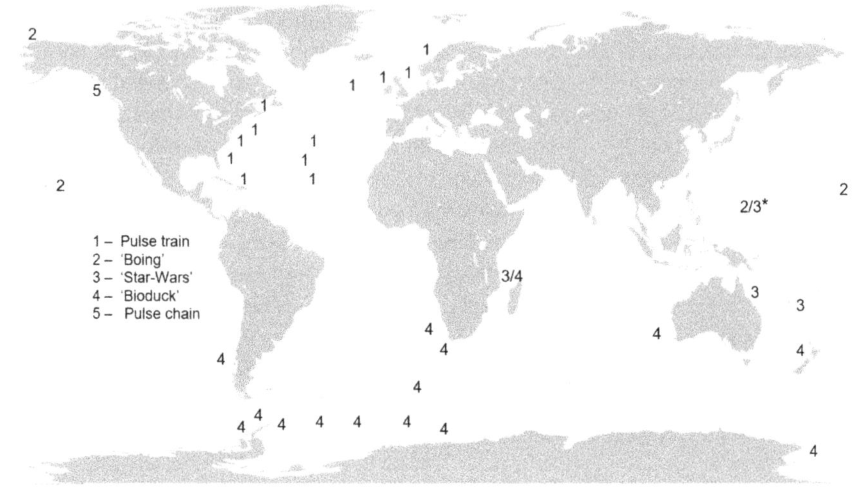

Fig. 14.1 Global distribution of the best described Antarctic and common minke whale vocalization types (Gedamke et al. 2001; Nieukirk et al. 2004, 2016; Matthews et al. 2004; Rankin and Barlow 2005; Dolman et al. 2005; Oswald et al. 2011; Martin et al. 2013; Risch et al. 2013, 2014a, 2019; b; Delarue et al. 2013; Dominello and Širović 2016; Norris et al. 2017; Cerchio et al. 2018, 2022; Nikolich and Towers 2018; Thomisch et al. 2019; Buchan et al. 2020; Shabangu et al. 2020; Filun et al. 2020). *Nieukirk et al. 2016 recorded a sound with similarities to the 'star-wars' vocalizations near the Marian Trench and hypothesized that it might be produced by minke whales

14.2.1 North Atlantic

On a summer feeding ground in the North Atlantic, a few sounds from a minke whale were first reported in the 1970s by Beamish and Mitchell (Beamish and Mitchell 1973). These authors used a calibrated hydrophone system to record a series of clicks in the 4–8 kHz range during a very close (20–80 m) encounter with a single animal. The clicks they described were 1–5 ms in duration and repeated at 6–7 clicks/second. About a decade later, in the Gulf of St Lawrence, Canada, Edds-Walton recorded sounds in the presence of minke whales; frequency-modulated (FM) downsweeps with a median start and end frequency of 118 Hz and 80 Hz, respectively (Edds-Walton 2000). These sounds had a median duration of 0.4 s and were produced by individual animals while traveling (Edds-Walton 2000).

In 1971, based on acoustic and visual observations conducted in the Caribbean Sea region, Winn and Perkins (1976) recorded low-frequency pulse trains with varying inter-pulse intervals (IPI) and peak frequencies from 55–150 Hz. In 1993–1996, with the advent of limited scientific access to the US Navy Sound Surveillance System (SOSUS), pulse trains described as songs and attributed unequivocally to minke whales in the Caribbean and throughout the Western North Atlantic were described and reported by Mellinger et al. (2000) and Clark and Gagnon (2002) (Fig. 14.2; Chap. 2). These pulse trains appear to be the most common in the repertoire of North Atlantic minke whales and have been recorded during migration and in several

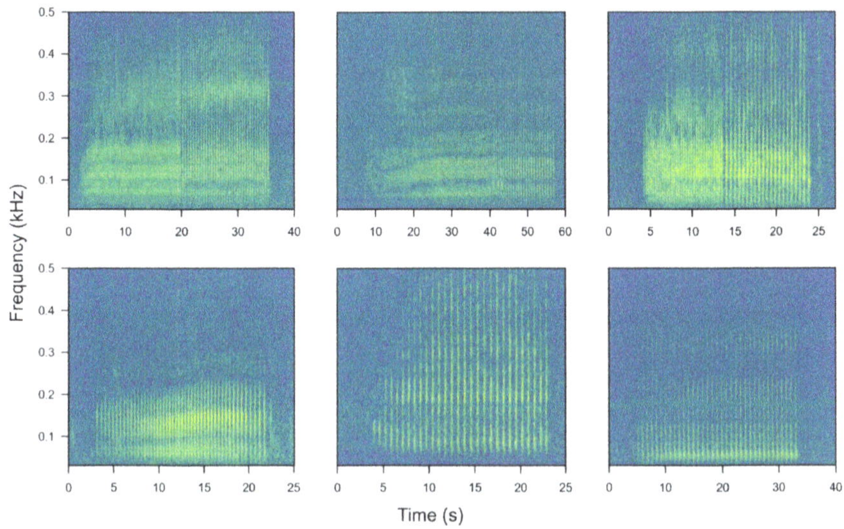

Fig. 14.2 Spectrograms of North Atlantic minke whale pulse train types identified by Risch et al. (2013) and recorded in the Stellwagen Bank National Marine Sanctuary, Eastern North Atlantic. Upper panel shows three different types of 'slow-down' pulse trains, and lower panel shows three types of 'constant' pulse trains. Not shown here is an example of a speed-up pulse train, which is mostly recorded in lower latitudes (Mellinger et al. 2000). Note the different time scales for these different pulse train types. Spectrogram parameters: 2000 Hz sample rate, 512 point FFT, 75% overlap = 75%, 3.9 Hz and 64 ms frequency and time resolution, respectively

summer feeding grounds (Folkow and Blix 1991; Nieukirk et al. 2004; Risch et al. 2013, 2019). Pulse trains are stereotypic, and three different pulse train types have been described (e.g., 'speed-up', 'slow-down', and 'constant') based on IPI and peak frequency (Fig. 14.3), while individual pulses are typically FM upsweeps and sometimes paired with a slight temporal overlap within a pair (Mellinger et al. 2000).

In addition to these stereotypic types of pulse trains, higher frequency (3–12 kHz) clicks described from initial Caribbean recordings (Winn and Perkins (1976) have more recently been confirmed to be regular components of some of these low-frequency pulse trains (Risch et al. 2015). These higher frequency clicks might also correspond to the ones originally described by Beamish and Mitchell (1973).

The durations of North Atlantic pulse trains appear to vary between pulse train types and geographically, with significantly longer pulse trains recorded in lower latitudes than in higher latitudes (Risch et al. 2014a). The average source level of 164–168 dB re 1 μPa measured by Risch et al. (2014c) gave an approximate active space, defined as the area within which a receiver might perceive the call of a signaler (Brenowitz 1982), of up to 10 km in a habitat dominated by shipping noise (Risch et al. 2014c).

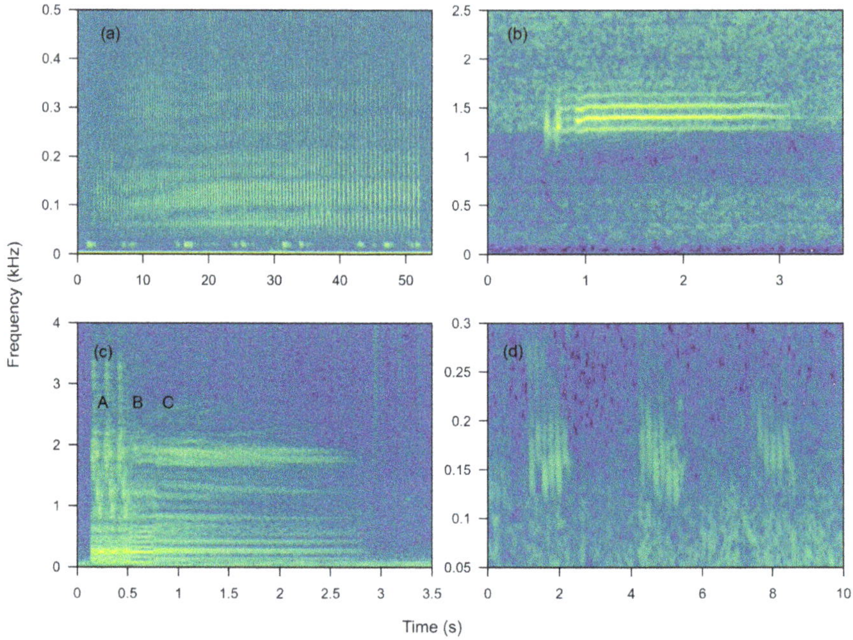

Fig. 14.3 Spectrogram of **a** North Atlantic slow-down pulse train type recorded during the spring migration season off Cape Cod; **b** North Pacific 'boing' sound; **c** South Pacific 'star-wars' sound. The sound is comprised of three sub-units: A, B, C (Gedamke et al. 2001); **d** Antarctic 'bio-duck' sound. Spectrogram parameters: **a** 2000 Hz sample rate, 512 point FFT, 75% overlap, 3.9 Hz and 64 ms frequency and time resolution, respectively; **b** 8000 Hz sample rate, 512 point FFT, 75% overlap, 15.6 Hz and 16 ms frequency and time resolution, respectively; **c** 16,000 Hz sample rate, 1024 point FFT, 90% overlap, 15.6 Hz and 6.4 ms frequency and time resolution, respectively; **d** 2000 Hz sample rate, 512 point FFT, 95% overlap, 3.9 Hz and 12.8 ms frequency and time resolution, respectively

14.2.2 North Pacific

A North Pacific sound type referred to as the 'boing' has been recorded in lower latitudes of the North Pacific since the late 1950s and was first described in the early 1960s (Wenz 1964). Although the sound was assumed to be biological, actual species attribution remained a mystery until 2003 when simultaneous visual observations and acoustic bearing angle localizations of individual animals were used to confirm that minke whales were the source of this mysterious sound (Rankin and Barlow 2005). Boing sounds consist of a relatively short introductory pulse followed by a longer amplitude modulation (AM) of a FM unit with decreasing amplitude over the course of the sound and a peak frequency of approximately 1.4 kHz (Fig. 14.3b; Thompson and Friedl 1982; Rankin and Barlow 2005; Oswald et al. 2011).

At least two types of the boing sound have been detected across the North Pacific, which may be indicative of geographic differences between genetically separate populations. Boings in the Eastern North Pacific have pulse repetition rates of about 92 s^{-1}, whereas in the central part of the North Pacific, boings have pulse repetition rates of 115 s^{-1} (Wenz 1964; Rankin and Barlow 2005). Eastern and central boings also differ in overall durations, lasting 3.6 and 2.6 s, respectively (Rankin and Barlow 2005). Boings are primarily recorded in low-latitude presumed breeding grounds (Norris et al. 2012) and are repeated in long sequences with varying inter-boing intervals (ICI). Several studies have described a bi-modal distribution of boing repetition rates, with shorter intervals between boings (28–30 s) when several animals are presumed to be in acoustic contact with one another and longer intervals (350–600 s) when animals are further apart (Wenz 1964; Thompson and Friedl 1982; Rankin and Barlow 2005). Apart from having been recorded in lower latitudes, a few 'boings' have also been recorded in a summer feeding ground, the Northeastern Chukchi Sea, during late summer and autumn (Delarue et al. 2013).

In addition to boing sounds, North Pacific minke whale vocalizations from a summer feeding ground off Vancouver Island, Canada, were recently described as low-frequency downsweeps starting at 142 Hz and ending at 38 Hz, a peak frequency of 105 Hz, and a duration of 0.7 s (Nikolich and Towers 2018). At the same recording site, a series of sounds similar to North Atlantic pulse trains was attributed to a minke whale and described as "pulse chains" with variable pulse rates in the 330–1400 Hz frequency range. Overall, pulse train occurrence in this summer feeding ground was very low, despite regular visual sightings of minke whales (Nikolich and Towers 2018).

14.2.3 South Pacific

One of the most fascinating baleen whale sound is produced by animals in the Australian population of the dwarf minke whale. Named the 'star-wars' vocalization (Fig. 14.3c), due to its distinct synthetic or metallic quality, this sound type was first scientifically described and unequivocally linked to the species in the Northern Great Barrier Reef sing a combination of dedicated visual observations and acoustic localization (Gedamke et al. 2001). The reef is a winter habitat and presumed breeding ground for the species. The star-wars sound is somewhat similar in structure to the boing sound. The sound is composed of three sub-units (A, B, C; Fig. 14.3c), each of which is an amplitude modulation of a FM unit, with AM rate and FM characteristics being different for each of the three components. This complex combination results in a sound that seems oddly unworldly to the human ear, hence, the reference to artificial sounds produced for the movie 'Star Wars'. Individual star-wars sounds are usually repeated in stereotyped sequences of equal intervals ranging from 1–2 s to 3–4 min. Source-level estimates for these sounds ranged from 150–165 dB re 1 μPa (Gedamke et al. 2001).

Similar to North Atlantic minke whales, dwarf minke whales in the Great Barrier Reef region also produce low-frequency, downswept vocalizations starting around 250 Hz and sweeping down to around 50 Hz. These sounds are typically 0.2–0.3 s in duration and have source levels of approximately 148–160 dB re 1 μPa (Gedamke et al. 2001). In addition, several 'noisy', broadband social sounds have been described for Great Barrier Reef dwarf minke whales (Gedamke 2004). As is the case in vocal repertoires of other baleen whales, the acoustic characteristics of these social sounds appear to be vary along a continuum, which makes their division into distinct call types difficult. They appear structurally similar to, but are generally less complex than, the typical star-wars sound (Gedamke 2004).

More recently, a complex call of unknown origin was recorded in the Mariana Trench, off the east coast of Guam and Saipan (Nieukirk et al. 2016; Fig. 14.1). This sound consisted of an AM 38 Hz moan, followed by a broadband sweep with energy up to 7.5 kHz. Due to the similarities to the star-wars sound, it was suggested that this call might be produced by either a dwarf or common minke whale (Nieukirk et al. 2016).

While dwarf minke whales are frequently visually observed in the South Atlantic (Zerbini et al. 1996; Acevedo et al. 2006), so far they have not been detected acoustically in the South Atlantic ocean basin.

14.2.4 Southern Ocean

For over five decades, a regular pulsed signal (Matthews et al. 2004) referred to as the 'bio-duck' has been recorded in the Southern Ocean, but the animal producing this mysterious sound remained unknown until recently. The signal was first described by submarine officers in the 1960s. Since then, it has been described off the west coast of Australia and in the Ross and Weddell Seas, as well as in sub-Antarctic waters (Poulter 1964; Matthews et al. 2004; McCauley 2004; Dolman et al. 2005; Klinck and Burkhardt 2008; Van Opzeeland 2010). It was not until 2013 that the 'bio-duck' was unequivocally linked to Antarctic minke whales based on analysis of multi-sensor acoustic recording tags deployed on two individuals of this species in Wilhelmina Bay, Western Antarctica (Risch et al. 2014b). The 'bio-duck' vocalization consists of one to six pulses in the 50–300 Hz frequency band (Fig. 14.3d), typically with harmonics up to 1 kHz, although recent descriptions of the different subtypes of the 'bio-duck' vocalization describe harmonics up to 2 kHz and beyond (Shabangu et al. 2020). These pulse trains are then often repeated in long sequences at very regular ICIs (Fig. 14.4).

Along with the first description of the 'bio-duck' signal from Antarctic minke whales, low-frequency downsweeps were also recorded from the same tagged individuals during several of their dives. These downsweeps were in the frequency range of 60–130 Hz with a mean duration of 0.2 s (Risch et al. 2014b). Similar downsweeps had previously been reported from the Ross Sea (Schevill and Watkins 1972; Leatherwood et al. 1981), as well as together with 'bio-duck' vocalizations that were recorded in the Weddell Sea and off Western Australia (Matthews et al. 2004; Klinck and Burkhardt 2008).

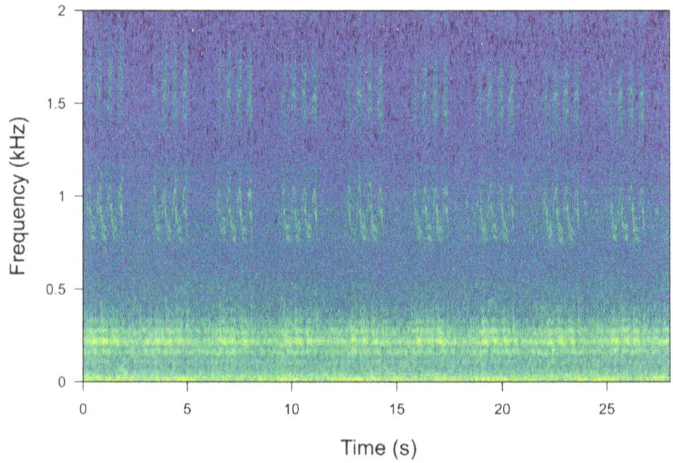

Fig. 14.4 Spectrogram of a partial Antarctic 'bio-duck' song sequence recorded during austral winter at PALAOA station, Antarctica, (acoustic sample courtesy of I. van Opzeeland). Spectrogram parameters: 48 kHz sample rate, 4096 point FFT, 75% overlap, 11.7 Hz and 21.3 ms frequency and time resolutions, respectively

A recent study of minke whale pulse trains collected near the Western Antarctic Peninsula distinguished at least four different variants of the bio-duck sound and several subtypes for one variant. These variants were based on frequency content, the number of pulses and IPIs within a train, and the pulse train repetition rate (Dominello and Širović 2016). Different bio-duck sound variants have also been described to co-occur in Western Australian and South African recordings (Matthews et al. 2004; Shabangu et al. 2020).

14.3 Commonalities Between Minke Whale Vocal Repertoires and a Definition of Minke Whale Song

Despite a remarkable amount of variability in the vocal repertoires currently described for common and Antarctic minke whales, there are some commonalities. Most minke whale populations produce series of relatively stereotypic low-frequency sounds that are often repeated in sequences lasting several minutes to hours. Although the sex of the caller is still unknown for most of these sequences (Sect. 14.5), the dwarf and North Atlantic minke whale vocalizations have been classified as song (Gedamke et al. 2001; Clark and Gagnon 2002; Clark and Ellison 2004; Gedamke 2004). This classification is based on the definition of song, commonly used to describe the patterned sequences produced by many birds species, where song is defined as: '*...long, complex, vocalizations produced by males in the breeding season*'. (Catchpole and Slater 2008) and by the description of male humpback whale song defined as *'...a series of notes, generally of more than one type, uttered in succession and so related as to form a recognizable sequence or pattern in time'* (Payne and McVay 1971; Chap. 11). Given that the North Pacific boing and the Antarctic bio-duck sounds are also produced in long sequences (Fig. 14.4) and both are produced on presumed breeding grounds (Sect. 14.5), they could similarly be classified as songs, making 'song' a common acoustic behavior among minke whale populations worldwide.

Next to humpback whale song (Payne and McVay 1971), baleen whale songs have been described for several species, including blue (*Balaenoptera musculus*; Chap. 9), fin (*Balaenoptera physalus*), Omura's (*Balaenoptera omurai*; Chap. 15), and bowhead (*Balaena mysticetus*; Chap. 12) whales, and singing is generally thought to function in mediating reproductive behavior (Cummings et al. 1987; Croll et al. 2002; Oleson et al. 2007a; Chap. 7).

In contrast to the highly dynamic characteristics of humpback and bowhead whale song (Cholewiak et al. 2013; Stafford et al. 2018; see Chaps. 10 and 11, respectively), minke whale songs appear more typical of the stereotypic songs of other balaenopterid whales such as blue and fin whales (Oleson et al. 2007b; Širović et al. 2013; Chap. 9), consisting of patterned phrases of a few notes with limited variability between years and individuals (Clark and Ellison 2004). That being said, individual song units can be structurally complex (e.g., the star-wars sound) and some variation has recently been shown in the Western North Atlantic where at least two different calling patterns or songs, consisting of three to four stereotyped pulse train types, have been described (Risch et al. 2014c). Across all populations, minke whale song units are typically low-frequency, pulsed FM-AM signals, with most energy below 1–2 kHz. Some geographic variation in boing song production has been described in the North Pacific (Rankin and Barlow 2005), while pulse train song units in the North Atlantic appear to increase in duration in lower latitude winter breeding grounds (Risch et al. 2014a).

The second sound type in most minke whale populations for which vocalizations have been partially described, are low-frequency (<500 Hz) downsweeps. These are usually not produced in patterned sequences and have most often been recorded in presumed summer feeding habitats (Schevill and Watkins 1972; Edds-Walton 2000; Shabangu et al. 2020).

14.4 Seasonal Distribution and Temporal Variation in Minke Whale Vocalizations

Marine environments are generally 'open' environments, with few natural barriers. Driven by wind, waves and upwellings, the distribution of available resources is variable over large spatial and temporal scales. Baleen whales have adapted to these dynamic conditions by capitalizing on large home ranges, with many species' annual migrations ranging over thousands of kilometers from high-latitude feeding to low-latitude breeding grounds (Whitehead 2001; Stevick et al. 2011; Corkeron and Connor 1999; Chaps. 4 and 7). Migratory routes are well described for some species such as humpback (Dawbin 1966; Clapham and Mead 1999; Calambokidis et al. 2001) and right whales (Kraus et al. 1986; Mate et al. 2011; Chap. 4).

Long longitudinal migrations have also been documented for common minke whales (Kasamatsu et al. 1995; Skaug et al. 2004; Risch et al. 2014a). It is during these migrations and in their presumed winter breeding habitats that minke whale songs are most often recorded (Gedamke et al. 2001; Nieukirk et al. 2004; Oswald et al. 2011; Delarue et al. 2013; Risch et al. 2014a). In contrast, while low-frequency downsweeps have been described from summer feeding and winter breeding grounds, they have been more commonly recorded in higher latitude feeding grounds (Schevill and Watkins 1972; Edds-Walton 2000; Gedamke et al. 2001; Dominello and Širović 2016; Shabangu et al. 2020).

As is the case in other baleen whales, results from long-term passive acoustic recordings of minke whales are increasingly providing evidence for complex migration and seasonal distribution patterns. For example, the bio-duck song of the Antarctic minke whale has been detected simultaneously in high and low latitudes during austral winter (Matthews et al. 2004; Buchan et al. 2020; Filun et al. 2020; Shabangu et al. 2020). This indicates that while part of the population migrates to warmer waters during winter, a large part of the population stays in high latitudes year-round. This finding was corroborated by satellite tag data, showing animals staying south of the Antarctic Circumpolar Current year-round (Lee et al. 2017). To understand these distribution patterns and migratory pathways, it is important to better describe the seasonality and behavioral context of minke whale vocalizations in different populations and habitats, by using long-term passive acoustic data in combination with direct visual observation and multi-sensor tags.

Distinct diel patterns in vocalizations have been described for several minke whale populations. In the North Atlantic, minke whale pulse trains are produced primarily at night (Risch et al. 2013, 2019). The same pattern was found in South African waters and near the Greenwich meridian in the Southern Ocean for 'bio-duck' song (Menze et al. 2017; Shabangu et al. 2020). The authors hypothesized that the bio-duck song might be related to feeding, as the observed diel pattern was closely linked to diel vertical migration of zooplankton, with high concentrations near the surface at night (Menze et al. 2017). However, in contrast to these findings, in the Western Antarctic Peninsula and other parts of Antarctica, the bio-duck song either showed no distinct diel pattern or was more commonly detected during daylight hours (Dominello and Širović 2016; Shabangu et al. 2020). Similarly, no significant diel patterns were found in Hawaii for North Pacific minke whale 'boing' song (Oswald et al. 2011).

Site-specific diel patterns are not uncommon in baleen whale vocalization behavior. For example, in North Atlantic right whales (*Eubalaena glacialis*), differences in diel patterns between two recording sites have been linked with prey availability (Mellinger et al. 2007), and the observed variation in diel patterns of the bio-duck song in different parts of Antarctica might be related to prey distribution as well. Since prey behavior and distribution will also vary seasonally, some of these diverging patterns might be linked to seasonal shifts in prey distribution. More concurrent passive acoustic and prey distribution data are needed to fully explain these patterns.

14.5 Behavioral Contexts and Potential Function of Minke Whale Vocalizations

Acoustic communication using high-amplitude, low-frequency (<1 kHz) signals with the potential to propagate over large distances is a common feature of baleen whale social systems, where individual whales are dispersed over a wide area. In such large acoustically mediated networks, individuals may employ eavesdropping to obtain information about conspecifics and maintain social cohesion (McGregor and Dabelsteen 1996). Although difficult to demonstrate in the field (McComb and Reby 2005), eavesdropping has been suggested to occur in humpback whales, where females could listen to vocal interactions between competing males (Cholewiak 2008; Chap. 11). In addition to true eavesdropping, it has been shown that many mammals attend to vocalizations even if these are not directed at them. Baleen whales might use passive listening to obtain information about a productive food source by attending to calls from conspecifics, and perhaps from heterospecifics, sharing the same acoustic habitat and possibly over very large distances. Finally, baleen whales might be alerted to the presence of predators such as killer whales by passive listening.

In the North Atlantic, minke whale song patterns, consisting of three to four different pulse train types, have been described through localization of individual animals. The localized individuals were in audible range of one another which suggests that these song patterns might play an important role in conspecific interactions. However, due to a lack of visual observation data, the actual behavioral function of these sequences remains unknown (Risch et al. 2014c). Uncertainty regarding the function of baleen whale vocalizations is a common feature across the different species, subspecies, and populations. Given that most known minke whale sounds have only been identified and described within the past 20 years, the biological significance of these sounds any associated acoustic behavior of these elusive species is still in its infancy.

In other baleen whales such as humpback, fin, and blue whales, only males produce song in a reproductive context (Glockner 1983; Croll et al. 2002; Oleson et al. 2007a). Although it is currently unknown whether this is similar in minke whales, the absence of North Atlantic minke whale song in habitats with a high proportion of females, such as the Gulf of St Lawrence, Canada, and west of Greenland, provides circumstantial evidence that it might be (Risch et al. 2014a). It has, therefore, been suggested that minke whale song, which has now been described to occur in all ocean basins (14.3), might also play a role in the reproductive behavior in this species complex (Risch et al. 2013). Another indication that minke whale song in the North Atlantic might be related to reproductive behavior is based on the observation that pulse trains recorded in presumed Caribbean winter breeding grounds are about 20 s longer and exhibit more than twice as many pulses compared to those recorded in the Gulf of Maine during spring and autumn when the whales are on migration (Mellinger et al. 2000; Risch et al. 2013). In other mammal species, such as baboons and giant pandas, it has been shown that call duration can increase with heightened arousal state during the breeding season (Rendall 2003; Charlton et al. 2011).

Minke whale boing sounds recorded in the North Pacific have so far mostly been recorded in the lower latitudes around Hawaii and the Mariana Trench region, and their occurrence appears to be seasonal, with most boings being recorded from October to February, which coincides with minke whale visual sightings in those areas and the presumed breeding period for the population (Rankin et al. 2007; Oswald et al. 2011; Norris et al. 2017). Similar to the situation in the North Atlantic, direct data on age or sex of calling animals are still missing. Based on analogous lines of evidence regarding the stereotypy of the signals, their seasonal and spatial distribution, and in comparison to song and its function described for other baleen whale species, it has been suggested that the boing song might be produced by males in a reproductive context (Norris et al. 2017).

In the South Pacific, given the stereotypy and regular repetition of the star-wars sound recorded on the Great Barrier Reef and its primary production during the breeding season, it has also been speculated that these sounds function as a reproductive advertisement display, although comparable data gaps to those for North Atlantic pulse trains and North Pacific boing sounds exist (Gedamke et al. 2001).

Compared to common and dwarf minke whale songs, the seasonal and spatial distribution patterns of occurrence of Antarctic minke whale bio-duck song appear to be more complex, with songs being produced in high- and low-latitude habitats (Fig. 14.4). It has been shown that bio-duck song in most areas peaks during austral spring and winter (June to August), although some sites show another peak later in the year (December). In addition, bio-duck song units appear to increase in duration during winter (Dominello and Širović 2016). Based on these data and reports of visually observed breeding behavior between August and October, it has been hypothesized that bio-duck song functions in a reproductive context, and mating in this species might take place across a wide range of habitats (Filun et al. 2020).

Compared to the song sequences, low-frequency downsweeps produced by most minke whale populations are less well described and their behavioral function remains also unclear. The most comprehensive study of minke whale downsweeps was carried out in the Gulf of St Lawrence, Canada, where no distinct temporal pattern of downsweep production was found (Edds-Walton 2000). For this summer foraging ground, it was suggested that downsweeps might function to maintain spacing between feeding individuals (Edds-Walton 2000). Downsweeps produced by Antarctic minke whales are positively correlated with sea ice presence, but more data from direct observation or multi-sensor acoustic tags are necessary to confirm the behavioral context of these downsweep vocalizations (Dominello and Širović 2016).

14.6 Minke Whale Hearing and Impacts of Noise Pollution

While information about minke whale hearing capabilities is limited, we assume they can hear one another. In the absence of field measurements, minke whale hearing sensitivity can either be inferred from species-specific vocalizations, behavioral responses to sounds, or anatomical measurements on which model predictions can be based (Mellinger et al. 2000; Tubelli et al. 2012; Yamato et al. 2012; Boisseau et al. 2021; Chap 3). While general estimates of audiograms for baleen whales assume highest sensitivities between 200 and 19,000 Hz (Southall et al. 2019), simulations of the minke whale auditory pathway suggests that the most sensitive frequency range is between approximately 30 and 25,000 Hz (Tubelli et al. 2012).

There are numerous natural biotic and abiotic sound sources in the world's oceans, including weather events such as lightening/thunder and waves, sediment movement due to tidal currents, ice noise in the Arctic and Antarctic, and earthquakes, as well as biological sounds from a variety of taxa from invertebrates to the large whales (Bass and Clark 2003). Like other baleen whales, minke whales have adapted to these natural soundscapes by producing sounds with lower fundamental frequencies and higheramplitudes that increase sound transmission range (Clark 1983).

Since the beginning of the twentieth century, noise levels in some ocean regions have been steadily increasing and have remained high since at least the 1960s (Andrew et al. 2002; McDonald et al. 2006). These increases in and continuing high levels of anthropogenic noise have been driven primarily by shipping noise and noise from seismic airgun arrays for offshore oil exploration. Other sources of man-made noise include offshore construction work, multi-beam echosounders, fish finders, and naval sonar (Duarte et al. 2021).

An 80% loss of communication space for minke whale pulse trains relative to historical 'quiet' conditions has recently been modeled for minke whale habitat in the Northwest Atlantic (Cholewiak et al. 2018). In other parts of the world, minke whale densities have been observed to decrease during naval exercises, and minke whale strandings occurred during at least two mass stranding events linked to military sonar activity (Parsons et al. 2000; Balcomb and Claridge 2001). Another study showed minke whales responding to mid-frequency active sonar by a reduced calling rate and avoidance of the sound source. This behavior was observed even at relatively low estimated received sound levels (Sivle et al. 2015; Martin et al. 2015; Kvadsheim et al. 2017). Minke whales have also been shown to strongly avoid acoustic deterrent devices (ADDs) that are frequently used to mitigate underwater noise of offshore construction activities (Boisseau et al. 2021).

North Pacific minke whales off Hawaii increase the amplitude of their boing sounds in association with increased background noise. However, despite this compensation, the detected increases in sound intensity were not of the same magnitude as increases in background noise, and thus, reductions of communication space with potentially negative impacts would still be expected (Helble et al. 2020).

It is likely that chronically elevated noise levels such as those found in increasingly industrialized ocean regions as a result of human activities can also increase stress levels in minke whales, as has been documented in right whales (Rolland et al. 2012), with unknown consequences for long-term health and reproduction. More research is needed to fully understand the impacts of anthropogenic noise on all baleen whale species, but it is likely that aggregate (e.g., vessel noise, seismic surveys, and sonar) and cumulative effects due to a variety of stressors including underwater noise, habitat loss, and pollution will eventually reach levels that could impact population survival. It is, therefore, important to address known and potential impacts as early as possible using effective mitigation and management actions (Andersen and Clubb 2013).

14.7 Conclusions

Minke whales live in large-scale, acoustically mediated social systems, and similar to other baleen whales, the combination of advanced vocal learning has given rise to some unique and intriguing social behaviors, much of which have yet to be discovered for the different minke populations and subspecies. Future research should focus on elucidating the functions of the different minke whale call and song types in different contexts, differences in age and sex of acoustically active individuals, and changes in their acoustic behaviors in relation to both natural and anthropogenically modified noise conditions. Effort should also be directed toward the identification and description of sounds of unknown but presumed biological origin, especially in regions, such as the South Atlantic, for which data are currently lacking. Finally, new machine learning techniques to enhance automated detection and classification of baleen whale vocalization (Bergler et al. 2019; Allen et al. 2021), coupled with a better description of individual calling rates and source levels using localization techniques, will enhance the application of density estimation methods to assess and monitor population size using passive acoustics (Marques et al. 2013; Martin et al. 2013; Norris et al. 2017).

Acknowledgements In memory of Thomas F. Norris, whose dedication and enthusiasm for studying marine mammal acoustics and minke whales were so infectious. Many shared discussions of minke whales sounds and their potential function have inspired large parts of this review. I thank Sofie van Parijs for her mentorship, support, and encouragement over the years, which ultimately enabled me to write this review. Finally, many thanks to Ellen Garland and Christopher W. Clark for their helpful comments on drafts of this chapter.

References

Acevedo J, Aguayo-Lobo A, Acuna P, Pastene L (2006) A note on the first record of the dwarf minke whale (*Balaenoptera acutorostrata*) in Chilean waters. J Cetacean Res Manag 8:293

Allen AN, Harvey M, Harrell L et al (2021) A convolutional neural network for automated detection of humpback whale song in a diverse, long-term passive acoustic dataset. Front Mar Sci. https://doi.org/10.3389/fmars.2021.607321

Andersen MS, Clubb DO (2013) Understanding and accounting for the costs of inaction. In: European Environment Agency (ed) Late lessons from early warnings: science, precaution, innovation. Rosendahls-Schultz Grafisk, Copenhagen, p 33

Andrew RK, Howe BM, Mercer JA, Dzieciuch MA (2002) Ocean ambient sound: Comparing the 1960s with the 1990s for a receiver off the California coast. Acoust Res Lett Online 3:65. https://doi.org/10.1121/1.1461915
Balcomb K, Claridge DE (2001) A mass stranding of cetaceans caused by naval sonar in the Bahamas. Bahams J Sci 3:2–12
Bass AH, Clark CW (2003) The physical acoustics of underwater sound communication. In: Simmons AM, Fay RR, Popper AN (eds) Acoustic communication. Springer, New York, pp 15–64
Beamish P, Mitchell E (1973) Short pulse length audio frequency sounds recorded in the presence of a Minke whale (Balaenoptera acutorostrata). Deep Sea Res Oceanogr Abstr 20:375–386. https://doi.org/10.1016/0011-7471
Bergler C, Schröter H, Cheng RX et al (2019) ORCA-SPOT: an automatic killer whale sound detection toolkit using deep learning. Sci Rep 9:10997. https://doi.org/10.1038/s41598-019-47335-w
Best PB (1985) External characters of southern minke whales and the existence of a diminutive form. Sci Rep Whales Res Inst 36:1–33
Boisseau O, McGarry T, Stephenson S et al (2021) Minke whales avoid a 15 kHz acoustic deterrent device. Mar Ecol Prog Ser. https://doi.org/10.3354/meps13690
Brenowitz EA (1982) The active space of red-winged blackbird song. J Comp Physiol 147:511–522. https://doi.org/10.1007/BF00612017
Buchan SJ, Balcazar-Cabrera N, Stafford KM (2020) Seasonal acoustic presence of blue, fin, and minke whales off the Juan Fernández Archipelago, Chile (2007–2016). Mar Biodivers 50:1–10. https://doi.org/10.1007/s12526-020-01087-3
Calambokidis J, Steiger G, Straley J et al (2001) Movements and population structure of humpback whales in the North Pacific. Mar Mammal Sci 17:769–794. https://doi.org/10.1111/j.1748-7692.2001.tb01298.x
Catchpole CK, Slater PJB (2008) Bird song: biological themes and variations, 2nd edn. Cambridge University Press, Cambridge
Cerchio S, Rasoloarijao T, Cholewiak DM (2018) Acoustic monitoring of blue whales (*Balaenoptera musculus*) and other baleen whales in the mozambique channel off the Northwest coast of madagascar. International Whaling Commission
Cerchio S, Laran S, Andrianarivelo N, et al (2022) Cetacean species diversity in Malagasy waters. In: Goodman S (ed) The new natural history of Madagascar. Princeton University Press, Princeton
Charlton BD, Keating JL, Kersey D et al (2011) Vocal cues to male androgen levels in giant pandas. Biol Lett 7:71–74. https://doi.org/10.1098/rsbl.2010.0582
Cholewiak D (2008) Evaluating the role of song in the humpback whale (*Megaptera novaeangliae*) breeding system with respect to intra sexual interactions. Cornell University
Cholewiak DM, Sousa-Lima RS, Cerchio S (2013) Humpback whale song hierarchical structure: historical context and discussion of current classification issues. Mar Mammal Sci 29:E312–E332. https://doi.org/10.1111/mms.12005
Cholewiak D, Clark CW, Ponirakis D et al (2018) Communicating amidst the noise: modeling the aggregate influence of ambient and vessel noise on baleen whale communication space in a national marine sanctuary. Endanger Species Res 36:59–75. https://doi.org/10.3354/esr00875
Clapham PJ, Mead JG (1999) Megaptera novaeangliae. Mamm Species 604:1–9
Clark CW (1983) Acoustic Communication and Behavior of the Southern right whale (Eubalaena australis). In: Payne R (ed) Communication and behavior of whales, vol. 76. Avalon Publishing, pp 163–198
Clark CW, Ellison WT (2004) Potential use of low-frequency sounds by baleen whales for probing the environment: evidence from models and empirical measurements. In: Thomas J, Moss CF, Vater M (eds) Advances in the study of echolocation in bats and dolphins. The University of Chicago Press, Chicago, pp 564–582

Clark CW, Gagnon GJ (2002) Low-frequency vocal behaviors of baleen whales in the North Atlantic: insights from integrated undersea surveillance system detections, locations, and tracking from 1992 to 1996. US Navy J Underw Acoust 52:609–640
Corkeron PJ, Connor RC (1999) Why do baleen whales migrate? Mar Mammal Sci 15:1228–1245. https://doi.org/10.1111/j.1748-7692.1999.tb00887.x
Croll D, Clark CW, Acevedo A et al (2002) Only male fin whales sing loud songs. Nature 417:809. https://doi.org/10.1038/417809a
Cummings WC, Diego S, Holliday DV (1987) Sounds and source levels from bowhead whales off Pt. Barrow, Alaska in the Canadian. 82
Dawbin WH (1966) The seasonal migratory cycle of humpback whales. In: Norris KS (ed) Whales, dolphins, and porpoises. University of California Press, Los Angeles, pp 145–170
Delarue J, Martin B, Hannay D (2013) Minke whale boing sound detections in the northeastern Chukchi Sea. Mar Mammal Sci 29:E333–E341. https://doi.org/10.1111/j.1748-7692.2012.00611.x
Dolman SJ, Swift RJ, Asmus K, Thiele D (2005) Preliminary analysis of passive acoustic recordings made in the Ross Sea during ANSLOPE III in 2004. Pap SC/57/E10 Present to Sci Comm Int Whal Com 1–8
Dominello T, Širović A (2016) Seasonality of antarctic minke whale (Balaenoptera bonaerensis) calls off the western antarctic Peninsula. Mar Mammal Sci 32:826–838. https://doi.org/10.1111/mms.12302
Duarte CM, Chapuis L, Collin SP et al (2021) The soundscape of the anthropocene ocean. Science (80-)371. https://doi.org/10.1126/science.aba4658
Edds PL, Macfarlane JAF (1987) Occurrence and general behavior of balaenopterid cetaceans summering in the St. Lawrence Estuary, Canada. Can J Zool 65:1363–1376. https://doi.org/10.1139/z87-216
Edds-Walton P (2000) Vocalizations of minke whales Balaenoptera acutorostrata in the St. Lawrence Estuary. Bioacoustics 11:31–50. https://doi.org/10.1080/09524622.2000.9753448
Filun D, Thomisch K, Boebel O et al (2020) Frozen verses: Antarctic minke whales (*Balaenoptera bonaerensis*) call predominantly during austral winter. R Soc Open Sci 7:192112. https://doi.org/10.1098/rsos.192112
Folkow LP, Blix AS (1991) Norwegian whale sighting and acoustic surveys in the Atlantic Ocean during the winter of 1989/90. Rep Int Whal Comm 41:531–538
Gedamke J, Costa DP, Dunstan A (2001) Localization and visual verification of a complex minke whale vocalization. J Acoust Soc Am 109:3038–3047. https://doi.org/10.1121/1.1371763
Gedamke J (2004) Minke whale song, spacing, and acoustic communication on the Great Barrier Reef, Australia
Glockner D (1983) Determining the sex of humpback whales (*Megaptera novaeangliae*) in their natural environment. In: Payne RS (ed) Communication and behavior of whales. Westview Press, Boulder, pp 447–464
Helble TA, Guazzo RA, Martin CR et al (2020) Lombard effect: Minke whale boing call source levels vary with natural variations in ocean noise. J Acoust Soc Am 147:698–712. https://doi.org/10.1121/10.0000596
Hoelzel AR, Dorsey EM, Stern SJ (1989) The foraging specializations of individual minke whales. Anim Behav 38:786–794. https://doi.org/10.1016/S0003-3472(89)80111-3
Horwood JW (1990) Biology and exploitation of the minke whale. CRC Press, Boca Raton
Kasamatsu F, Nishiwaki S, Ishikawa H (1995) Breeding areas and southbound migrations of southern minke whales *Balaenoptera acutorostrata*. Mar Ecol Prog Ser 119:1–10. https://doi.org/10.3354/meps119001
Klinck H, Burkhardt E (2008) Marine mammal automated perimeter surveillance—MAPS. Rep Polar Mar Res 580:114–121
Kraus SD, Prescott JH, Knowlton AR, Stone GS (1986) Migration and calving of right whales (*Eubalaena glacialis*) in the western North Atlantic. Rep Int Whal Comm 10:139–144

Kvadsheim PH, DeRuiter S, Sivle LD et al (2017) Avoidance responses of minke whales to 1–4kHz naval sonar. Mar Pollut Bull 121:60–68. https://doi.org/10.1016/j.marpolbul.2017.05.037
Leatherwood S, Thomas JA, Awbrey FT (1981) Minke whales oft northwestern ross Island. Antarct J 154–156
Lee JF, Friedlaender AS, Oliver MJ, DeLiberty TL (2017) Behavior of satellite-tracked Antarctic minke whales (*Balaenoptera bonaerensis*) in relation to environmental factors around the western Antarctic Peninsula. Anim Biotelemetry 5:23. https://doi.org/10.1186/s40317-017-0138-7
Marques TA, Thomas L, Martin SW et al (2013) Estimating animal population density using passive acoustics. Biol Rev 88:287–309. https://doi.org/10.1111/brv.12001
Martin SW, Marques TA, Thomas L et al (2013) Estimating minke whale (*Balaenoptera acutorostrata*) boing sound density using passive acoustic sensors. Mar Mammal Sci 29:142–158. https://doi.org/10.1111/j.1748-7692.2011.00561.x
Martin SW, Martin CR, Matsuyama BM, Henderson EE (2015) Minke whales (*Balaenoptera acutorostrata*) respond to navy training. J Acoust Soc Am 137:2533–2541. https://doi.org/10.1121/1.4919319
Mate BR, Best PB, Lagerquist BA, Winsor MH (2011) Coastal, offshore, and migratory movements of South African right whales revealed by satellite telemetry. Mar Mammal Sci 27:455–476. https://doi.org/10.1111/j.1748-7692.2010.00412.x
Matthews D, Macleod R, McCauley RD (2004) Bio-duck activity in the perth canyon. An Automatic Detection Algorithm. Proc Acoust 1–4
McCauley R (2004) Western Australian exercise area blue whale project. C Rep R2004–29. Proj - 350:1–73
McComb K, Reby D (2005) Vocal communication networks in large terrestrial mammals. In: McGregor PK (ed) Animal communication networks. Cambridge University Press, Cambridge, pp 372–389
McDonald MA, Hildebrand JA, Wiggins SM (2006) Increases in deep ocean ambient noise in the Northeast Pacific west of San Nicolas Island, California. J Acoust Soc Am 120:711. https://doi.org/10.1121/1.2216565
McGregor PK, Dabelsteen T (1996) Communication networks. In: Kroodsma DE, Miller HE (eds) Ecology and evolution of acoustic communication in birds. Cornell University Press, Ithaca, pp 409–425
Mellinger DK, Carson CD, Clark CW (2000) Characteristics of minke whale (*Balaenoptera acutorostrata*) pulse trains recorded near Puerto Rico. Mar Mammal Sci 16:739–756. https://doi.org/10.1111/j.1748-7692.2000.tb00969.x
Mellinger DK, Nieukirk SL, Matsumoto H et al (2007) Seasonal occurrence of North Atlantic right whale (*Eubalaena glacialis*) vocalizations at two sites on the Scotian Shelf. Mar Mammal Sci 23:856–867. https://doi.org/10.1111/j.1748-7692.2007.00144.x
Menze S, Zitterbart DP, van Opzeeland I, Boebel O (2017) The influence of sea ice, wind speed and marine mammals on Southern Ocean ambient sound. R Soc Open Sci 4
Nieukirk SL, Stafford KM, Mellinger DK et al (2004) Low-frequency whale and seismic airgun sounds recorded in the mid-Atlantic Ocean. J Acoust Soc Am 115:1832–1843. https://doi.org/10.1121/1.1675816
Nieukirk SL, Fregosi S, Mellinger DK, Klinck H (2016) A complex baleen whale call recorded in the Mariana Trench Marine National Monument. J Acoust Soc Am 140:EL274–EL279. https://doi.org/10.1121/1.4962377
Nikolich K, Towers JR (2018) Vocalizations of common minke whales (*Balaenoptera acutorostrata*) in an eastern North Pacific feeding ground. Bioacoustics 29:97–108. https://doi.org/10.1080/09524622.2018.1555716
Norris TF, Dunleavy KJ, Yack TM, Ferguson EL (2017) Estimation of minke whale abundance from an acoustic line transect survey of the Mariana Islands. Mar Mammal Sci 33:574–592. https://doi.org/10.1111/mms.12397
Norris TF, Martin S, Thomas L et al (2012) Acoustic ecology and behavior of minke whales in the Hawaiian and marianas Islands: localizations, abundance estimation, and characterization of

minke whale "Boings." In: Popper AN, Hawkins T (eds) The effects of noise on aquatic life. pp 149–153

Oleson EM, Calambokidis J, Burgess WC et al (2007a) Behavioral context of call production by eastern North Pacific blue whales. Mar Ecol Prog Ser 330:269–284. https://doi.org/10.3354/meps330269

Oleson EM, Wiggins SM, Hildebrand JA (2007b) Temporal separation of blue whale call types on a southern California feeding ground. Anim Behav 74:881–894. https://doi.org/10.1016/j.anbehav.2007.01.022

Oswald JN, Au WWL, Duennebier F (2011) Minke whale (*Balaenoptera acutorostrata*) boings detected at the Station ALOHA Cabled Observatory. J Acoust Soc Am 129:3353–3360. https://doi.org/10.1121/1.3575555

Parsons ECM, Birks I, Evans PGH et al (2000) The possible impacts of military activity on cetaceans in West Scotland. Eur Res Cetaceans 14:185–190

Pastene LA, Acevedo J, Goto M et al (2010) Population structure and possible migratory links of common minke whales, *Balaenoptera acutorostrata*, in the Southern Hemisphere. Conserv Genet 11:1553–1558. https://doi.org/10.1007/s10592-009-9944-7

Payne RS, McVay S (1971) Songs of humpback whales. Science (80-) 173:585–597. https://doi.org/10.1126/science.173.3997.585

Poulter T (1964) Recording # 120342 (December 26th, 1964). Macaulay Libr

Rankin S, Barlow J (2005) Source of the North Pacific "boing" sound attributed to minke whales. J Acoust Soc Am 118:3346–3351. https://doi.org/10.1121/1.2046747

Rankin S, Norris TF, Smultea MA et al (2007) A visual sighting and acoustic detections of minke whales, balaenoptera acutorostrata (*Cetacea: Balaenopteridae*), in Nearshore Hawaiian Waters. Pacific Sci 61:395–398. https://doi.org/10.2984/1534-6188

Rendall D (2003) Acoustic correlates of caller identity and affect intensity in the vowel-like grunt vocalizations of baboons. J Acoust Soc Am 113:3390–3402. https://doi.org/10.1121/1.1568942

Rice DW (1998) Marine mammals of the world: systematics and distribution. In: Special publication number 4, the society for marine mammology. Lawrence, Kansas

Risch D, Clark CW, Dugan PJ et al (2013) Minke whale acoustic behavior and multi-year seasonal and diel vocalization patterns in Massachusetts Bay, USA. Mar Ecol Prog Ser 489:279–295. https://doi.org/10.3354/meps10426

Risch D, Castellote M, Clark C et al (2014a) Seasonal migrations of North Atlantic minke whales: novel insights from large-scale passive acoustic monitoring networks. Mov Ecol 2:24. https://doi.org/10.1186/s40462-014-0024-3

Risch D, Gales NJ, Gedamke J et al (2014b) Mysterious bio-duck sound attributed to the Antarctic minke whale (*Balaenoptera bonaerensis*). Biol Lett 10:20140175. https://doi.org/10.1098/rsbl.2014.0175

Risch D, Siebert U, van Parijs SM (2014c) Individual calling behaviour and movements of North Atlantic minke whales (*Balaenoptera acutorostrata*). Behaviour 151:1335–1360. https://doi.org/10.1163/1568539X-00003187

Risch D, Baumann-Pickering S, Norris T et al (2015) High-frequency clicks associated with mink whale pulse trains at three locations in the Western North Atlantic. In: 21st Biennial conference on the biology of marine mammals. San Francisco

Risch D, Wilson SC, Hoogerwerf M et al (2019) Seasonal and diel acoustic presence of North Atlantic minke whales in the North Sea. Sci Rep 9:1–11. https://doi.org/10.1038/s41598-019-39752-8

Rolland RM, Parks SE, Hunt KE et al (2012) Evidence that ship noise increases stress in right whales. Proc R Soc B Biol Sci 279:2363–2368. https://doi.org/10.1098/rspb.2011.2429

Rychel AL, Reeder TW, Berta A (2004) Phylogeny of mysticete whales based on mitochondrial and nuclear data. Mol Phylogenet Evol 32:892–901. https://doi.org/10.1016/j.ympev.2004.02.020

Schevill WE, Watkins WA (1972) Intense low-frequency sounds from an antarctic minke whale, Balaenoptera acutorostrata. Breviora 388:1–8

Shabangu FW, Findlay K, Stafford KM (2020) Seasonal acoustic occurrence, diel-vocalizing patterns and bioduck call-type composition of Antarctic minke whales off the west coast of South Africa and the Maud Rise, Antarctica. Mar Mammal Sci 36:658–675. https://doi.org/10.1111/mms.12669
Širović A, Williams LN, Kerosky SM et al (2013) Temporal separation of two fin whale call types across the eastern North Pacific. Mar Biol 160:47–57. https://doi.org/10.1007/s00227-012-2061-z
Sivle LD, Kvadsheim PH, Curé C et al (2015) Severity of expert-identified behavioural responses of humpback whale, minke whale, and Northern Bottlenose whale to Naval Sonar. Aquat Mamm 41:469–502. https://doi.org/10.1578/AM.41.4.2015.469
Skaug HJ, Øien N, Schweder T, Bøthun G (2004) Abundance of minke whales (*Balaenoptera acutorostrata*) in the Northeast Atlantic: variability in time and space. Can J Fish Aquat Sci 61:870–886. https://doi.org/10.1139/F04-020
Southall BL, Finneran JJ, Reichmuth C et al (2019) Marine mammal noise exposure criteria: updated scientific recommendations for residual hearing effects. Aquat Mamm 45:125–232. https://doi.org/10.1578/AM.45.2.2019.125
Stafford KM, Lydersen C, Wiig Ø, Kovacs KM (2018) Extreme diversity in the songs of Spitsbergen's bowhead whales. Biol Lett 14:20180056. https://doi.org/10.1098/rsbl.2018.0056
Stevick PT, Neves MC, Johansen F et al (2011) A quarter of a world away: female humpback whale moves 10,000 km between breeding areas. Biol Lett 7:299–302. https://doi.org/10.1098/rsbl.2010.0717
Thomisch K, Boebel O, Bachmann J et al (2019) Temporal patterns in the acoustic presence of baleen whale species in a presumed breeding area off Namibia. Mar Ecol Prog Ser 620:201–214. https://doi.org/10.3354/meps12952
Thompson PO, Friedl WA (1982) A long term study of low frequency sounds from several species of whales off Oahu, Hawaii. Cetology 45:1–19
Tubelli AA, Zosuls A, Ketten DR et al (2012) A prediction of the minke whale (*Balaenoptera acutorostrata*) middle-ear transfer function. J Acoust Soc Am 132:3263. https://doi.org/10.1121/1.4756950
Valentine PS, Birtles A, Curnock M et al (2004) Getting closer to whales—passenger expectations and experiences, and the management of swim with dwarf minke whale interactions in the Great Barrier Reef. Tour Manag 25:647–655. https://doi.org/10.1016/j.tourman.2003.09.001
Van Opzeeland IC (2010) Acoustic ecology of marine mammals in polar oceans. Reports Polar Mar Res 619:332
Wenz GM (1964) Curious noises and the sonic environment in the ocean. In: Tavolga WN (ed) Marine bio-acoustics. Pergamon, New York, pp 101–119
Whitehead H (2001) Analysis of animal movement using opportunistic individual identifications: application to sperm whales. Ecology 82:1417–1432. https://doi.org/10.1890/0012-9658
Winn HE, Perkins PJ (1976) Distribution and sounds of the minke whale, with a review of mysticete sounds. Cetology 19:1–12
Yamato M, Ketten DR, Arruda J et al (2012) The auditory anatomy of the minke whale (*Balaenoptera acutorostrata*): a potential fatty sound reception pathway in a baleen whale. Anat Rec Adv Integr Anat Evol Biol 295:991–998. https://doi.org/10.1002/ar.22459
Zerbini AN, Secchi ER, Siciliano S, Simoes-Lopes P (1996) The dwarf form of the minke whale Balaenoptera acutorostrata Lacépède, 1804, in Brazil. Report: International Whaling Commission 46

Chapter 15
The Omura's Whale: Exploring the Enigma

Salvatore Cerchio

Abstract The rapidly growing understanding of the Omura's whale (*Balaenoptera omurai*) is an example of how the drive to explore and discover has been enhanced by advances in technology, particularly in bioacoustics. This combination of inspiration and technology has allowed exploration of topics that had been out of reach for many scientists. The Omura's whale was first described in 2003 from a handful of specimens, and for a decade remained an enigmatic "ghost" species without confirmation of live sightings, let alone any understanding of its ecology. After the chance discovery of a population off northwest Madagascar, the stage was set for the exploration of a new species. In five short years, through the application of recent advancements in bioacoustics, molecular genetics, and satellite telemetry, we may have a better understanding of the Omura's whale (or at least one of its populations) than of many other cetaceans. In this chapter, I describe new information gathered on Omura's whales off Madagascar during 2013–2017, with a focus on its singing behavior. Combined with growing information from other sites around the world, I make inferences about the species' behavioral patterns and describe the potential for innovative research that could be conducted on this enigmatic species.

Keywords Omura's whale · *Balaenoptera omurai* · Balaenopteridae · Passive acoustic monitoring · Spatiotemporal distribution · Song · Singing behavior · Geographic variation · Breeding ecology

---

S. Cerchio (✉)
African Aquatic Conservation Fund, Chilmark, MA 02535, USA
e-mail: scerchio@africanaquaticconservation.org

C.W. Clark and E. C. Garland (eds.), *Ethology and Behavioral Ecology of Mysticetes*, Ethology and Behavioral Ecology of Marine Mammals,
https://doi.org/10.1007/978-3-030-98449-6_15

An Omura's whale mother with her calf near Banc du Goliath, north of Nosy Be, Madagascar, in November 2018. This female was first seen on Nosy Be without a calf in November 2015. Drone photo collected under research permit 266/18/MEEF/SG/DGF/DSAP/SCB issued by la Direction Générale de l'Environnement et des Forêts, and la Direction de la Gestion des Ressources Naturelles Renouvelables et des Écosystèmes, Madagascar. (*Photo credit* and ©, Salvatore Cerchio)

15.1 Introduction

During February and March 1984, the *r/v Regina Maris* set anchor each day on Silver Bank, Dominican Republic, in the fading light of the late afternoon. As a young student under the tutelage of Ken Balcomb, I took that cue to hang a small hydrophone from the course yard of the foremast, and plug it into the latest advancement in commercially available acoustic recording technology, a Sony Walkman Pro. Through the night, I lay on the deck of the Regina, listening to the cacophony of a dense chorus of humpback whale singers, waiting for a single singer to come into the foreground, and then push the REC button to record his song onto a 45 m cassette audio tape. Since portable computers had yet to be invented, and digital signal processing software was not yet accessible to non-specialists, analysis of these recordings started with hand notation taught to a few students like me by Katy Payne, to scroll the sequences of song phrases onto a piece of lined engineering paper. Later at a farmhouse lab in Lincoln, Massachusetts, I followed a laborious process to create hardcopy spectrograms on long strips of 35 mm photographic paper, from which I, and other students and researchers, would qualitatively define the structure of the songs for each individual and make quantitative measurements of duration and frequency using a metric ruler. Archaic as these practices seem today, they represented remarkable innovations in the 1980s, leaps in technology from previous decades that facilitated new research outside the walls of exclusive, highly funded labs. These allowed young students like me to take part in rapidly progressing research that was laying down the foundations of what we know about cetacean bioacoustics. This was fertile ground for the city-boy that I was, who had been obsessed with nature since a child, and was mesmerized by the sounds of humpback whales when first played off a flexible little vinyl record pulled from an issue of National Geographic. My natural curiosity and penchant for discovery were ignited by these experiences on the *Regina Maris* and in the Paynes' lab. And so, it was this combination of the romance of the field and the sheer exhilaration of discovery that captured my soul and set me on my career's path. That path would ultimately lead me to a humbling discovery and the rare opportunity to explore and describe the behavior of an enigmatic species of baleen whale, the Omura's whale, which in turn would be enhanced by a new wave of technological advancements.

15.2 Background

Discovery and leaps in technology progress together and feed into each other in a self-reinforcing loop, continuously advancing our understanding of the natural world. Perhaps nowhere has this been as evident in the past two decades, as with the discovery of a new species of baleen whale, an animal the size of a bus, which for centuries had been overlooked in the tropical waters of mostly remote exotic regions. The Omura's whale (*Balaenoptera omurai*) was only described and established as a distinct species in 2003, historically misidentified as the "small form" of the Bryde's whale (*Balaenoptera edeni*), when it was first encountered during Japanese scientific whaling operations in the late 1970s (Ohsumi 1980; Wada et al. 2003). In fact, it was not even part of, nor the nearest relative to the Bryde's whale complex (which currently includes Bryde's whale, *Balaenoptera brydei*; Eden's whale, *B. edeni*; and most recently Rice's whale, *Balaenoptera ricei*). Early research on the Omura's whale established that it is an ancient lineage, phylogenetically distinct with a divergence time of ca. 17 million years ago, and basal to a large clade containing the three species of the Bryde's whale complex and the sei whale (*Balaenoptera borealis*) (Sasaki et al. 2006). This was not the simple splitting of two closely related groups that had recently diverged, which has become quite common with advances in phylogenetic research (such as is the case with Rice's whale, Rosel et al. 2021), but a novel discovery in the spirit of nineteenth century exploration. Until 2013, the species remained an enigma, known only from a handful of historic whaling and stranding specimens from the western Pacific and eastern Indian Oceans (Wada et al. 2003, Yamada et al. 2006a, b, 2008; reviewed in Cerchio et al. 2019). With no confirmed observations or detailed ecological studies in its natural habitat, there was a complete absence of information on species range, conservation status, ecology, acoustic repertoire, and behavior.

Enter Madagascar, an isolated island continent in the southwest Indian Ocean, off the southeastern coast of Africa. At the time of the 2003 description of the Omura's whale in the Indo-Pacific region, little was known about the diversity of cetaceans in the coastal waters of Madagascar. More recently, field surveys and passive acoustic monitoring conducted from 2009 through 2019 off the island of Nosy Be established the northwest waters of Madagascar as a cetacean biodiversity hotspot. Research off northwest Madagascar has now documented a minimum of 22 different cetacean species with an additional seven species likely to occur (Cerchio et al. in press). Among these was the unexpected discovery in 2013 of a previously unknown population of Omura's whale (Cerchio et al. 2015). Prior to this discovery, the nearest documentation of Omura's whale came from a place over 5000 km away off the Coco (Keeling) Islands in the eastern Indian Ocean, where Japanese scientific whaling operations had hunted Bryde's whales (Ohsumi 1980; Wada et al. 2003). Thus, the Omura's whales discovered off northwest Madagascar were the first members of the species identified from the western Indian Ocean and represented a substantive expansion of the known species range. Extensive field observations during 2013 and 2014 in the Nosy Be region resulted in an initial study that in just two years established some fundamental ecological aspects of the species. This work revealed the region to be an important habitat for Omura's whales and provided the first *in situ* observations of this poorly known species globally (Cerchio et al. 2015). Phylogenetic analysis of tissue samples from 18 animals confirmed that the Madagascar population was definitively Omura's whale, and also suggested low genetic variation across its range. This was itself the result of remarkable leaps in molecular genetic techniques since the first recognition of the taxon in the early 1990s. Wada and Numachi (1991) first suggested a species-level classification for Omura's whales based on allozyme analysis, a laborious time-consuming technique of identifying protein polymorphisms that predated the polymerase chain reaction (PCR) and wide use of mitochondrial DNA sequencing to assess taxonomy. Once the Nosy Be samples were imported from Madagascar (a laborious bureaucratic process that took a few years), species identification following PCR and mtDNA sequencing actually took only two weeks; so ironically, the confirmation that the medium-sized balaenopterids of Nosy Be were Omura's whale was not delayed because of technological limitations, but rather modern bureaucracy.

Once the suspected species attribution had been confirmed, detailed photographic evidence allowed the first accurate description of external appearance and scientific depiction of the species (Cerchio et al. 2015; Cerchio and Yamada 2018). The Omura's whale had previously been confused with Bryde's whales, primarily due to its similar size, body shape, and tropical distribution; however, its external appearance was quite distinct, with a different dorsal fin shape, and diagnostic pigmentation that includes extreme asymmetry between left and right sides (similar to a fin whale, though bolder and more striking) (Fig. 15.1). Sighting data included documentation of mothers with young calves, as indicated by their small size and bent dorsal fins, an indication that calving occurred off Madagascar. There was also extensive documentation of foraging behavior, including surface lunge feeding on swarms of euphausiids and likely other zooplankton. Acoustic recordings during these initial years, from a boat in the presence of a whale and from archival recorders, indicated a form of acoustic behavior that included stereotyped and rhythmically repetitive sounds that were at times heard in choruses of multiple individuals (Cerchio et al. 2015). These aspects of acoustic behavior are indicative of singing in other balaenopterids (e.g., fin whales, *Balaenoptera physalus*, Watkins et al. 1987, 2000; and blue whales,

Fig. 15.1 Omura's whale, *Balaenoptera omurai*, northwest Madagascar population, showing asymmetric pigmentation between right side (top) and left side (bottom) and typical external characters; illustration by Frédérique Lucas, executed with photographic and video materials from Nosy Be, Madagascar. Reprinted with permission from F. Lucas & S. Cerchio

Balaenoptera musculus, MacDonald et al. 2006) and are therefore suggestive of a male breeding display. Nosy Be ecotourism operators reported sighting these "petite rorquals" while working from April through December. Taken together, these observations led to the conclusion that Nosy Be and the northwest waters of Madagascar support both breeding and feeding habitat for a potentially small, resident population of Omura's whales.

Since the initial publication of the Madagascar observations in Cerchio et al. (2015), multiple accounts of Omura's whales have been reported throughout the globe, at least in part due to an elevated awareness of the species and descriptions of their diagnostic field characters and acoustic behavior. Cerchio et al. (2019) compiled accounts from all available sources, verifying records based on a descending hierarchy of genetic, morphological, photographic, acoustic, and/or detailed sighting data. It was found that the Omura's whale is widely distributed around the globe and has currently been documented in tropical and/or warm-temperate latitudes of all ocean basins between 35°N and 35°S, although primarily in the true tropics (Fig. 15.2). There are currently no confirmed reports in the central and eastern Pacific Ocean (both north and south), even in areas that have been extensively visually surveyed such as the Eastern Tropical Pacific (ETP). This suggests that the central Pacific may represent a real barrier to dispersal as it does for other highly coastal marine mammal taxa (e.g., genera *Sousa* and *Neophocaena*). Most accounts of Omura's whales are from neritic continental shelf areas; however, there are several deep water oceanic accounts, suggesting that preferred habitat varies among populations. Cerchio et al. (2019) suggested that some populations may have exclusive shallow water shelf habits, whereas others may utilize both shallow and nearby deep water habitats, likely related to the availability and distribution of food resources. Among the accounts reported were several cases of acoustic documentation, illustrating that Omura's whale vocalizations are stereotyped and recognizable across a broad range of geographic locations and ocean conditions from the central Atlantic Ocean to the southeastern Indian Ocean.

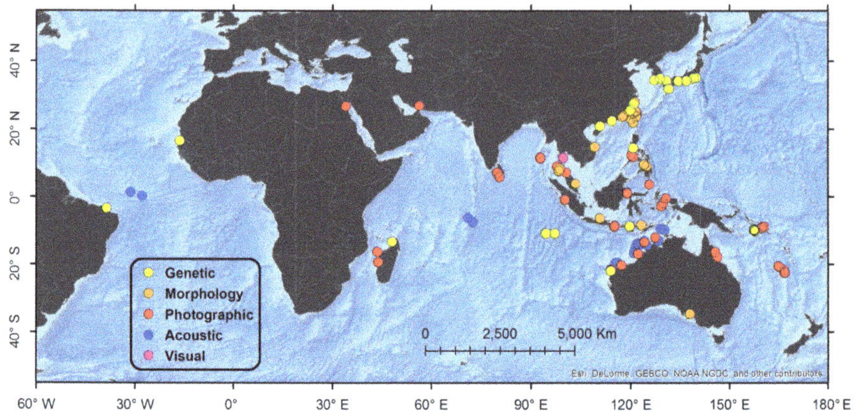

Fig. 15.2 Map showing global distribution of reported Omura's whales based on a diagnostic verification method. Legend: "Genetic" refers to typing of mtDNA markers; "Morphology" refers specifically to diagnostic characters of the skull; "Photographic" refers to photographs or videos illustrating diagnostic external physical characters and pigmentation; "Acoustic" refers to recordings of vocalizations; "Visual" refers to a personal report congruent with external appearance when also substantiated by a separate account from the same general locale verified by a higher level method. When multiple methods of verification are available from a single account, the highest level is reported in order of presentation, with "Genetic" being the highest. (From Cerchio et al. (2019, Fig. 1) Global distribution of Omura's whales (*Balaenoptera omurai*) and assessment of range-wide threats. Front. Mar. Sci., https://doi.org/10.3389/fmars.2019.00067 (CC BY)

15.3 Insights from Passive Acoustic Monitoring off Madagascar

15.3.1 Omura's Whale Sightings off Nosy Be

During six field seasons of coastal boat surveys between July and November (2011–2017) within an approximately 80 km coastal range off Nosy Be, Madagascar, there was a total of 286 encounters consisting of 311 Omura's whales, not accounting for re-sightings of individuals between groups (Fig. 15.3). Most of these encounters occurred during the 2015–2017 field seasons, with only 44 sightings documented during 2011–2014. This increase in sightings during the latter three years is believed to be a combination of both greater focus on Omura's whales and a higher encounter rate due to a real increase in density/abundance of animals during 2015 and 2016 (also suggested by local tour operators and fishers). During boat surveys, Omura's whales showed a strong preference for shallow continental shelf or neritic habitat, avoiding the near-coastal habitat (<20 m depth) and the deeper waters of the off-shelf pelagic habitat (>100 m depth); accordingly, the mean bottom depth among 60 encounters measured in situ was 53.8 m (range 23–81 m).

The social structure of the species, at least within this study region, may be best characterized by "loose aggregations" of individuals. When applying a traditional

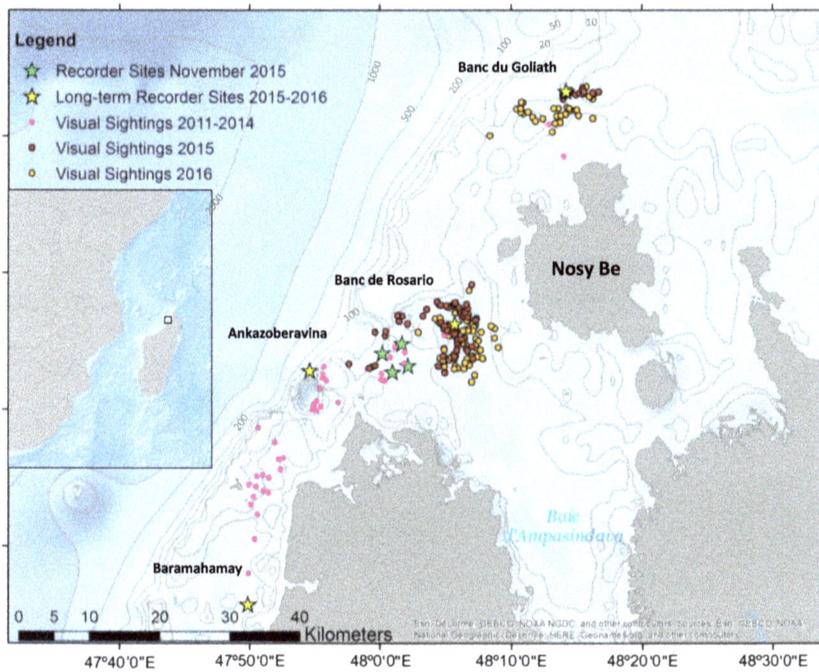

Fig. 15.3 Map indicating sightings of Omura's whales in the Nosy Be, Madagascar study area and positions of passive acoustic recorder sampling sites (stars) during the 2015–2016 study period. Sightings are shown separately for 2011–2014, 2015 and 2016 (colored dots).

definition of "group" typically used for humpback whales (*Megaptera novaeangliae*) and other large baleen whales (i.e., multiple individuals within a few body lengths of each other), Omura's whales were predominantly encountered as singletons (mode = 1) with a mean group size of 1.1 (s.d. 0.3, range 1–2, $n = 261$; Cerchio et al. unpublished). In only three cases were two adults considered ephemerally associated as a "group"; the only other "groups" encountered were mother-calf pairs ($n = 22$). However, despite this seemingly solitary social structure based on a restrictive definition of "group", when one individual was encountered there were often several others within 100 s of meters to several kilometers, likely within acoustic communication range of one another. Therefore, the term "aggregation" may be a more appropriate term than "group" when considering the social structure of the species (Chap. 7). The largest aggregation of Omura's whale documented to date was encountered in 2015, with photographic identification indicating a minimum of 12 individuals within a 2 × 5 km area that was surveyed over a 3 h period (Fig. 15.4). This aggregation was documented 10 km off the north coast of Nosy Be just south of a submerged bathymetric feature called Banc du Goliath.

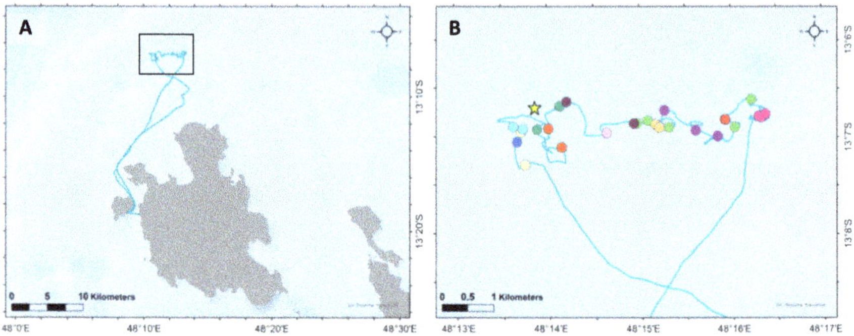

Fig. 15.4 Boat survey track (**a**) north of Nosy Be, Madagascar, on November 19, 2015, illustrating an example of an Omura's whale aggregation (**b**) in an approximately 2 × 5 km area that was surveyed over 3 h, within which a minimum of 12 different individual Omura's whales were photographically identified. Each different color circle indicates the position(s) of a different individual, and the yellow star indicates the position of the long-term recorder that was thereafter deployed at Banc du Goliath.

15.3.2 Acoustic Monitoring Effort

Omura's whale vocalizations were initially described from boat-based recordings in the presence of individuals off Nosy Be, Madagascar (Cerchio et al. 2015). Thereafter, autonomous archival recorders were deployed in the waters of the Nosy Be region that had been documented as Omura's whale habitat. Initially, a pilot study was conducted over a three week period in 2014, in which three SoundTrap-202 recorders (oceaninstruments.co.nz), sampling continuously at a sample rate (SR) of either 96 kHz or 144 kHz, were deployed at approximately 20 m depth at four different sites for periods of 7–14 days. These recordings were useful for obtaining long, uninterrupted sequences of Omura's whale sounds, and proved the effectiveness of using remote archival recorders to document spatiotemporal occurrence of vocalizing Omura's whales. Following on from the pilot study, four SoundTrap-300 recorders were used to monitor for Omura's whale and other cetacean vocalizations from 1 December 2015 to 10 November 2016, recording on a 33% duty-cycle (10 min every 30 min) at a SR of 24 kHz. The recorders were deployed in water depths of approximately 40 m and distributed across an approximately 80 km range within the study area (Fig. 15.3). Two recorders were placed at the extreme north (Banc du Goliath) and south (Baramahamay) of the study area, respectively, and two were placed in the core of the study area, separated by approximately 20 km (Banc de Rosario and Ankazoberavina). The southern site and the two core sites were previously monitored during the pilot deployment in 2014, during which many Omura's whale vocalizations were recorded. The Banc du Goliath site was chosen to monitor the area near the site of the large aggregation encountered in 2015 (Fig. 15.4). The recorders operated for an aggregate of 1161 recorder-days, yielding a total of 9287 h of acoustic data. These data were used to define the spatiotemporal occurrence of Omura's whale throughout the study area over a one year period (Cerchio et al. 2018).

15.3.3 Description of Omura's Whale Song

As initially described in Cerchio et al. (2015, 2018), vocalizations recorded in the presence of Omura's whale were classified as songs. An Omura's whale "song," as recorded off northwest Madagascar, consists of a single stereotyped phrase comprised of one to two unit types; this is repeated at a regular interval of approximately two to three minutes for extended periods of time in long sequences (Figs. 15.5 and 15.6). In northwest Madagascar to date, a single "song type" has been positively attributed to Omura's whales through simultaneous visual observations and acoustic recordings, referred to as Type1 song. Type1 song occurs in two similar forms (or two similar "phrase types"), which are referred to as Type1A and Type1B. Type1A is composed from a single, partially amplitude-modulated (AM), low-frequency sound (Unit A, Fig. 15.5A). Unit A has an average duration of 10.2 (s.d. 0.31) sec in the ~15–50 Hz frequency band, its main energy in an initial 8.2 (s.d 0.27) sec wideband section with a slightly bimodal energy distribution (20.8 and 40.0 Hz) (Fig. 15.5C), and terminates with a short multi-narrowband "tail" of 1.9 (s.d. 0.37) sec. Type1B is composed from Unit A and a distinct constant frequency sound (Unit B, Fig. 15.5B), which has a duration of approximately 4 s in the 16–17 Hz band (Cerchio et al. 2018; Moreira et al. 2020; Fig. 15.5B). Songs are repeated at regular intervals, with a singer varying between phrase types in its song sequence (Fig. 15.5D). Frequently, overlapping series with different received levels and acoustic quality were recorded, indicating choruses of several individual singers (Fig. 15.5E).

Uninterrupted sequences of songs can, at times, last for periods in excess of several hours, with the longest recorded uninterrupted sequence lasting for at least 11.6 h, during which time the whale was within detection range of the recorder. Figure 15.6 illustrates an 8 h example observation of singing, in which the song sequence fades in at the start and then fades out at the end, suggesting that the whale was singing prior to moving within detection range of the recorder, sung for many hours within detection range, and then continued to sing as it moved out of the detection range. An interesting aspect of this observation is that the singer had to be actively maintaining its relative position in the vicinity of the recorder to be detected for this period of time. The currents in this area range from approximately 1–4 km/hr depending on tidal regime, and given the relatively short detection range of the recorder in this habitat (likely <5 km, see below) a passively drifting whale likely would have drifted out of detection range during this period; similarly, if moving even at less than the average documented movement speed (3.9 km/hr from four satellite tagged individuals; Cerchio et al. 2018), the whale would have traversed through the detection range in less time. Recordings of long sequences in excess of several hours are not uncommon, suggesting that some singers may remain relatively stationary by actively maintaining position in a specific area while singing.

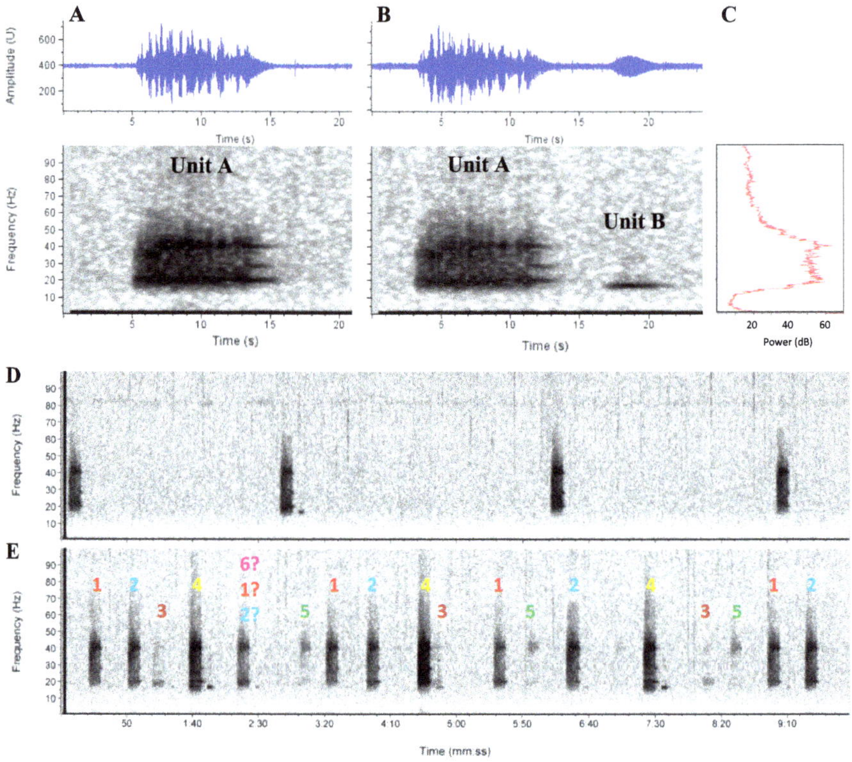

Fig. 15.5 Northwest Madagascar Omura's whale song. Sound spectrograms illustrating **A** phrase Type1A, and **B** phrase Type1B (2000 Hz SR; 2048 pt FFT, 95% overlap, Hann window); **C** power spectral density level of Unit A (averaged across 22 units from 6 different series; 2000 **Hz** SR, 32,768 pt FFT, 0% overlap, Hann window); and spectrograms of **D** a sequence of four songs from a single singer, and **E** a chorus of an estimated five or six Omura's whale singers with 18 separate songs audible (with a subjective assessment of individual attribution indicated by colored numbers; 2000 Hz SR; 4096 pt FFT, 95% overlap, Hann window)

To assess song repetition rate, spectrograms (2000 Hz SR, 4096 pt FFT, 50% overlap) of 963 h of recordings from the four locations monitored in 2014 were manually scanned to define unambiguous, continuous song sequences. Each sequence was presumed to be a single individual singer based upon an uninterrupted, consistent song repetition rate throughout the sequence (Fig. 15.6), and defined either as a solo singer or as a prominent foreground singer when more than one singer was audible. A total of 161 individual sequences were isolated ranging from 10 to 244 repeated songs, typically a mix of the two phrase types. As a measure of song repetition rate, the interval from the onset of one song to the onset of the next (hereafter referred to as the song repetition interval) was measured from 111 individual series with $\geq$20 consecutive songs. Mean repetition interval for individual series ranged from 141.6 s (s.d. 34.5, $n = 44$ intervals) to 219.7 s (s.d. 45.73, $n = 32$ intervals) with a population mean across the 111 individual means of 177.5 s (s.e. 15.3) (Fig. 15.7).

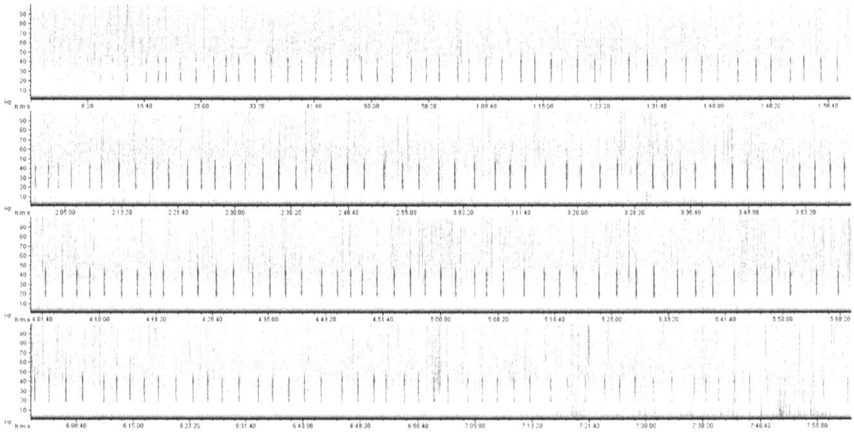

Fig. 15.6 Spectrogram illustrating an uninterrupted 8-h sequence of Omura's whale songs (n = 199). Due to its uninterrupted and relatively regular repetition rate, this sequence is presumed to be from a single singing whale

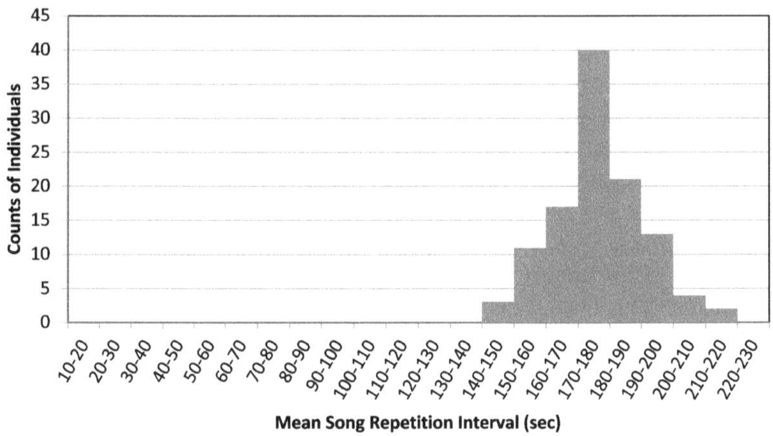

Fig. 15.7 Variation in song repetition rate, as expressed by the distribution of mean song repetition intervals, measured from 111 individual singers. Data included are for continuous series of $\geq$ 20 songs, presumed to be from a single individual and typically with mixed phrase types (Type1A and Type 1B) repeated uninterrupted at a relatively consistent repetition rate

Since these 111 song sessions indicated that the range of song repetition intervals is consistent and typically 150–200 s, the summation of the number of detected song phrases in a 10-min sample of the duty-cycled data (33%, 10 min every 30 min) should be a function of the number of singing whales. Therefore, total number of songs across an entire day of samples provides a reliable index of daily singing activity and the relative number of singing whales within detection range of the recorder. The actual detection ranges of the recorders were undetermined, but short deployments of pairs of recorders separated by 2, 3, and 5 km, suggested that in this shallow water environment songs can be detected at ranges > 2km, but likely <5 km (unpublished data). Under the assumption that singing is a display limited to reproductively mature males, the number of singers only reflects a portion of the population, and therefore, summation of songs, or acoustic presence, represents only a general proxy for Omura's whale daily presence.

15.3.4 Spatiotemporal Occurrence of Omura's Whale Singing

There was apparent spatiotemporal variation in the occurrence and relative prominence of Omura's whale singing activity among the four recorder sites during the 2015/2016 recording period (Cerchio et al. 2018). Initial review of data from 1 December 2015 to 10 November 2016 indicated that singing was clearly most prominent at the northernmost site, and relatively rare at the two most southern sites after December. An automated detector for song Unit A was developed using the Band-limited Energy Detector of Raven Pro 1.5 (Bioacoustics Research Program 2014) in the 15–50 Hz frequency band in order to detect occurrences of songs composed of either phrase Type1A or Type1B. Several SNR thresholds for automatic detection were tested on one week of data from Banc du Goliath. A detection threshold of 5 dB was chosen, yielding a detection rate of 81.3% for all phrases and a false detection rate of only 1.9%, thus providing a conservative estimate of the total number of true detections. The detector was run on all data from all sites, and examination of one week from each month (ca. 25% of data) indicated widely varying rates of "false positive" detections due to particularly high levels of ambient noise at some recorders during certain months. Therefore, detector results were exhaustively browsed manually to verify all detections, removing detections of other sounds and all false positives. The resultant output provided confident and conservative counts of songs (recognizing an estimated 81% detection rate since missed detections, or "false negatives," were not logged during the elimination of false positives). Songs were recorded at all sites during December on a daily basis, sometimes in choruses of multiple individuals; singing activity subsided in January, particularly at the three southern sites where it was thereafter sporadic and uncommon from February through April, and varied substantially among the four sites from May through November (Fig. 15.8) (Cerchio et al. 2018, unpublished data).

Whereas singing activity declined during January in the three southern sites, song was recorded extensively at the most northern site, Banc du Goliath, in all months with choruses of multiple individuals occurring throughout the year (Fig. 15.8A). Moreover, there was no indication of strict seasonality despite fluctuation in singing activity throughout the year. Singing was present daily from 1 December to the end of January and then was intermittently present throughout the year with several strong peaks in activity, the most prominent being late May to late June, and late October

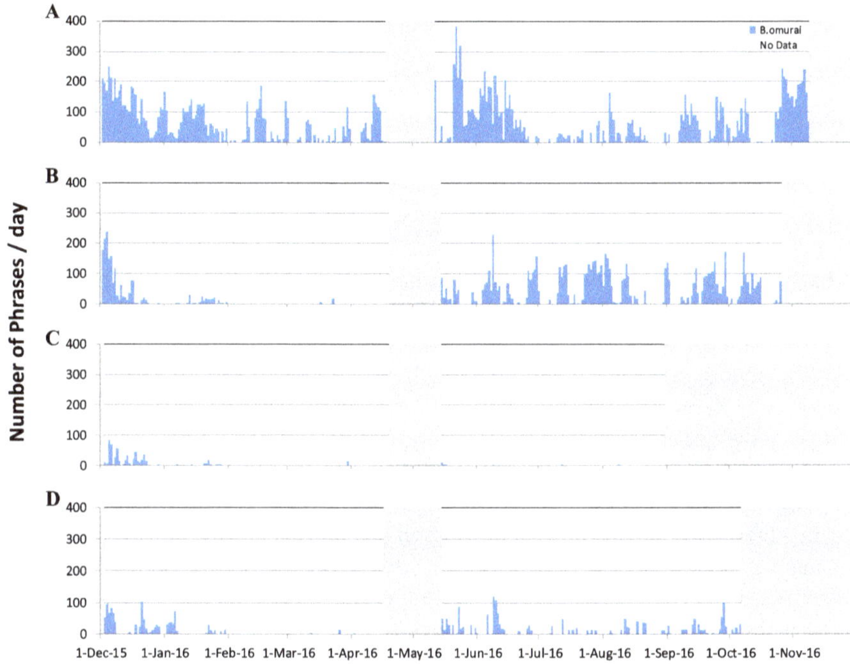

Fig. 15.8 Daily counts of songs from December 01, 2015, to November 30, 2016, at each of the four recording sites in the Nosy Be region (see Fig. 15.3): **A** Banc du Goliath; **B** Ankazoberavina; **C** Banc de Rosario; and **D** Baramahamay. Data were recorded on a 33% duty-cycle of 10 min every 30 min (on the hour and half hour, local time), so totals represent the number of songs detected during a total of 8 h of recording each day. Gray indicates periods with no data or incomplete data

to mid-November when our recording effort ended (Fig. 15.8A). It is assumed that singing activity remained high throughout November, as this was the period of boat surveys when many animals were sighted in the vicinity of the recorder. The longest stretch of days without any song detections was eight days in late August; however, even this hiatus was shortly followed by a period of high singing activity in mid-September (Fig. 15.8A). The presence of singing year-round off Banc du Goliath indicates definitively that this population of Omura's whales is non-migratory, in that it does not seasonally move out of northwest Madagascar waters. This was a key discovery of the acoustic monitoring effort, since residency of the population and a non-migratory life history of the species had been previously suspected, but not confirmed.

The Banc Rosario recording site, 35 km south of Goliath, had the second highest levels of singing activity, with relatively high levels of singing during December and from May to November, but very little from late December through mid-May (Fig. 15.8B). These two sites are considered hotspots for Omura's whale as indicated by boat-based sightings (Fig. 15.3), and also indicated by the long-term experience of local ecotourism companies. Singing activity at the Ankazoberavina recorder, 20 km to the southwest of Banc Rosario, and Baramahamay recorder, another 30 km south of Ankazoberavina, was markedly less than at Goliath and Rosario. At Ankazoberavina, singing was heard consistently only in December, but even in that month at less than half the occurrence rate as during the same days at Goliath (Fig. 15.8C). After December, song was recorded at Ankazoberavina on less than one day per month overall, with no activity from late January to late March, or late May to late August (when the recorder prematurely stopped recording). At Baramahamay, singing was recorded more frequently, during December into early January, and from late May to early November when our recording effort ended, but at a far lower rate than Goliath or Rosario (Fig. 15.8D). This indicates that despite whales being resident year-round in the Nosy Be region, there was distinct micro-geographic variation in their distribution within the region throughout at least the monitored year. Moreover, the variation throughout the year at Goliath and Rosario suggests influxes of singing animals moving throughout the area, into and out of detection range of the recorders. This result underscores the importance of a broad sampling strategy throughout a region; e.g., a very different conclusion regarding local residency would have been reached if acoustic monitoring had only been conducted at Ankazoberavina and not at Banc du Goliath. Further monitoring effort in additional years will be required to determine if this pattern is consistent across years, suggesting stable micro-habitat preferences, or conversely if spatial heterogeneity changes across years, perhaps suggesting the influence of fluctuating and variable oceanographic conditions on distribution of prey resources, and consequently on the distribution of whales.

Multiple simultaneous singers were heard throughout the recording period, at times in choruses of several individuals within detection range, as indicated by the presence of multiple overlapping sequences of songs. The frequency of occurrence of multiple singers was assessed at the Banc du Goliath site, since it had the highest measured levels of singing activity (Fig. 15.8A). From 1 December 2015 to 10 November 2016, a total of 15,349 10-min samples were evaluated, of which 40.5% (n = 6,213 samples) had at least one song indicating the presence of a singing Omura's whale. A total of 26,438 songs were counted, with the number of songs per sample ranging from 0 to 23. The daily average number of songs per sample closely tracked the total number of songs per day (Fig. 15.8A), suggesting little to no seasonal variation in tendency to chorus. Based on a mean song repetition interval of 177.5 s (Fig. 15.7), on average a single singer produces 3.4 songs per 10 m sampling period. Considering this interval $\pm$ 2 s.d. (i.e., an approximated 95% C.I.), the number of individual singing males can be estimated based on number of songs per sample. There were no singers detected in 59.5% of all samples, 1 singer estimated present in 20–27% of all samples, and $\geq$2 singers estimated present in 14–20% of all samples (Fig. 15.5E for a 10-min sample with 18 songs from an estimated 5–6 different singers). This indicated that when singing whales were present, there were multiple singing males within detection range of the recorder from one-third to one-half of the time, suggesting that choruses were a common phenomenon.

Assessment of hourly rates of song phrases/10-min (using Goliath site data) gave no indication of a cyclic diel trend in singing activity across the complete sampling period (Cerchio et al. 2018), unlike other baleen whales for which diel trends have been repeatedly found. For example, the tendency of blue whales to sing more frequently at twilight or night than during the day in several locales (Stafford et al. 2005; Wiggins et al. 2005; Chap. 9), humpback whales to increase singing activity during dark hours in multiple ocean basins (Cholewiak 2008; Cerchio et al. 2014; Chap. 11), North Atlantic right whales to produce "gunshot" sounds, a signal that has been associated with a male display, more often at night in spring and summer (Matthews et al. 2001; Parks et al. 2005; Chap. 13), and minke whales to sing more at night in the North Atlantic and off South Africa, but either have no distinct pattern or be more common during daylight hours in the western Antarctic Peninsula (Chap. 14).

15.3.5 Acoustic Detections of Other Song Types

Two additional novel forms of song were detected off Nosy Be with lower occurrence rates than the confirmed Omura's whale song (Type1A and 1B). Both were in the 20–60 Hz frequency band, with stereotyped units repeated in a series at regular intervals, and therefore considered a form of balaenopterid singing. They were never produced within the same series as an Omura's whale Type1 song; therefore, it is unclear if they represent a different form of Omura's whale song or were produced by Omura's whales at all. Each shared similarities with the Type1 Omura's whale song, in terms of AM unit structure, unit duration and repetition rate. To our knowledge, these specific forms have not been previously reported in the literature, and since they were never recorded in the presence of a whale, species attribution can only be speculated. The first novel song was recorded sporadically at three of the sites (Banc du Goliath, Banc de Rosario and Ankazoberavina) from late February to mid-September, and at times a series of high SNR novel songs overlapped with a series of Omura's whale Type1 songs; however, during its peak occurrence, this first novel song was never as common as Omura's whale Type1 song. The second novel song form was very rare and documented only twice on the Banc du Goliath recorder.

It is possible that these novel forms of song are from another baleen whale species, such as Bryde's whale (*B. edeni* or *B. brydei*); however, Bryde's whales have only been confidently documented far to the south of Madagascar during Japanese scientific whaling operations and other visual boat surveys (Ohsumi 1980; Best 2001), and never off the west coast of Madagascar, nor off Nosy Be despite the extensive effort to study Omura's whales (Cerchio et al. in press). Alternatively, these may be songs of Omura's whale, representing different song types that are distinct from the common Type1 song. If these songs were from Omura's whales, it is not obvious why they are less prevalent than Type1 song. The comingling of Omura's Type1 song and the first novel form of song in close proximity (likely within a few kilometers) suggests that they may both be from males of the same species singing in a chorus. Little is known about the range and dispersal of Omura's whale populations, but our data from Nosy Be, including photographic identification and satellite telemetry data (Cerchio et al. 2018), suggest that this Nosy Be population is resident and has a restricted range along a ca. 400 km stretch of the northwest coast of Madagascar.

Nothing is known about population structure, long distance movements or population mixing of Omura's whale, nor is it clear whether there exist multiple populations in the southwest Indian Ocean. It is conceivable that there are other as yet undetected populations in the region, that populations have diagnostic song types, and the presence of these two song types represents males from nearby populations temporarily moving into the range of the northwest Madagascar population. In the next section, we explore what is currently known about the geographic variation in Omura's song across its global range.

15.4 Global Geographic Variation of Omura's Whale Song

It is now recognized that Omura's whales have a nearly pan-tropical distribution, with documented accounts in all ocean basins except the central and eastern Pacific Ocean (Cerchio et al. 2019). Among the accounts that have been reported are several acoustic documentations of Omura's whale song occurring widely throughout the global range: in the equatorial Atlantic Ocean off the São Pedro and São Paolo Archipelago, Brazil (Moreira et al. 2020), and Ascension Island (Leroy et al. 2021); in the central Indian Ocean off the Chagos Archipelago (Sousa and Harris 2015); in the South China Sea (G. C. Gagnon, pers. comm.); and broadly along the northwestern coast of Australia (McCauley 2009, 2011; McPherson et al. 2016; Erbe et al. 2017). All of these cases of acoustic identification were based upon strong similarities in acoustic features with the published description of Omura's whale song from Madagascar (Cerchio et al. 2015). Only some were accompanied by independent photographic verification of the presence of the species in the general region (only off Australia), but none was recorded in the presence of Omura's whales, as was done off Madagascar. That the similarities of these songs with Madagascar Omura's whale song were strong enough to infer the occurrence of the species in these other regions, indicates that Omura's whale song has diagnostic, species-specific acoustic features across an extended range (Fig. 15.9).

The first reports of Omura's whale songs came from the northwest coast of Australia and initially were attributed to Bryde's whales (McCauley 2009), before the description of Madagascar Omura's whale song in 2015. The song was documented across an extensive range, from at least ten recording sites from Exmouth northeast into the Timor Sea to north of Darwin (McCauley 2009, 2011). After the description of Omura's whale song from Madagascar, the songs in these first reports were independently attributed to Omura's whales in three separate studies (McPherson et al. 2016; Erbe et al. 2017; Cerchio et al. 2019). Off the Chagos Archipelago, Sousa and Harris (2015) reported two vocalizations of unknown origin off Diego Garcia, attributing them to a baleen whale and speculating that they could represent undescribed blue whale vocalizations. One of these, the "Diego Garcia Croak" (DGC), was recognized as an Omura's whale song by Cerchio et al. (2019). In the equatorial Atlantic, sounds recognized as Omura's whale song were reported from deep water recorders near the São Pedro and São Paulo Archipelago on the mid-Atlantic

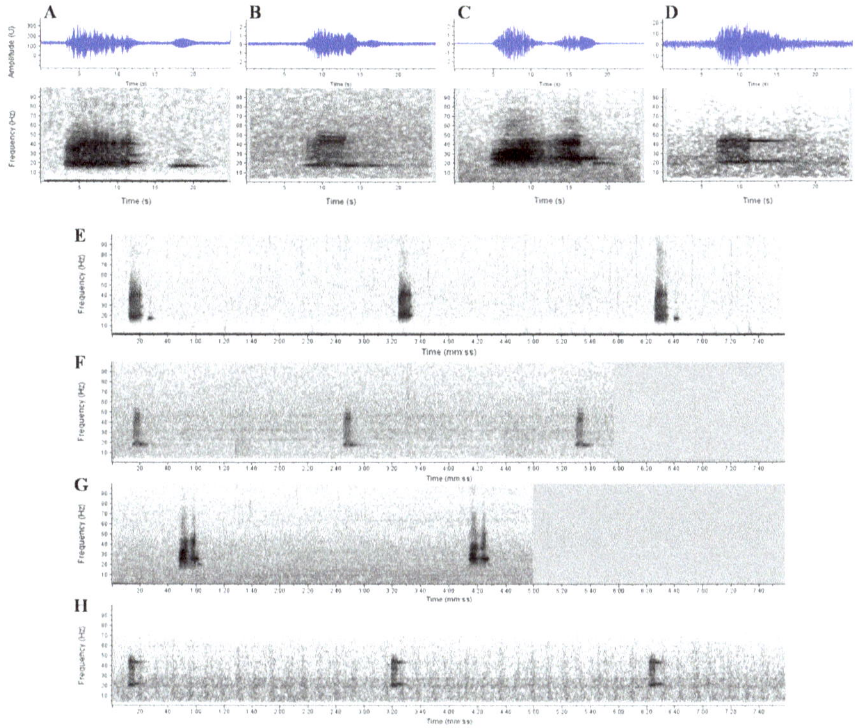

Fig. 15.9 Waveforms and spectrograms illustrating similarities for songs attributed to Omura's whales from four disparate sites in the species' range as presently documented. Top panel: spectrogram examples of four song types from: **A** northwest Madagascar (Type1B), (Cerchio et al. 2015, 2018) with confirmed attribution to Omura's whales; **B** Diego Garcia, Chagos Archipelago in the central Indian Ocean originally described as a baleen whale of unknown species (Sousa and Harris 2015, sound contributed by D. Harris from Diego Garcia CTBT acoustic data); **C** the northwest coast of Australia (IMOS acoustic data, Kimberley site); and **D** São Pedro and São Paulo Archipelago on the mid-Atlantic Ridge (from Moreira et al. 2020). Lower four panels show short sequences of songs from: **E** northwest Madagascar, **F** Diego Garcia, Chagos Archipelago, **G** northwest coast of Australia and **H** São Pedro and São Paulo Archipelago, Brazil. All four of these song types have been attributed to Omura's whales due to their common acoustic features and repetition rates compared to song features from other baleen whale species (2 kHz SR, FFT 4096 DFT, 95% overlap, frequency resolution 0.5 Hz, time resolution of 200 ms)

Ridge (Moreira et al. 2020), and very similar songs have more recently been reported from Ascension Island (Leroy et al. 2021), using hydrophones of the Comprehensive Nuclear Test Ban Treaty Organization, CTBTO).

The songs from four disparate locations for which acoustic descriptions are available (Madagascar, the Chagos Archipelago, Australia and the Equatorial Atlantic) broadly share structural characteristics and acoustic features that make them unambiguously identifiable as Omura's whale song and distinguish them from all other described baleen whale vocalizations (Table 15.1, Fig. 15.9). All combine wideband, noisy, AM, and narrowband components with predominant energy in the 15–50 Hz

Table 15.1 Comparison of frequency and temporal characteristics of Omura's whale songs from four locations across the species distributional range, as illustrated in Fig. 15.9

Parameter	Madagascar[a]	Chagos[b,c]	Australia[c, d]	São Pedro and São Paulo Archipelago[a]
Sample size	22	10[b], 78–80[c]	not reported[d], 100[c]	36
Low bounding Freq	16.1 (1.2) Hz	16.9 (0.1) Hz[b]	ca. 26 Hz[d]	17.0 (1.3) Hz
High bounding Freq	47.8 (4.7) Hz	49.6 (0.1) Hz[b]	ca. 40 Hz[d]	46.9 (1.1) Hz
Bimodal peak Freq	20.8 (4.8) Hz 40.0 (1.6) Hz	19.5 (0.08) Hz[c] 43.5 (0.11) Hz[c]	25.3 (0.18) Hz[c] 41.2 (0.18) Hz[c]	20.6 (0.0) Hz 43.6 (0.9) Hz
Ending tonal Freq	16–17 Hz	17.6 (0.04) Hz[c]	19.8 (0.02) Hz[c]	n/a
Duration	10.2 (0.3) s (Type1A) 19.4 s (Type1B)[c]	13.1 (0.6) s[b] 10.6 (0.14) s[c]	range of 11–16 s[d] 16.4 (0.08) s[c]	10.4 (1.2) s
Repetition rate interval	177.5 (14.3) s	186 (18) s[b]	ca. 200 s[c]	172.2 (32.5) s

Values are mean (s.d.), when reported by the original source, except for: "Peak Freq" values reported by Moreira et al. (2020), which are median (IQR); all Chagos values reported by Sousa and Harris (2015) and values reported by Leroy et al. (2021), which are mean (s.e.). Sample sizes varied across studies and between measured variables, so the reader is directed to the original literature for details
Sources: [a]Moreira et al. (2020); [b]Sousa and Harris (2015); [c]Leroy et al. (2021); [d]McPherson et al. (2016)

band, a bimodal energy distribution and spectral energy peaks at approximately 20 and 40 Hz. The three Indian Ocean songs all end with a low-frequency narrowband component or a separate low-frequency narrowband unit, at approximately 16–20 Hz, which appears optional in at least the Madagascar population (e.g., Type1A vs Type1B song). Song phrase duration is between approximately 10–18 s, with lower variability when considering only the first unit or sub-unit (duration between 10-11 s). Repetition intervals are consistent with averages of one song every approximately 170-200 s. Despite these broad similarities, there are distinct geographic differences in the structure and acoustic features of the songs from the different regions. Such geographic variation often develops between populations that have been at least semi-isolated and undergone vocal or dialectical divergence (Fig. 15.9; Cerchio et al. 2019; Moreira et al. 2020; Leroy et al. 2021).

In each of these regions, long-term monitoring has indicated that singing occurs throughout the year with little to no seasonality, as described above for Madagascar. Additional song types have also been described in multiple regions, some of which may also be attributed to Omura's whales. Off northwest Australia, McCauley (2009, 2011) reports that song was detected in all months of the year from at least three sites (Scott's Reef, northwest of Broome, and Joseph Bonaparte Gulf). In the eastern

Timor Sea, north of Darwin, Australia, McPherson et al. (2016) report that song was detected in all months from August 2014 through July 2015 except November and December, describing apparent movement of whales in and out of the area over the course of several months. McPherson et al. (2016) also described two distinct types of Omura's whale song in the Timor Sea with overlapping but slightly varying temporal distributions: a two-part song type (shown in Fig. 15.9c), and a single-part broadband song type that is much more similar to the Madagascar song Type1A (Fig. 15.5a). Off the Chagos Archipelago, Sousa and Harris (2015) reported peaks of occurrence in both the Austral winter and summer of 2002 and 2003, with hiatuses of no occurrence during spring and autumn, and it was concluded that the population may reside year-round at Diego Garcia. Sousa and Harris (2015) also reported two different song types, only one of which has been positively attributed to Omura's whale based upon similarity with Madagascar song; the other song type, the "Diego Garcia Downsweep" may represent a second Omura's whale song type for the region, although some researchers argue it to be from a blue whale (Leroy et al. 2021). In the Equatorial Atlantic, Omura's whale song was detected off the São Pedro and São Paolo Archipelago during nine of 11 monitored months between February and December 2013, with a strong peak in singing activity during April through June, and no occurrence during September and October (Moreira et al. 2020). Currently, there is no evidence for additional song types off São Pedro and São Paulo, and the song recorded off Ascension Island appears similar, but not identical (Leroy et al. 2021). The observations of temporal occurrence for each of these regions suggest local movements within a region, without a strong seasonal signal, similar to what has been more fully described off Nosy Be, Madagascar (Fig. 15.8).

15.5 Synthesis and Implications for Breeding Ecology

Acoustic displays, referred to as song, have been described in many Mysticetes, particularly Balaenopteridae, and for those for which data exist, the trait is sex-limited to the males of the population. It is widely accepted that these vocal displays are therefore a product of sexual selection, represent a male breeding tactic, and are an integral part of each species' breeding system. As in other taxa (e.g., passerine birds), the displays take the form of stereotyped songs that are repeated with a consistent rhythm, and singing sometimes occurs in aggregations of male singers (choruses, although not the synchronized choruses of some anuran amphibians or insects). Communal display in aggregations is a common characteristic of lek mating systems (Höglund and Alatalo 1995), and a form of lekking has been proposed to describe the breeding system for at least humpback whales, the best studied singers among the Balaenopteridae (Herman and Tavolga 1980; Clapham 1996). As we learn more about the acoustic behaviors of *Balaenoptera* species, it is becoming clearer that they share the general trait of having male vocal displays, and that song, communal singing (i.e., displaying in aggregations), and perhaps even lek mating systems might eventually be considered ancestral balaenopterid behavioral traits.

The Omura's whale has only recently been added to the list of *Balaenoptera* species; however, because of the advancements in technology and methodology in the twenty-first century, along with the identification of a tractable population off Madagascar, we are learning about this species at a relatively rapid rate compared to previous work on other Mysticetes. The current understanding of Omura's whale behavioral ecology and vocal display behavior fits well within the context of what is known about other balaenopterids. Omura's whales produce stereotyped songs that are uttered in consistent, highly repetitive forms. Singing occurs uninterrupted over extended periods of time (several hours without pause) and therefore represents a time-intensive vocal display. There are currently no data on the sex of Omura's whale singers, but given what is known about sexually selected breeding displays, it seems highly likely that singing is a male-limited vocal display important in the breeding system. Social structure appears to be characterized by loose aggregations, with groups of individuals spread over a few kilometers in a patchy and ephemeral spatial distribution. Observations in Madagascar suggest that these aggregations probably are related to both feeding and social behavior, sometimes associated with patchy food resources and other times not. Similarly, long-term acoustic monitoring indicates that chorusing by singers is a prominent part of singing behavior. Micro-geographic variation in spatial distribution is high (i.e., there is a high degree of spatial and temporal heterogeneity over an 80 km stretch of coastal habitat in Madagascar revealed by both acoustic monitoring and satellite telemetry), with apparently preferred areas for aggregation and singing. Some degree of the spatial heterogeneity is probably in response to prey distribution, but aggregating and displaying in preferred "arenas" is also characteristic of lekking behavior (Höglund and Alatalo 1995).

Whereas some characteristics of Omura's whale behavior appear to be general to Balaenopteridae as a whole, several other emerging characteristics for the species are unusual for the family. Populations of Omura's whales appear to be non-migratory, with likely restricted ranges and small sizes, and the species has an exclusively tropical and warm-temperate distribution, all of which are rare among balaenopterids. There is no apparent seasonality to singing behavior, which is also apparently rare among balaenopterids. Like other balaenopterids, their general song structure is diagnostic to the species across a broad global range, allowing it to be readily recognized in many cases. Currently recognized macro-geographic variation in songs types appears moderate to low on a global scale (Fig. 15.9), in comparison to some balaenopterids, for which there is markedly greater variation in song types among populations (e.g., among the song types of different populations of pygmy blue whales, Chap. 9; or among songs of non-interacting populations of humpback whales, Chap. 11).

There are indications of different song types within regions, but in most cases, the attribution of the additional song types to Omura's whales is speculative. Thus, the perception of song similarity across their global range may be inflated by ascertainment bias, in that researchers are currently only able to identify those regional song types that are most similar to the only one that has been confidently described for the species off Madagascar (Cerchio et al. 2015). When we consider songs of unknown origin already documented, then there may be two to three possible song types off northwest Madagascar (described above), two or more off the Chagos Archipelago (Sousa and Harris 2015), and at least two off northwest Australia (McPherson et al. 2016). Therefore, it is possible that neighboring populations have distinctive song dialects, and migrants/wanderers move between populations; this would explain the presence of a "minority" song type among a much more common song type (as may be the case off Madagascar). However, there is as yet no evidence of "cultural diffusion" (Mundinger 1980; Whiten et al. 2016) of song types from one population to another, as would be indicated by the replacement of a local song by another different one (but more time for observation is needed). Alternatively, different song types may serve different roles, and each population may have a song that is similar to those currently recognized as Omura's whale song, as well as more divergent song type(s). These observations elicit wide open questions that require expanded research on multiple populations and further interpretation of new results.

In addition to the novelty of new species-specific discoveries, Omura's whales may present a unique opportunity for the study of balaenopterid vocal/breeding behavior. Humpback whales have been the most intensively studied baleen whale, in large part due to their tendency to densely aggregate in accessible coastal areas; however, their song and singing behavior is complex, presenting challenges to interpretation and inference (Chap. 11). Conversely, blue and fin whales have comparatively simple songs and singing behaviors (Chap. 9); however, they are challenging to study due to their dispersed pelagic distributions. The relatively simple songs and spatial distribution characteristics of Omura's whale populations appear to combine some of the most tractable characteristics of other balaenopterids, and therefore might make them favorable for research compared to other species within the family.

Blue and fin whales are pelagic and inhabit large oceanic ranges, and their spatial distribution is likely strongly influenced by the long distance propagation of their low-frequency songs (Chap. 2). Singing males are consequently spread over large regions of deep water, and if in fact their mating systems resemble leks, their display arenas may spread over substantial portions of an ocean basin. So far, Omura's whales appear to be neritic and coastal, with relatively small core ranges, based upon what has been learned off Madagascar. Their song propagation is limited by shallow water stripping of low frequencies and likely moderate source levels due to their small size; pairs of recorders deployed on the shelf habitat off Madagascar suggested that individual singers were simultaneously audible on hydrophones separated by 2–3 km but not on hydrophones separated by 5 km. This limited acoustic range is congruent with the visual observations of aggregations that are spatially tight in comparison to larger pelagic balaenopterids, and therefore, chorusing individuals may be separated by only 100 s of meters to several kilometers, compared to likely 10 s to 100 s of kilometers for blue and fin whales. Consequently, the lekking grounds of Omura's whales may be of a tractable size for study.

Population characteristics of Omura's whales might also be favorable for long-term latitudinal studies of individuals within a population. Considering humpback whales in comparison, most modern populations are large, on the order of >10,000 individuals, requiring prohibitively large samples of individuals (i.e., genetic and behavioral sampling) to study the breeding system. Additionally, long-range movements of individuals result in mixing and substantive immigration/emigration between populations. There is still much to be learned about Omura's whale population characteristics, but based on the initial information from northwest Madagascar, the species may be characterized by small resident populations. This presents the potential for long-term latitudinal study of individuals, including for example, assessments of genetic relatedness relative to individual associations and male reproductive success.

Lastly, song characteristics of Omura's whales are relatively conducive to study. In comparison, humpback whales (the most widely studied species in terms of song and mating behavior) have complex singing behaviors, representing the most highly derived balaenopterid in terms of evolutionary exaggerated traits. Humpback whale song is organized into a relatively complex hierarchal structure, with substantive variation on several interacting levels, including intra-individual, inter-individual, temporal, and inter-population (Chap. 11). Such complexity makes it difficult to identify unambiguous differences among individual males. This presents great challenges in attributing variation to consequential behavioral metrics, such as reproductive success, thus making it difficult to correlate song features with selection. In contrast, Omura's whale song is comparatively simple, with a single phrase and non-hierarchal structure, uttered in a consistent repetitive rhythm. Inter-individual variation is not readily apparent, nor has it yet been properly assessed, but under the assumption that it is a male sexually selected trait, such variation must exist between individuals and should be measurable. Possible acoustic features that may contain measurable variation include the pattern of pulses in the AM units, quantitative details of phrase structure, or the repetition rates of songs. The salient feature is that the simplicity of the song increases the likelihood that consistent variation between individuals can be reliably measured and regressed against important physical and behavioral variables, such as size, age, and ultimately male mating success. It may therefore be possible to correlate song acoustic features with selection and make some inference on the evolutionary process of this species' singing behavior and mating system.

There are several challenges that also need to be acknowledged. Perhaps most complicating is the difficulty in identifying singing individuals in the field, due to the low-frequency nature of the song, the long intervals between songs, and the difficulty in sighting and tracking individuals at the surface due to small size and an inconspicuous blow. Therefore, obtaining focal animal song samples paired with genetic samples is not straight forward. This may be surmountable using combinations of emergent technology, such as acoustic tags deployed on known (recognizable) males in an area that is being monitored by a synchronized array of hydrophones with localizing capability. Also, there are challenges inherent to working with a species that feeds throughout the year, unlike breeding humpback whales that are fasting and therefore not linked to food resources. Dependence on patchy and ephemeral prey resources results in somewhat unpredictable temporal variability in Omura's whale distribution within their range, limiting the reliability of finding aggregations within and across years. Observations off Madagascar suggest that such variability complicates field work, but some areas exist where aggregations are more consistently found. These challenges aside, Omura's whales may embody the simplicity and tractability of singing systems that are apparent in species such as fin and blue whales, in a species that has an accessible coastal distribution like the humpback whale, and furthermore may have small tractable populations. Therefore, the species may represent a potential model for assessing questions about general balaenopterid vocal and mating behavior, aside from species-specific questions.

15.6 Afterward: Consequences of Leaps for Improving Our Understanding of Poorly Known Species and Habitats

Over the past four to five decades, scientific technology has steadily developed on all fronts, and passive acoustic monitoring devices have become increasingly more powerful, portable, and affordable. These technological advancements have enabled us to expand our study of the marine environment, in regions around the world where resources and technology have been limiting. The technological advancements of the 1970s made it possible for small-scale research groups to acquire the means to conduct cutting-edge research, and make available learning experiences to undergraduate students, like myself, at local universities that would in turn motivate careers in bioacoustics. Remarkable as this was, these opportunities were still limited to developed countries and the privilege of relatively wealthy societies when considered in a global context. Advancements continue, and during the past 5–10 years in particular, there has been another major technological leap that is again revolutionizing the field of cetacean bioacoustics. There are now commercially available archival recorders the size of a soda can, at a modest price, and these recorders can record continuously for months at sample rates covering the frequency range of all mysticetes and many odontocetes. Analysis methods have shifted from analog to digital, and modestly priced high-powered computers and open-source software have pushed the limits beyond what was conceivable in the 1980s and 1990s. Today's advancements allow researchers in remote locations, on limited budgets, access to technology that will revolutionize our understanding of previously poorly studied regions of the world. This ability opens the door to conduct studies that only ten years ago would have had orders of magnitude higher cost, if they were logistically possible at all. In addition, the discoveries that this technology facilitates will in turn synergistically inspire the wonder and passion of a new generation of explorers, as individuals discover in themselves the drive to push the boundaries of discovery itself. The case study of the Omura's whale of Madagascar, and related work applying passive acoustic technologies to study marine mammals in Madagascar, the Mozambique Channel, and other parts of coastal Africa, are examples of what could be considered a new age of discovery in the twenty-first century. In the coming decades, as technology continues to advance, costs become more affordable, and capacity around the globe is increased through access to education and global networking, we should expect to see the inspiration of new explorers and continued discovery in regions that previously have been poorly studied.

Acknowledgements This chapter is a synthesis of work from Madagascar presented elsewhere in peer-reviewed papers and scientific reports, with other studies conducted around the globe, in an effort to cast our current understanding of Omura's whales as a species in the context of baleen whale behavior and biology. The long-term study off Madagascar would not be possible without the dedication of my Malagasy colleagues, staff, and students, including most prominently Norbert Andrianarivelo, Boris Andrianantenaina, and Tahina Rasoloarijao, and partnership with the Malagasy institutions, Institut Halieutique et des Sciences Marines, and Center National de Recherches Océanographiques, in particular Directors Thierry Lavitra, Jamal Angelot Mahafina, and Mamy Nirina Rajaonarivelo. Danielle Cholewiak, Liam Mueller-Brennan and Sandra Dorning assisted with acoustic analysis, and Tanguy Guillemain d'Echon and Mada Megafauna contributed to field work. Funding was provided by generous grants from the US Marine Mammal Commission, the National Geographic Society Committee for Research and Exploration, the New England Aquarium Marine Conservation Action Fund, and the Woods Hole Oceanographic Institution (through the Dalio Foundation). Lastly, the work presented here has greatly benefited by the support—intellectual, motivational and inspirational—and the friendships of Robert L. Brownell, Christopher W. Clark, and Tadasu K. Yamada.

References

Bioacoustics Research Program (2014) Raven pro: Interactive sound analysis software (Version 1.5) [Computer software]. The Cornell Lab of Ornithology, Ithaca, NY. Available from http://ravensoundsoftware.com/

Best PB (2001) Distribution and population separation of Bryde's whale *Balaenoptera edeni* off southern Africa. Mar Ecol Prog Ser 220:277–289

Cerchio S, Andrianantenaina B, Lindsay A, Rekdahl M, Andrianarivelo N, Rasoloarijao T (2015) Omura's whales (*Balaenoptera omurai*) in northwest Madagascar: ecology, behavior and conservation needs. Roy Soc Open Sci 2:150301. https://doi.org/10.1098/rsos.150301

Cerchio S, Andrianantenaina B, Zerbini A, Pendleton D, Rasoloarijao T, Cholewiak D (2018) Residency, feeding ecology, local movements and potential isolation of the Madagascar Omura's Whale (*Balaenoptera omurai*) population. Paper SC/67B/NH/09 presented to the International Whaling Commission Scientific Committee. Available from: https://archive.iwc.int

Cerchio S, Laran S, Andrianarivelo N, Saloma A, Andrianantenaina B, Van Canneyt O, Rasoloarijao T (In press) Cetacean species diversity in Malagasy waters. In: Goodman SM (ed) The new natural history of Madagascar. Princeton University Press, Princeton

Cerchio S, Strindberg S, Collins T, Bennett C, Rosenbaum H (2014) Seismic surveys negatively affect humpback whale singing activity off northern Angola. PLoS ONE 9(3):e86464. https://doi.org/10.1371/journal.pone.0086464

Cerchio S, Yamada TK (2018) Omura's whale: *Balaenopteridae omurai*. In: Würsig B, Thewissen JGM, Kovacs K (eds) Encyclopedia of Marine Mammals, 3rd ed, Academic Press, San Diego, CA, USA, pp 656–659

Cerchio S, Yamada TK, Brownell Jr RL (2019) Global distribution of Omura's whales (*Balaenoptera omurai*) and assessment of range-wide threats. Front Mar Ecol 6:67

Cholewiak DM (2008) Evaluating the role of song in the humpback whale (*Megaptera novaeangliae*) breeding system with respect to intra-sexual interactions. Doctoral thesis. Cornell University

Clapham PJ (1996) The social and reproductive biology of humpback whales: an ecological perspective. Mammal Rev 26:27–49

Erbe C, Dunlop R, Jenner KCS, Jenner MN, McCauley RD, Parnum I, Parsons M, Rogers T, Salgado-Kent C (2017) Review of underwater and in-air sounds emitted by Australian and antarctic marine mammals. Acoust Aust 45:179–241
Herman LM, Tavolga WN (1980) The communication systems of cetaceans. In: Herman LM (ed) Cetacean behavior: mechanisms and functions. Wiley, New York, pp 149–209
Höglund J, Alatalo RV (1995) Leks. Princeton University Press
Leroy EC, Royer JY, Alling A, Maslen B, Rogers TL (2021) Multiple pygmy blue whale acoustic populations in the Indian Ocean: whale song identifies a possible new population. Sci Rep 11:1–21
Matthews JN, Brown S, Gillespie D, Johnson M, McLanaghan R, Moscrop A, Nowacek D, Leaper R, Lewis T, Tyack P (2001) Vocalization rates of the North Atlantic right whale (*Eubalaena glacialis*). J Cetac Res Manage 3:271–281
McCauley RD (2009) Sea noise logger deployment Scott Reef, 2006–2008—whales, fish and seismic surveys. Report prepared for Woodside Energy, CMST R2009–15, p 88, 61 Fig
McCauley RD (2011) Woodside kimberley sea noise logger program, September 2006 to June 2009: Whales, fish and man made noise. Report R2010–50_3 prepared for Woodside Energy, Ltd., p 80
McDonald MA, Hildebrand JA, Mesnick SL (2006) Biogeographic characterization of blue whale song worldwide: using song to identify populations. J Cetac Res Manage 8:55–65
McPherson C, Kowarski K, Delarue J, Whitt C, MacDonnell, J, Martin B (2016) Passive acoustic monitoring of ambient noise and marine mammals—barossa field: JASCO document 00997, Version 1.0. Technical report by JASCO Appl Sci Jacobs. pp. 99
Moreira S, Weksler M, Sousa-Lima RS, Maia M, Sukhovich A, Royer JY, Macondes MCC, Cerchio S (2020) Occurrence of Omura's whale, *Balaenoptera omurai* (Cetacea: Balaenopteridae), in the equatorial Atlantic Ocean based on passive acoustic monitoring. J Mammal 101:1727–1735
Mundinger PC (1980) Animal cultures and a general theory of cultural evolution. Ethol Sociobiol 1(3):183–223
Ohsumi S (1980) Population study of the Bryde's whale in the southern hemisphere under scientific permit in the three seasons 1976/77–1978/79. Rep Int Whal Comm 30:319–331
Parks SE, Hamilton PK, Kraus SD, Tyack PL (2005) The gunshot sound produced by male North Atlantic right whales (*Eubalaena glacialis*) and its potential function in reproductive advertisement. Mar Mamm Sci 21:458–475
Rosel PE, Wilcox LA, Yamada TK, Mullin KD (2021) A new species of baleen whale (*Balaenoptera*) from the Gulf of Mexico, with a review of its geographic distribution. Mar Mam Sci 2021:1–34. https://doi.org/10.1111/mms.12776
Sasaki T, Nikaido M, Wada S, Yamada TK, Cao Y, Hasegawa M, Okada N (2006) *Balaenoptera omurai* is a newly discovered baleen whale that represents an ancient evolutionary lineage. Mol Phylogenet Evol 41:40–52. https://doi.org/10.1016/j.ympev.2006.03.032
Sousa AG, Harris D (2015) Description and seasonal detection of two potential whale calls recorded in the Indian Ocean. J Acoust Soc Am 138:1379–1388. https://doi.org/10.1121/1.4928719
Stafford KM, Moore SE, Fox CG (2005) Diel variation in blue whale calls recorded in the eastern tropical pacific. Anim Behav 69:951–958
Wada S, Numachi K (1991) Allozyme analyses of genetic differentiation among the populations and species of the Balaenoptera. Rep Int Whaling Comm Spec Issue 13:125–154
Wada S, Oishi M, Yamada TK (2003) A newly discovered species of living baleen whale. Nature 426:278–281
Watkins WA, Daher MA, Reppucci GM, George JE, Martin DL, DiMarzio NA, Gannon DP (2000) Seasonality and distribution of whale calls in the North Pacific. Oceanography 13:62–67
Watkins WA, Tyack PL, Moore KE, Bird JE (1987) The 20-Hz signals of finback whales (*Balaenoptera physalus*). J Acoust Soc Am 82:1901–1912
Wiggins SM, Oleson EM, McDonald MA, Hildebrand JA (2005) Blue whale (*Balaenoptera musculus*) diel call patterns offshore of southern California. Aquat Mamm 31:161–168
Whiten A, Caldwell CA, Mesoudi A (2016) Cultural diffusion in humans and other animals. Curr Opin Psychol 8:15–21

Yamada TK, Chou LS, Chantrapornsyl S, Adulyanukosol K, Chakravarti SK et al (2006a) Middle-sized balaenopterid whale specimens (Cetacea: Balaenopteridae) preserved at several institutions in Taiwan, Thailand, and India. Mem Natl Sci Mus Tokyo 44:1–10

Yamada TK, Kemper C, Tajima Y, Umetani A, Janetzki H, Pemberton D (2006b) Marine mammal collections in Australia. National Sci Museum Monogr 34:117–126

Yamada TK, Kakuda T, Tajima Y (2008) Middle sized balaenopterid whale specimens in the Philippines and Indonesia. Mem National Museum Natl Sci Tokyo 45:75–83

Index

C.W. Clark and E. C. Garland (eds.), *Ethology and Behavioral Ecology of Mysticetes*, Ethology and Behavioral Ecology of Marine Mammals,
https://doi.org/10.1007/978-3-030-98449-6

GPSR Compliance
The European Union's (EU) General Product Safety Regulation (GPSR) is a set of rules that requires consumer products to be safe and our obligations to ensure this.

If you have any concerns about our products, you can contact us on

ProductSafety@springernature.com

In case Publisher is established outside the EU, the EU authorized representative is:

Springer Nature Customer Service Center GmbH
Europaplatz 3
69115 Heidelberg, Germany

www.ingramcontent.com/pod-product-compliance
Ingram Content Group UK Ltd.
Pitfield, Milton Keynes, MK11 3LW, UK
UKHW022215051026
15624UKWH00001B/178
* 9 7 8 3 0 3 0 9 8 4 4 8 9 *